HANDBOOK OF

Compressed Gases

Among the most dramatic of many essential uses of the compressed and liquefied gases in modern life are their applications in man's first exploratory voyages in outer space, which are powered by oxygen liquefied at cryogenic temperatures. Above, liquefied oxygen helps to launch one of the Saturn missions of the National Aeronautics and Space Administration. Inset shows use of liquefied oxygen and liquefied hydrogen as fuels.

HANDBOOK OF

Compressed Gases

COMPRESSED GAS ASSOCIATION, INC.
New York, New York

VNR VAN NOSTRAND REINHOLD COMPANY
New York Cincinnati Toronto London Melbourne

Van Nostrand Reinhold Company Regional Offices:
New York Cincinnati Chicago Millbrae Dallas

Van Nostrand Reinhold Company International Offices:
London Toronto Melbourne

Library of Congress Catalog Card Number 66-26993
ISBN 0-442-15033-4

 Manufactured in the United States of America.

Published by Van Nostrand Reinhold Company
450 West 33rd Street, New York, N.Y. 10001

Published simultaneously in Canada by
Van Nostrand Reinhold Ltd.

10 9 8 7 6

Illustration credits: **Frontispiece, National Aeronautics and Space Administration; pp. 338, 349, 350, 351 (upper illus.), 352 (lower illus.), Air Reduction Inc.; pp. 340, 341, 346, 347, Pressed Steel Tank Co.; p. 343, Olin Mathieson Chemical Corporation; p. 344, Suburban Propane Gas Corporation; p. 345 (upper illus.), The Chlorine Institute; p. 345 (lower illus.), Air Products & Chemicals, Inc.; pp. 348, 353, Phillips Petroleum Co.; pp. 351 (lower illus.), 352 (upper illus.), Union Carbide Corporation.**

TO
FRANKLIN R. FETHERSTON

Foreword

This *Handbook* was prepared by the Compressed Gas Association, Inc. under the guidance of Franklin R. Fetherston, who for thirty-eight years served as Secretary-Treasurer and Managing Director of the Association. A substantial part of the volume's initial planning and preparation was carried out by John Henry Kunkel, long active in editorial capacities in the compressed gas industry. The book has been compiled and edited principally by Gene R. Hawes, author, Editor of *Columbia Research News* at Columbia University, and Consultant to Columbia University Press. Material covering technical information was furnished by many members of the Association, highly expert in their respective fields, who most generously cooperated in providing knowledge and assistance.

G. C. CUSACK, CHAIRMAN
Handbook Committee
Compressed Gas Association, Inc.

Preface

For many years, most men working in any way with compressed, liquefied and cryogenic gases have agreed that a handbook like the present volume was seriously needed. Their daily experience made the need crystal clear. Surely the field involved a burgeoning wealth of unique technical data, safety and regulatory practice, and production, shipping and handling art that would be extraordinarily helpful to have summarized between the covers of a single book. Doing such a volume well has long been recognized as a very extensive task compounded in difficulty by an increasingly rapid pace of significant innovation in the field.

Finally, rather than have the field continue indefinitely without the advantages of such a work, the Compressed Gas Association, Inc. (CGA) embarked on preparing a handbook several years ago. This "Handbook of Compressed Gases" is the result. Criticisms and suggestions for improvement in future editions will be most welcome.

The character and authority of this first edition stem, essentially, from the CGA itself. Just as the Association is concerned with the broadest of interests relating to gases, so is the book designed to serve the needs of the broadest groups of users—groups including businessmen, public officials and students no less than engineers and scientists seeking basic guideposts in unfamiliar terrain. And just as the Association is a technical organization interested in both adequate data and sound utilization for gases, so do the two main parts of the book, II and III, deal primarily with information about individual gases on the one hand, and general information about their safe handling, containment and shipment on the other.

CGA, a nonprofit, membership corporation chartered in New York, represents all segments of the compressed gas industry in the United States, Canada and Mexico. Among its membership are some 250 companies—including North America's largest firms devoted primarily to producing and distributing compressed, liquefied and cryogenic gases, as well as leading chemical industry corporations which have compressed and liquefied gas divisions. Equipment, container and valve manufacturers and distributors of gases and equipment constitute others especially important types of members. Many of the largest producers and equipment manufacturers in Latin America, the British Commonwealth, Continental Europe, Africa and Asia are represented in another particularly significant group of members. Any interested businesses, corporations, associations and individuals may join.

More than 30 technical committees conduct the most fundamental work of the CGA. Some, like the Ammonia Committee or the Hydrogen Committee, attack problems connected with individual gases; others, like the committees on the medical gases and the halogenated hydrocarbons, deal with gas families and their uses; still other groups of committees are responsible for kinds of containers, such as cylinders or tank cars, or for relevant areas of expert knowledge like metallurgy or low-temperature phenomena. Committees are organized and take on assignments flexibly, as befits a field which cuts across many professional and business specialities.

Primarily, the CGA's technical committees work to develop standards for the guidance of the industry, regulating agencies, and the public. In this work they frequently collaborate with other organizations concerned with safety standards for compressed gases, such as the National Fire Protection Association, the American Standards Association, the American Society for Testing and Materials, and international standards organizations. For a company to have its seasoned executives serve on CGA committees represents both a vital contribution to the industry and to the public, and a prized honor.

In a particularly significant part of its work, the CGA acts as an advisor to regulatory authorities that include the U.S. Interstate Commerce Commission and the U.S. Coast Guard, Canada's Board of Transport Commissioners and Department of Transport, and state, provincial, municipal and local agencies concerned with safe handling of compressed gases. The Association has also long maintained a very close working relationship with the Bureau of Explosives of the Association of American Railroads, which serves as the official representative of all American and Canadian railroads to the ICC and the Board of Transport Commissioners for Canada. CGA recommendations have consequently played an essential role, down through the years, in developing the whole system of practices and regulations for transporting compressed gases throughout North America—the continent which benefits from the world's most highly advanced compressed gas technology. The CGA has thus served industry and the public through more than a half century, since the time of its founding as the Compressed Gas Manufacturers Association in 1913.

Through the decades of its past accomplishments, the CGA has seen profound changes develop increasingly within the industry as the march of technological progress has steadily accelerated. Fluorocarbon gases, for example, were unknown in their common air-conditioning and aerosol-packaging applications of today when the Association began early in the century. Another even more portentous development has been the rise of the entire field of cryogenics, based unavoidably on gases, and crucial to progress in such divergent fields as surgery, food preservation and the exploration of space. It is indicative of today's widening horizons that the CGA's Constitution currently interprets "compressed gas" to mean gases in their gaseous, liquid or solid state.

To report in full on this field, now expanding more rapidly than in any past era, would require nothing less than loose-leaf binding with hourly bulletins and acres of shelf space. However, in the field, there is a core of essential knowledge that persons need to refer to most frequently and most widely, and that underlies advances unfolding today and tomorrow. This book is a first attempt to put that essential core in the hands of the many and diverse people to whom it can be most helpful. As such, this Handbook is offered to further the CGA's very first Constitutional purpose:

"To promote, develop and coordinate technical activities in the compressed gas industries, including end uses of products, in the interest of safety and efficiency, and to the end that they may serve to the fullest extent the best interest of the public."

NORMAN A. EVANS, PRESIDENT
Compressed Gas Association, Inc.

Contents

PART I

An Introduction to Using Compressed Gases

CHAPTER 1

The Compressed and Liquefied Gases Today

Gases in compressed or liquefied form play countless indispensable roles in modern technology. Oxygen and acetylene often cut the physical shape in steel of our structures and machinery, after oxygen has first helped to produce stronger and cheaper steels. Acetylene welding and brazing of familiar metals have long been very common; other flammable gases, like hydrogen, are equally essential for the welding of certain metals, and some of the newer metals and alloys (like stainless steels, titanium and zirconium) can be welded only under an inert gas atmosphere. Propane and other fuel gases are also widely used in metal cutting.

Anhydrous ammonia is used to fertilize croplands pressed for greater productivity with lower costs and higher yields in the face of a growing world food shortage. Carbon dioxide hangs in special containers on thousands of plant and laboratory walls as the most widely used dry extinguisher for chemical and electrical fires.

Nitrous oxide, cyclopropane and ethylene are the workhorse anesthetic gases of present-day surgery, while giving oxygen is standard first aid in many emergencies as well as standard treatment for maladies including heart ailments; also, oxygen given in the new "hyperbaric" chambers (ones in which pressures exceed atmospheric pressure) have begun bringing patients literally back from the dead.

Propane and butane provide precisely controllable heat for innumerable processes in industry; similarly, low temperatures used in industrial processing and scientific research are produced by refrigerant gases and gases liquefied into cryogenic fluids below temperatures of about −200 F.

Many chemicals essential to the making of plastics, synthetic rubber and modern drugs are compressed or liquefied gases—among them, butadiene, chlorine, vinyl chloride, acetylene, ammonia and the methylamines.

Compressed gases make it possible to explore vast unknown realms underseas, within and without submarines. And not only do gases keep space-voyaging astronauts alive; liquefied gases produce the power that drives rocket vehicles out into space.

Even everyday life in industrialized nations today depends on gases for technological advantages so common that they are taken for granted. All refrigerating and air conditioning machines, in homes, stores, offices, plants, schools, hospitals, warehouses, autos, trucks and trains, use refrigerant gases—often, the fluorocarbon gases—while the "dry ice" that solid carbon dioxide represents has largely replaced its natural namesake in the bulk handling of perishable foods. It is, of course, bubbles of carbon dioxide dissolved under pressure that make up the "pop" in all kinds of soda pop or "carbonated" beverages.

In a like manner, the versatile fluorocarbons, nitrous oxide and some of the LP-Gases provide the push behind all kinds of new pressure-packaged products which foam or spray out of hand-size containers—foams of whipped dessert toppings and shaving lathers, and sprays of perfumes, shampoos, suntan lotions, paints, insecticides and many other useful liquids. Millions of children and adults instantly recognize the traces of chlorine gas in the water of swimming pools that guard them against infection. Propane and butane—the chief "liquefied petroleum" or "LP" gases—again are the compact portable fuels in millions of homes that lie beyond gas mains. LP-Gas powers whole fleets of trucks and buses instead of gasoline or diesel fuels, especially in the mid-west and western U. S., while LP-Gas is used heavily on farms for powering vehicles, burning weeds and many other purposes. And in the heart of cities, signs lighting the way to refreshment or entertainment contain in their scripts of tubular glass the bright, glowing colors of neon, argon, helium or krypton.

In sum, to a steadily increasing degree, many of the powers of modern technology, as well as many necessities and conveniences of modern life, stem from mastery of the production and use of compressed and liquefied gases.

TRENDS TOWARD LARGER USAGE

Current trends point to an accelerating development of the uses of compressed gas. Many such growing uses apply to human physiology. Use of gases to sustain life in hostile environments outside the earth's atmosphere has barely begun to be developed, and will surely mushroom with widening exploration. In medicine, cryosurgery with liquefied nitrogen is becoming a standard operating technique and oxygen therapies keep multiplying, while hyperbaric oxygenation treatments have been introduced only by more advanced hospitals.

Gases are handled more and more as cold liquids and cryogenic fluids with steadily improved equipment and materials for low-temperature work, largely because many gases are most compact and exert little or no pressure when cooled down to liquids. Methane or natural gas liquefied at about −260 F, already referred to today as "LNG" (liquefied natural gas), may be expected to become as commonplace as LP-Gas. Compatible shipping, handling and receiving facilities for low-temperature liquid transportation of gases appear with growing frequency. A host of new materials and methods for containing and piping cryogenic fluids have been originated in recent years, and will certainly be augmented in years ahead by many new materials now completely unknown.

The present era of burgeoning demand for gases in ever larger quantities has made the search for means to transport and store them in greater quantities one of the most conspicuous present trends in the compressed gas industry. "Pregnant whale" and "jumbo" tank cars with two or three times the capacity of conventional tank cars have recently gone into service, as have larger highway "tankers" and double-trailer tank carriers. Truck trailers and large removable truck cargo tanks ride railroad flatcars "piggyback" for flexibility in large-volume, low-cost shipment. Pipeline transport in bulk to quantity industrial users near or even at some distance from gas producers is on the rise. Shipment of gases in tank barges and tankships continues to mount, and both water and land shipments are made increasingly at low temperatures and low or no pressures. Storage tanks of unprecedented size and design have been developed at a rising rate; one striking new type consists of a huge lined and domed pit in the ground, around which the earth is kept frozen.

PURPOSE OF THIS HANDBOOK

Today's very large and growing uses of compressed gas in almost every area of industry, in medicine and even in many facets of daily life, give rise to widespread needs for authoritative information which the *Com-*

pressed Gas Handbook attempts to meet. The *Handbook* is designed primarily for use by the many persons in business, industry, government, education, transportation, the health professions and the armed services who deal with gases. In order to be of the greatest utility, it is written to be understood by the person without professional training in the sciences or engineering—to serve for reading and reference beyond just the engineering office, the research laboratory and the college library. The book gives basic information about each gas having present commercial importance, and general information about all gases that is essential for persons who are working in any way with gases.

This first part of the book introduces persons who deal with gases, as well as students, to the whole compressed gas field and to the American and Canadian regulatory bodies active within it. Part II presents detailed information about individual major gases. The standards for safe handling and for containers that have been developed by the compressed gas industry itself are given in Part III. Reading the two chapters of Part I should prove helpful to at least all persons new to the field. Parts II and III are mainly for reference, though it is essential for anyone to read the Part II introduction before using any of the subsequent information on individual gases. Reference use of Parts II and III is facilitated by their organization into many sections and subsections—a section on each gas in Part II, and in Part III, a chapter on each specific area of standardization (such as cylinder valve connections, storage tank safety relief devices, safe handling of cylinders, hydrostatic retesting, etc.).

WHAT THE COMPRESSED AND LIQUEFIED GASES ARE

Whether or not we consider any compound or element to be a gas depends, of course, on the cosmic coincidence of temperatures and pressures stable within fortunately narrow limits that are found on the surface of the planet, Earth. Accordingly, to us a gas is any substance that boils at atmospheric pressure and any temperature between the absolute zero of outer space through some 40 to perhaps 80 degrees Fahrenheit (40 to 80 F, as abbreviated in this book). Eleven of the 92 elements (not including transuranium elements) happen to have such boiling points, as do apparently unlimited numbers of compounds and of mixtures like air. (The 11: hydrogen, nitrogen, oxygen, fluorine, chlorine, and the 6 inert gases—helium, neon, argon, krypton, xenon and radon.)

Gases are defined as "compressed" for practical reasons of transportation, storage and use. The definition carrying greatest weight in the U. S. is the one that has been adopted by the Interstate Commerce Commission, the federal regulatory body empowered by act of Congress to prescribe regulations for their safe transportation in interstate commerce. The ICC definition currently reads:

". . . any material or mixture having in the container an absolute pressure exceeding 40 psi (pounds per square inch) at 70 F or, regardless of the pressure at 70 F, having an absolute pressure exceeding 140 psi at 130 F;

"or any flammable material having a vapor pressure exceeding 40 psi absolute at 100 F . . ." (the latter, as determined by the Reid method specified in the "Method of Test for Vapor Pressure of Petroleum Products" of the American Society for Testing and Materials).

Absolute pressure, incidentally, is the pressure in a container that would appear on an ordinary gage plus the local atmospheric pressure of some 15 psi (14.696 psi at sea level and 32 F is the generally accepted standard value).

In America, then, a compressed gas is generally taken to be any substance which, when enclosed in a container, gives a pressure reading of at least:

(1) either 25 psig (pounds per square inch, gage pressure) at 70 F; or 125 psig at 130 F; or

(2) if the contained substance is flammable, 25 psig at 100 F.

Two major groups of gases that differ in

physical state when contained result from this definition, and from the range of boiling points among gases. They are:

(1) Gases which do not liquefy in containers at ordinary terrestial temperatures and under pressures attained in commercially used containers, which range up to 2000 to 2500 psig.

(2) Gases which do become liquids to a very large extent in containers at ordinary temperatures and at pressures from 25 to 2500 psig.

The first group, commonly called the *nonliquefied gases*, are elements or compounds that have relatively low boiling points, say, from −150 F on down. However, the nonliquefied gases, of course, become liquids if cooled to temperatures below their boiling points. Those that liquefy at "cryogenic" temperatures (temperatures from absolute zero, or −459.7 F, up to around −200 F—though the upper limit set for the cryogenic range varies widely from one authority to another), are therefore also known as *cryogenic fluids*.

The second group, called for some years the *liquefied gases*, are elements or compounds that have boiling points relatively near atmospheric temperatures, from about −130 F to 25 F or 30 F. The liquefied gases solidify at cryogenic temperatures, and only one of them—carbon dioxide—has as yet come into wide use in solid form.

Oxygen, helium and nitrogen are examples of gases in wide use both as nonliquefied gases and cryogenic fluids. With respective boiling points of −297 F, −425 F and −320 F, they are charged into high-pressure steel cylinders at more than 2000 psig at 70 F for shipment and use as nonliquefied gases. However, when shipped as cryogenic fluids, they are cooled down to liquid form and charged into special insulated containers that keep them below their boiling points and operate at gage pressures of less than 75 psig or even less than 1 psig.

Examples of widely used liquefied gases and their boiling points are anhydrous ammonia (−28 F), chlorine (−29 F), propane (−44 F) and carbon dioxide (which sublimes directly from a solid to a gas at −109 F and one atm, and liquefies only under pressure). When charged into cylinders at typical pressures ranging from 85 psig for chlorine to 860 psig for carbon dioxide, these are all largely liquids. In cylinders filled with them to the maximum amounts permitted for shipment, only a small vapor space is left inside to allow for expansion in case of heating.

A third physical state in the container is represented by only one widely used gas—acetylene, which is sometimes referred to as a "dissolved" gas (in contrast to a "nonliquefied" or "liquefied" gas). The industry recommends that free acetylene should not ordinarily be handled at pressures greater than 15 psig because, if handled at higher pressures without special equipment, it can decompose with explosive violence. In consequence, acetylene cylinders are packed with an inert porous material which is then saturated with acetone. Acetylene charged into the cylinder dissolves in the acetone and in solution will not decompose at or below the maximum authorized shipping pressure of 250 psig at 70 F. Some consumers of acetylene in bulk for chemical processing make acetylene at their plants by reacting calcium carbide and water rather than having it supplied in the special acetylene cylinders.

THE MAJOR FAMILIES OF COMPRESSED GASES

Compressed and liquefied gases are also often described according to loosely knit families to which they belong through common origins, properties or uses.

Atmospheric gases comprise one of their leading families. Its bulkiest member is nitrogen, constituting 78 per cent air by volume, and oxygen (21 per cent of the air) is its second most abundant member. All animal life, of course, utterly depends on these two elements, needing the one in nutrition and the other in respiration. Most of the remaining 1 per cent of the atmosphere consists of a sub-family of gases sharing the property of chemical inertness, the inert gases: chiefly argon, with minute amounts of helium, neon, krypton, xenon and radon. The last four are frequently called the "rare gases" of the atmosphere, due to their scarcity. Hydrogen also occurs minutely in the atmosphere, as do a large variety of trace constituents, small amounts of carbon dioxide and large amounts of water vapor.

Nitrogen, oxygen, argon and the rare gases are commercially produced by cooling the air down to liquid form and then distilling off "fractions" of it having different boiling points, much as petroleum fractions are distilled at higher temperatures. The process is called fractionation. Some high-purity helium is obtained in fractionation, but most helium used today comes from wells of natural gas in which it occurs in concentrations of a few per cent.

Fuel gases burned in air or with oxygen to produce heat make up a large family of gases related through the major use. Its members are notably the hydrocarbon gases—especially the leading LP-gases, propane and butane. Methane, the largest component of natural gas, is another leading representative of the family, and such welding gases as acetylene and hydrogen are somewhat special representatives. New members now joining the fuel gas family in a current development in the field are the inhibited methylacetylene-propadiene mixtures, which resemble LP-gas in a number of ways.

An opposite application relates members of another extensive family, that of refrigerant gases. A refrigerant gas should be one that liquefies easily under pressure, for it works by being compressed to a liquid mechanically and then by absorbing large amounts of heat as it circulates in cooling coils and vaporizes back into gas. Dry or anhydrous ammonia does liquefy under low pressure, and was the earliest widely used refrigerant. Any liquefied gas, though, is a strong candidate for membership in the refrigerant family; even some gases bordering on cryogenic fluids, like methane, have been classified as refrigerant gases by one of the main organizations in the field, ASHRAE (the American Society of Heating Refrigerating and Air Conditioning Engineers).

Among the most popular refrigerant gases today are the fluorocarbons, a family of almost indefinitely large size since they are any of the endless series of hydrocarbons which have been fluorinated. Fluorocarbons serve well as refrigerants because most of them are chemically inert to a large extent and they can be selected, mixed or compounded to provide almost any physical properties desired in particular refrigerant applications.

Aerosol propellant gases constitute a family related by use that has come into commercial importance in recent years, with the introduction of pressure-packaged products used in the form of a foam or a spray. The insecticides, paints, whipped toppings, perfumes, shaving soaps and other liquids with which propellants are used are packaged with propellants so that a dip tube runs from the finger-tripped valve on top of the container down to the bottom of the liquid. When the valve is opened, the vapor pressure of the propellant forces the liquid up the dip tube and out through the valve in spray or foam form. The propellant, liquefied or dissolved under moderate pressure, then evaporates into the enlarged vapor space, and thus continues to exert its constant vapor pressure until the product liquid is consumed. Such constant container pressure assures uniform consistency of foam or spray through the life of the package. Propellant gases have moderate vapor pressures at room temperatures—70 psig down to 35 to 40 psig, and still lower in some cases. It is usually thought that a good propellant should also be nontoxic, chemically stable, noncorrosive and, of course, inexpensive.

Nitrous oxide illustrates how confused family ties among the gases can be, for it is a prominent member of the family of gases used in medicine, as well as a respected propellant gas and a reliable refrigerant gas. Cyclopropane and ethylene are also gases that, like nitrous oxide, serve widely in medicine as anesthetics. Oxygen enjoys large and growing application in medicine, where it is employed alone as well as mixed with carbon dioxide or helium for many kinds of inhalation therapy.

Gases considered to be members of the poison gas family in the U. S. are generally those that the ICC has classified as poison gases to insure public safety in interstate shipment. Two of these gases that have commercial importance are included in Part II of this volume, hydrogen cyanide and phosgene. Both are shipped as liquefied gases, and their inhalation hazards are incidental to their use as intermediates in the chemical industry. The two gases serve as intermediates in the production of wide varieties of compounds, some of the most familiar being "Plexiglas" and other acrylic plastics (in the case of hydrogen cyanide) and barbiturate drugs (in the case of phosgene). As is generally the case with gases designated by the ICC to be poisons of the class A or "extremely dangerous" group, the potential danger posed by hydrogen cyanide and phosgene is chiefly through inhalation. Hydrogen cyanide concentrations of from 100 to 200 parts per million (ppm) in air, if inhaled for 30 to 60 minutes, can be fatal. Inhaling phosgene concentrations of some 700 ppm can be fatal within several minutes. Extensive safety requirements and practices have been developed for shipping and handling these and other poison gases, however, and accidents do not occur with them when all recommended protective measures are taken.

Still other gases that are important commercially have no marked family ties; many of these are used in chemical industry processing. The methylamines, for example, serve as chemical intermediate sources of reactive organic nitrogen, and methyl mercaptan helps synthesize insecticides. They are liquefied gases. Another liquefied gas, sulfur dioxide, is widely employed as a preservative and bleach in processing foods, as a bleach in manufacturing sulfite papers and artificial silk, and as an additive to irrigating water to improve the yield of alkaline soils in the American Southwest. Carbon monoxide, a nonliquefied gas, serves chiefly in the making of other chemicals, such as ethylene, and also in refining high-purity nickel metal. Fluorine, another nonliquefied gas, is used to make certain fluorides—among them sulfur hexafluoride, which has high dielectric strength and is widely employed as an insulating gas in electrical equipment.

BASIC SAFETY PRECAUTIONS TAKEN AGAINST COMPRESSED GAS HAZARDS

The very powers that make the compressed and liquefied gases widely useful in modern life—high heat output in combustion for some gases, high reactivity in chemical processing with others, extremely low temperatures available from some, and the economy of handling them all in compact form at high pressure or low temperature—these powers often can also represent hazards if the gases are not handled with full knowledge and care.

Part III of this volume presents the specific and detailed standards that have been developed by the compressed gas industry to insure safety, and readers are urged to consult the full recommendations there for any aspect of the field with which they may be concerned. At this point, however, it may be helpful to outline briefly the chief hazards posed by compressed gases and the main precautions taken against them.

Practically all gases can act as simple asphyxiants by displacing the natural oxygen in the air. The chief precaution taken against this potential hazard is adequate ventilation of all enclosed areas in which unsafe concentrations might build up. A second precaution is to avoid entering unventilated areas that might contain high concentrations of gas without first putting on breathing apparatus with a self-contained or hose-line air supply. A number of gases do have characteristic odors which can warn of their presence in air; others, however, like the atmospheric gases, have no odor or color whatever. Warning labels are required for compressed and liquefied gas shipping containers; similar warning signs are placed at the approaches to areas in which the gases are regularly stored and used, and unauthorized persons are kept away from such areas.

Some gases can also have a toxic effect on the human system, either through being inhaled, or through having high vapor concentrations or by liquefied gas coming in contact with the skin or the eyes. Adequate ventilation of enclosed areas similarly serves as the chief precaution against high concentrations of gas which can exert toxic effects. In addition, for unusually toxic gases, automatic devices can be purchased or built to monitor the gas concentration constantly and set off alarms if the concentration should approach a danger point. Precautions against skin or eye contact with liquefied gases that are toxic or very cold, or both, include thorough knowledge and training for all personnel handling such gases, the development of foolproof procedures and equipment for handling them and special protective clothing and equipment (such as protective garments, gloves and face shields).

With the flammable gases, it is necessary to guard against the possibility of fire or explosion. Ventilation again represents a prime precaution against these hazards, together with safe procedures and equipment to detect possible leaks. Should fire break out, suitable fire extinguishing apparatus and preparation are also developed to limit damage. Care is also taken to keep any flammable gas from reaching any source of ignition or heat, like sparking electrical equipment or sparks struck by ordinary tools, boiler rooms or unenclosed flames.

Oxygen poses a combustible hazard of a special kind; though it does not itself ignite, it lowers the ignition point of flammable substances and greatly accelerates combustion. It should not be allowed closer than 10 ft to any flammable substance, including grease and oil, and should be stored no closer than that to cylinders or tanks containing flammable gases.

Hazard resulting from the possible rupture of a cylinder or other vessel containing gas at high pressure is protected against by careful and secure handling of containers at all times. For example, cylinders should never be struck nor allowed to fall, for, should the cylinder valve be broken off and the cylinder is charged at high pressure, it could become a projectile. Under most circumstances, the relatively small opening through the remaining portion of the valve is not large enough to cause "rocketing." Cylinders should not be dragged or rolled across the floor; they should be moved by a hand truck. Also, when they are upright on a hand truck or floor or vehicle, they should be chained securely to keep them from falling over. Moreover, cylinders should not be heated to the point at which any part of their outside surface exceeds a temperature of 125 F, and never with a torch or other open flame. Similar precautions are taken with larger shipping and storage containers. Initial protection against the possibility of vessel rupture is provided by the demanding requirements and recommendations which compressed gas containers fulfill in their construction, testing and retesting.

The compressed and liquefied gases are handled today in the U. S. and Canada in enormous volumes and in a great variety of applications with a very high record of industrial and public safety. The compressed gas industry itself has led in the development of safe equipment and practices, acting both out of public interest and enlightened self interest.

In addition, the public and the industry are protected by many government authorities which make regulations about shipping, storage, labelling and other matters in the interest of public safety. The industry cooperates fully with these bodies, and has often served them by helping develop regulatory codes that are sound and thoroughly safe for all concerned. The main regulatory authorities with responsibilities in the compressed gas field are described in the next chapter.

CHAPTER 2

Regulatory Authorities for Compressed Gases in the United States and Canada

Persons who produce, supply or use compressed gases, as well as those involved in compressed gas transportation, must comply with a variety of governmental safety regulations in the United States and Canada. These regulations are issued and enforced by regulatory bodies on the federal, state or provincial, and local levels of government in the two countries.

Federal regulation applies chiefly to land shipment of compressed gas between states or provinces and to shipment by water and by air, and sets detailed specifications for shipping containers and various shipping practices. State and provincial and local regulation applies mainly to the storage and use of compressed gases, though it extends in some cases to certain transportation matters within the locality, province or state.

The major regulatory bodies or kinds of bodies on each level of government in the two nations, and important aspects of their powers, are as follows. Regulation of shipping rates or charges by various governmental bodies is not treated here, although compressed gas shipment is subject to rate regulations of certain federal and state or provincial authorities.

FEDERAL REGULATORY AUTHORITIES

ICC in the U. S.; BTC in Canada

The two most influential agencies regulating compressed gas shipments in North America are the Interstate Commerce Commission (ICC) of the U. S. Government and the Board of Transport Commissioners (BTC) for Canada, an arm of the Canadian Government. Both issue requirements for shipping compressed gas by rail in the case of gases which they classify as "dangerous" articles or commodities in interstate or interprovince commerce. Their codes require that the designated gases be shipped in containers which comply with certain specifications, and that the containers be equipped with safety devices as stipulated, tested by methods identified, filled within listed maximum amounts, and in some instances be boxed or transferred in particular ways.

It was at the turn of the century that the U. S. Congress charged the Interstate Commerce Commission with the promulgation of regulations, including container specifications, for the safe transportation of hazardous commodities by rail and authorized the ICC to utilize the services of the organization now known as the Bureau of Explosives of the Association of American Railroads. The Bureau had already been functioning for many years as a source of safety requirements for the railroads; it began working in close cooperation with the ICC, and has continued to do so down through the present time. It has long worked in cooperation with the BTC as well, for the Association of American Railroads (AAR) includes Canadian as well as U. S. rail carriers.

Today, the Bureau acts as a central point of coordination and communication for the railroads on the one hand and the ICC and the BTC on the other. It also serves all three groups as a central office for registration, licensing and inspection. Its director and chief inspector also holds the title of "Agent," and he serves in that capacity as the official agent and attorney for all AAR member rail carriers with the ICC and the BTC (and similarly acts on behalf of some steamship lines and freight forwarders as well).

In part because the Bureau of Explosives represents both the Canadian and U. S. railroads, substantially identical requirements have been adopted by the BTC for Canada and by the ICC for the U. S. concerning compressed gas shipment by rail across state or province lines. Canadian needs and interests are thus taken into account in the setting of ICC regulations, as are U. S. needs and interests in the setting of BTC regulations. In consequence, compressed gas cylinders, tank cars and portable tanks can move freely between the two countries, and the shipping, charging, labelling and other practices authorized for the one country largely meet the requirements of the other country. The result represents a remarkable achievement in international cooperation that realizes substantial economies and increased economic activity for both nations.

One major difference between the ICC and the BTC is that the ICC regulates interstate motor vehicle transport of designated gases as well as rail transport in the U. S., while the BTC regulates only rail shipment of gases in Canada. It is the Canadian Provincial Governments rather than the Canadian Federal Government that set regulations on shipping gases by motor vehicles in Canada.

The complete regulations of the ICC, including those applying to compressed gases, are published in their current form under the title:

"Agent T. C. George's Tariff No. 15, Publishing Interstate Commerce Commission Regulations for Transportation of Explosives and Other Dangerous Articles by Land and Water, in Rail Freight Service, and by Motor Vehicle (Highway) and Water, including Specifications for Shipping Containers."

This document, and supplements revising parts of it since it appeared in 1963, is issued by the previously mentioned agent. It may be obtained from the Bureau of Explosives at 63 Vesey Street, New York, N.Y., 10007. Complete revisions or reissues of the ICC regulations are published under new tariff numbers every few years.

The complete regulations of the Board of Transport Commissioners for Canada are similarly published as now in force under the title:

"Board of Transport Commissioners for Canada, Regulations for the Transportation of Dangerous Commodities by Rail."

These regulations are also Agent E. R. Evans' C.T.C. No. E.T. 4600 (the tariff of the Express Traffic Association of Canada), and Agent G. A. Richardson's C.T.C. No. 5 (the tariff of the Railway Association of Canada, part of the AAR).

The regulations and supplements subsequent to their publication in 1962 are issued by, and available from, Roger Duhamel, Printer to the Queen's Most Excellent Majesty, Ottawa, Ontario.

U. S. Coast Guard; Department of Transport of Canada

Water shipment of compressed gases is regulated in the U. S. by the Coast Guard, and in Canada by the Department of Transport, both of which are federal government organizations in their respective countries. The regulations of each of these bodies generally provide for compressed gas shipment by water in the same kinds of containers that meet ICC and BTC requirements for rail shipment. As the Department of Transport notes, however, the packing requirements for shipment by sea are in some cases more stringent than those set for shipment by rail.

Additional regulations of the two bodies give requirements and specifications for shipping certain compressed gases by tankships and tank barges. Ammonia, chlorine and liquefied petroleum gas are the main gases that have been shipped in bulk aboard such tank vessels (though others for which tank-barge shipment in particular is growing include inhibited butadiene, anhydrous dimethylamine, liquefied hydrogen, liquefied oxygen, methyl chloride and vinyl chloride).

Current regulations of the U. S. Coast Guard are published in full as:

"Agent T. C. George's Water Carrier Tariff No. 17, Publishing United States Coast Guard Regulations Governing the Transportation or Storage of Explosives or Other Dangerous Articles or Substances, and Combustible Liquids on Board Vessels."

This tariff and supplements issued after its publication in 1965 are available from the Bureau of Explosives.

Two Coast Guard publications giving regulations for shipping compressed gases in tank vessels are, "Rules and Regulations for Cargo and Miscellaneous Vessels, Subchapter I," (CG-257, Sept. 1, 1964), and "Rules and Regulations for Tank Vessels, Subchapter D," (CG-123, April 1, 1964). Both of these publications of the Treasury Department, United States Coast Guard, may be obtained from the Superintendent of Documents in Washington, D. C.

The complete regulations of the Department of Transport of the Canadian government are issued under the title:

"Dangerous Goods Shipping Regulations."

The document, and amendments made since its publication in 1954, are available from the Queen's Printer, Ottawa, Ontario.

International Air Transport Association

Regulations for the air transport of compressed gases made by Federal agencies in the U. S. and Canada, with still further joint requirements of almost all major airlines, are incorporated in an annual publication of the International Air Transport Association, to which airlines belong. The volume reflects in addition the national regulations of many other countries, and is published in several different languages. Its tenth annual edition was effective April 1, 1965. The publication is entitled, "IATA Regulations Relating to the Carriage of Restricted Articles by Air," and may be obtained from IATA at 1060 University Street, Montreal 3, Quebec.

OTHER AREAS OF FEDERAL REGULATION

A few special areas of the compressed gas field are subject to further regulation by other federal government bodies in the U. S. and Canada, as in the labelling and purity requirements prescribed for gases used in medicine under national laws concerning drugs. Another special area of labelling regulation in the U. S. arose with the recent enactment of the Hazardous Substances Labelling Act; the provisions of this act apply only to those gases under pressure which go into the home, as with aerosol propellants. The act does not apply generally to compressed gases, as sometimes has been assumed mistakenly. Information about regulation in these special areas is available from the various appropriate authorities in the two countries, such as the Federal Food and Drug Administration in the U. S.

REGULATION BY STATES AND PROVINCES

States of the U. S. and the Canadian Provinces vary widely in their regulatory provisions for compressed gases. These governments generally have developed regulations applying to compressed gas through their chief fire safety officer, pressure vessel authorities or industry and labor commission. However, in some instances, it may prove to be the Railway Commission or the Industrial Accident Commission which has developed the main body of applicable regulations. Certain states have adopted the ICC regulations in full or in part. These are the ICC regulations concerning the shipment of dangerous articles, as well as the Motor Carrier Safety Regulations of the ICC.

Also, many of the states have based their fire codes for precautions with liquefied petroleum gas on the recommendations of the National Fire Protection Association given in NFPA Pamphlet No. 58, "Storage and Handling of Liquefied Petroleum Gases." Also widely reflected in state as well as local regulations is NFPA Pamphlet No. 51, "Standard for the Installation and Operation of Oxygen-Fuel Gas Systems for Welding and Cutting."

State or provincial regulation of compressed gas concerns primarily its storage and use, though some state regulation in the U. S. governs its transportation. In Canada, as noted previously, the provinces (rather than the national government) regulate shipment by motor vehicle.

Producers, distributors and users of compressed gases should obtain current and full information on state or province regulations directly from the appropriate governmental offices. A summary of certain aspects of current compressed gas regulation by states of the U.S. appears in Appendix A to this volume.

LOCAL REGULATION

Municipalities, towns and other local governments in the U.S. and Canada have also adopted a host of regulations applying to compressed gas storage, use and transportation. These will frequently be issued under the authority of officials like the fire commissioner, the head of the department of buildings or the zoning commissioner. Inquiries about local regulations should again be made with the governments concerned.

PART II

Individual Compressed Gases: Properties, Manufacture, Uses, Safe Handling, Shipping, Containers

Introduction: Essential Reference Material

Basic information on compressed and liquefied gases that are of present commercial importance appears in this part of the *Handbook*. Each gas is treated individually in a separate section (except in the cases of four groups of closely related gases dealt with in group sections). Gases produced and used in only small laboratory quantities are not included.

KEY TO ABBREVIATIONS AND TERMS USED

Abbreviations and terms given in the gas sections (and elsewhere in the volume) have these meanings:

atm—atmospheres (or pressure; 1 atm = 14.7 psi)
Btu—British thermal units
C—degrees Centigrade (Celsius)
cargo tank—large unitary container fixed to a motor vehicle or motor vehicle trailer or semitrailer
C_p—specific heat at constant pressure
C_v—specific heat at constant volume
cu—cubic
cm—centimeters
F—Farenheit, degrees Farenheit
filling density—the container charging ratio for liquefied gases. It is the maximum limit to which containers are authorized to be charged under ICC and BTC regulations. (See also "SERVICE PRESSURE.") It is given in ICC and BTC regulations as "The per cent ratio of the weight of gas in a container to the weight of water that the container will hold at 60 F (1 lb of water = 27.737 cu in. at 60 F)."
ft—feet
g—grams
gal—U. S. gallons (not Imperial gallons)
in.—inches
kg—kilograms
lb—pounds avoirdupois
portable tank—large transferrable container originally designed to transport compressed gas by motor vehicle
psi—pounds per square inch pressure
psia—pounds per square inch, absolute pressure
psig—pounds per square inch, gage pressure
service pressure—under ICC and BTC regulations it is the authorized pressure marking on the container. For example, for cylinders marked "ICC-3A1800," the service pressure is 1800 psig.
sq—square
TMU—ton multi-unit; a TMU container is a large transferrable cylinder originally designed to transport a ton of liquefied chlorine in multiple units on railroad flatcars.
tube trailer—motor vehicle trailer fitted with long, multiple, fixed tubes (usually ones built to meet ICC or BTC cylinder specifications 3A, 3AX, 3AA or 3AAX) for transporting nonliquefied gas at high pressure.

Chemical symbols and other standard abbreviations not shown above are also employed in Part II.

STANDARD INFORMATION GIVEN ABOUT THE GASES

Standard topics covered in the gas sections of Part II are as follows, and in the following order:

- Generally accepted chemical name of the gas.
- Chemical formula (the empirical formula, or structural formula in some cases)
- Synonyms (other names by which the gas is or has been known)
- ICC classification of the gas
- Physical constants
  - International symbol (most often, the empirical formula)
  - Molecular weight
  - Vapor pressure
  - Density of the gas
  - Specific gravity of the gas compared to air
  - Specific volume of the gas
  - Density or specific gravity as a liquid
  - Boiling point
  - Melting point
  - Critical temperature
  - Critical pressure
  - Critical density
  - Latent heat of vaporization
  - Latent heat of fusion
  - Specific heat
  - Ratio of specific heats
  - Heat of combustion (gross and net)
  - Solubility
  - Weight per gallon of liquid
- Properties
- Materials of construction
- Manufacture
- Commercial uses
- Physiological effects (followed in some cases by subsections on leak detection, special precautions for handling, protective equipment, etc.)
- Containers
  - Filling limits
- Shipping methods; regulations
  - by rail
  - by highway
  - by water
  - by air
- Cylinders
  - Valve outlet and inlet connections
  - Cylinder requalification
  - Safe handling (in some cases)
- Single-unit tank cars and TMU tanks
- Cargo tanks and portable tanks
- Storage containers and handling equipment
- References

QUALIFICATIONS ON PHYSICAL CONSTANTS AND ALL MATERIAL

Data given on the physical constants in the gas sections are based on authoritative scientific and industrial sources, as are all other matters of factual information and recommended practice. In presenting all this material, the publisher and Compressed Gas Association, Inc., assume no legal responsibility whatever for any losses or injuries sustained, or liabilities incurred, by persons or organizations acting in any way on the basis of any part of the material. However, the information and recommendations are believed to be accurate and sound to the best knowledge of the CGA.

The physical constants data that are given generally represent the properties of pure commodities rather than those of commercial grades of the gases. The properties of commercial grades should be expected to differ somewhat from the values for pure grades presented here.

NECESSARY REFERENCE TO SECTIONS IN PART III

Concerning safety with individual gases, readers should refer to certain chapters in Part III of this book for essential detailed information. These chapters are:

Chapter 1, especially Section A, on the general rules for the safe handling of gases, and Section B, on the safe handling of gases used medicinally. The sections directly and very significantly supplement all Part II gas sections.

Chapter 2, on safety relief device standards

for different gases and different types of containers.

Chapter 3, on safety practices with compressed gas cylinders—marking and labeling, testing and inspecting for requalification, repairing and disposition.

Chapter 4, on insuring safety through standard cylinder valve connection systems in general and for gases used in medicine.

Sections of Part III also directly supplement the gas sections of Part II, as follows: (1) when hydrostatic retesting is identified as required for cylinder requalification in a gas section, the testing meant is that described in Sections D and F of Chapter 3 in Part III; (2) when visual inspection is identified as required for cylinder requalification, the present standards for such visual inspection are those given in Section C of Chapter 3, Part III; and (3) when standard cylinder valve outlet and inlet connections are identified in a gas section, the standard connections cited are those about which further information is given in Chapter 4, Part III.

Persons unfamiliar with the compressed gas field will also find it helpful to refer to Chapter 5 in Part III for descriptions and illustrations of the different types of containers that are merely identified in the following gas sections. Similarly, Chapter 6 of Part III will be helpful for those who need to know how to safely unload bulk shipments of liquefied gas.

VITAL ASPECTS OF ICC AND BTC REGULATIONS

Shipping regulations summarized in the gas sections are those of both the BTC and the ICC; however, for the sake of simplicity, ICC has been given consistently in the sections but it means ICC or BTC whenever it appears in the sections (as pertaining to rail shipment or cylinders, of course, for the BTC does not regulate motor vehicle shipment or water shipment in Canada). It is of course the ICC and BTC regulations in effect as the book is being prepared that are summarized here; for the full and current regulations, readers should see the current editions of their published forms as cited in the preceding chapter.

It is also important for readers to understand that, when specifications of cylinders and tank cars authorized for an individual gas are identified in the gas sections, cylinders or tank cars having the same service pressures as those given *and also those having higher service pressures* are authorized. The required service pressures thus shown in the gas sections, and similarly given in ICC and BTC regulations, are the *minimum* service pressures (not the sole service pressures) authorized for containers of the specified type. For example, if cylinders meeting ICC or BTC specifications 3A150 are noted as authorized for a given gas, then any other 3A cylinders with higher service pressures are also authorized (such as 3A1000, 3A2000, etc.).

SOURCES OF WATER AND AIR SHIPPING REGULATIONS

Shipping regulations which apply specifically to shipment by water and to shipment by air that are summarized in the gas sections are the respective regulations of the U. S. Coast Guard and of the International Air Transport Association. As with the ICC and BTC regulations, it is the Coast Guard and IATA regulations in force as the *Handbook* is being prepared which are summarized. Readers should see the full current regulations of the Coast Guard and of IATA in the current editions of their published forms as cited in the previous chapter, and in addition should see the regulations cited there of the Canadian Department of Transport for requirements concerning shipment by water from Canadian ports.

Explanation of one detail concerning summarization of current U. S. Coast Guard regulations in the individual gas sections that follow may be helpful to readers. Many gas sections carry, under the subheadings on U. S. regulations for shipment of the gases by water, notations indicating that the gas is authorized for shipment in cylinders on barges of indicated U. S. Coast Guard classes of barges, such as Class A, Class BA, BB or BC, or Class CA or CB. These classes of barges are identified in section 146.10–3 of the U. S. Coast Guard regulations in effect as this volume went to press ("Agent T. C. George's Water Carrier Tariff No. 17, Publishing United States Coast Guard Regulations Governing the Transportation or Storage of Explosives or Other Dangerous Articles or Substances, and Combustible Liquids on Board Vessels," available from the Bureau of Explosives, 63 Vesey St., New York, N. Y. 10007).

CONVERSION TO METRIC AND OTHER UNITS

For the convenience of readers who would like to convert the values given for physical constants to metric units, conversion formulas are as follows (the conversion formulas and factors shown will give values correct to at least three significant figures):

$°F = (1.8 \times °C) + 32$

$°C = 5/9\,(°F - 32)$

$°K = °C + 273.16$

kg = 2.205 lb

$$\text{liters} = \frac{\text{cu ft}}{28.32}$$

$$\text{g/liter} = \frac{\text{lb/cu ft}}{0.0624}$$

$$\text{calories/g} = \frac{\text{Btu/lb}}{1.80}$$

calories/(g)(°C) = Btu/(lb)(°F)

millimeters Hg = 0.193 psia

Imperial gal = 0.833 U. S. gal

1 U. S. gal = 0.133681 cu ft = 231 cu in.

Acetylene

C_2H_2
Synonym: Ethine, ethyne
ICC Classification: Flammable compressed gas, red label

PHYSICAL CONSTANTS

International symbol	C_2H_2
Molecular weight	26.04
Specific gravity of gas at 32 F and 1 atm (air = 1)	0.906
Specific gravity of liquid at −112 F	0.613
Specific volume of gas at 1 atm, cu ft/lb*	
at 60 F	14.5
at 70 F	14.7
Sublimation point at 1 atm	−118 F
Boiling point at 10 psig**	−103 F
Melting point at 10 psig**	−116 F
Triple point	−116 F at 3 psig
Critical temperature	96.8 F
Critical pressure, psia	907
Latent heat of vaporization at triple point, Btu/lb	264
Specific heat, C_p, at 60 F and 1 atm, Btu/(lb)(°F)	0.383
Specific heat, C_v, at 60 F and 1 atm, Btu/(lb)(°F)	0.304
Ratio of specific heats, C_p/C_v, at 60 F and 1 atm	1.26
Heat of combustion, gross, Btu/cu ft	1457
Heat of combustion, net, Btu/cu ft	1407
Solubility in water at 1 atm, vol/vol	
at 32 F	1.7
at 60 F	1.1

Properties. Acetylene is a compound of carbon and hydrogen in proportions by weight of about 12 parts carbon to one part hydrogen (92.3 to 7.7 per cent). A colorless flammable gas, it is slightly lighter than air. Acetylene of 100 per cent purity is odorless, but gas of ordinary commercial purity has a distinctive, garlic-like odor.

Acetylene burns in air with an intensely hot, luminous and smoky flame. The ignition temperatures of acetylene and of acetylene-air and acetylene-oxygen mixtures vary according to composition, initial pressure, initial temperature, and water-vapor content. As a typical example, an air mixture containing 30

* Based on 1.171 grams/liter at 32 F and 1 atm.

** Reported at 10 psig instead of at 1 atm because, under atmospheric pressure, acetylene sublimes directly from the solid to the gaseous state without entering the liquid state.

per cent acetylene by volume at atmospheric pressure can be ignited at about 581 F. The flammable limits of acetylene-air and acetylene-oxygen mixtures similarly depend on initial pressure, temperature and water-vapor content. In air at atmospheric pressure the upper flammable limit is about 80 per cent acetylene by volume, and the lower limit is 2.5 per cent acetylene. If an ignition source is present, 100 per cent acetylene will decompose with violence under certain conditions of pressure and container size and shape.

Acetylene can be liquefied and solidified with relative ease. However, in both the liquid and solid states, acetylene explodes with extreme violence when ignited unless special conditions of confinement are employed. A mixture of gaseous acetylene with air or oxygen in certain proportions explodes if ignited. Gaseous acetylene under pressure may also decompose with explosive force under certain conditions, but experience indicates that 15 psig is generally acceptable as a safe upper pressure limit. Generation, distribution through hose or pipe, or utilization of acetylene at pressures in excess of 15 psig gage pressure or 30 psig absolute pressure for welding and allied purposes should be prohibited.

Pressure exceeding 15 psig can be employed provided specialized equipment is used. Where acetylene is to be utilized for chemical synthesis at pressures in excess of 15 psig, or transported through large diameter pipelines, means to prevent propagation, should ignition occur, must be employed. Packing large-diameter pipe with small-diameter pipes or with metallic Raschig rings are methods that have been successfully used. Insulating of large diameter pipes as a protection against exposure to fires is recommended.[1]

Acetylene cylinders avoid the decomposition characteristics of the gas by providing a porous-mass packing material having minute cellular spaces so that no pockets of appreciable size remain where "free" acetylene in gaseous form can collect. This porous mass is saturated with acetone in which the acetylene actually dissolves. The combination of these two features—porous filler and acetone solvent—allows acetylene to be contained in such cylinders at moderate pressure without danger of explosive decomposition (the maximum authorized cylinder pressure is 250 psig at 70 F, with a variation of about 2.5 psig rise or fall per degree of temperature change).

Materials of Construction. Only steel or wrought iron pipe should be used for acetylene. Joints in piping must be welded or made up of threaded or flanged fittings. The materials for fittings can be rolled, forged or cast steel, or malleable iron. Cast iron fittings are not permissible.

Under certain conditions, acetylene forms readily explosive compounds with copper, silver, and mercury. For this reason, contact between acetylene and these metals, or their salts, compounds and high-concentration alloys, is to be avoided. It is generally accepted that brass containing less than 65 per cent copper in the alloy, and certain nickel alloys, are suitable for use in acetylene service under normal conditions. In normal service, conditions involving contact with highly caustic salts or solutions, or contact with other materials corrosive to copper or copper alloys, can render the above generally acceptable alloys unsatisfactory for this service. The presence of moisture, certain acids, or alkaline materials tends to enhance the formation of copper acetylides. Further information on metallic acetylides is given in the references listed in the concluding "Additional References" section.

For recommendations on acetylene cylinder manifolds and shop piping, users should consult their supplier and recognized safety authorities such as the Underwriters' Laboratories, Inc., the Associated Factory Mutual Fire Insurance Companies, and Compressed Gas Association, Inc.

Manufacture. In the United States and Canada calcium carbide is the principal raw material for acetylene manufacture. Calcium carbide and water may be made to react by several methods to produce acetylene, with calcium hydroxide as a by-product. Acetylene is also manufactured by the thermal or arc cracking of hydrocarbons and by a process employing the partial combustion of methane with oxygen.

Acetylene manufactured from carbide made in the United States and Canada normally contains less than 0.4 per cent impurities other than water vapor. Apart from water, the chief impurity is air, in concentrations of approximately 0.2 to 0.4 per cent. The remainder is mostly phosphine, ammonia, hydrogen sulphide, and in some instances, small amounts of carbon dioxide, hydrogen, methane, carbon monoxide, organic sulphur compounds, silicon hydrides, and arsine. Purified cylinder acetylene is substantially free from phosphine, ammonia, hydrogen sulphide, organic sulphur compounds, and arsine. The other impurities are nearly the same as in the original gas.

Commercial Uses. Approximately 80 per cent of the acetylene produced annually in the United States is used for chemical synthesis. It is possible to use acetylene for an almost infinite number of organic chemical syntheses, but this use in North America has been less extensive than in Europe owing to the ready availability of petroleum from which competitive syntheses are often possible. Nevertheless, acetylene has come into increasing prominence as the raw material for a whole series of organic compounds, among them acetaldehyde, acetic acid, acetic anhydride, acetone, and vinyl chloride. These compounds may be used in turn to produce a diverse group of products including plastics, synthetic rubber, dyestuffs, solvents and pharmaceuticals. Acetylene is also utilized to manufacture carbon black.

The remaining 20 per cent of the annual U. S. acetylene production is used principally for oxyacetylene welding, cutting, heat treating, etc. Small amounts are utilized for lighting purposes in buoys, beacons and similar devices.

Physiological Effects. Acetylene can be inhaled in rather high concentrations without chronic harmful effects, and it has, in fact, been used as an anesthetic. It is of course a simple asphyxiant if present in concentrations high enough to deprive the lungs of oxygen and produce suffocation. However, the lower flammable limit of acetylene in air would usually be reached long before suffocation could occur as the result of an acetylene leak.

ACETYLENE CONTAINERS

Only cylinders are authorized for shipping acetylene. Acetylene producers using the gas for chemical synthesis store acetylene in low-pressure gas holders for which the recommended material is carbon steel.[2]

Filling Limits. The maximum filling density authorized for acetylene in cylinders that meet the specifications and solvent filling requirements of the ICC is 250 psig at 70 F, or lower maximum charging pressures at 70 F for cylinders marked with such lower maximum pressures.

SHIPPING METHODS: REGULATIONS

Under the appropriate regulations and tariffs, acetylene is authorized for shipment as follows:

By Rail: In cylinders (freight or express).

By Highway: In cylinders on trucks.

By Water: In cylinders via cargo vessels, passenger vessels, passenger or vehicle ferry vessels, and passenger or vehicle railroad car ferry vessels. On barges in cylinders for barges of U. S. Coast Guard classes A and C only.

By Air: Aboard cargo aircraft only, in cylinders as required up to 300 lb (140 kg) maximum net weight per cylinder.

ACETYLENE CYLINDERS

Only cylinders that meet ICC specifications 8 or 8-AL, and that also meet requirements for fillings of a porous material and a suitable solvent, are authorized by the ICC for acetylene service.

Important Warnings. Do not attempt to charge acetylene into any cylinders except those constructed for acetylene.

Do not charge any other gas but acetylene into an acetylene cylinder.

Do not mix any other gas with acetylene in an acetylene cylinder.

Failure to observe these warnings may result in an explosion.

ICC regulations prohibit shipment of cylinders containing acetylene gas unless they were charged by or with the consent of the owner.

Valve Outlet and Inlet Connections. Standard connection, U. S. and Canada—No. 510. Alternate standard connection, U. S. and Canada—No. 300. Additional alternate standard connection, Canada only—No. 410. Small valve series standard connections, U. S. and Canada—Nos. 200 and 520.

Cylinder Requalification. Under present regulations, no periodic hydrostatic retest is required for cylinders 8 and 8-AL which are authorized for acetylene service.

Cylinder Capacities. Acetylene is most commonly available in cylinders of capacities of 10, 40, 60, 100, 225 and 300 cu ft. "Lighthouse" type cylinders—those generally used in acetylene-operated automatic aids to marine navigation—are available in larger sizes, the biggest having a capacity of approximately 1400 cu ft.

Marking and Labeling. The following marks are required by the ICC to be plainly stamped on or near the shoulder or top head of all acetylene cylinders as follows: (a) the ICC specification number—ICC-8 or ICC-8AL; (b) a serial number and the user's, purchaser's or maker's identifying symbol (the symbol must be registered with the Bureau of Explosives, 63 Vesey St., New York, N. Y., 10007); (c) the date of the test to which it was subjected in manufacture; and (d) the tare weight of the cylinder in pounds and ounces.

The markings on cylinders must not be changed except as provided in ICC regulations, which require that serial numbers and ownership marks may be changed only when a detailed report is filed with the Bureau of Explosives. Markings on cylinders must be kept in a readable condition.

SAFE HANDLING OF ACETYLENE CYLINDERS

Special precautions that must be observed for the safe handling of acetylene cylinders (in addition to the general requirements set forth in Part III, Chapter 1, "Safe Handling of Compressed Gases") are presented in the following sections.

STORING ACETYLENE CYLINDERS

In storing acetylene cylinders, the user should comply with all local, municipal, and state regulations, and with Standard No. 51 of the National Fire Protection Association.[3]

Inside all buildings, acetylene cylinders should not be stored near oxygen cylinders. Unless they are well separated, there should be a fire-resistant partition between acetylene cylinders and oxygen cylinders.

Acetylene cylinders stored inside a building must be limited to a total capacity of 2000 cu ft of gas exclusive of cylinders in use or attached for use. Quantities exceeding this total must be stored in a special building or in a separate room as required by NFPA Standard No. 51.

Conspicuous signs must be posted in the storage area forbidding smoking or the carrying of open lights.

While storage in a horizontal position does not make the acetylene in cylinders less stable or less safe, it does increase the likelihood of solvent loss, which will result in a lower flame quality when the cylinder is used. Therefore, it is always preferable to store acetylene cylinders in an upright position.

HANDLING ACETYLENE CYLINDERS

Always call acetylene by its proper name, "acetylene," to promote recognition of its hazards and the taking of proper precautions. Never refer to acetylene merely as "gas."

Never attempt to repair or alter cylinders or valves. This should be done only by the cylinder manufacturer. If a cylinder is leaking, follow the recommendations made in the closing section on "Handling Leaking Cylinders."

Never tamper with safety devices in valves or cylinders. Keep sparks and flame away from acetylene cylinders and under no circumstances allow a torch flame to come in contact with safety devices. Should the valve outlet of an acetylene cylinder become clogged by ice, thaw with warm—*not boiling*—water.

Never, under any circumstances, attempt to transfer the acetylene from one cylinder to another, to refill acetylene cylinders, or to mix any other gas with acetylene in a cylinder.

In welding shops and industrial plants using both oxyacetylene and electric welding apparatus, care must be taken to avoid the handling of this equipment in any manner which may permit the compressed gas cylinders to come in contact with the electric welding apparatus or electrical circuits.

Never use acetylene cylinders as rollers or supports, or for any other purpose than storing acetylene.

MOVING CYLINDERS

Cylinders must be protected against dropping when being unloaded from a truck or platform. One method of protection is to use a V-shaped trough as a skid with a wooden or rope bumper at the bottom if the skid is steep. Another method is to use a welded angle iron cradle or rocker-rack so constructed that the cylinder may be slid horizontally into a steel trough at the top of the cradle. The cradle is then tilted so that the cylinder is upended and can be lowered to the ground. A heavy iron counterweight attached to the cradle will help balance the cylinder during the tilting operation.

Special caution is necessary in transporting acetylene cylinders by crane or derrick. Lifting magnets, slings or rope or chain, or any other device in which the cylinders themselves form a part of the carrier, must never be used for hoisting acetylene cylinders. Instead, when a crane is used, a platform, cage or cradle should be provided to protect the cylinders from being

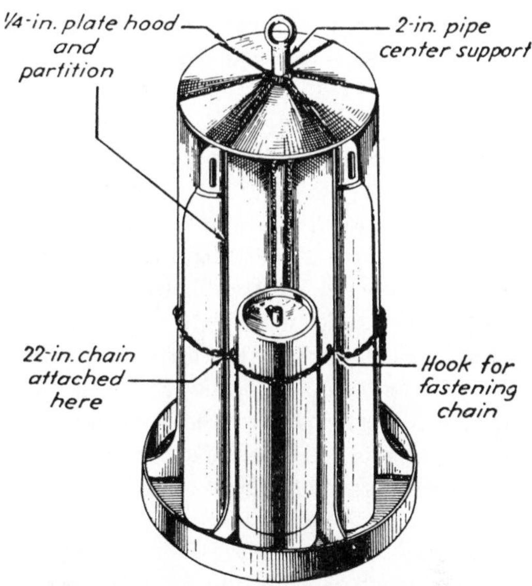

FIG. 1. Recommended type of cradle to hold acetylene cylinders when moved with a crane or derrick.

damaged by slamming against obstructions, and to keep them from falling out. A recommended type of cradle to build for this purpose is shown in Fig. 1.

Horizontal movement of cylinders is easily accomplished by the use of a hand truck; however, when a hand truck is used some positive method, such as chaining, should be used to secure a cylinder standing upright in the truck. Cylinders must not be transported lying horizontally on trucks with the valve overhanging in a position to collide with stationary objects. Cylinders should never be dragged from place to place.

Valves should always be closed before cylinders are moved. Unless cylinders are to be moved while secured in an upright position to a suitable truck, pressure regulators should be removed and valve protecting caps, if provided for in the design, should be attached.

WITHDRAWING ACETYLENE FROM CYLINDERS

Never use acetylene through blowpipes or other devices equipped with shut-off valves on the acetylene supply connections without reducing the pressure through a suitable regulator attached to the cylinder valve. As previously explained, acetylene should never be used in equipment outside the cylinder at pressures exceeding 15 psig. It is always preferable to use acetylene cylinders in an upright position to avoid loss of solvent and accompanying reduction in flame quality. However, use in a horizontal position, with or without the loss of solvent, does not make the acetylene less stable or less safe.

In preparing to withdraw acetylene from cylinders, never use wrenches or other tools for opening cylinder valves except those provided or approved by the manufacturer of the gas.

After removing the valve protection cap, slightly open the valve an instant in order to clear the opening of particles of dust or dirt, being careful to stand so that the valve points away from the body. Avoid blowing dangerous amounts of the gas in confined spaces. Do not "crack" an acetylene cylinder valve near welding work, sparks, open flame, or any other possible sources of ignition.

Be sure that all connections are gas-tight and remain so, and that the connected hose is in good condition and does not have any leaks.

Always open the acetylene cylinder valve slowly. Never use a hammer or mallet in attempting to open or close a valve.

Do not open an acetylene cylinder valve more than $1\frac{1}{2}$ turns. Do not stand in front of the regulator and gage faces when opening the valve.

Do not pile hose, tools or other objects on top of an acetylene cylinder where they might interfere with quick closing of the valve.

The wrench used for opening the cylinder valve should always be kept on the valve spindle when the cylinder is in use.

Always close the cylinder valve when the work is finished. Be sure the cylinder valve is closed and all gas is released from the regulator before removing the regulator from a cylinder.

Never apply a torch to the side of a cylinder to raise the pressure. Serious accidents have resulted from violation of this rule.

DETERMINING ACETYLENE CONTENTS IN CYLINDERS

Since acetylene gas is in solution in acetone, the contents of an acetylene cylinder cannot be determined accurately by pressure gage readings. A pressure gage indicates what is known as "solution pressure" in the cylinder. This pressure is greatly affected by changes in cylinder temperature. For example, the gage pressure of a cylinder may be 230 psig at a temperature of 70 F and less than 100 psig at 0 F without any acetylene having been withdrawn from the cylinder. Acetylene cylinder contents are instead measured by weight. Weight of acetylene can be converted to cubic feet of gas by the factor of 14.5 cu ft per lb.

HANDLING LEAKING CYLINDERS

Because acetylene and air in certain proportions are explosive, care should be taken to prevent acetylene leakage. Connections should be kept tight and hose maintained in good condition. Points of suspected leakage should be tested by covering them with soapy water. A leak will be indicated by bubbles of escaping acetylene passing through the soap film. *Never test for leaks with an open flame!*

If acetylene leaks around the valve spindle when the valve is open, close the valve and tighten the gland nut. This compresses the packing around the spindle. If this does not stop the leak, close the valve, and attach to the cylinder a tag stating that the valve is unserviceable. Notify the gas supplier and follow his instructions for the cylinder's return.

If acetylene leaks from the valve even when the valve is closed, or if rough handling should cause any fusible safety plugs to leak, move the cylinder to an open space well away from any possible source of ignition and plainly tag the cylinder as having an unserviceable valve or fuse plug. Open the valve slightly to let the acetylene escape slowly. Place a sign at the cylinder warning persons against approaching the cylinder with cigarettes or other open lights. When the cylinder is empty, close the valve. Notify the manufacturer immediately the serial number of the cylinder and the particulars of its defect, as far as known, and await shipping instructions.

REFERENCES

1. "Acetylene Transmission for Chemical Synthesis" (Pamphlet G-1.3), Compressed Gas Association, Inc.
2. *Op. cit.*, p. 9.
3. "Gas Systems for Welding and Cutting," Standard No. 51, National Fire Protection Association, 60 Batterymarch St., Boston, Mass. 02110 (Also available as NBFU Standard No. 51, American Insurance Association, 85 John St., New York, N. Y., 10007.)

ADDITIONAL REFERENCES

"Recommendations for Chemical Acetylene Metering" (Pamphlet G-1.2), Compressed Gas Association, Inc.

"Standard for Acetylene Cylinder Charging Plants" (Pamphlet G-1.4), Compressed Gas Association, Inc.

"Safe Practices for Installation and Operation of Oxy-Acetylene Welding and Cutting Equipment" (Pamphlet OA-5), Compressed Gas Association, Inc.

*Metallic Acetylides**

"Copper Acetylides," V. F. Bramfeld, M. T. Clark, and A. P. Seyfang; *J. Soc. Chem. Ind.* (*London*), **66**, 346–53 (October 1947).

"The Formation and Properties of Acetylides," paper presented by G. Benson, Shawinigan Chemicals, Ltd., at the Compressed Gas Association Canadian Section, September 17, 1950.

"The Chemistry of Acetylene," ACS Monograph No. 99, Nieuwland and Vogt, New York, Reinhold.

"Conditions of Formation and Properties of Copper Acetylide," unpublished research paper by L'Air Liquide, Paris, France.

"Ueber Bildung und Eigenschaften der Kupferacetylide," von H. Feitnecht and L. Hugi-Carmes', *Schweizer Archiv Angew. Wiss. Tech.*, **10**, 23 (1957).

* Most of the above articles contain additional bibliographies which are also informative on this subject.

Air

Synonyms: Compressed air, atmospheric air, the atmosphere (of the Earth)
ICC Classification: Nonflammable compressed gas, n. o. s. (not otherwise specified); green label

PHYSICAL CONSTANTS

International symbol	AIR
Molecular weight	28.9752
Density of gas at 70 F and 1 atm, lb/cu ft	0.07493
Density of liquid at boiling point and, 1 atm, lb/cu ft	54.56
Liquid/gas ratio (liquid at boiling point, gas at 70 F and 1 atm), vol/vol	1/728.1
Boiling point at 1 atm	−317.8 F
Freezing point at 1 atm	−357.2 F to −363.8 F
Critical temperature	−220.3 F
Critical pressure, psia	547
Critical density, lb/cu ft	21.9
Latent heat of vaporization at boiling point, Btu/lb	88.2
Specific heat, C_p, at approx. 70 F, Btu/(lb)(°F)	0.241
Specific heat, C_v, at approx. 70 F, Btu/(lb)(°F)	0.1725
Ratio of specific heats, C_p/C_v, at approx. 70 F	1.40
Thermal conductivity, Btu/(hr)(sq ft)(°F/ft)	
at −148 F	0.0095
at 32 F	0.0140
at 212 F	0.0183
Solubility in water at 1 atm, vol/1 vol of water	
at 32 F	0.032
at 68 F	0.020
at 212 F	0.012

Properties. Air is the natural atmosphere of the Earth—a nonflammable, colorless, odorless gas that consists of a mixture of gaseous elements (with water vapor, a small amount of carbon dioxide and traces of many other constituents). Liquefied air containing carbon dioxide has a milky color, and, with the carbon dioxide removed, it appears transparent with a bluish cast.

Air is of course very commonly compressed at the point of use for most practical applications. However, to meet needs for air of special purity or specified composition (as in certain medical, scientific, industrial, fire protection, undersea and aerospace uses), it is purified or compounded synthetically and shipped in cylinders as a nonliquefied gas at high pressures.

The composition of dry air at sea level, according to one scientific source is as follows:

Component	*% by Volume*	*% by Weight*
Nitrogen	78.03	75.5
Oxygen	20.99	23.2
Argon	0.94	1.33
Carbon dioxide	0.03	0.045
Hydrogen	0.01	—
Neon	0.00123	—
Helium	0.0004	—
Krypton	0.00005	—
Xenon	0.000006	—

Air as naturally encountered also contains varying amounts of water vapor. For most practical purposes, the air composition is taken to be 79 per cent nitrogen and 21 per cent oxygen by volume, and to be 76.8 per cent nitrogen and 23.2 per cent oxygen by weight.

Air which meets specified purity requirements is important in certain applications, particularly those involving human respiration. The CGA specifications for air, which define 9 grades of gaseous air and 2 grades of liquid air according to differing maximum limits for particular trace constituents, is therefore given as an appendix to this section.

Materials of Construction. Dry air is noncorrosive, and may be contained in equipment constructed with any common, commercially available metals.

Manufacture. Air of known purity and composition is either compressed from the atmosphere and purified by chemical and mechanical means, or is made synthetically from its already purified major components, chiefly nitrogen and oxygen.

Commercial Uses. Liquefied air is the chief source of the atmospheric gases, which are removed from it by fractionation, but practically all air separation plants collect and liquefy air at the plant itself. Air meeting particular specifications of purity serves various uses. It has important applications in medical, undersea, aerospace and atomic energy fields; it is needed in tunnel construction; it is very widely employed in self-contained breathing apparatus used by industrial personnel and firemen; and it is needed in some kinds of pneumatic equipment.

Physiological Effects. Air is of course nontoxic and nonflammable, but cylinders and other containers charged with air at high pressure must be handled with all the precautions necessary for safety with any nonflammable compressed gas.

AIR CONTAINERS

Compressed air is shipped under ICC regulations in authorized cylinders. Liquefied air may be shipped in special insulated and vented containers like the cryogenic containers used for liquefied nitrogen, hydrogen or helium without being subject to ICC regulations if it is shipped at pressures below 25 psig.

Filling Limits. The maximum filling limits at 70 F authorized for compressed air under present regulations are the authorized service pressures marked on the cylinders (also, in the *that meet special requirements*, up to 10 per cent in excess of their marked service pressures).

SHIPPING METHODS; REGULATIONS

Under the appropriate regulations and tariffs, air is authorized for shipment as follows:

By Rail: In cylinders as a compressed gas, and as a liquid in special cryogenic containers at pressures below 25 psig.

By Highway: In cylinders as a gas, and as a liquid in special cryogenic containers at pressures below 25 psig.

By Water: In cylinders on passenger vessels, cargo vessels, and ferry and railroad-car ferry vessels (passenger or vehicle). In cylinders on barges for barges of U. S. Coast Guard classes A, BA, BB, CA and CB.

By Air: In cylinders as a compressed gas, aboard passenger aircraft up to 150 lb (70 kg), or aboard cargo aircraft up to 300 lb (140 kg), maximum net weight per cylinder. In insulated containers meeting specified requirements, liquid air, either low-pressure or pressurized, in cargo aircraft only up to 300 lb (140 kg) maximum net weight per container. Nonpressurized liquid air is not accepted for shipment aboard cargo or passenger aircraft.

AIR CYLINDERS

Compressed air may be shipped in any cylinders authorized by the ICC for nonliquefied compressed gas. (These include cylinders meeting ICC specifications 3A, 3AA,

3B, 3C, 3D, 3E, 4, 4A, 4B, 4BA and 4C; in addition, continued use of cylinders meeting ICC specifications 3, 7, 25, 26, 33 and 38 is authorized, but new construction is not authorized.

Valve Outlet and Inlet Connections. Industrial Air, Standard Connection, U. S. and Canada—No. 590. Air for Human Respiration, Standard Connections, U. S. and Canada—No. 950, No. 1310, No. 1340. Industrial Air or Air for Human Respiration, Alternate Standard Connection, Canada only —No. 400 (obsolete effective 1/1/67).

Cylinder Requalification. All cylinders authorized for compressed air service must be requalified by hydrostatic retest every 5 years under present regulations, with the following exceptions: ICC-4 cylinders, every 10 years; and no periodic retest is required for cylinders of types 3C, 3E, 4C and 7.

APPENDIX

Air Specification[1] *Compressed Gas Association, Inc.*

1. SCOPE

1.1. This document describes the specification requirements for air, including atmospheric air and air synthesized by blending oxygen and nitrogen in the proper proportions. It is different from other gas specifications because atmospheric air is not a manufactured product but is naturally-occuring. Atmospheric air contains a large variety of trace constituents on many of which it is impractical to set individual limits. However, this specification qualifies certain grades of air by limiting the concentrations of specific trace constituents.

2. CLASSIFICATION

2.1. Types—Gaseous air is denoted as Type I and liquid air as Type II.

2.2. Grades—Table 1 presents the component maxima, in ppm (*v*/*v*) unless shown otherwise, for the types and grades of air. A blank indicates no maximum limiting characteristic.

2.3. Quality Tests—The supplier will assure, by his standard practice, the quality of the specified type and grade of air. If otherwise required, alternative control procedures are described in 2.3.1—analytical procedures; 3.2.1—lot definitions; 3.2.2 —samples per lot; and 4.2, 4.3 and 4.4—samples.

2.3.1. Analytical Procedures. Table 2 lists the subsequent sections which cover the determination of the characteristics listed in 2.2.

3. QUALITY VERIFICATION SYSTEMS

3.1. Production qualification tests are a single or series of analyses performed on the product to assure the reliability of the production facility to supply air of the required type and grade. This production qualification can be verified by the analytical records of product from the supplier, or, if required, by the analysis of representative samples of the product from the facility at appropriate intervals as agreed between the supplier and the customer. Production qualification tests may be performed by the supplier or by a laboratory agreed upon between the supplier and the customer.

3.1.1. Sampling and analytical requirements of the production qualification tests will include the determination of all the limiting characteristics of the required type and grade of air.

3.2. Lot acceptance tests are those analyses performed on the air in the shipping container, or a sample thereof, which is representative of the lot.

3.2.1. Lot Definitions. One of the following is to be used for lot acceptance tests.

3.2.1.1. No specific quantity.

3.2.1.2. All of the air supplied during the contract period.

3.2.1.3. All of the air supplied or containers charged during a calendar month.

3.2.1.4. All of the air supplied or containers charged during 7 consecutive days.

3.2.1.5. All of the air supplied or containers charged during a consecutive 24-hour period.

3.2.1.6. All of the air supplied or containers charged during one 8-hour shift.

3.2.1.7. All of the air supplied in one shipment.

3.2.1.8. All of the air in the container(s) charged on one manifold at the same time.

3.2.1.9. Any quantity of air agreed upon between the supplier and the customer.

[1] This Specification is the first in a series which the CGA plans to develop and publish. It attempts to qualify in a single document all of the types and grades of product available from the compressed gas industry and for which a requirement has been established.

3.2.2. Number of samples per lot is to be in accordance with one of the following:

3.2.2.1. One sample per lot.

3.2.2.2. For Type I container lots, 4 per cent of the containers filled but not less than one container.

4. SAMPLING

4.1. Sample Volume—The volume of air in a single sample container shall be sufficient to perform the analysis for at least one limiting characteristic. If a single sample does not contain a sufficient quantity of air to perform all of the analyses for the specified type and grade due to the size of the sample container, physical state of the sample, etc., additional samples from the same container shall be taken under similar conditions.

4.2. Type I air samples shall be representative of the air supply. Sampling shall be performed in accordance with one of the following:

TABLE 1.

	TYPE I (Gaseous) GRADES									TYPE II (Liquid) GRADES	
Limiting Characteristics	A	B	C	D	E	F	G	H	J	A	B
% O_2 (v/v) Balance predominantly N_2 (Note 1)	atm	atm	atm/ 19–23	atm/ 19–23	atm/ 19–23	atm/ 19–23	atm/ 19–23	atm/ 19–23	atm/ 19–23	19–25	19–23
Water		None condensed (per 5.4.1)	Note 2	Note 2	Note 2	Note 2	Note 2	Note 2	1		Note 2
Hydrocarbons (condensed) in Mg/m³ of gas at NTP		None per 5.5.1	5	5	5						
Carbon monoxide			50	20	10	5	5	5	1		5
Odor			see 5.1.5	see 5.1.5	see 5.1.5	see 5.1.5	see 5.1.5	see 5.1.5	see 5.1.5		none
Carbon dioxide				1000	500	500	500	500	0.5		500
Gaseous Hydrocarbons (as methane)						25	15	10	0.5		10
Nitrogen dioxide							2.5	0.5	0.1		0.5
Nitrous oxide									0.1		
Sulfur dioxide							2.5	0.5	0.1		0.5
Halogenated solvents							10	1	0.1		1
Acetylene									0.05		0.5
Permanent particulates											see 5.16

Note 1: The term "atm" (atmospheric) denotes the oxygen content normally present in atmospheric air; the numerical values denote the oxygen limits for synthesized air.

Note 2: The water content of compressed air required for any particular grade may vary with the intended use from saturated to very dry. If a specific water limit is required, it should be specified as a limiting dew point or concentration in ppm (v/v). Dew point is expressed in temperature (°F) at 1 atm absolute pressure (760 mm Hg). To convert dew point (°F) to °C, ppm (v/v), or mg/liter, see 7.1.

TABLE 2.

Limiting Characteristics	Procedures	Type I (Gaseous)									Type II (Liquid)	
		A	B	C	D	E	F	G	H	J	A	B
Oxygen, %	5.3.1			x	x	x	x	x	x	x	x	x
	5.3.2			x	x	x	x	x	x	x	x	x
	5.3.3			x	x	x	x	x	x	x	x	x
	5.3.4			x	x	x	x	x	x	x	x	x
	5.3.5			x	x	x	x	x	x	x	x	x
	5.3.6			x	x	x	x	x	x	x	x	x
Water	5.4.1		x									
	5.4.2			x	x	x	x	x	x	x		x
	5.4.3			x	x	x	x	x	x	x		x
	5.4.4			x	x	x	x	x	x	x		x
	5.4.5			x	x	x	x	x	x	x		x
Condensed hydrocarbons	5.5.1		x									
	5.5.2			x	x	x						
	5.5.3			x	x	x						
Carbon monoxide	5.6.1			x	x	x	x	x	x			x
	5.6.2			x	x	x	x	x	x			x
	5.6.3			x	x	x	x	x	x	x		x
	5.6.4			x	x	x	x	x	x	x		x
Odor	5.7			x	x	x	x	x	x	x		
	5.7.1											x
Carbon dioxide	5.8.1				x	x	x	x	x			x
	5.8.2				x	x	x	x	x			x
	5.8.3				x	x	x	x	x	x		x
	5.8.4				x	x	x	x	x	x		x
Gaseous hydrocarbons	5.9.1						x	x	x	x		x
	5.9.2						x	x	x	x		x
	5.9.3						x	x	x	x		x
	5.9.4						x	x	x	x		x
Nitrogen dioxide	5.10.1							x	x	x		x
	5.10.2							x	x	x		x
	5.10.3							x	x	x		x
Nitrous oxide	5.11.1									x		
Sulfur dioxide	5.12.1							x	x	x		x
	5.12.2							x	x	x		x
	5.12.3							x	x	x		x
	5.12.4							x	x	x		x
Halogenated solvents	5.13.1							x	x	x		x
	5.13.2							x	x	x		x
	5.13.3							x	x	x		x
Acetylene	5.14.1									x		x
	5.14.2									x		x
	5.14.3									x		x
Permanent particulates	5.16.1											x

4.2.1. By charging the sample container at the same time delivery containers are charged, on the same manifold and in the same manner.

4.2.2. By withdrawing a sample from the supply container through a suitable connection into the sample container. No regulator or external valve is to be used between the supply and the sample containers (a suitable purge valve is permissible). *Note: For safety reasons the sample container must have a rated service pressure at least equal to the pressure in the supply container.*

4.2.3. By connecting the container being sampled directly to the analytical equipment.

4.2.4. By any other method agreed upon between the supplier and the customer.

4.3. Type II air samples shall be representative of the liquid air supply. Sampling shall be performed in accordance with one of the following:

4.3.1. By withdrawing a sample into an open container for odor and acetylene tests; see 5.14.1 and 5.14.2.

4.3.2. By withdrawing a sample directly through an analytical filter system; see 5.16.1.

4.3.3. By any other method agreed upon between the supplier and the customer.

4.4. Type II air samples gasified from liquid shall be representative of the liquid air supply. Sampling shall be performed in accordance with one of the following.

4.4.1. By vaporizing liquid air from the supply container in the sampling tubing.

4.4.2. By flowing liquid air from the supply container into or through a suitable container in which a representative sample is vaporized; see 8.1.

4.4.3. By any other method agreed upon between the supplier and the customer.

5. Analytical Procedures

5.1. The parameters of analysis for analytical techniques contained in this section are:

5.1.1. Per cent (v/v) = molar per cent.

5.1.2. Ppm (v/v) = molar parts per million.

5.1.3. "As Methane" for the purposes of this specification is defined as the single carbon atom equivalent.

5.1.4. Calibration gas standards are required to calibrate (zero and span) some of the analytical instruments used to determine some of the constituents of air. If required by the customer, the accuracy of measurement equipment used in preparing these standards is to be traceable to the National Bureau of Standards.

5.1.5. Specific measurement of odor in Type I air is impractical. Air normally may have a slight odor. The presence of a pronounced odor should render the air unsatisfactory for breathing purposes. This concept is applied to the odor limit in Table 2.2.

5.1.6. Dew point is expressed in temperature (degrees F) at one atmosphere absolute pressure (760 mmHg). To convert dew point (°F) to °C, ppm (v/v), or mg/l, see 7.1.

5.1.7. Analytical equipment is to be operated in accordance with the manufacturer's instructions.

5.2. Analytical methods not listed in this specification are acceptable if agreed upon between the supplier and the customer.

5.3. The per cent oxygen shall be determined by any one of the following procedures:

5.3.1. By an apparatus employing a comparison tube filled with a color-reactive chemical.

5.3.2. By a volumetric (Orsat type) or manometric gas analysis apparatus using a suitable oxygen-absorbing reagent.

5.3.3. By a paramagnetic-type analyzer. The accuracy should be at least 0.5 per cent. The analyzer is to be calibrated (zeroed and spanned) at appropriate intervals by the use of calibration gas standards (see 5.1.5), using nitrogen as the base gas.

5.3.4. By a thermal conductivity-type analyzer. The accuracy should be at least 0.5 per cent. The analyzer is to be calibrated (zeroed and spanned) at appropriate intervals by the use of calibration gas standards (see 5.1.4), using nitrogen as the base gas.

5.3.5. By an electrochemical-type analyzer containing a solid or a gaseous electrolyte. The accuracy should be at least 0.5 per cent. The analyzer is to be calibrated (zeroed and spanned) at appropriate intervals by the use of calibration gas standards (see 5.1.4), using nitrogen as the base gas.

5.3.6. By a gas chromatograph in accordance with 5.15. The technique used must be specific for oxygen.

5.4. The water content as specified for any particular grade of air shall be determined by one of the following procedures:

5.4.1. By supporting the cylinder in an inverted position (valve at the bottom) for 5 minutes. The cylinder and content should be at room temperature or above 32 F. The cylinder valve is then opened slightly (USE CAUTION) while the cylinder remains inverted, and the air vented with a barely audible flow into an open dry container for one minute. *Note: A rapid gas flow may cause any liquid to disperse and not collect in the container.* This procedure will detect condensed hydrocarbons (oil) as well as water.

5.4.2. By an electrolytic hygrometer having an indicator graduated in ppm (v/v) on a range which is no greater than 10 times the specified maximum moisture content.

5.4.3. By a frost-point analyzer in which the temperature of a viewed surface, at the time frost first begins to form, is measured or indicated.

5.4.4. By a piezoelectric sorption hygrometer on a range which is no greater than 10 times the specified maximum moisture content.

5.4.5. By an analyzer built and operated in accordance with NBS Research Paper No. 1865 (see 8.2).

5.5. The condensed hydrocarbon content of the specified type and grade of air shall be determined by one of the following procedures:

5.5.1. By supporting the cylinder in an inverted position (valve at the bottom) for 5 minutes. The cylinder and content should be at room temperature or above 32 F. The cylinder valve is then opened slightly (USE CAUTION) while the cylinder remains inverted, and the air vented with a barely audible flow into an open dry container for one minute. *Note: A rapid gas flow may cause any liquid to disperse and not collect in the container.* This procedure will detect condensed (liquid) water as well as oil.

5.5.2. By scrubbing the air sample with spectral grade carbon tetrachloride which is then examined by an infrared technique for condensible hydrocarbons; see 8.3.

5.5.3. By passing a sample of air through an adequate filter medium and measuring the increase in weight of the filter or noting presence of visible discoloration.

5.6. The carbon monoxide content of the specified type and grade of air shall be determined by one of the following procedures:

5.6.1. By an apparatus employing a comparison tube filled with a color-reactive chemical.

5.6.2. By a catalytic combustion analyzer. The analyzer is to be calibrated (zeroed and spanned) at appropriate intervals with calibration gas standards (see 5.1.4), using air as the base gas. The limit of detectability shall be no greater than the specified maximum amount of carbon monoxide.

5.6.3. By a gas cell-equipped dispersive or nondispersive infrared analyzer. The analyzer is to be calibrated (zeroed and spanned) at appropriate intervals with calibration gas standards (see 5.1.4) at approximately 4.6 microns. The analyzer should be operated so that its sensitivity for carbon monoxide is either 0.5 ppm or 10 per cent of the specified maximum amount, whichever is greater.

5.6.4. By a gas chromatograph in accordance with 5.15.

5.7. Odor in Type I is checked directly by smelling of a moderate flow of air from the container being tested; see 5.1.5. (USE CAUTION.)

5.7.1. Odor in Type II is checked by evaporating to dryness 200 cc of liquid in a loosely covered 400 cc beaker with a fresh filter paper in the bottom. The cover is removed at the point of complete evaporation and the beaker is odor tested several times until it has warmed to above the freezing point of condensed water on the outside.

5.8. The carbon dioxide content of the specified type and grade of air shall be determined by one of the following procedures:

5.8.1. By an apparatus employing a comparison tube filled with a color-reactive chemical.

5.8.2. By a volumetric or manometric gas-absorption (Orsat type) analysis apparatus, using a suitable carbon dioxide absorbing reagent. Precision of the apparatus should be at least 10 per cent of the specified maximum amount.

5.8.3. By a gas cell-equipped dispersive or nondispersive infrared analyzer. The analyzer is to be calibrated (zeroed and spanned) at appropriate intervals with calibration gas standards (see 5.1.4) at approximately 4.3 microns. The analyzer should be operated so that its sensitivity for carbon dioxide is 0.5 ppm or 10 per cent of the specified maximum amount, whichever is greater.

5.9. The gaseous hydrocarbon content shall be determined by one of the following procedures:

5.9.1. By a flame ionization type analyzer. The analyzer is to be calibrated (zeroed and spanned) at appropriate intervals by the use of calibration gas standards (see 5.1.4). The range used should be no greater than 10 times the specified maximum gaseous hydrocarbon content expressed as methane.

5.9.2. By a gas cell-equipped infrared analyzer.

The analyzer is to be calibrated (zeroed and spanned) at appropriate intervals with methane standards (see 5.1.4), at approximately 3.5 microns (the characteristic absorption wavelength for C—H stretching). The analyzer should be operated so that its sensitivity for methane is either 0.5 ppm or 10 per cent of the specified maximum amount, whichever is greater.

5.9.3. By oxidizing the hydrocarbon which may be determined by classical chemical methods or infrared equipment prepared as specific for carbon dioxide.

5.9.4. By a gas chromatograph in accordance with 5.15.

5.10. The nitrogen dioxide content of the specified type and grade of air shall be determined by one of the following procedures:

5.10.1. By an apparatus employing a comparison tube filled with a color-reactive chemical.

5.10.2. By a suitable wet chemical method; see 8.4.

5.10.3. By a gas cell-equipped infrared analyzer. The analyzer is to be calibrated (zeroed and spanned) at appropriate intervals with calibration gas standards (see 5.1.4) at approximately 6.2 microns. The analyzer should be operated so that its sensitivity for nitrogen dioxide is at least the specified maximum amount.

5.11. The nitrous oxide content of the specified type and grade of air shall be determined by one of the following procedures:

5.11.1. By a gas cell-equipped infrared analyzer. The analyzer is to be calibrated (zeroed and spanned) at appropriate intervals with calibration gas standards (see 5.1.4) at approximately 4.5 microns. The analyzer should be operated so that its sensitivity for nitrous oxide is at least the specified maximum amount.

5.12. The sulfur dioxide content of the specified type and grade of air shall be determined by one of the following procedures:

5.12.1. By an apparatus employing a comparison tube filled with a color-reactive chemical.

5.12.2. By a suitable wet chemical method; see 8.4.

5.12.3. By a gas cell-equipped infrared analyzer. The analyzer is to be calibrated (zeroed and spanned) at appropriate intervals with calibration gas standards (see 5.1.4) at approximately 7.3 microns. The analyzer should be operated so that its sensitivity for sulfur dioxide is at least the specified maximum amount.

5.12.4. By a gas chromatograph in accordance with 5.15.

5.13. The halogenated solvent content of the specified type and grade of air shall be determined by one of the following procedures:

5.13.1. By an electronic-type halide detector. The analyzer is to be calibrated (zeroed and spanned) at appropriate intervals with gaseous calibration standards or with air in equilibrium with standard solutions of chlorinated solvent (e.g., trichloroethylene) in mineral oil. This procedure must be carried out only in an atmosphere free from contamination with halogens or their compounds. The analyzer should be operated so that its sensitivity for halogenated solvents is at least the maximum specified amount.

5.13.2. By a gas-cell equipped infrared analyzer appropriately calibrated with a halogenated solvent. The analyzer should be operated so that its sensitivity for halogenated solvents is at least the specified maximum amount.

5.13.3. By a gas chromatograph in accordance with 5.15.

5.14. The acetylene content of the specified type and grade of air shall be determined by one of the following procedures:

5.14.1. By a wet chemical (Illosvay-colorimetric) method in which the sensitivity for acetylene is at least the maximum specified amount. This procedure may be adapted for either Type I or Type II air.

5.14.2. By a gas cell-equipped infrared analyzer. The analyzer is to be calibrated (zeroed and spanned) at appropriate intervals with calibration gas standards (see 5.1.4) at approximately 13.7 microns. The analyzer should be operated so that its sensitivity for acetylene is at least the maximum amount specified.

5.14.3. By a gas chromatograph in accordance with 5.15.

5.15. A gas chromatograph may be used to determine many of the limiting characteristics listed in 2.2 and above in this section. The analyzer must be capable of separating and determining the specified component with a sensitivity of 0.1 ppm or 20 per cent of the specified maximum amount of the component, whichever is greater. Appropriate impurity concentrating techniques may be used to attain this sensitivity. The analyzer is to be calibrated at appropriate intervals by the use of calibration gas standards (see 5.1.4) which contain the applicable limiting characteristic gaseous components of air.

5.16. The permanent particulate matter content of Type II shall be determined by one of the following procedures. This specification recognizes that it is not possible to qualify the permanent particulates in Type I air.

5.16.1. By passing liquid air sample through a low micron-rated tared analytical filter disc contained in a suitable holder. The filtered liquid is collected and measured in an open container. The assembly is warmed and the disc reweighed.

6. CONTAINERS

6.1. Air containers shall comply with Interstate Commerce Commission specifications and shall be maintained, filled, packaged, marked, labeled, and shipped to comply with current ICC regulations Title 49 (C.F.R. 71 to 90) and American Standard Z48.1; see 7.2, 8.5 and 8.6.

6.2. Container preparation shall be as necessary to assure that the container contents meet the requirements of the specified type and grade of air.

6.3. Containers for air intended for human respiration shall be processed by a method which encompasses evacuation, purging or cleaning procedures to assure that the container and component parts are not reactive, additive or absorptive to an extent that significantly affects the identity, quality or purity of the specified type and grades of air being supplied.

6.4. Valves on air containers shall conform to American Standard B57.1 (see 8.7) unless otherwise agreed upon between the supplier and the customer.

7. SUPPLEMENTAL SPECIFICATION DATA

7.1. Moisture Conversion Data (all referred to 70 F and 14.7 psig).

Dew Point (°F)	Dew Point (°C)	ppm (*v*/*v*)	mg/lit
−110	−78.9	0.58	0.00045
−105	−76.1	0.94	0.00070
−100	−73.3	1.5	0.0011
−95	−70.5	2.3	0.0017
−90	−67.8	3.2	0.0024
−85	−65.0	5.0	0.0037
−80	−62.2	7.1	0.0055
−75	−59.4	10.6	0.0079
−70	−56.7	16.1	0.012
−65	−53.9	24.2	0.018
−60	−51.1	30.9	0.023
−55	−48.3	43.0	0.032
−50	−45.6	60.5	0.045
−45	−42.8	87.3	0.065
−40	−40.0	121	0.09
−35	−37.2	161	0.12
−30	−34.4	229	0.17
−25	−31.6	382	0.21
−20	−28.9	403	0.30
−15	−26.1	538	0.40
−10	−23.3	685	0.51
−5	−20.5	900	0.67
0	−17.8	1180	0.88

7.2. Pressure Temperature Conversion Chart

Settled Temperature (°F)	Type I Container Service Pressure (expressed in psig)						
	1800	2000	2200	2265	2400	2490	2640
−50	1267	1396	1523	1564	1649	1705	1799
−48	1276	1406	1534	1576	1662	1719	1813
−46	1285	1416	1546	1588	1674	1732	1827
−44	1293	1426	1557	1600	1687	1745	1841
−42	1302	1436	1569	1611	1700	1758	1856
−40	1311	1447	1580	1623	1712	1772	1870
−38	1320	1457	1592	1635	1725	1785	1884
−36	1329	1467	1603	1647	1738	1798	1898
−34	1338	1477	1614	1659	1750	1811	1912
−32	1347	1487	1626	1670	1763	1824	1926
−30	1356	1497	1637	1682	1776	1838	1941
−28	1365	1508	1648	1694	1788	1851	1955
−26	1374	1518	1660	1706	1801	1864	1969
−24	1383	1528	1671	1718	1813	1877	1983
−22	1392	1538	1683	1729	1826	1890	1997
−20	1401	1548	1694	1741	1839	1903	2011
−18	1410	1558	1705	1753	1851	1917	2025
−16	1419	1568	1717	1765	1864	1930	2039

7.2. Pressure Temperature Conversion Chart—continued

Settled Temperature (°F)	Type I Container Service Pressure (expressed in psig)						
	1800	2000	2200	2265	2400	2490	2640
−14	1428	1578	1728	1776	1876	1943	2053
−12	1437	1589	1739	1788	1889	1956	2068
−10	1446	1599	1751	1800	1901	1969	2082
− 8	1455	1609	1762	1811	1914	1982	2096
− 6	1464	1619	1773	1823	1927	1995	2110
− 4	1472	1629	1784	1835	1939	2008	2124
− 2	1481	1639	1796	1847	1952	2022	2138
0	1490	1649	1807	1858	1964	2035	2152
+ 2	1499	1659	1818	1870	1977	2048	2166
4	1508	1669	1830	1882	1989	2061	2180
6	1517	1679	1841	1893	2002	2074	2194
8	1526	1689	1852	1905	2014	2087	2208
10	1535	1699	1863	1917	2027	2100	2222
12	1544	1710	1875	1928	2039	2113	2236
14	1552	1720	1886	1940	2052	2126	2250
16	1561	1730	1897	1952	2064	2139	2264
18	1570	1740	1909	1963	2077	2152	2278
20	1579	1750	1920	1975	2089	2165	2292
22	1588	1760	1931	1987	2102	2178	2306
24	1597	1770	1942	1998	2114	2191	2320
26	1606	1780	1954	2010	2127	2204	2334
28	1615	1790	1965	2021	2139	2217	2348
30	1623	1800	1976	2033	2152	2230	2362
32	1632	1810	1987	2045	2164	2243	2376
34	1641	1820	1998	2056	2176	2256	2390
36	1650	1830	2010	2068	2189	2269	2404
38	1659	1840	2021	2080	2201	2282	2418
40	1668	1850	2032	2091	2214	2295	2431
42	1676	1860	2043	2103	2226	2308	2445
44	1685	1870	2055	2114	2239	2321	2459
46	1694	1880	2066	2126	2251	2334	2473
48	1703	1890	2077	2138	2264	2347	2487
50	1712	1900	2088	2149	2276	2360	2501
52	1721	1910	2099	2161	2288	2373	2515
54	1729	1920	2111	2172	2301	2386	2529
56	1738	1930	2122	2184	2313	2399	2543
58	1747	1940	2133	2196	2326	2412	2557
60	1756	1950	2144	2207	2338	2425	2571
62	1765	1960	2155	2219	2350	2438	2584
64	1774	1970	2166	2230	2363	2451	2598
66	1782	1980	2178	2242	2375	2464	2612
68	1791	1990	2189	2253	2388	2477	2626
70	1800	2000	2200	2265	2400	2490	2640
72	1809	2010	2211	2277	2412	2503	2654
74	1818	2020	2222	2288	2425	2516	2668
76	1826	2030	2233	2300	2437	2529	2682
78	1835	2040	2245	2311	2450	2542	2695
80	1844	2050	2256	2323	2462	2555	2709
82	1853	2060	2267	2334	2474	2568	2723
84	1862	2070	2278	2346	2487	2580	2737
86	1870	2080	2289	2357	2499	2593	2751
88	1879	2090	2300	2369	2511	2606	2765
90	1888	2100	2312	2380	2524	2619	2779
92	1897	2110	2323	2392	2536	2632	2792
94	1906	2120	2334	2404	2548	2645	2806

7.2. Pressure Temperature Conversion Chart—continued

Settled Temperature (°F)	Type I Container Service Pressure (expressed in psig)						
	1800	2000	2200	2265	2400	2490	2640
96	1914	2129	2345	2415	2561	2658	2820
98	1923	2139	2356	2427	2573	2671	2834
100	1932	2149	2367	2438	2585	2684	2848
102	1941	2159	2378	2450	2598	2697	2861
104	1949	2169	2389	2461	2610	2710	2875
106	1958	2179	2401	2473	2622	2722	2889
108	1967	2189	2412	2484	2635	2735	2903
110	1976	2199	2423	2496	2647	2748	2917
112	1985	2209	2434	2507	2659	2761	2931
114	1993	2219	2445	2519	2672	2774	2944
116	2002	2229	2456	2530	2684	2787	2958
118	2011	2239	2467	2542	2696	2800	2972
120	2020	2249	2478	2553	2709	2813	2986
122	2028	2259	2490	2565	2721	2825	3000
124	2037	2269	2501	2576	2733	2838	3013
126	2046	2278	2512	2588	2746	2851	3027
128	2055	2288	2523	2599	2758	2864	3041
130	2063	2298	2534	2611	2770	2877	3055
132	2072	2308	2545	2622	2783	2890	3068
134	2081	2318	2556	2634	2795	2903	2082
136	2090	2328	2567	2645	2807	2915	3096
138	2098	2338	2578	2657	2819	2928	3110
140	2107	2348	2589	2668	2832	2941	3124
142	2116	2358	2600	2680	2844	2954	3137
144	2125	2368	2612	2691	2856	2967	3151
146	2133	2377	2623	2702	2869	2980	3165
148	2142	2387	2634	2714	2881	2992	3179
150	2151	2397	2645	2725	2893	3005	3192

8. Referenced Documents

8.1. Liquefied Gas Sampler.

8.1.1. Military Specification MIL-S-27626A "Sampler, Liquid Oxygen, TTU-131/E."

8.1.2. Navy liquid oxygen sampler, Drawing (Alameda) G-276; also MIL-O-27210B (ASG) section 4.4.3.2.2.

8.2. National Bureau of Standards Research Paper No. 1865, "Measurement of Water in Gases by Electrical Conduction in a Film of Hygroscopic Material and the Use of Pressure Changes in Calibration." Sup't. of Documents, Government Printing Office, Washington, D.C. 20402.

8.3. Marshall Space Flight Center Procedure MSFC-PROC-245 "Carbon Tetrachloride Scrubber Method for Analysis of Condensible Hydrocarbons in Compressed Gases."

8.4. Procedure: "Calorimetric Micro-determination of Nitrogen Dioxide in the Atmosphere" by B. E. Saltzman at the USDHEW at Cincinnati, Ohio. Reported in *Analytical Chemistry*, **26**, 1949–1955 (1954).

8.5. CFR 71-90, "Interstate Commerce Commission Regulations for the Transportation of Explosives and other Dangerous Articles." Available as Agent T. C. George's Tariff (current edition), Agent T. C. George, Bureau of Explosives, 63 Vesey Street, New York, New York, 10007.

8.6. See Section A of Chapter 3 in Part III (American Standard Z48.1), "Marking Portable Compressed Gas Containers."

8.7. See Section A of Chapter 4 in Part III (portion of American Standard B57.1), "Compressed Gas Cylinder Valve Outlet and Inlet Connections."

Ammonia (Anhydrous)

NH_3 ICC Classification: Nonflammable compressed gas: green label

PHYSICAL CONSTANTS

International symbol	NH_3
Molecular weight	17.032
Vapor pressure at 70 F, psig (liquefied, in cylinders)	114.1
Specific gravity of gas at 32 F and 1 atm (air = 1)	0.5970
Specific volume of gas at 32 F and 1 atm, cu ft/lb	20.78
Vapor density at −28 F and 1 atm, lb/cu ft	0.05555
Specific gravity of liquid at −28 F (compared to water at 4 C)	0.6819
Liquid density at −28 F and 1 atm, lb/cu ft	42.57
Boiling point at 1 atm	−28 F
Melting point at 1 atm	−107.9 F
Critical temperature	271.4 F
Critical pressure, psia	1657
Latent heat of vaporization at boiling point and 1 atm, Btu/lb	589.3
Flammable limits (% in air, by volume)	16–25%
Weight per gallon, liquid, at 60 F, lb	5.147

Properties. Anhydrous ammonia consists of 1 part nitrogen to 3 parts hydrogen by volume, and of about 82 per cent nitrogen to 18 per cent hydrogen by weight. Only anhydrous ammonia, which is shipped as a liquefied compressed gas, is treated here. Ammonia readily dissolves in water, and is also shipped in aqueous solution.

At atmospheric temperatures and pressures, anhydrous ammonia is a pungent, colorless gas. It may be easily compressed to a colorless liquid. At atmospheric pressure ammonia has a boiling point of −28 F and it freezes to a white crystalline mass at −107.9 F. When heated above its critical temperature of 271.4 F, it exists only as a vapor, regardless of the pressure. When liquid ammonia in a closed container is in equilibrium with ammonia vapor, the pressure within the container bears the relationship to temperature shown in Fig. 1. Figure 1 also shows the density-temperature relationship under those conditions, while Table 1 reports the relationship of temperature to the vapor pressure, density, specific gravity, and latent heat for liquefied anhydrous ammonia.

It takes excessive temperatures (about 840 to 930 F) to cause the elements of anhydrous ammonia to dissociate slightly at atmospheric pressure. Ammonia gas burns in a mixture with air within flammable limits (at atmospheric pressure) of 16 to 25 per cent by volume of anyhdrous ammonia. Experiments conducted by a nationally recognized laboratory indicate that an anhydrous ammonia-air mixture in a standard quartz bomb will not ignite at a temperature below 1562 F (850 C). When an iron bomb having catalytic effect was used, the ignition temperature of the anhydrous ammonia-air mixture was 1203.8 F (651 C).

Materials of Construction. Most common metals are not affected by dry ammonia. However, when mixed with very little water or water vapor, ammonia will vigorously attack copper, silver, zinc, and many alloys, especially those containing copper. It is this last property that is most likely to concern the bulk storage operator because absolutely no brass or bronze fittings must be used in connection with the storage or handling of ammonia. Tanks, valves, and fittings used with ammonia must be made of iron or steel since

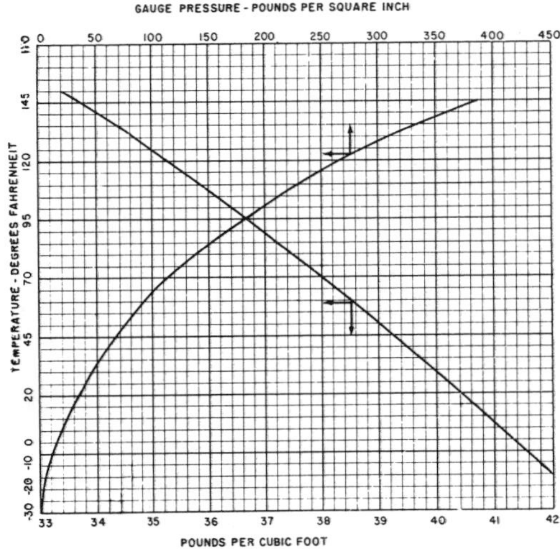

FIG. 1. Pressure-temperature and density-temperature curves of anhydrous ammonia when liquefied under compression.

ammonia will not attack these materials. For details of materials of construction for piping, tubing, fittings, and hose, refer to "American Standard Safety Requirements for the Storage and Handling of Anhydrous Ammonia."[1] For pipe schedules, refer to "American Standard Wrought-Steel and Wrought-Iron."[2]

Ammonia is a highly reactive chemical. It forms ammonium salts in reactions with inorganic and organic acids; amides in reactions with esters, acid anhydrides, acylhalides, carbon dioxide, or sulfonyl chlorides; and amines in reactions with halogen compounds or oxygen-containing compounds such as polyhydric phenols, alcohols, aldehydes and aliphatic ring oxides.

Manufacture. Ammonia was first produced in commercial quantities as a by-product of the destructive distillation of coal in the manufacture of coal-gas and coke.

TABLE 1. PROPERTIES OF LIQUID AMMONIA AT VARIOUS TEMPERATURES
(data for Columns 1, 2 and 6 taken from Bureau of Standards Circular No. 142. Values for Columns 3, 4 and 5 calculated from Column 2)

Temperature (F)	Vapor Pressure (psig) (1)	Liquid Density			Specific Gravity of Liquid (compared to water at 4 C) (5)	Latent Heat (Btu/lb) (6)
		Pounds per Cubic Foot (2)	Pounds Per U. S. Gallon (3)	Pounds Per Imperial Gallon (4)		
−28	0.0	42.57	5.69	6.83	.682	589.3
−20	3.6	42.22	5.64	6.77	.675	583.6
−10	9.0	41.78	5.59	6.70	.669	576.4
0	15.7	41.34	5.53	6.63	.663	568.9
10	23.8	40.89	5.47	6.56	.656	561.1
20	33.5	40.43	5.41	6.49	.648	553.1
30	45.0	39.96	5.34	6.41	.641	544.8
40	58.6	39.49	5.28	6.33	.633	536.2
50	74.5	39.00	5.21	6.26	.625	527.3
60	92.9	38.50	5.14	6.18	.617	518.1
65	103.1	38.25	5.11	6.14	.613	513.4
70	114.1	38.00	5.08	6.10	.609	508.6
75	125.8	37.74	5.04	6.06	.605	503.7
80	138.3	37.48	5.01	6.01	.600	498.7
85	151.7	37.21	4.97	5.97	.596	493.6
90	165.9	36.95	4.94	5.93	.592	488.5
95	181.1	36.67	4.90	5.89	.588	483.2
100	197.2	36.40	4.87	5.84	.583	477.8
105	214.2	36.12	4.83	5.80	.579	472.3
110	232.3	35.84	4.79	5.75	.573	466.7
115	251.5	35.55	4.75	5.71	.570	460.9
120	271.7	35.26	4.71	5.66	.565	455.0
125	293.1	34.96	4.67	5.61	.560	448.9
130	315.6	34.66	4.63	5.56	.555	443
135	339.4	34.35	4.59	5.51	.550	436
140	364.4	34.04	4.55	5.46	.545	430

In 1913, the first successful synthetic ammonia plant was put into operation in Germany. It utilized the Haber-Bosch process in which a preheated mixture of nitrogen and hydrogen was subjected to pressure in the presence of a suitable contact catalyst.

Most commercial ammonia is made today by processes which are modifications of the original Haber-Bosch process. They differ only in the sources for the hydrogen and nitrogen, pressures, temperatures, and composition catalysts. In a typical modern plant a mixture of natural gas, air and steam are passed at high temperature through catalysts. After the oxides of carbon have been removed from the reaction products, the remaining hydrogen-nitrogen mixture is heated, compressed and passed through catalysts to produce synthesis anhydrous ammonia.

Commercial Uses. Anhydrous ammonia is one of the oldest commercial refrigerants known. It has been used in both the absorption and compression type of systems.

Its most extensive use from the standpoint of volume is in soil fertilization. In this application it is used in the form of ammonia, ammonium salts, nitrates and urea. It is also used to ammoniate fertilizers containing superphosphates and in making nitrogen solutions which are water solutions of ammonia and ammonium nitrate or urea or both. Anhydrous ammonia is applied to the soil by direct injection or by addition to irrigation water.

Ammonia or dissociated ammonia is used in such metal-treating operations as nitriding, carbo-nitriding, bright annealing, furnace brazing, sintering, sodium hydride descaling, and atomic hydrogen welding and other applications for which protective atmospheres are required.

The petroleum industry utilizes anhydrous ammonia in neutralizing the acid constituents of crude oil and in protecting equipment such as bubble towers, heat exchangers, condensers and storage tanks from corrosion.

Controlled combustion of dissociated ammonia in air provides a source of pure nitrogen.

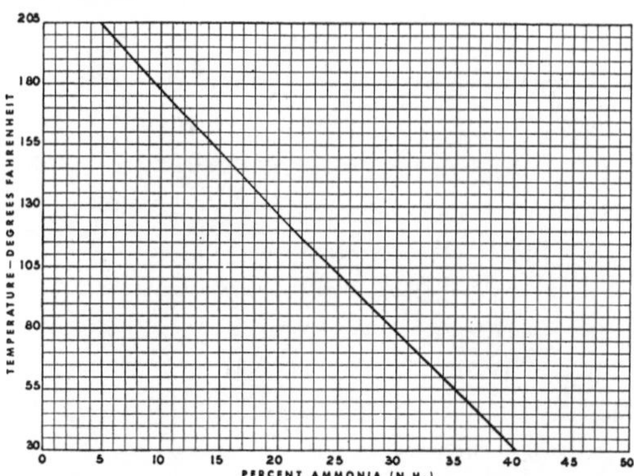

FIG. 2. Solubility of ammonia in water (per cent ammonia by weight).

Dissociated ammonia serves as a convenient source of hydrogen for hydrogenation of fats and oils.

Ammonia is used in extracting from their ores certain metals such as copper, nickel and molybdenum.

Ammonia can be oxidized to nitric acid or it can furnish nitrogen oxides for the production of sulphuric acid by the chamber process. Most industrial and military explosives of the conventional types contain nitrogen, and ammonia is the basic source of nitrogen in their manufacture.

As a processing agent ammonia is used in the manufacture of alkalies, ammonium salts, dyes, pharmaceuticals and cuprammonium rayon.

Where a solution of ammonia in water (aqua ammonia) is required, it is sometimes more economical to buy anhydrous ammonia and unload it directly into water under controlled conditions to form solutions of the desired strength. Details of the necessary equipment and process can be obtained from ammonia manufacturers. Solubility of ammonia in water at various temperatures is shown in Fig. 2.

Physiological Effects. Ammonia, either in the gaseous state or in aqueous (water) solution, acts like an alkali. Depending upon the concentration, its effects will vary from mild irritation to severe corrosion of the sensitive membranes of the eyes, nose, throat and lungs. It is readily detected in the air at very low concentrations by its sharp, pungent odor. The maximum allowable exposure for an 8-hour daily exposure is generally considered to be 100 ppm. Inhalation of air containing from 5000 to 10,000 ppm may cause sudden death from spasm or inflammation of the larynx. Concentrations exceeding 700 ppm of vapor will cause irritation of the eyes, and permanent eye injury may result from exposure to 700 ppm concentrations for longer than ½ hour. Ammonia's high solubility in water causes it to irritate moist skin surfaces.

Liquid ammonia vaporizes rapidly when released to atmospheric pressure and will absorb heat from any substance it contacts. It has a great affinity for water; therefore, if it should come in contact with the skin, it may cause severe injury by freezing the tissue and subjecting it to caustic action.

Further information on the physiological effects, safety equipment, and first aid in connection with anhydrous ammonia is given in Pamphlet G-2 of Compressed Gas Association, Inc.[3]

Leak Detection. A leak in ammonia valve connections or feed lines is quickly noticed by odor. The exact location of the leak may be detected by allowing the fumes from an open bottle of hydrochloric acid (or from a squeeze bottle of sulfuric acid, or an SO_2 aerosol container) to come in contact with the ammonia vapor, which will then combine with the detector to produce a dense, white fog. Other means of leak detection consist of moist phenolphthalein paper or red litmus paper, either of which will change color in ammonia vapor. Phenolphthalein paper may be obtained free of charge from ammonia manufacturers upon request.

ANHYDROUS AMMONIA CONTAINERS

Anhydrous ammonia is transported as a liquefied compressed gas in cylinders, insulated and uninsulated cargo and portable tanks, insulated and uninsulated single-unit tank cars, and TMU (ton multi-unit) tank cars. It is stored in bulk in large-capacity containers installed above or below ground. Large above-ground containers are often insulated and refrigerated.

Filling Densities. The maximum allowable filling densities for anhydrous ammonia shipping containers are as follows:

Container Type	ICC Specification	Max. Allowable Filling Density* (%)
Cylinders	Any authorized type**	54
Portable tanks	ICC-51	56
Cargo tanks	MC-330, MC-331	56
TMU tanks	106A500-X	50
Single-unit tank cars	105A300-W	57
	112A400-F	57***
	112A340-W	57***

* Per cent of water capacity by weight.

** See the following section, "Anhydrous Ammonia Cylinders."

*** 58.8 per cent permitted November to March, inclusive. Storage in transit prohibited.

SHIPPING METHODS: REGULATIONS

Under the appropriate regulations and tariffs, anhydrous ammonia is authorized for shipment as follows:

By Rail: In cylinders (via freight, express, or baggage) and portable tanks, in TMU tank cars, and in single-unit tank cars (insulated or uninsulated).

By Highway: In cylinders, portable tanks, and TMU tanks on trucks, and in cargo tank motor vehicles (including semi- and full trailers).

By Water: In bulk in cargo barges, tank vessels, or cargo vessels as specified in U. S. Coast Guard regulations.[4] Via cargo vessels in cylinders, portable tanks (ICC-51 specification, maximum 20,000 lb gross weight), TMU tanks, and railroad tank cars and tank motor vehicles complying with ICC regulations. On passenger vessels in cylinders; on ferry vessels (passenger or vehicle) in cylinders and tank motor vehicles authorized by the ICC; and on railroad car ferry vessels (passenger or vehicle) in cylinders, portable tanks (ICC-51, maximum 20,000 lb gross weight), TMU tanks, and tank cars and tank trucks authorized by the ICC. In cylinders on barges of U. S. Coast Guard classes A, BA, BB, CA, and CB.

By Air: Aboard cargo aircraft only in appropriate cylinders up to 300 lb (140 kg) maximum net weight per cylinder.

ANHYDROUS AMMONIA CYLINDERS

There are two types of anhydrous ammonia cylinders in current use, the tube-type (ICC-4), and the bottle-type. Each type is available in a number of convenient sizes, those most widely used having capacities of 50, 100, and 150 lb. Dimensions and weights of typical cylinders are shown in Table 2.

Cylinders that meet the following ICC specifications are authorized for anhydrous ammonia service: 4, 3A480, 3A480X, 3AA480, 4A480, 4AA480, and 3E1800. Specification ICC-3 cylinders are also authorized for use, but new construction is prohibited.

Withdrawing Liquid or Gas. Most ammonia cylinder valves have a dip tube connected to them. This makes it possible to withdraw either liquid or vapor from a cylinder in a horizontal position by rolling it until the open dip tube terminates in the desired phase. In bottle-type cylinders, the dip tube terminates in liquid when the valve *outlet* points up and in vapor when the valve *outlet* points down. In tube-type cylinders, the dip tube terminates in liquid when the valve *outlet* or valve *stem* points up, and terminates in vapor when the valve *outlet* or valve *stem* points down.

Valve Outlet and Inlet Connections. Standard connection, U. S. and Canada—No. 240, No. 800.

Safety Relief Devices. Cylinders containing less than 165 lb of anhydrous ammonia are not required to be equipped with safety relief devices.

Cylinder Requalification. Under present regulations, the following cylinders may be requalified by complete external visual inspection at the indicated intervals if used exclusively in service with anhydrous ammonia of 99.95 per cent purity: ICC-4 (10 years); ICC-3A480 (5 years); ICC-3AA480 (5 years); ICC-3A480X (5 years); ICC-4A480 (5 years);

TABLE 2. APPROXIMATE DIMENSIONS AND WEIGHTS OF TYPICAL AMMONIA CYLINDERS

ICC Cylinder Spec.	Ammonia Capacity (lb)	Average Tare Weight (pounds)		Overall Length (in.)	Outside Diameter (in.)	Wall Thickness (in.)	Minimum Volume (cu ft)
		Less Cap*	With Cap*				
ICC-4**	50	108	112	48	10	.190	1.5
ICC-4**	100	167	171	86	10	.190	3.0
ICC-4**	150	237	242	86	12	.210	4.5
3A480	50	74	77	43	10½	.150	1.5
3A480	100	132	135	57	12½	.175	3.0
3A480	150	193	196	58	15	.210	4.5
3A480X	100	88	91	57	12¼	.120	3.0
3A480X	150	135	138	58	14¾	.125	4.5
4AA480	100	100	103	55	12¼	.145	3.0
4AA480	150	143	146	56	14¾	.175	4.5

* Valve protecting cap
** Tube-type Cylinder

and ICC-4AA480 (5 years). In addition, the retest intervals may be 10 years instead of 5 years for cylinders 3A480, 3A480X, and 4AA480 if they are used exclusively for anhydrous ammonia commercially free from corroding components, and if they are protected externally by suitable corrosion resistant coatings (such as painting, etc.).

With respect to the two remaining cylinder types authorized for ammonia service, periodic hydrostatic retest is required every 5 years for ICC-3 cylinders, and no periodic retest is required for ICC-3E cylinders.

Charging Cylinders From a Refrigerating Unit. The charging of cylinders with anhydrous ammonia withdrawn from a refrigerating unit is a hazardous operation which normally cannot be recommended. If a careful study of all conditions makes such an operation necessary, it must be performed only with the approval and according to the instructions of the cylinder owner.

Storing Cylinders. Anhydrous ammonia bottle-type cylinders should be stored upright; tube-type cylinders, horizontally, and blocked to prevent rolling.

Manifolding Cylinders. For methods, flow diagrams, equipment and piping arrangements for manifolding cylinders, see Pamphlet G-2 of Compressed Gas Association, Inc.[3] For specifications of piping, tubing, and fittings for use with manifolding and other equipment, see "American Standard Safety Requirements for the Storage and Handling of Anhydrous Ammonia."[1]

CARGO TANKS AND PORTABLE TANKS

Anhydrous ammonia is authorized for shipment in motor vehicle cargo tanks complying with ICC specifications MC-330 and MC-331, and in steel portable tanks meeting ICC-51 specifications (sometimes called "skid" tanks). The minimum design pressure required for tanks of both types is 265 psig. Shipment of ammonia in TMU tanks (ICC-106A500-X type) on trucks is also authorized.

SINGLE-UNIT TANK CARS AND TMU TANK CARS

Anhydrous ammonia is authorized for shipment in single-unit tank cars complying with ICC specifications 105A300-W (an insulated car) and 112A400-F or 112A340-W (uninsulated cars). Regulations permit the unloading of railroad car shipments on the carrier's tracks (rather than private tracks) under certain conditions.

FARM VEHICLES FOR ANHYDROUS AMMONIA

Two main types of special farm vehicles for carrying anhydrous ammonia used as a soil fertilizer have been developed. One type applies ammonia to the ground from portable or "skid" tanks. Vehicles of the second type transport ammonia in bulk to replenish storage tanks at various locations on the farm or skid tanks on ammonia-applying vehicles. These vehicles usually operate only within individual states and are hence not subject to ICC regulations. However, state or local regulations may establish requirements for them. Sound standards for ammonia containers and equipment associated with these farm vehicles appear in the previously cited pamphlet, "American Standard Safety Requirements for the Storage and Handling of Anhydrous Ammonia."[1]

STORAGE CONTAINERS, PIPING AND EQUIPMENT

Relatively large quantities of anhydrous ammonia are often stored in refrigerated containers, which maintain it in its compact liquid state. Systems for such refrigerated storage must be properly engineered. The advice of the gas manufacturer or of experienced engineering firms should be obtained before proceeding with anhydrous ammonia storage installations. Standards for the safety requirements of these installations are given in the pamphlet cited above.[1]

REFERENCES

1. "American Standard Safety Requirements for the Storage and Handling of Anhydrous Ammonia" (ASA K61-1966), (Pamphlet G-2.1), Compressed Gas Association, Inc.
2. "American Standard Wrought-Steel and Wrought-Iron" (ASA B36.10-1959). American Standards Association, 10 E. 40th St., New York, N. Y., 10017.
3. "Anhydrous Ammonia" (Pamphlet G-2), Compressed Gas Association, Inc.
4. "Rules and Regulations for Cargo and Miscellaneous Vessels," Subchapter 1, Section 98.25 (Sept. 1, 1964). Superintendent of Documents, U. S. Government Printing Office, Washington, D. C. 20402.

ADDITIONAL REFERENCES

"Federal Specification O-A-445 on Anhydrous Ammonia in Cylinders." Superintendent of Documents, U. S. Government Printing Office, Washington 25, D. C.

"Joint Army-Navy Specification JAN-A-182 on Anhydrous Ammonia for Explosives." Superintendent of Documents.

U. S. Bureau of Standards, Scientific Paper No. 501, "Specific Heat of Superheated Ammonia Vapor," March, 1925.

Department of Commerce, Circular No. 142, "Tables of Thermodynamic Properties of Ammonia," April, 1923.

Wilson, T. A., "The Total and Partial Pressures of Aqueous Ammonia Solutions," Bulletin No. 146, University of Illinois, 1925.

Underwriters' Laboratories, "The Comparative Life, Fire and Explosion Hazards of Common Refrigerants," November 1933.

Y. Henderson and H. W. Haggard, *Noxious Gases*, New York: Chemical Catalog Co., 1927.

Compressed Gas Association, Inc., "Low Pressure Storage of Anhydrous Ammonia." Results of CGA ammonia research project—1966.

Argon

Ar ICC Classification: Nonflammable compressed gas: green label

PHYSICAL CONSTANTS

International symbol	Ar
Molecular weight	39.944
Density at 70 F and 1 atm, lb/cu ft	0.1034
Density, gas, at boiling point, lb/cu ft	0.356
Density, liquid, at boiling point, lb/cu ft	86.98
Boiling point at 1 atm	−302.6 F
Freezing point at 1 atm	−308.7 F
Critical temperature	−188.5 F
Critical pressure, psia	705.4
Latent heat of vaporization at melting point, Btu/lb	70.2
Latent heat of fusion at melting point, Btu/lb	12.1
Specific heat, C_p at 70 F and 1 atm, Btu/(lb)(°F)	0.1252
Specific heat, C_v, at 70 F and 1 atm, Btu/(lb)(°F)	0.075
Ratio of specific heats, C_p/C_v, at 70 F and 1 atm	1.67

Properties. Argon belongs to the family of inert, rare gases of the atmosphere. It is plentiful compared to the other inert atmospheric gases, 1 million cu ft of dry air containing 9300 cu ft of argon. Argon is colorless, odorless, tasteless and nontoxic. It is extremely inert, and forms no known chemical compound. It is slightly soluble in water.

Argon Mixtures. Mixtures of argon and helium, oxygen, hydrogen, nitrogen, or carbon dioxide are used for lamp-bulb filling and other purposes, and are available in any desired combinations. Small users may obtain these in high-pressure cylinders, mixed according to specifications. Large users purchase the individual gases in proportionate amounts and mix them at point of use.

Materials of Construction. Argon gas may be contained in any commonly available metals since it is inert and noncorrosive. Among materials suitable for use with argon liquefied at low temperatures are 18-8 stainless steel and other austenitic nickel-chromium alloys, copper, monel, brass and aluminum. Ordinary carbon steels and most alloy steels lose their ductility at the temperatures of liquid argon, and hence are usually considered unsafe for use with it.

Manufacture. Argon is manufactured in oxygen-nitrogen plants by means of fractional distillation after the liquefaction of air. In the distillation columns, liquid nitrogen is the product from the bottom of the high pressure column, followed by liquid oxygen containing argon and some krypton and xenon, gaseous oxygen, gaseous nitrogen, and crude neon gas. Crude gas is drawn off in the middle of the column for further processing to obtain high purity.

Commercial Uses. Argon is extensively used in filling incandescent and fluorescent lamps, and electronic tubes; as an inert gas shield for arc welding and cutting; as a blanket in the production of titanium, zirconium, and other reactive metals; to flush molten metals to eliminate porosity in castings; and to provide a protective shield for growing silicon and germanium crystals.

Physiological Effects. Argon is nontoxic. Due to its ability to displace air, it is a simple asphyxiant. Liquid argon must be handled with all the precautions required for safety with a gas at extremely low temperatures. In particular, severe burn-like injuries will result if liquid argon remains in contact with the skin for more than a few seconds. Delicate tissues, such as those of the eyes, can be damaged by a contact with liquid argon too brief to affect the skin of the hands or face.

ARGON CONTAINERS

While argon is available in gaseous form, it is increasingly being shipped as a liquid for reasons of economy. Argon gas is authorized for shipment in cylinders and tank cars. Liquid argon is shipped in cylinders, and in tank trucks and tank trailers. Most trucks used for shipping liquid argon maintain it at pressures below 25 psig; argon at such pressures is not classified or regulated as a "dangerous article" by the ICC, and hence is not subject to any ICC regulations. Trucks for shipping liquid argon at pressures at or above 25 psig are authorized by special permit of the ICC. Some liquid argon has also been shipped in tank cars at pressures below 25 psig or by special ICC permit.

Filling Limits. The maximum filling limits authorized for argon in shipping containers are as follows:

Gaseous argon in cylinders—up to their marked service pressure at 70 F and, in the case of cylinders meeting ICC specifications 3A and 3AA and special requirements, up to 10 per cent in excess of their marked service pressure.

Pressurized liquid argon in ICC-4L cylinders —115 per cent (per cent water capacity by weight).

Argon in uninsulated tank cars—up to 7/10 of the marked test pressure at 130 F.

SHIPPING METHODS; REGULATIONS

Under the appropriate regulations and tariffs, argon is authorized for shipment as follows:

By Rail: Gaseous and liquid argon in cylinders (by freight and express), and in tank cars.

By Highway: Gaseous and liquid argon in cylinders on trucks, and liquid argon in tank trucks (by special ICC permit for trucks operating at pressures at or above 25 psig; otherwise, as a commodity not regulated by the ICC).

By Water: Gaseous and liquid argon in cylinders in cargo vessels, passenger vessels, passenger ferries, and railroad car ferries. Gaseous argon also in tank cars via cargo vessels and railroad car ferries. Gaseous argon in cylinders on barges of U. S. Coast Guard classes A, BA, BB, CA and CB.

By Air: Gaseous argon in cylinders up to 150 lb (70 kg) in passenger aircraft and in cylinders up to 300 lb (140 kg) in cargo aircraft. Low-pressure liquid argon in cylinders up to 300 lb (140 kg) in cargo aircraft only. Non-pressurized liquid argon in containers as specified up to 13.2 gal (50 liters) in passenger and cargo aircraft. (Quantities are maximum net weight per container.)

ARGON CYLINDERS

Gaseous argon may be shipped in any cylinders authorized by the ICC for non-liquefied compressed gas. (These include cylinders meeting ICC specifications 3A, 3AA, 3B, 3C, 3D, 3E, 4, 4A, 4B, 4BA, and 4C; in addition, continued use of cylinders meeting ICC specifications 3, 7, 25, 26, 33, and 38 is authorized, but new construction is not authorized.)

Pressurized liquid argon may be shipped in cylinders meeting ICC specifications 4L200 (and in cylinders of the 4L type with a higher marked service pressure).

Valve Outlet and Inlet Connections. Standard connection, U. S. and Canada—No. 580.

Cylinder Requalification. All cylinders authorized for gaseous argon service must be requalified by hydrostatic retest every 5 years under present regulations, with the following exceptions: ICC-4 cylinders, every 10 years; and no periodic retest is required for cylinders of types 3C, 3E, 4C and 7.

TANK CARS

Argon is authorized for shipment in uninsulated multi-tube tank cars meeting ICC specification 107A. No ICC regulations apply to liquid argon shipped in insulated single-unit tank cars at pressures below 25 psig.

OTHER SHIPPING METHODS; STORAGE

Bulk shipments of liquid argon are also made in tank trucks and tank trailers to which ICC regulations do not apply (pressure below 25 psig) or by special ICC permit, as previously noted.

For large-scale argon users, bulk-storage systems for low-temperature liquid argon are available through suppliers. The systems are equipped with units for converting the liquid to gas at ordinary temperatures unless the liquid argon is drawn through insulated piping for special use at low temperatures.

REFERENCE

"Standard Density Data, Atmospheric Gases and Hydrogen" (Pamphlet P-6), Compressed Gas Association, Inc.

Boron Trifluoride

BF_3
Synonym: Boron fluoride
ICC Classification: Nonflammable compressed gas; green label

PHYSICAL CONSTANTS

International symbol	BF_3
Molecular weight	67.82
Specific gravity of gas at 70 F and 1 atm (air = 1)	2.38
Specific gravity of liquid at −148.7 F	1.57
Boiling point at 1 atm	−148.7 F
Melting point at 1 atm	−196.8 F
Triple point temperature (approx.)	−196.8 F
Critical temperature	10.5 F
Critical pressure, psia	723
Latent heat of vaporization, Btu/lb	1.18×10^{-4}
Latent heat of fusion, Btu/lb	2.68×10^{-5}
Specific heat, C_p, at 78 F, Btu/(lb)(°F)	0.178
Solubility in water at 32 F and 1 atm, by weight	322%

Properties. Boron fluoride is a colorless gas which has a persistent, irritating, acidic odor and which hydrolyzes in moist air to form dense white fumes. It is shipped as a non-liquefied compressed gas at varying pressures in the neighborhood of 2000 psig. Boron trifluoride reacts readily with water with the evolution of heat to form the hydrates, $BF_3.H_2O$ and $BF_3.2H_2O$—which are relatively strong acids. Inhalation of the gas irritates the respiratory system, and high concentrations in contact with the skin can cause a dehydrating type of burn (similar to burns from sulfur trioxide). Boron trifluoride readily dissolves in water and in organic compounds containing oxygen or nitrogen. The reaction is so rapid that, if boron trifluoride from a cylinder is fed beneath the surface of these liquids, there is danger of suck-back into the cylinder unless a vacuum break or trap is provided in the feed line. Boron trifluoride catalyzes a variety of reactions, and forms a great many addition compounds.

Materials of Construction. Dry boron trifluoride does not react with the common metals of construction, but if moisture is present, the hydrate acids identified above can corrode all common metals rapidly. In consequence, lines and pressure reducing valves in boron trifluoride service must be well protected from the entrance of moist air between periods of use. Cast iron must not be used because active fluorides attack its structure. If steel piping is used for boron trifluoride, forged steel fittings must be used with it instead of cast iron fittings. Among materials suitable for gaskets are "Teflon" and other appropriate fluorocarbon or chlorofluorocarbon plastics. Most plastics become embrittled in boron trifluoride service, but tubing of neoprene, butyl rubber or "Tygon" can be used temporarily where pressure is not involved.

Manufacture. In one major method of commercial preparation, boron trifluoride is produced from boric oxide and hydrofluoric acid (in the reaction $B_2O_3 + 6HF \rightarrow 2BF^3 + 3H_2O$), the product being purified and compressed before packaging in cylinders. Among other methods are ones which employ reactions between boron trichloride or borax and hydrofluoric acid.

Commercial Uses. Boron trifluoride is used as a catalyst for polymerizations, alkalations, and condensation reactions; as a gas flux for internal soldering or brazing; as an extinguisher for magnesium fires; and as a source of B^{10} isotope.

Physiological Effects. Boron trifluoride irritates the nose, mucous membranes and other parts of the respiratory system, and concentrations as low as 1 ppm in air can be detected by the sense of smell. The irritating sensation and white fumes produced by boron trifluoride in air give easily noticed warning of the escape of even small amounts of the gas; moreover, personnel do not get used to its odor, and tend to seek fresh air if traces of it are being inhaled. Its toxicity to humans is unknown, but one producer has found no indications of chronic effects in 30 years of experience with contact between personnel and low-exposure concentrations of boron trifluoride.

High concentrations are not only injurious if inhaled but in contact with the skin can cause dehydrating burns similar to those inflicted by acids.

In case of burns or other serious exposures, call a physician immediately and administer first aid measures that have been previously provided for in consultation with medical authorities. As a first aid treatment for skin burns, some users apply a paste made of milk of magnesia, magnesium oxide and glycerine.

Protective clothing and equipment required for personnel working with boron trifluoride includes at least rubber gauntlets, goggles and face shields worn as minimum protection. It is advisable as well that long sleeved shirts buttoned at the wrists be worn. Full-protection rubber or plastic garments and breathing apparatus with self-contained air supplies must be available for emergencies that may make it necessary for personnel to enter an area containing a high concentration of boron trifluoride.

BORON TRIFLUORIDE CONTAINERS

Boron trifluoride is authorized for shipment in cylinders, and is also shipped in motor vehicle tube trailers with tubes built to comply with cylinder specifications.

Filling Limits. The maximum filling limit at 70 F permitted for boron trifluoride is the service pressure of the container.

SHIPPING METHODS; REGULATIONS

Under the appropriate regulations and tariffs, boron trifluoride is authorized for shipment as follows:

By Rail: In cylinders.

By Highway: In cylinders on trucks, and in tube trailers with tubes meeting cylinder specifications.

By Water: In cylinders aboard cargo vessels only. In cylinders on barges of U. S. Coast Guard classes A, BA, BB, CA, and CB only.

By Air: In appropriate cylinders aboard cargo aircraft only up to 300 lb (140 kg) maximum net weight per cylinder.

BORON TRIFLUORIDE CYLINDERS

Boron trifluoride is frequently shipped and used in cylinders meeting ICC specifications 3A and 3AA and having service pressures ranging from 1800 to 2400 psig. Common commercial cylinder sizes have capacities from 6 to 62 lb (about 345 cu ft) of gas. Under ICC regulations, boron trifluoride is also authorized for shipment in any other cylinders specified as appropriate for nonliquefied compressed gas (which include cylinders meeting ICC specifications 3B, 3C, 3D, 3E, 4, 4A, 4BA, and 4C; cylinders meeting specifications 3, 7, 25, 26, 33, and 38 may also be continued in boron trifluoride service, but new construction is not authorized).

Valve Outlet and Inlet Connections. Standard connection, U. S. and Canada—No. 330.

Cylinder Requalification. Cylinders of types 3A and 3AA used in boron trifluoride service must be requalified by hydrostatic restest every 5 years under present regulations. (All other cylinders authorized for boron trifluoride must similarly be requalified by hydrostatic retest every 5 years, with the following exceptions: ICC-4 cylinders, every 10 years; and no periodic retest is required for cylinders of types 3C, 3E, 4C, and 7.)

TUBE TRAILERS

Bulk shipment of boron trifluoride is made in high-pressure tube trailers, for which the tubes have been built to comply with ICC cylinder specifications 3A or 3AA and to have service pressures of around 2000 psig. Common capacities of these trailers vary from 9000 to 15,000 lb of gas.

STORAGE AND PIPING EQUIPMENT

Boron trifluoride users rarely if ever have storage facilities for the gas, and instead draw supplies from cylinders or tube trailers as needed. Users of smaller quantities usually transfer the gas to process through seamless tubing, with a needle valve for control. Materials may be steel or stainless steel, with compression or flare couplings for junctions. The entire system should be capable of withstanding minimum pressures of 3000 psig.

Large-quantity users usually employ systems of extra-heavy-duty steel pipe with forged steel fittings and large valves of special-alloy bronze for control. In the systems, pressure gages must have all-steel internal parts, pressure regulators must have only steel or nickel-alloy internal parts, and any brass interior parts used must be silver-plated. Large-volume systems must also be built to minimum design pressures of 3000 psig throughout. Cast-iron lines or fittings must not be used, since active fluorides cause deterioration of the structure of cast iron.

Any boron trifluoride transfer system must include vacuum breaks or effective check valves to prevent suck-back of process materials into cylinders or tubes supplying the gas.

GENERAL REFERENCES

Booth, H. S., and Martin, D. R., "Boron Trifluoride and Its Derivatives," New York, John Wiley & Sons, 1949.

———, *J. Am. Chem. Soc.*, **64**, 2198–2205 (1942).

Eucken, A., and Schröder, E., *Z. Physik. Chem.*, **B41**, 307–319 (1938).

Fischer, W., and Weidemann, W., *Z. Anorg. Allgem. Chem.*, **213**, 106–114 (1933).

LeBoucher, L., Fischer, W., and Blitz, W., *Z. Anorg. Allgem. Chem.*, **207**, 61–72 (1932).

Topchiev, A. V., Zavgorodnii, S. V., and Paushkin, Y. M. (Greaves, J. T., trans.), "Boron Fluoride and Its Compounds as Catalysts in Organic Chemistry," New York, Perachemon Press, 1959.

Butadiene (1,3-Butadiene)

$H_2C{:}CHCH{:}CH_2$ (or $CH_2{:}CHCH{:}CH_2$)
Synonyms: Vinylethylene, biethylene, erythene, bivynl, divynl B
ICC Classification (butadiene, inhibited): Flammable compressed gas; red label

PHYSICAL CONSTANTS
(for Research Grade product)

International symbol	—
Molecular weight	54.088
Vapor pressure, psia	
at 70 F	36.05
at 100 F	59.30
at 120 F	80.11
at 140 F	105.93
Boiling point at 1 atm	24.06 F
Freezing point at 1 atm	−164.05 F
Critical temperature	306 F
Critical pressure, psia	628
Critical volume, cu ft/lb	0.0654
Specific gravity of gas at 60 F and 1 atm (air = 1)	1.9153
Density of liquid at saturated pressure (apparent value from weight in air of the air-saturated liquid), lb/cu ft	
at 60 F	39.05
at 70 F	38.69
at 105 F	37.0
at 115 F	36.57
at 130 F	35.69
Specific gravity of liquid, 60 F/60 F, at saturation pressure (absolute value from weights in vacuum for the air-saturated liquid)	0.6272
Specific heat of ideal gas at 60 F, C_p, Btu/(lb)(°F)	0.3412
Specific heat of ideal gas at 60 F, C_v, Btu/(lb)(°F)	0.3045
Ratio of specific heats, C_p/C_v, ideal gas at 60 F	1.121
Specific heat of liquid at 1 atm	0.5079
Latent heat of vaporization at boiling point and 1 atm, Btu/lb (estimated)	174
Gross heat of combustion	
Gas at 60 F and 1 atm, Btu/cu ft of the real gas	2954.8
Liquid at 60 F and saturation pressure, Btu/lb	20095.
Liquid at 60 F and saturation pressure, Btu/gal	104898.
Net heat of combustion	
Gas at 60 F and 1 atm, Btu/cu ft of the real gas	2800.
Liquid at 60 F and saturation pressure, Btu/lb	19035.
Liquid at 60 F and saturation pressure, Btu/gal	99364.

Air required for combustion at 60 F and 1 atm	
cu ft of air per cu ft of gas	26.92
lb of air per lb	14.06
Flammable limits in air, by volume	2.0–11.5%
Flash point	−105 F

Properties. Butadiene (1,3-butadiene) is a flammable, colorless gas with a mild aromatic odor. It is highly reactive and readily polymerizes, and is authorized for shipment only if inhibited. (Among inhibitors often used are tertiary butyl, di-*n*-butylamine, and phenyl-beta-naphthylamine.) Inhibited butadiene is shipped as a liquefied compressed gas under its own low vapor pressure of about 21 psig at 70 F. (Only 1,3-butadiene is treated in this section; the information given here should not be assumed to apply to 1,2-butadiene nor to other butadienes.)

Materials of Construction. Butadiene is noncorrosive, and may be used with any common metals. Steel is recommended for tanks and piping in butadiene service by some authorities. Welded rather than threaded connections are similarly recommended because butadiene tends to leak through even extremely small openings. Before being exposed to butadiene that is not inhibited, iron surfaces should be treated with an appropriate reducing agent like sodium nitrite because polymerization is accelerated by oxygen (even if present in ferrous oxide) as well as by heat.

Manufacture. Butadiene is made commercially by dehydrogenating butanes or butenes in the presence of a catalyst, by reacting ethanol and acetaldehyde, and by the cracking of naphtha and light oil. It is also derived as a by-product in ethylene production.

Commercial Uses. One major use of butadiene has been in the making of synthetic rubber (styrene-butadiene and nitrile-butadiene rubbers, to a large extent; *cis*-polybutadiene is also an extender and substitute for rubber, and *trans*-polybutadiene is a type of rubber with unusual properties). Butadiene is also used extensively for various polymerizations in manufacturing plastics. Copolymers with high proportions of styrene have found applications as stiffening resins for rubber, in water-base and other paints, and in high-impact plastics. Butadiene also serves as a starting material for nylon 66 (adiponitrile) and an ingredient in rocket fuels (butadiene-acrylonitrile polymer).

Physiological Effects. If inhaled in high concentrations, butadiene has an anesthetic or mild narcotic action which appears to vary with individuals. Inhalation of a 1 per cent concentration in air has been reported to have had no effect on the respiration or blood pressure of individuals, but such exposures may cause the pulse rate to quicken and give a sensation of prickling and dryness in the nose and mouth. Inhalation in higher concentrations has brought on blurring of vision and nausea in some persons. Inhalation in excessive amounts leads to progressive anesthesia, and exposure to a 25 per cent concentration for 23 minutes proved fatal in one instance. No cumulative action on the blood, lungs, liver or kidneys has been evidenced. The maximum allowable concentration for an 8-hour exposure, as generally accepted in the U. S., is 1000 ppm.

Contact with excessive concentrations of butadiene vapors also irritates the eyes, lungs and nasal passages. Contact between liquid butadiene and the skin causes freezing of the tissues, if prolonged, for butadiene liquid evaporates rapidly, and delayed skin burns may result if liquid butadiene is allowed to remain trapped in clothing or in shoes.

All the precautions necessary for the safe handling of a flammable compressed gas must also be observed with butadiene, and special precautions must be taken against its possible polymerization.

Detailed suggestions concerning safety precautions, protective equipment and first aid are given in Safety Data Sheet SD-55 of the Manufacturing Chemists' Association.[1]

BUTADIENE CONTAINERS

Inhibited butadiene is authorized for shipment in cylinders, single-unit tank cars, TMU tank cars, and motor vehicle cargo tanks and portable tanks.

Filling Limits. The maximum filling limits authorized for inhibited butadiene are as follows:

In cylinders—Not in excess of the cylinder service pressure at 70 F nor in excess of 5/4 of the service pressure at 130 F.

As with liquefied petroleum gas in other authorized containers; these maximum filling densities are prescribed according to the specific gravity of the liquid material at 60 F in detailed tables that are part of the ICC regulations. Producers and suppliers who charge inhibited butadiene containers other than cylinders should consult these tables in the current regulations. The lower and upper limits of the maximum filling densities authorized in the present regulations for such containers are as follows (per cent water capacity by weight):

In single-unit tank cars and TMU tank cars—from 45.500 per cent (insulated tanks, April through October) for 0.500 specific gravity, to 61.57 per cent (uninsulated tanks, November through March with no storage in transit) for 0.635 specific gravity. (In addition, filling must not exceed various specified limits of pressure and liquid content at temperatures of 105 F, 115 F or 130 F, as given in ICC regulations.)

In cargo tanks and portable tanks—from 38 per cent (tanks of 1200-gal capacity or less) for 0.473 to 0.480 specific gravity, to 60 per cent (tanks of over 1200-gal capacity) for 0.627 specific gravity and over (except when using fixed length dip tubes or other fixed maximum liquid level indicators). Moreover, the tank must be not liquid full at 105 F if insulated nor at 115 F if uninsulated, and the gage vapor pressure at 115 F must not exceed the tank's design pressure.

SHIPPING METHODS; REGULATIONS

Under the appropriate regulations and tariffs, inhibited butadiene is authorized for shipment as follows:

By Rail: In cylinders, and in single-unit tank cars and TMU (ton multi-unit) tank cars.

By Highway: In cylinders on trucks, and in cargo tanks and portable tanks.

By Water: In cylinders and portable tanks (tanks meeting ICC-51 specifications and not over 20,000 lb gross weight) aboard cargo vessels only. In authorized tank cars aboard trainships only, and in authorized motor vehicle tank trucks aboard trailerships and trainships only. In cylinders on barges of U. S. Coast Guard classes A, CA and CB only. In cargo tanks aboard tankships and tank barges (to maximum filling densities by specific gravity as stated in Coast Guard regulations).

By Air: In cylinders aboard cargo aircraft only up to 300 lb (140 kg) maximum net weight per cylinder.

BUTADIENE CYLINDERS

Inhibited butadiene is authorized by the ICC for shipment in any cylinders specified for liquefied compressed gas; such cylinders include those that meet the following ICC specifications: 3A, 3AA, 3B, 3BN, 3D, 3E, 4, 4A, 4B, 4BA, 4B-ET, 9, 40 and 41 (cylinders complying with ICC specifications 3, 25, 26 and 38 may also be continued in butadiene service, but new construction is not authorized).

Valve Outlet and Inlet Connections. Standard connection, U. S. and Canada—No. 510.

Cylinder Requalification. Cylinders of all types authorized for inhibited butadiene service must be requalified by hydrostatic retest every 5 years under present regulations with the following exceptions:

(1) no periodic retest is required for 3E cylinders;

(2) 10 years is the required retest interval for type 4 cylinders; and

(3) external visual inspection may be used in lieu of hydrostatic retest for cylinders that are used exclusively for inhibited butadiene which is commercially free from corroding components and that are of the following types (including cylinders of these types with higher service pressures): 3A480, 3AA480, 3A480X, 3B, 4B, 4BA, 26-240 and 260-300.

SINGLE-UNIT TANK CARS AND TMU TANK CARS

Single-unit tank cars and TMU tank cars are authorized by the ICC for the shipment of inhibited butadiene as follows:

At pressures not exceeding 75 psig at 105 F —in single-unit tank cars which comply with ICC specifications 105A100W or 111A100W-4 (provided that they have excess flow valves as required; shipment may also be continued in ICC-105A100 cars, but new construction is not authorized); also, in TMU tank cars which meet ICC specifications 106A500X.

At pressures not exceeding 255 psig at 115 F —in single-unit tank cars which comply with ICC specifications 112A340W and 114A340W.

CARGO TANKS AND PORTABLE TANKS

Inhibited butadiene may be shipped by motor vehicle under ICC regulations in cargo tanks meeting ICC specifications MC-330 or MC-331, and in portable tanks complying with ICC-51 specifications. The minimum design pressure required for these tanks is 100 psig.

STORAGE AND HANDLING EQUIPMENT

Butadiene must be kept inhibited in storage to prevent polymerization and the formation of spontaneously flammable peroxides. The inhibitor content of butadiene stored for any appreciable period should be regularly measured and maintained at safe levels. Ignition within a storage tank can be prevented by diluting the vapor phase with a sufficient proportion of inert gas.

Because of its high volatility, butadiene is usually stored under pressure, or in insulated tanks at reduced temperatures, preferably below 35 F.

All precautions necessary for storage tanks containing flammable compressed gas must be taken with butadiene storage installations. Tanks should be located outdoors, isolated from boilerhouses and other possible sources of ignition, and provided with adequate diking or drainage to confine or discard the content should the tank rupture. Processes employing butadiene should be designed so that personnel are not exposed to butadiene vapor or liquid. Installations must of course comply with all local regulations, and should be designed with the help of authorities thoroughly familiar with butadiene.

REFERENCE

1. "Butadiene" (SD-55), Manufacturing Chemists' Association, 1825 Connecticut Ave., N. W., Washington, D. C., 20009.

Butane

See **Liquefied Petroleum Gases**

Butene (1-Butene, Cis-2-Butene, Trans-2-Butene)

See **Liquefied Petroleum Gases**

Butylenes

See **Liquefied Petroleum Gases**

Carbon Dioxide

CO_2 ICC Classification: Non-flammable compressed gas; green label

PHYSICAL CONSTANTS

International symbol	CO_2
Molecular weight	44
Specific gravity of gas at 32 F and 1 atm (air = 1)	1.5290
Vapor density at 32 F and 1 atm, lb/cu ft	0.12341
Liquid density at 2 F, lb/cu ft	63.3
Liquid density at 80 F, lb/cu ft	42.2*
Triple point	−69.9 F at 60.4 psig
Sublimation temperature at 1 atm	−109.4 F
Critical temperature	87.8 F
Critical pressure, psig	1057.4
Latent heat of sublimation at −109.4 F and 1 atm, Btu/lb	246.3
Latent heat of liquid at 2 F and 301.5 psig, Btu/lb	119.1

Properties. Carbon dioxide is a compound of carbon and oxygen in proportions by weight of about 27.3 per cent carbon to 72.7 per cent oxygen. A gas at normal atmospheric temperatures and pressures, carbon dioxide is colorless, odorless and about 1.5 times as heavy as air. A slightly acid gas, it is felt by some persons to have a slight pungent odor and biting taste. It is normally inert and nontoxic. It does not burn.

Carbon dioxide exists simultaneously as a solid, liquid and gas at a temperature of −69.9 F and a pressure of 60.4 psig, its triple point. Figure 1 shows the triple point and full equilibrium curve for carbon dioxide.

At temperatures and pressures below the triple point, carbon dioxide may be either a solid ("dry ice") or a gas, depending upon conditions. Solid carbon dioxide at a temperature of −109.4 F and atmospheric pressure transforms directly to a gas (sublimes) without passing through the liquid phase. Lower temperatures will result if solid carbon dioxide sublimes at pressures less than atmospheric or in moving atmosphere (partial pressure effect).

At temperatures and pressures above the triple point and below 87.8 F, carbon dioxide liquid and gas may exist in equilibrium in a closed container. Within this temperature range the pressure in a closed container holding carbon dioxide liquid and gas in equilibrium bears a definite relationship to the temperature. Above the critical temperature, which is 87.8 F, carbon dioxide cannot exist as a liquid regardless of the pressure.

Carbon dioxide is produced and sold in liquid form as a compressed gas, and in solid form.

Manufacture. Unrefined carbon dioxide gas is obtained from the combustion of coal, coke, natural gas, oil or other carbonaceous fuels; from by-product gases from ammonia

* This approximates the "liquid full" condition in a container charged to 68 per cent filling density.

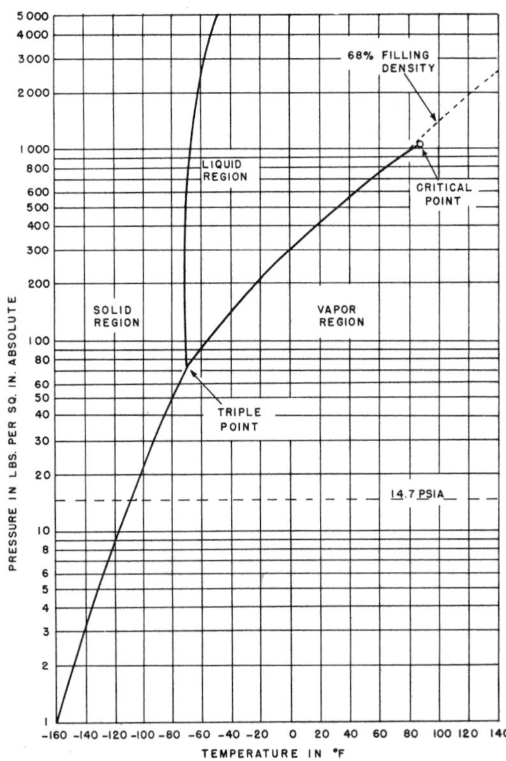

FIG. 1. Equilibrium curve for carbon dioxide.

plants, lime kilns, etc.; from fermentation processes; and from gases found in certain natural springs and wells. The gas obtained from these sources is liquefied and purified by several different processes to a purity of about 99.9 per cent or better.

In general the process involved in producing solid carbon dioxide is as follows: First, cold liquid carbon dioxide is piped into a special hydraulic press. As the liquid boils and evaporates, the vapors are pumped off. The remaining liquid cools until it finally freezes into solid carbon dioxide crystals. After the vapor pressure has been reduced to atmospheric pressure by pumping or bleeding the vapors away, the solid carbon dioxide crystals are pressed into a block by the hydraulic press.

Materials of Construction. The common commercially available metals can be used for dry carbon dioxide installations. Any carbon dioxide system at the user's site must be designed to contain safely the pressures involved, and must conform with all state and local regulations. For low-pressure carbon dioxide systems (up to 400 psig), containers and related equipment should have design pressures rated at least 10 per cent above the normal maximum operating pressure. For such systems, schedule 80 threaded steel pipe with 2000-lb forged steel fittings are recommended; alternate recommendations include stainless steel, copper or brass pipe, and seamless carbon steel, stainless steel or copper tubing. Special materials and construction are required for containers operating at temperatures below −20 F. Since wet carbon dioxide forms carbonic acid, systems handling carbon dioxide in aqueous solutions must be fabricated from such acid-resistant materials as certain stainless steels, "Hastelloy" metals, or "Monel" metal.

Commercial Uses. Solid carbon dioxide is used quite extensively to refrigerate dairy products, meat products, frozen foods, and other perishable foods while in transit. It is also used as a cooling agent in many industrial processes, such as grinding heat-sensitive dyes and pigments, rubber tumbling, cold-treating metals, shrink fitting of machinery parts, vacuum cold traps, etc. Solid carbon dioxide placed in liquefiers provides a source of gaseous carbon dioxide.

Gaseous carbon dioxide, obtained from liquid or solid CO_2, is used to carbonate soft drinks in chemical processing, as a food preservative, as a chemically inert "blanket" in food processing and metal welding, for hardening molds and cores in foundries, and for pressure pumping.

Liquid carbon dioxide is used as a fire extinguishing agent in portable and built-in fire extinguishing systems. It is also used as as expendable refrigerant for low-temperature testing of aviation, missile, and electronic components, for pre- and post-chilling trucks, railroad cars, containers, etc., for rubber tumbling, and for controlling chemical reactions.

Physiological Effects. Carbon dioxide is present in the atmosphere to the extent of

0.03 per cent by volume and is normal to body processes in limited amounts. However, high concentrations become dangerous in their effect of diluting or depleting the oxygen content in the air. A concentration of 5000 ppm is generally accepted as the maximum allowable for a daily eight-hour exposure.

Being denser than air, under certain conditions of use or storage, carbon dioxide gas may accumulate in low or confined areas. Precautions with regard to ventilation are required.

When entering low or confined areas where a high concentration of carbon dioxide gas is present, *do not* use air-breathing or filter-type gas masks. Gas masks of the self-contained type, or the type which feeds clean outside air to the breathing mask are required.

Appropriate warning signs should be affixed outside of those areas where high concentrations of carbon dioxide gas may accumulate. Suggested wording for such a sign is:

"CAUTION—CARBON DIOXIDE GAS
Ventilate before entering. A high
CO_2 gas concentration may occur
in this area."

Contact between the skin and liquid carbon dioxide can result in frostbite, and must be avoided. Solid carbon dioxide must be handled with dry ice tongs or heavy gloves, for careless handling of it leading to contact with the skin may cause injuries like severe burns.

CARBON DIOXIDE CONTAINERS

Carbon dioxide is contained, shipped, and stored in either liquefied or solid form. Applications using gaseous carbon dioxide are supplied by gas converted from liquid or solid carbon dioxide.

Liquefied Carbon Dioxide Containers. Liquefied carbon dioxide is shipped in cylinders, insulated portable tanks, insulated tank trucks, and in insulated tank cars. In high-pressure supply systems, it may be stored and used from single or manifolded cylinders, or high-pressure receiver-tube assemblies. In low-pressure systems, it is stored in insulated pressure vessels with controlled heating and refrigeration systems.

Solid Carbon Dioxide Containers. Normally, solid carbon dioxide is packaged in 50-lb blocks wrapped in heavy Kraft paper or heavy, insulated bags. It is also shipped in insulated containers and storage boxes of varying size. Solid carbon dioxide is stored in heavily insulated, top-opening boxes.

Filling Limits. The maximum allowable filling densities authorized for carbon dioxide are:

In cylinders—50 lb except for cylinders rated for 75 lb or 100 lb (the 50-lb limit results in less filling than the overriding maximum of 68 per cent water capacity by weight; for special carbon dioxide mining devices, 85 per cent is allowed if other requirements are met).

In single-unit tank cars—so that the liquid portion of the gas does not completely fill the tank at 0 F.

In cargo tanks and portable tanks on trucks —95 per cent by volume.

SHIPPING METHODS: REGULATIONS

Under the appropriate regulations and tariffs, carbon dioxide is authorized for shipment as follows:

By Rail: In cylinders (by freight, express or baggage), and in insulated single-unit tank cars.

By Highway: In cylinders, in insulated tank trucks, and in insulated portable tanks.

By Water: In cylinders aboard passenger vessels, cargo vessels, and all types of ferry vessels. In authorized tank cars, tank trucks, and portable tanks (maximum 20,000 lb gross weight) aboard cargo vessels and railroad car ferry vessels (passenger or vehicle); and in tank trucks aboard passenger or vehicle ferry vessels. In appropriate cylinders on barges of U. S. Coast Guard classes A, BA, BB, CA, and CB.

By Air: In cylinders aboard passenger aircraft up to 150 lb (70 kg) maximum net weight per cylinder; and in cylinders aboard cargo aircraft up to 300 lb (140 kg) maximum net weight per cylinder.

Shipment of solid carbon dioxide by rail or highway is not subject to ICC regulations (under which it is not designated as a dangerous article), while shipment by water and air must meet only certain labeling and packaging requirements.

CYLINDERS

There are two kinds of liquefied carbon dioxide cylinders, the standard type and the syphon type. The standard cylinder, in an upright position, discharges gas; inverted, it discharges liquid. The syphon cylinder is equipped with a dip tube. It discharges liquid only. With the exception of fire extinguisher cylinders, all syphon-type cylinders are clearly identified by the word "SYPHON."

Cylinders that meet the following ICC specifications are authorized for liquefied carbon dioxide service: 3A1800, 3AA1800, 3E1800 and 3HT2000. ICC-3 cylinders may also be continued in carbon dioxide service, but new construction is not authorized.

Valve Outlet and Inlet Connections. Standard connection U. S. and Canada—No. 320, No. 940. Alternate standard connection, Canada only: No. 420 (obsolete, effective Sept. 1, 1966).

Cylinder Requalification. Under present regulations, cylinders authorized for liquefied carbon dioxide service must be requalified by hydrostatic retest every 5 years, with two exceptions: 3HT cylinders must be requalified by retest every 3 years, and no periodic retest is required for 3E cylinders.

SINGLE-UNIT TANK CARS

Liquefied carbon dioxide is authorized for shipment in single-unit tank cars that meet ICC specifications 105A500-W, and are fitted as required with insulation and pressure-regulating valves.

CARGO TANKS AND PORTABLE TANKS

Liquefied carbon dioxide is authorized for shipment in cargo tanks on trucks complying with ICC specifications MC-330 and MC-331, and in portable tanks conforming to specifications ICC-51. The minimum design pressure for these tanks must be 200 psig (or, if built to requirements in "Low Temperature Operation of the ASME Boiler and Pressure Vessel Code, Section VIII, Unfired Pressure Vessels," the design pressure may be reduced to 100 psig or the controlled pressure, whichever is greater).

STORING AND HANDLING LIQUEFIED CARBON DIOXIDE

Users of liquefied carbon dioxide must comply with all state, municipal and other local regulations.

Storage containers of liquefied carbon dioxide are noninsulated and nonrefrigerated (except for bulk low-pressure containers, discussed below). The contained carbon dioxide is, therefore, at ambient temperatures and relatively high pressures.

Cylinders and high-pressure tubes charged with liquid carbon dioxide must never be allowed to reach a temperature exceeding 125 F. Storage should never be near furnaces, radiators, or any other source of heat.

Cylinder Handling Precautions. At an average room temperature of 60 F, a cylinder containing liquid carbon dioxide under balanced thermal conditions has a pressure of 733 psig. No attempt should be made to use carbon dioxide gas without a pressure regulator of the correct design and in good condition.

Piping from cylinders to point of use must be of correct high-pressure design, of at least Schedule 80 piping or high-pressure copper tubing, with proper provisions made in the piping for adequate safety relief devices. Piping should be adequately braced.

Transfer of liquid carbon dioxide from one high-pressure carbon dioxide cylinder to another may be accomplished by several means, including direct transfer by pressure differential, or, more usually, by means of a pump. For refilling fire extinguishers, or any other carbon dioxide cylinder, consult the gas supplier and follow his recommendations.

When a depleted carbon dioxide cylinder is removed from a manifold supply line, close the valve first and leave it closed to prevent air from entering the so-called "empty" cylinder.

High-pressure Tubes. High-pressure tubes are filled from bulk liquid carbon dioxide cargo tank trailers. Their location, therefore, should provide ready access to driveways capable of taking these delivery units.

Tubes should be in an area where unauthorized persons cannot tamper with fittings and valves. Adequate protection should be provided to prevent heavy objects from shearing off piping, valves, or safety relief devices.

Outlets from safety relief valves should be piped to the outside to prevent accumulation of heavy carbon dioxide vapors. Such piping should be provided with drain holes at low points and must not be capped on the end or equipped with valves or other means of stopping the flow of gas.

High-pressure Tube Handling Precautions. The same handling precautions given above with respect to cylinders apply equally to high-pressure tubes.

Bulk Low-pressure Liquid Containers. Bulk containers for storing liquid carbon dioxide at low pressures are well insulated and equipped with a means, usually mechanical refrigeration, to control and limit internal temperatures and pressures. Storage temperatures are maintained well below ambient, usually in the range of −20 to 4 F, with corresponding carbon dioxide pressures of 200 to 312 psig.

The storage container should preferably be located in an area that is not subject to unduly high temperatures. If the ambient temperature is above 110 F for long periods of time, it may be necessary to provide additional refrigeration capacity. If the ambient temperature falls below 0 F for a prolonged period, no harm will result but the carbon dioxide pressure may fall below the desired range.

Dusty, oily locations should be avoided because of the tendency of dust and oil to collect on the refrigerator condenser and thus reduce its efficiency. A dry, well ventilated location is preferred.

The storage container should not be located in an area where it might be struck by heavy moving or falling objects. A break or tear in the outer shell or covering of the insulation will destroy the vapor seal and allow water vapor to enter the insulation with eventual losses of insulation efficiency.

The storage container should be protected from tampering by unauthorized individuals. If a small enclosed location is used, the outlet from the safety relief valves should be piped to the outside or other point where a discharge of carbon dioxide vapor will not result in a high concentration of carbon dioxide. Such piping should be provided with drain holes at low points and must not be equipped with valves or other means of stopping the flow of gas.

Bulk Low-pressure Liquid Container Handling Precautions. No attempt should be made to use carbon dioxide vapor without a pressure-reducing regulator of suitable design and in good condition.

Whenever liquid carbon dioxide is discharged directly to the atmosphere, as in those cases where sudden cooling is desired, extreme caution should be exercised to guard against and counteract the heavy recoil inherent with the discharge of a dense liquid (weighing more than water) under high pressure. The liquid carbon dioxide hose should be anchored firmly against this recoil by means of chains or other positive mechanical devices installed prior to use of the hose.

All lines from the bulk carbon dioxide receiver should be of Schedule 80 steel pipe or high-pressure copper tubing, anchored in such a fashion that shrinkage of the piping or tubing due to the passage of the sub-cooled liquid carbon dioxide through them will not tear them loose. Such piping must be protected with adequate safety relief devices to prevent undue pressure buildup from entrapped liquid carbon dioxide. Valves used in such lines should have a design pressure not less than 350 psig.

The rapid discharge of liquid carbon dioxide through a line which is not grounded will result in a buildup of static electricity potential, potentially dangerous to operating personnel. Such lines should, therefore, be grounded before use.

Flexible hoses used with liquid carbon dioxide should have a minimum design pressure of 500 psig for low-temperature operation, with added wire reinforcement. Attachments at each end should be mechanically sound to at least the same pressure as the hose itself. Coupling devices used with the hose should be of extra heavy grade.

After use of such flexible hose, it becomes quite rigid and may contain loose dry ice snow. Do not fold, bend or distort the hose, or point it in any direction where pressure buildup within the hose will eject the dry ice snow so as to endanger personnel. These hoses should be hydrostatically tested to 1.5 times the design pressure semiannually, and rejected if found unsound.

Whenever liquid carbon dioxide is discharged into confined spaces to reduce temperature rapidly, large volumes of carbon dioxide gas will be evolved, amounting to some 8.5 cu ft of gas per lb. These spaces must therefore have some provision incorporated to vent this vapor to the atmosphere to prevent pressure rise within the container.

STORING AND HANDLING SOLID CARBON DIOXIDE

Solid carbon dioxide (dry ice) has a temperature of −109.4 F, and must be protected during storage with thermal insulation in order to minimize loss through sublimation.

Dry ice should be stored in well-insulated storage containers preferably in a cool, nonconfined, or ventilated area.

Handling Precautions. Do not handle dry ice with bare hands. Use heavy gloves or dry ice tongs.

Handle dry ice carefully as injuries can occur if it is accidentally dropped on the feet. Make certain that dry gloves or tongs are used at all times.

A suggested wording for a caution label for dry ice follows:

"Warning—Extremely cold. May cause severe burns in contact with skin. Liberates heavy gas which may cause suffocation. Do not enter confined areas where used or stored until adequately ventilated. Do not taste. Do not put in stoppered glass jars or bottles or other sealed containers. Handle with gloves. Keep out of children's reach."

Dry Ice Converters. Dry ice converters are devices for transforming supplies of solid carbon dioxide into gas, which is then charged into cylinders for use. These converters should be located in areas where they will never be subjected to temperatures of more than 125 F. Converter locations must also be chosen or protected so that unauthorized persons cannot tamper with fittings and valves. Adequate protection should be provided to prevent heavy objects from shearing off piping, valves, or safety relief devices.

Outlets from safety relief valves should be piped to the outside to prevent accumulation of heavy carbon dioxide vapors. Such piping should be provided with drain holes at low points and must not be capped on the end or equipped with valves or other means of stopping the flow of gas.

Converter Handling Precautions. The same precautions for handling high-pressure gas or liquid carbon dioxide in cylinders apply equally to converters. The only differences between the two types of storage are the method of filling and the amount of liquefied carbon dioxide contained. Dry ice converters are charged with dry ice and the amount charged must not exceed the rated capacity of the converter. *Do not overfill.*

Dry ice converters have a screwed or flanged lid designed to be removed when the units are emptied so that they can be recharged with blocks of dry ice. It is virtually impossible to remove the lid when any pressure remains. Even so, be sure that all valves of the converter venting to the atmosphere are opened when the lid is removed and left open during the recharging period.

Prior to loading dry ice into a converter, remove all paper wrappings from the dry ice. Be sure that the converter lid is positioned correctly before closing valves, and that the screwed lid is fitted all the way into the converter.

When venting converters prior to opening, be sure that venting is made to the outside atmosphere in order to prevent accumulation of heavy carbon dioxide vapors in enclosed areas.

Carbon Monoxide

CO ICC Classification: Flammable compressed gas, red label

PHYSICAL CONSTANTS

International symbol	CO
Molecular weight	28.01
Specific gravity of gas at 1 atm and 70 F (air = 1)	0.9678
Specific volume at 1 atm and 70 F, cu ft/lb	13.8
Boiling point at 1 atm	−312.7 F
Melting point at 1 atm	−340.6 F
Triple point	2.2 psia at −337.1 F
Critical temperature	−220 F
Critical pressure, psia	507.5
Specific heat, vapor, C_p, at 60 F, Btu/(lb)(°F)	0.2478
Specific heat, vapor, C_v, at 60 F, Btu/(lb)(°F)	0.1766
Specific heat ratio, C_p/C_v	1.405
Latent heat of vaporization at boiling point, Btu/lb	92.79
Latent heat of fusion, at melting point, Btu/lb	12.85
Net heat of combustion at 77 F, Btu/lb	4343.6
Flammable limits in air, by volume	12.5–74%
Solubility in water at 32 F, by volume	3.5%
Weight per gallon, liquid, at −317.2 F, lb	6.78

Properties. Carbon monoxide is a toxic, flammable gas with no color and no odor. If inhaled, concentrations of 0.4 per cent prove fatal in less than an hour, while inhalation of high concentrations can cause sudden collapse with little or no warning. Pure carbon monoxide has a negligible corrosive effect on metals at pressures below 500 psig, although it is corrosive to many metals at higher pressures or when containing sulfur compounds as impurities. Chemically, carbon monoxide is stable with respect to decomposition. At temperatures of 570–2700 F it reduces many metal oxides to lower metal oxides, metals or metal carbides. Hydrogenation of carbon monoxide yields products varying according to catalysts and conditions, which include methane, benzene, olefins, paraffin waxes, hydrocarbon high polymers, methanol, higher alcohols, ethylene glycol, glycerol and other oxygenated products. Carbon monoxide also combines with the alkali and alkaline earth metals, reacts with chlorine, bromine, sulfur, Grignard reagents, and sodium alkyls, adds to alcohols, and enters into many other reactions.

Materials of Construction. Steels and other common metals are satisfactory for use with sulfur-free carbon monoxide at pressures below 500 psig. Iron, nickel and other metals react with carbon monoxide at pressures above 500 psig to form carbonyl liquids or vapors in small quantities, and the presence of sulfur-containing impurities in carbon monoxide appreciably increases its corrosive action on steel at any pressure. High-pressure plant equipment is often lined with copper for increased resistance to carbon monoxide attack,[1] and very highly alloyed chrome steels are sufficiently resistant to corrosion by carbon monoxide with small amounts of sulfur-bearing impurities.[2] Users are strongly urged to make corrosion tests of samples of proposed construction materials in order to select ones which will withstand high-pressure use of carbon monoxide under actual conditions.

Manufacture. Pure carbon monoxide is made commercially from synthesis gas, blast-furnace gas or coke-oven gas by two methods: (1) absorption of carbon monoxide by an ammoniacal cuprous salt solution at elevated pressure, followed by pressure release; (2) low-temperature condensation and fractionation.

Commercial Uses. Carbon monoxide is used in the chemical industry to produce such commodities as methanol and ethylene, and in organic synthesis; and in metallurgy, to recover high-purity nickel from crude ore in the Mond process, and for special steels and reducing oxides. It is also used to obtain powdered metals of high purity, such as zinc white pigments; to form certain metal catalysts applied in synthesizing hydrocarbons or organic oxygenating compounds; and in hydrogenating fats and oils. It is used as well in the manufacture of acids, esters and hydroxy acids, such as acetic and propionic acids and their methyl esters and glycolic acid.

Physiological Effects.[3] A chemical asphyxiant, carbon monoxide acts toxically by combining with the hemoglobin of the red blood cells to form the stable compound, carbon monoxide-hemoglobin. It thus prevents the hemoglobin from taking up oxygen and cuts off needed oxygen from the body. The affinity of carbon monoxide for hemoglobin is about 300 times the affinity of oxygen for hemoglobin. The inhalation of concentrations as low as 0.04 per cent will result in headache and discomfort within 2 to 3 hours, and, as noted, inhalation of a 0.4 per cent concentration proves fatal in less than one hour. Lacking odor and color, carbon monoxide gives no warning of its presence, and inhalation of heavy concentrations can cause sudden, unexpected collapse.

The maximum concentration allowable for a daily 8-hour exposure, according to the American Conference of Governmental Industrial Hygienists, is 100 ppm.

First-Aid Suggestions. Personnel accidentally overcome by carbon monoxide should be given first aid prior to a physician's arrival. The first aid measures presented here are based upon what is believed to be common practice in industry, but they should be reviewed and amplified into a complete first-aid program by a competent medical advisor before adoption in any specific case.

It is extremely important to hasten the elimination of carbon monoxide from the blood stream, should poisoning occur. Such elimination is best effected by inhalation of a mixture of oxygen containing 7 to 10 per cent carbon dioxide by volume. The inhalation of pure oxygen will eliminate carbon monoxide much more rapidly than inhalation of fresh air, and oxygen alone could be used instead of the oxygen-carbon dioxide mixture. The mixture, however, acts as a powerful respiratory and cardiac stimulant, inducing deep breathing and rapid ventilation of the lungs. Administering oxygen alone after a prolonged and severe asphyxia does not prove very effective since the victim generally breathes poorly, although spontaneously.

Any person showing symptoms of carbon monoxide poisoning must be moved immediately to fresh, but not cold, air. Keep him warm and place him on his stomach with his face turned to one side. If he is breathing, he should be given oxygen containing 7 to 10 per cent carbon dioxide to inhale. If he is not breathing, start manual artificial respiration at once with simultaneous administration of the oxygen-carbon dioxide mixture. Inhalation of the mixture should continue for 15 to 30 min after spontaneous respiration returns. The after-treatment consists of general measures to prevent pneumonia from developing.

Drugs are of little value in the treatment of carbon monoxide poisoning. Coffee may be given if the patient is able to hold a cup. Alcohol should not be given.

Detailed information on the treatment of carbon monoxide asphyxia is presented in the volume, "Noxious Gases and the Principles of Respiration Influencing Their Action."[4]

Leak Detection. As in the case of any flammable gas, never use a flame in trying to detect carbon monoxide leaks. Soapy water painted over the suspected area will indicate leaks by the formation of bubbles. Carbon monoxide alarm detectors must be installed in all indoor areas in which the gas is regularly used in more than small laboratory amounts. Alarm detector units based on infrared absorption or measurement of the heat of reaction occuring in the catalytic conversion of CO to CO_2 are available.

CARBON MONOXIDE CONTAINERS

Carbon monoxide is authorized for shipment only in cylinders.

Filling Limits. The maximum pressure authorized for carbon monoxide in cylinders having a minimum service pressure 1800 is 1000 psig at 70 F except on special authorization of the ICC.

SHIPPING METHODS; REGULATIONS

Under the appropriate regulations and tariffs, carbon monoxide is authorized for shipment as follows:

By Rail: In cylinders (freight, or express in one outside container to a maximum of 150 lb).

By Highway: In cylinders on trucks.

By Water: In cylinders via cargo vessels only. In cylinders on barges of U. S. Coast Guard classes A and C only.

By Air: Aboard cargo aircraft only in appropriate cylinders up to 150 lb (70 kg) maximum net weight per cylinder.

CARBON MONOXIDE CYLINDERS

Cylinders that meet the following ICC specifications are authorized for carbon monoxide service: 3A1800, 3AA1800, and 3E1800. ICC-3 cylinders are also authorized for continued use, but not for new construction.

Valve Outlet and Inlet Connections: Standard connection, U. S. and Canada—No. 350.

Cylinder Requalification. Under present regulations, the cylinders authorized for carbon monoxide service must be requalified by hydrostatic test every 5 years with the exception of type 3E (for which periodic hydrostatic retest is not required).

Precautions in Handling Cylinders. All precautions necessary for the safe handling of any flammable gas must be observed with carbon monoxide. Among these, special care should be taken to avoid storing carbon monoxide cylinders with cylinders containing oxygen or other highly oxidizing or flammable materials. It is recommended that carbon monoxide cylinders in use be grounded, and protected by check valve to prevent suck-back of reaction contents into the cylinders. Areas in which cylinders are being used must be free of all ignition sources and hot surfaces.

REFERENCES

1. W. M. Newitt, "The Design of High Pressure Plant and Properties of Fluids at High Pressure," p. 20, Oxford, England, Clarendon Press, 1940.
2. O. Sunden and E. Bohm, *Chem.-Ing.-Tech.*, **23**, 30 (1951).
3. M. B. Jacobs, "The Analytical Chemistry of Industrial Poisons, Hazards, and Solvents," 2nd Ed., p. 403, New York, Interscience Publishers, 1949.
4. Henderson and Haggard, "Noxious Gases and the Principles of Respiration Influencing Their Action," (2nd Ed.), pp. 169–170, New York, Reinhold Publishing Corporation, 1943.

Chlorine

Cl_2 ICC Classification: Nonflammable compressed gas; green label

PHYSICAL CONSTANTS*

International symbol	Cl_2
Molecular weight	70.914
Vapor pressure, psig	
at 70 F	85.464
at 105 F	151.12
at 115 F	174.69
at 130 F	213.89
Density, dry gas, at 32 F and 1 atm, lb/cu ft	0.2003
Specific gravity of dry gas (air = 1), both at 32 F and 1 atm	2.482
Specific volume, dry gas, at 32 F and 1 atm, cu ft/lb	4.992
Density, liquid, at 32 F, lb/cu ft	91.67
Specific gravity, liquid, at 32 F and 53.155 psia (3.617 atm)	1.468
Density, saturated liquid, lb/cu ft	
at 70 F	87.72
at 105 F	83.82
at 115 F	82.65
at 130 F	80.91
Liquid/gas ratio at 32 F and 1 atm, vol/vol	1/457.6
Boiling point	−29.29 F
Melting point at 1 atm	−149.76 F
Critical temperature	291.2 F
Critical pressure, psia	1118.4
Latent heat of vaporization at boiling point, Btu/lb	123.7
Latent heat of fusion at melting point, Btu/lb	38.86
Specific heat, liquid, at 32–75 F, Btu/(lb)(F)	0.226
Specific heat, gas, at 59 F	
C_p, Btu/(lb)(F)	0.115
C_v, Btu/(lb)(F)	0.0849
Ratio, C_p/C_v	1.355
Thermal conductivity, gas, at 32 F, Btu/(hr)(sq ft)(F/ft)	0.0043
Viscosity at 68 F	
liquid, centipoises	0.3518
gas, centipoises	0.0132
Weight per gallon, liquid, at 32 F, lb	12.25

Properties.[1] Chlorine is a greenish-yellow, nonflammable gas with a distinctive, disagreeable odor; it is almost 2.5 times as heavy as air. The gas acts as a severe irritant if inhaled. Chlorine liquid has the color of clear amber and is about half again as heavy as water. It is shipped as a compressed liquefied gas with a vapor pressure of 85.5 psig at 70 F. Chlorine is nonflammable and nonexplosive in both gaseous and liquid states. However, like oxygen, it is capable of supporting the combustion of certain substances. Many organic chemicals react readily with chlorine, in some cases with explosive violence.

* See concluding "References" section for sources of physical constants.

Chlorine usually forms univalent compounds, but it can combine with a valence of 3, 4, 5 or 7.

Chlorine is only slightly soluble in water. When it reacts with pure water, weak solutions of hydrochloric and hypochlorous acids are formed. Chlorine hydrate ($Cl_2.8H_2O$), may crystallize below 49.3 F.

Chlorine unites, under specific conditions, with most of the elements; these reactions may be extremely rapid. At the boiling point of chlorine it reacts with sulfur. It does not react directly with oxygen or nitrogen; the oxides and nitrogen compounds are well known, but can be prepared only by indirect methods. Mixtures of chlorine and hydrogen composed of more than 5 per cent of either component can react with explosive violence, forming hydrogen chloride.

The preparation of soda and lime bleaches (sodium and calcium hypochlorite) are typical reactions of chlorine with the alkalies and alkaline earth metal hydroxides; the hypochlorites formed are powerful oxidizing agents. Because of its great affinity for hydrogen, chlorine removes hydrogen from some of its compounds, such as the reaction with hydrogen sulfide to form hydrochloric acid and sulfur. Chlorine reacts with ammonia or ammonium compounds to form various mixtures of chloramines, including the explosive, nitrogen trichloride, depending on the conditions.

Chlorine reacts with organic compounds much the same as with inorganics to form chlorinated derivatives and hydrogen chloride. Some of these reactions can be explosive if not controlled, including those with hydrocarbons, alcohols and ethers.

Commercial Uses. The largest quantities of chlorine produced are used in manufacturing chemicals which do not contain chlorine, such as ethylene glycol, tetraethyl lead, and ethylene oxide, and in manufacturing chlorine-containing chemicals—among them: such solvents as carbon tetrachloride, trichloroethylene, perchloroethylene, and methylene chloride; pesticides and herbicides, including DDT, benzene hexachloride, and toxaphene; such plastics and fibers as vinyl chloride and vinylidene chloride; and refrigerants and propellants like the halocarbons and methyl chloride. It is also an ingredient in the widely used bleach, deodorizer and disinfectant, chlorinated lime. Chlorine is also widely employed in bleaching pulp, paper, and textiles; for drinking and swimming water purification; in the sanitation of industrial and sewage wastes; and in the degassing of aluminium melts.

Materials of Construction. Below 230 F, the maximum temperature at which chlorine is normally handled, steel, iron, copper, nickel and lead are among metals resistant to dry chlorine in liquid or gas states. Nickel and certain nickel alloys resist corrosion by dry chlorine at temperatures of 600 to 1000 F. Even small amounts of moisture mixed with chlorine form hypochlorous and hydrochloric acids that are very corrosive to most metals. Tantalum and titanium resist attack by moist chlorine and chlorinated water of any concentration. Tantalum is not affected by wet or dry chlorine at temperatures up to 250 F, but titanium can burn in dry chlorine. Silver and platinum resist the action of wet chlorine gas fairly well, and have been used in special pieces of chlorine equipment. Certain nonmetals withstand corrosion by wet or dry chlorine. Glass or glass-lined steel has proven serviceable to temperatures of 200 to 300 F, and some ceramics and plastics are used at lower temperatures. Seamless carbon steel pipe, Schedule 80, is among materials recommended for piping at temperatures from −20 to 300 F; materials recommended for other purposes include lead and fluorocarbon plastics for gaskets, and forged steel for valves.

Dry chlorine reacts with aluminum, arsenic, gold, mercury, selenium, tellurium, tin and titanium. Potassium and sodium burn in heated chlorine gas, and carbon steel ignites in it at 483 F. Chlorine reacts rapidly with most metals at elevated temperatures, and at lower temperatures if the metal is in powdered, sponge or wire form.

Manufacture. Chlorine is produced largely

by the electrolysis of salt. The two leading devices for electrolysis are diaphragm cells and mercury (amalgam) cells. Together, these account for about 95 per cent of world chlorine production, the balance being produced with sodium cells and nonelectrolytic processes (with the latter accounting for about 1 per cent of the total production). Raw material availability and energy economics influence the selection of the diaphragm and mercury cell methods. While the largest percentage of U. S. chlorine production is by the diaphragm method, the mercury cell method, which accounts for the major part of production in Europe, is gaining increasing acceptance in this country.[2,3]

Physiological Effects.[1] Chlorine gas is primarily a respiratory irritant. Concentrations above 3 to 5 parts per million (by volume) in air are readily detectable by the normal person. In higher concentrations the severely irritating effect of the gas makes it unlikely that any person will remain in a chlorine contaminated atmosphere unless he is unconscious or trapped.

Liquid chlorine may cause skin and eye burns upon contact.

Acute Toxicity; Systemic Effects. When a sufficient concentration of chlorine gas is present, it will irritate the mucous membranes, the respiratory system and the skin. Large amounts cause irritation of eyes, coughing and labored breathing. If the duration of exposure or the concentration of chlorine is excessive, it will result in general excitement of the person affected and will be accompanied by restlessness, throat irritation, sneezing and copious salivation. The symptoms of exposure to high concentrations include retching and vomiting followed by difficult breathing. In extreme cases, the difficulty of breathing may increase to the point where death can occur from suffocation. The physiological effects of various concentrations of chlorine gas are shown in Table 1. Chlorine produces no known cumulative effects.

Liquid chlorine produces no demonstrated systemic effects but, when exposed to normal atmospheric pressure and temperature, it vaporizes to gas which poses the hazards described above.

TABLE 1. PHYSIOLOGICAL RESPONSES TO VARIOUS CONCENTRATIONS OF CHLORINE GAS*

Effects	Parts of Chlorine Gas per Million Parts of Air, by Volume, (ppm)
Least amount required to produce slight symptoms, after several hours' exposure	1
Least detectable odor	3.5
Maximum amount that can be inhaled for 1 hr without serious disturbances	4
Noxiousness, impossible to breathe several minutes	5
Least amount required to cause irritation of throat	15.1
Least amount required to cause coughing	30.2
Amount dangerous in 30 min to 1 hr	40–60
Kills most animals in very short time	1000

* *Source:* U. S. Bureau of Mines Technical Paper 248, "Gas Masks for Gases Met in Fighting Fires" (1921).

Chronic Toxicity. Careful examination of workers exposed daily to detectable concentrations has reportedly shown no chronic sytemic effects. Local chronic effects due to chlorine have not been clinically demonstrated. Sensitization has not been a problem with chlorine.

Employee Protection and Training. Persons afflicted with asthma, bronchitis and other chronic lung conditions or irritations of the upper respiratory tract should not be employed in areas where chlorine is handled. All employees working with or around chlorine should be given physical examinations, including chest x-rays.

All employees handling or working around chlorine should be trained to handle it properly and safely with special emphasis placed on actions to be taken in case of emergencies such as leaks. Each employee should be trained in the proper use of the several types of respiratory equipment and be familiar with the conditions under which each type must be used. Each employee should also be trained in first aid procedures, particularly in administering artificial respiration. Quiz sessions on actions to be taken in emergencies, the proper use of respiratory equipment, and first aid measures should be held at regular intervals. (See the following separate section on "Chlorine Leaks" for actions to take in emergencies created by leaking chlorine. See also the concluding section on "Chlorine Protective Equipment and First Aid Measures.")

CHLORINE CONTAINERS

Chlorine is stored and shipped as a liquid in cylinders, single-unit tank cars, TMU (ton multi-unit) tank cars, TMU tanks on trucks, and tank barges. ICC regulations permit shipment of chlorine in cargo tank trucks complying with specification MC-330 but few if any are currently in service in the U. S. or Canada.

Filling Limits. The maximum filling densities authorized for chlorine are as follows (per cent water capacity by weight):

In cylinders—125 per cent (the maximum content permitted for cylinders purchased after Nov. 1, 1935, is 150 lb of gas).

In single-unit tank cars—125 per cent (with a maximum of 60,000 lb, except that specially insulated cars tested in full compliance with ICC specification 105A500-W may be loaded with between 107,800 and 110,000 lb of chlorine).

In TMU tanks—125 per cent.

SHIPPING METHODS: REGULATIONS

Under the appropriate regulations and tariffs, chlorine is authorized for shipment as follows:

By Rail: In cylinders (by freight, express or baggage), single-unit tank cars, and TMU tank cars.

By Highway: In cylinders and TMU tanks on trucks. (Shipment in cargo tank trucks meeting ICC specifications MC-330 is also authorized but is now little used, as noted above.)

By Water: In cylinders, single-unit tank cars, and TMU tanks aboard cargo vessels only. In bulk in steel tank barges and steel cargo barges. (In an emergency involving life or health, and on application made to the Commandant of the Coast Guard, limited shipments of chlorine may be made under conditions authorized by the Commandant aboard passenger vessels and ferry vessels.) In cylinders on barges of U. S. Coast Guard classes A, BA, BB, CA, and CB.

By Air: In cylinders aboard cargo aircraft only up to 150 lb (70 kg) maximum net weight per cylinder.

CHLORINE CYLINDERS

Chlorine cylinders are of the "foot ring" type or the "bumped bottom" type. Cylinders that meet the following ICC specifications are authorized for chlorine service: 3A480, 3AA480, 3BN480, and 3E1800. Cylinders of the ICC-3 and the ICC-25 types may be continued in chlorine service, but new construction is not authorized. Cylinders of the same types with higher service pressures are also authorized. (For cylinders purchased after Oct. 1, 1944, it is required that they contain no aperture other than the one provided in the neck of the cylinder for attachment of a valve equipped with an approved safety relief device.)

Valve Outlet and Inlet Connections. Standard connection, U. S. and Canada—No. 820. Alternate standard connections, U. S. and Canada—No. 660, No. 840.

Cylinder Requalification. All cylinders authorized for chlorine service must be requalified by hydrostatic retest every 5 years, except that no periodic retest is required for ICC-3E cylinders.

Storing Cylinders. In addition to the precautions required for the safe handling and storage of any compressed gas cylinders, the following practices must be observed for chlorine cylinders. Never store chlorine cylinders next to cylinders containing other compressed gases. Similarly, never store chlorine containers near turpentine, ether, anhydrous ammonia, finely divided metals, hydrocarbons such as oil, grease and gasoline, or any flammable materials. The storage area must be well ventilated, and storage below ground should be avoided.

SINGLE-UNIT TANK CARS AND TMU TANK CARS

Insulated single-unit tank cars complying with ICC specifications 105A300-W are authorized for bulk shipment of chlorine.

Chlorine is also authorized for shipment in TMU tank cars conforming to ICC specifications 106A500-X.

TMU TANKS ON TRUCKS

TMU tanks conforming to ICC specifications 106A500-X are also authorized for the shipment of chlorine on motor vehicles.

STORAGE FACILITIES

Users of chlorine in bulk quantities who receive shipment in single-unit tank cars often withdraw the chlorine direct to process, without transferring it to bulk storage facilities of their own. This practice avoids both the expense of storage facilities and such requirements of safe storage as having specialized, experienced personnel available at all times for possible emergencies.

Bulk chlorine users supplied by tank barges, however, usually unload into their own storage facilities. Very few chlorine barges are built with capacities of less than 600 tons, and capacities this large make it impractical to hold the barge while unloading direct to process. Piping and unloading systems for use with tank barges must be approved by the U. S. Coast Guard. Whether supplied by tank car or tank barge, storage facilities must be designed by experienced engineers and must conform to all applicable state and local regulations.

CHLORINE LEAKS[1]

Immediate steps should be taken to find and stop chlorine leaks as soon as there is any indication of the presence of chlorine in the air. Chlorine leaks never get better; they always get worse, unless promptly corrected. Authorized, trained personnel equipped with suitable gas masks should investigate whenever a chlorine leak occurs. All other persons must be kept away from the affected area until the cause of the leak has been found and remedied. If the leak is extensive, all persons in the path of the fumes must be warned. If outdoors, keep all persons upwind from the leak. Also, if possible, keep all persons in locations higher than the leak. At sites involving the handling of chlorine outdoors, it is advisable to have a wind sock or weathervane installed in a prominent location. Gaseous chlorine tends to lie close to the ground or floor because it is approximately $2\frac{1}{2}$ times as heavy as air.

Finding Leaks. To find a leak tie a cloth to the end of a stick, soak the cloth with ammonia water, and hold close to the suspected area. (Avoid contact of ammonia water with brass.) A white cloud of ammonium chloride will result if there is any chlorine leakage. A supply of strong ammonia water (commercial 26 Bé) always should be available (household ammonia is not strong enough). Containers, piping and equipment should be checked for leaks daily.

Emergency Assistance. If a chlorine leak cannot be handled promptly by the user's personnel, the nearest office or plant of the supplier should be called for assistance. If the supplier cannot be reached, the nearest chlorine-producing plant at which help is available should be called. Chlorine-producing plants operate around the clock and can be reached by telephone at any time. The telephone numbers of the supplier and of the nearest chlorine producer who is able to provide assistance in an emergency should be posted in suitable places so that they will be quickly available if needed; these should be checked periodically to be sure that the numbers are correct. When phoning for assistance the following should be given:

(1) Name of chlorine supplier.

(2) Your company name, address, telephone number, and the person or persons to contact for further information.

(3) Type (and serial number if possible) of container or other equipment which is leaking.

(4) Nature, location and extent of the leak.

(5) Corrective measures that are being applied.

In Case of Fire. In case of fire, chlorine containers should be removed from the fire zone immediately. Tank cars or barges should be disconnected and pulled out of the danger area. If no chlorine is escaping, water should be applied to cool any containers that cannot be moved. All unauthorized persons should be kept at a safe distance.

Do Not Use Water with Leaks. Never use water on a chlorine leak. Chlorine is only slightly soluble in water; also, the corrosive action of chlorine and water *always* will make a leak worse. In addition, the heat supplied by even the coldest water applied to a leaking container will cause liquid chlorine to evaporate faster. Never immerse or throw a leaking chlorine container into a body of water; the leak will be aggravated and the container may float when still partially full of liquid chlorine, allowing gas evolution at the surface.

Equipment and Piping Leaks. If a leak occurs in equipment in which chlorine is being used, the supply of chlorine should be shut off and the chlorine which is under pressure at the leak should be disposed of.

Valve Leaks. Leaks around valve stems usually can be stopped by tightening the packing nut or gland by turning clockwise. If this does not stop the leak, the container valve should be closed, and the chlorine which is under pressure in the outlet piping should be disposed of. If a container valve does not shut off tight, the outlet cap or plug should be applied. Ton containers have two valves; in case of a valve leak, the container should be rolled so the valves are in a vertical plane with the leaky valve on top.

Container Body Leaks. If confronted with container leaks other than at the valves, one or more of the following steps should be taken:

(1) If possible turn the container so that gas instead of liquid escapes. The quantity of chlorine that escapes from a gas leak is about 1/15 the amount that escapes from a liquid leak through the same size hole.

(2) Apply appropriate emergency kit device, if available.

(3) Call the chlorine supplier for emergency assistance.

(4) If practical, reduce pressure in the container by removing the chlorine as gas (not as liquid) to process or a disposal system.

(5) Move the container to an isolated spot where it will do the least harm.

Leaks in Transit. If a chlorine leak develops in transit through a populated area, it is generally advisable to keep the vehicle or tank car moving until open country is reached in order to minimize the hazards of the escaping gas. Appropriate emergency measures should then be taken as quickly as possible.

If a motor vehicle is wrecked, leaking chlorine containers should be positioned, if possible, so that only gas escapes, and, if necessary, moved to an isolated area before attempts are made to stop the leaks. If a tank car is wrecked and chlorine is leaking, the danger area should be evacuated and emergency clearing operations should not be started until safe working conditions have been restored.

Preparations for Handling Emergencies. List at least several physicians who could be summoned in the event of an emergency. Request all physicians listed to familiarize themselves with the treatment of persons exposed to chlorine. List the phone number of these physicians with the plant telephone operator and post similar lists in areas where chlorine is handled. Also, list and post the phone number of the nearest hospital, fire, and police departments. Arrange for a telephone extension in areas where chlorine is handled and keep these extensions open at night, on weekends and on holidays.

Alkali Absorption. At regular points in areas of chlorine storage and use, provisions should be made for emergency disposal of chlorine from leaking containers. Chlorine may be absorbed in solutions of caustic soda or soda ash, or in agitated hydrated lime slurries. Caustic soda is recommended as it absorbs chlorine most readily. The proportions of alkali and water recommended for this purpose, in the amounts needed to absorb indicated quantities of chlorine, are given in Table 2.[1] A suitable tank to hold the solution should be provided in a convenient location. Never immerse any chlorine container. Chlorine should be passed into the solution through an iron pipe or rubber hose properly weighted to hold it under the surface.

Emergency Kits. Most chlorine suppliers have emergency kits and skilled technicians to use them. These kits can be used to stop most leaks in a chlorine cylinder, TMU container, tank car or tank barge, and can usually be delivered to consumer plants within a few hours in an emergency. Some consumers find it advisable to purchase kits and to train employees in their use.

CHLORINE PROTECTIVE EQUIPMENT AND FIRST AID MEASURES

Protective Equipment. For maximum bodily protection against accidental contact with chlorine, suits, aprons and shoes of protective material and design should be provided for personnel handling chlorine. Rigid helmets should be supplied for locations which have possible overhead liquid chlorine leaks. Rubber gloves must be worn for work on piping connections. Respiratory masks must be used wherever there is danger of eye contact with liquid and gaseous chlorine, or danger of gas inhalation. Safety harnesses and attached life lines are required for persons entering chlorine tanks. All masks and equipment must be of types approved by the U. S. Bureau of Mines for chlorine service. All such equipment should be used and maintained in accord with the manufacturer's instructions. All respiratory equipment should be carefully maintained, inspected and cleaned after each use and at regularly scheduled intervals.

Oxygen administration equipment should be available and personnel trained in its use.

Shower baths and bubble-type fountains should be installed where contact of the skin, eyes, or clothing with liquid chlorine or chlorinated water is a possibility.

First-aid Measures. Immediately give first aid and summon a physician if a person is injured by exposure to chlorine. When calling the physician state the location of the affected person and the nature of the exposure.

Inhalation.[1] Anyone overcome by or seriously exposed to chlorine gas should be moved at once to an uncontaminated area. If breathing has not ceased, the person should be placed on his back, with head and back elevated. He should be kept quiet and warm, with blankets if necessary.

Artificial Respiration. If breathing has apparently ceased, artificial respiration should be started immediately. The mouth-to-mouth method is preferred; the Nielson arm lift-back pressure method and the Schaefer prone-

TABLE 2. RECOMMENDED ALKALINE SOLUTIONS FOR ABSORBING CHLORINE

Chlorine Container Capacity (lb net)	Casutic Soda (100%)		Soda Ash		Hydrated Lime *	
	Alkali (lb)	Water (gal)	Alkali (lb)	Water (gal)	Alkali (lb)	Water (gal)
100	125	40	300	100	125	125
150	188	60	450	150	188	188
2000	2500	800	6000	2000	2500	2500

* Hydrated lime solution must be continuously and vigorously agitated while chlorine is to be absorbed.

pressure method may also be used—but, with the latter, do not exceed 18 cycles per minute.

Oxygen Administration. If oxygen inhalation apparatus is available, oxygen should be administered, preferably by a physician or a person authorized for such duty. The instructions which come with the equipment must be followed carefully.

Stimulants. Stimulants rarely will be necessary where adequate oxygenation is maintained and any such drugs for shock treatment should be given only by the attending physician.

Milk may be given in mild cases as a relief from throat irritation. Never give anything by mouth to an unconscious patient.

Contact with Skin or Mucous Membranes.[1] If the patient has inhaled chlorine in addition to having gotten it on his skin or mucous membranes, first aid for inhalation should be given first.

If liquid chlorine or chlorinated water has contaminated skin or clothing, the emergency shower should be used immediately. Contaminated clothing should be removed under the shower and the chlorine should be washed off with very large quantities of water. Skin areas should be washed with large quantities of soap and water. Never attempt to neutralize the chlorine with chemicals. No salves or ointments should be applied for 24 hours.

Contact With Eyes.[1] If even minute quantities of liquid chlorine enter the eyes, or if the eyes have been exposed to strong concentrations of chlorine gas, they should be flushed immediately with copious quantities of running water for at least 15 minutes. Never attempt to neutralize with chemicals. The eyelids should be held apart during this period to insure contact of water with all accessible tissues of the eyes and lids. Call a physician, preferably an eye specialist, at once. If a physician is not immediately available, the eye irrigations should be continued for a second period of 15 minutes. After the first period of irrigation is complete, it is permissible as a first aid measure to instill into the eye two or three drops of 0.5 per cent solution of Pontocaine (tetracaine hydrochloride) or other equally effective aqueous topical anesthetic. No oils or oily ointment should be instilled unless ordered by the physician.

REFERENCES

1. "Chlorine Manual," (3rd Ed., 1959), The Chlorine Institute, 342 Madison Ave., New York, N. Y., 10017.
2. Macmullan, Robert B. "Diaphragm vs. Amalgam Cells for Chlorine Caustic Production," Chem. Industries, July 1947.
3. Hardie, D. W. F., "Electrolytic Manufacture of Chemicals from Salt," Oxford Univ. Press, 1959.

Sources for Physical Constants

The "Chlorine Manual," above, is the source of the physical constants except as follows:

Boiling point, melting point; latent heat of fusion—Giaque, W. F., and Powell, T. M., "Chlorine; The Heat Capacity, Vapor Pressure, Heats of Fusion and Vaporization and Entropy," *J. Am. Chem. Soc.*, **61**, 1970 (1939).

Critical temperature, critical pressure—Pellaton, M., "Constantes Physiques du Chlor," *J. Chem. Phys.*, **13**, 426 (1915).

Vapor pressure, density of saturated liquid, latent heat of vaporization, liquid/gas ratio, specific gravity of gas, specific gravity of liquid—Kapoor, R. M., and Martin, J. J., *Thermodynamic Properties of Chlorine*, Engineering Research Institute, Univ. of Michigan (1957).

Specific heat, liquid—Hodgman, C. D., *Handbook of Chemistry and Physics*, 30th Ed., Cleveland, Ohio, Chemical Rubber Publishing Co., 1946.

Specific heat, gas—Lange, N. A., *Handbook of Chemistry*, Sandusky, Ohio, Handbook Publishers, Inc., 1949.

Thermal conductivity, gas—Eucken, A., *Physik. Z.*, **14**, 324–32 (1913).

Viscosity, liquid, gas—Steacie, E. W. R., and Johnson, F. M. G., *J. Am. Chem. Soc.*, **47**, 754–62 (1925).

Cyclopropane

C_3H_6 [or $(CH_2)_3$]
Synonym: Trimethylene
ICC Classification: Flammable compressed gas; red label

PHYSICAL CONSTANTS

International symbol	Cyclopropane
Molecular weight	42.08
Specific gravity of gas at 70 F and 1 atm (air = 1)	1.48
Specific gravity of liquid at −110 F	0.720
Vapor pressure at 68 F, psig	67.7
Boiling point at 1 atm	−27.15 F
Melting point at 1 atm	−197.7 F
Critical temperature	256 F
Critical pressure, psia	797
Latent heat of vaporization at boiling point, Btu/lb	205
Flammable limits, by volume	
in air	2.40–10.3%
in oxygen	2.48–60.0%
Autoignition temperature	
in air	928 F
in oxygen	849 F

Properties. Cyclopropane is a colorless, flammable gas with a sweet, distinctive odor resembling that of petroleum benzin or solvent naphtha. It is shipped and used as a liquefied compressed gas in cylinders, with a vapor pressure of about 68 psig at 68 F. Cyclopropane is an anesthetic and is used as a medical gas. Chemically, it reacts with hydrogen iodide or bromide and with bromine to give the corresponding halopropanes or dihalopropanes, and with chlorine to form chlorocyclopropanes.

Materials of Construction. A noncorrosive gas, cyclopropane may be contained by any commercially available metals.

Manufacture. One major method used to produce cyclopropane commercially is the reduction of a water solution of trimethylene-chlorobromide in the presence of zinc at 200 F. It is also made by the progressive thermal chlorination of propane.

Commercial Uses. Cyclopropane is used chiefly as an inhalant anesthetic in medicine. It is also employed for organic synthesis in the chemical industry.

Physiological Effects. A general anesthetic, cyclopropane can produce all levels of inhalation anesthesia and maximum muscle relaxation in patients of all ages and body types. It is not altered or combined in the body; the major part is exhaled within 10 minutes, while full desaturation takes several hours. It tends to irritate the circulatory system; laryngeal spasm, emergence delirium and nausea after anesthesia with it are common. Concentrations (by volume) of 6 to 8 per cent result in unconsciousness; of 7 to 14 per cent, in moderate anesthesia; and of 14 to 23 per cent, in deep anesthesia. Concentrations ranging from 23 to 40 per cent are lethal through respiratory failure.

The principal hazard met in handling cyclopropane stems from its high flammability, and all the precautions necessary for the safe handling of any flammable gas must be observed in its use.

CYCLOPROPANE CONTAINERS

Cyclopropane is shipped in cylinders as a liquefied compressed gas at gage pressures which range from about 53 at 50 F to 199 at 130 F.

Filling Limit. The maximum filling density authorized by the ICC for cyclopropane cylinders is 55 per cent (per cent of water capacity by weight).

SHIPPING METHODS; REGULATIONS

Under the appropriate regulations and tariffs, cyclopropane is authorized for shipment as follows:

By Rail: In cylinders (freight, express or baggage).

By Highway: In cylinders on trucks.

By Water: In cylinders on cargo or passenger vessels, and on ferry or railroad ferry vessels (either passenger or vehicle). In cylinders on barges of U. S. Coast Guard classes A and C only.

By Air: Aboard cargo aircraft only in appropriate cylinders up to 300 lb (140 kg) maximum net weight per cylinder.

CYCLOPROPANE CYLINDERS

Cylinders that meet the following ICC specifications are authorized for cyclopropane service: 3A225, 3A480X, 3AA225, 3B225, 4A225, 4AA480, 4B225, 4BA225, 4B240ET, and 3E1800 (cylinders manufactured under the now obsolete specifications ICC-3 and ICC-7-300 may be continued in service but new construction is not authorized).

Valve Outlet and Inlet Connections. Standard connections—U. S. and Canada—No. 510, No. 920.

Cylinder Requalification. Requalification by hydrostatic retest is required every 5 years under present regulations for all types of cylinders authorized for cyclopropane, except ICC-3E and ICC-7 for which no periodic retests are required.

However, the following types of cylinders can be requalified by visual inspection every 5 years under present ICC regulations if cyclopropane is shipped in them as a liquefied hydrocarbon gas and if the cylinders are used exclusively for liquefied hydrocarbon gas which is commercially free from corroding components: 3A480, 3A480X, 3B225, 4B225, and 4BA225. Cylinders of the same types but with higher service pressures that are used as specified can also be requalified by visual inspection.

Medical Gas Cylinders. For use as an anesthetic, cyclopropane is supplied in medical gas cylinders of standard styles. See Section B of Chapter 1 in Part III, on gases used medicinally, for data on cyclopropane in these standard cylinder styles and an explanation of their safe handling.

Dichlorodifluoromethane

See **Fluorocarbons**

Dichlorodifluoromethane-Difluoroethane Mixture

See **Fluorocarbons**

Dimethylamine

See **Methylamines**

Dimethyl Ether

CH_3OCH_3 [or $(CH_3)_2O$]
Synonyms: Methyl ether, methyl oxide, wood ether
ICC Classification: Flammable compressed gas; red label

PHYSICAL CONSTANTS

International symbol	—
Molecular weight	46.07
Vapor pressure at 70 F, psig	60
Specific gravity of gas (air = 1)	1.59
Specific gravity of liquid, at 68 F	0.661
Density, liquid, lb/cu ft	
at 70 F	41.2
at 105 F	39.2
at 115 F	38.6
at 130 F	37.7
Boiling point at 1 atm	−12.8 F
Freezing point at 1 atm	−223 F
Critical temperature	264 F
Critical pressure, psia	772
Specific heat, at −18 F, Btu/(lb)(°F)	0.96
Latent heat of vaporization, at −12.6 F, Btu/lb	201
Flammable limits in air, by volume	3.4–26.7%
Flash point (closed cup)	−42 F
Autoignition temperature	662 F
Solubility in water at 75 F and 5 atm, by weight	35%

Properties. Dimethyl ether is a colorless, flammable gas easily compressed to a colorless liquid. It has a faint sweetish odor and leads to anesthesia when inhaled in fairly large concentrations. It readily forms complexes with inorganic compounds and acts as a methylating agent.

Materials of Construction. Any commercially available metals may be used with dimethyl ether, as is it noncorrosive.

Manufacture. Dimethyl ether is produced by the dehydration of methanol, either with sulfuric acid or over alumina at high temperatures and pressures.

Commercial Uses. Dimethyl ether is used as a propellant in aerosol sprays and as a refrigerant (mixed with a fluorocarbon to reduce flammability). It is also used as a methylating agent in the dye industry, and as a chemical reaction medium, a solvent, and a catalyst and stabilizer in polymerization.

Physiological Effects. Studies of dimethyl ether have found that inhalation of a 7.5 per cent concentration for 12 minutes resulted in a feeling of intoxication and some lack of attention; of a 10 per cent concentration for 64 minutes, in nauseous sickness; and of a 20 per cent concentration for 17 minutes, in unconsciousness. Early experiments investigating dimethyl ether as an anesthetic with animals and humans found no permanent residual effects.[1]

Prolonged contact of liquid dimethyl ether with the skin causes freezing or frostbite of the skin.

Precautions necessary for the safe handling of any flammable gas must be observed with dimethyl ether.

DIMETHYL ETHER CONTAINERS

Dimethyl ether is authorized for shipment by the ICC in cylinders, in single-unit tank cars, in TMU (ton multi-unit) tank cars, and in TMU tanks on trucks.

Filling Limit. The maximum filling density authorized for dimethyl ether in cylinders is the maximum cylinder service pressure at 70 F. The maximum filling densities authorized for dimethyl ether in other containers are (per cent water capacity by weight): for single-unit tank cars, 62 per cent for TMU tanks, 59 per cent.

SHIPPING METHODS; REGULATIONS

Under the appropriate regulations and tariffs, dimethyl ether is authorized for shipment as follows:

By Rail: In cylinders (freight or express), and in single-unit tank cars and TMU tank cars.

By Highway: In cylinders on trucks, and in TMU tanks on trucks.

By Water: In cylinders via cargo vessels only, and in authorized tank cars via trainships only. In cylinders on barges of U. S. Coast Guard classes A and C only.

By Air: Aboard cargo aircraft only in appropriate cylinders up to 300 lb (140 kg) maximum net weight per cylinder.

DIMETHYL ETHER CLYINDERS

Dimethyl ether is authorized for shipment of cylinders of any type currently approved by the ICC for liquefied compressed gases (these are cylinders that meet ICC specifications 3A, 3AA, 3B, 3BN, 3D, 3E, 4, 4A, 4B, 4BA, 4B-ET, 9, 40, and 41; cylinders meeting ICC specifications 3, 25, 26 and 38 may be continued in dimethyl ether service, but new construction is not authorized).

Valve Outlet and Inlet Connections. Standard connection, U. S. and Canada—No. 510.

Cylinder Requalification. All cylinders authorized for dimethyl ether service must be requalified by hydrostatic test every 5 years with the exceptions of: type 4, for which the retest period is 10 years; type 3E, for which periodic retest is not required; and types 40 and 41, which are small inside containers that it is illegal to refill.

SINGLE-UNIT TANK CARS AND TMU TANK CARS

Dimethyl ether is authorized for shipment in single-unit tank cars meeting ICC specifications 105A300W (provided that they have properly fitted loading and unloading valves). Rail shipment is also authorized for TMU tank cars of ICC specifications 106A500X and 110A500W.

TMU TANKS ON TRUCKS

Shipment of TMU tanks of ICC specifications 106A500X and 110A500W on trucks is also authorized.

STORAGE AND HANDLING EQUIPMENT

Storage and handling equipment for dimethyl ether must be designed to keep it from contact with the air, as it may form peroxides when exposed to atmospheric oxygen. Unloading and storage systems must be purged of all air before dimethyl ether is introduced into them. Compressed nitrogen is among substances recommended for such purging; carbon dioxide must not be used because it is highly soluble in dimethyl ether. Indoor storage areas must be located only in fire-resistive buildings and fitted with sprinkler systems to keep the storage container cool, should fire occur, because dimethyl ether exerts extreme pressure when heated. Ventilation must be provided for the floor level, to which dimethyl ether vapors sink. Should dimethyl ether ignite, recommended extinguishing agents for firefighting equipment include dry chemical, carbon dioxide and carbon tetrachloride.

REFERENCE

1. Brown, W. E., *J. Pharmacol. Exp. Therapeutics*, **23**, 485–96 (1924); Davidson, B. M., **26**, 43–8 (1925).

Ethane

C_2H_6 (or CH_3CH_3)
Synonyms: Bimethyl, dimethyl, ethyl hydride, methyl-methane
ICC Classification: Flammable compressed gas; red label

PHYSICAL CONSTANTS
(for Research Grade product)

International symbol	C_2H_6
Molecular weight	30.068
Vapor pressure at 70 F, psig	544
Pressure in typical full cylinder at 130 F, psig (approx.)	2474.
Boiling point at 1 atm	−127.53 F
Freezing point at saturation pressure (triple point)	−297.89 F
Critical temperature	90.32 F
Critical pressure, psia	709.08
Critical volume, cu ft/lb	0.0788
Specific volume of gas at 60 F and 1 atm, cu ft/lb of the real gas	12.5151
Specific gravity of gas at 60 F and 1 atm (air = 1)	1.0469
Density of liquid at saturation pressure (apparent value from weight in air of the air-saturated liquid), lb/cu ft	
at 60 F	23.52
at 70 F	22.40
Specific gravity of liquid, 60 F/60 F, at saturation pressure (absolute value from weights in vacuum for the air-saturated liquid)	0.3771
Specific heat of ideal gas at 60 F, C_p, Btu/(lb)(°F)	0.4097
Specific heat of ideal gas at 60 F, C_v, Btu/(lb)(°F)	0.3436
Ratio of specific heats, C_p/C_v, ideal gas at 60 F	1.192
Specific heat of liquid at 1 atm	0.9256
Latent heat of vaporization at boiling point and 1 atm, Btu/lb	210.41
Latent heat of fusion, Btu/lb	0.09014
Gross heat of combustion	
Gas at 60 F and 1 atm, Btu/cu ft of the real gas	1783.7
Liquid at 77 F and saturation pressure, Btu/lb	22169.
Net heat of combustion	
Gas at 60 F and 1 atm, Btu/cu ft of the real gas	1631.5
Liquid at 77 F and saturation pressure, Btu/lb	20281.
Air required for combustion, cu ft of air per cu ft of the real gas, at 60 F and 1 atm	16.845
Air required for combustion, lb of air per lb	16.090
Flammable limits in air, by volume	2.9–13%
Flash point	−211 F

Properties. Ethane is a colorless, odorless, flammable gas that is relatively inactive chemically and is considered nontoxic. Slightly heavier than air, it is shipped as a liquefied compressed gas under its vapor pressure of 545 psig at 70 F.

Materials of Construction. Ethane is noncorrosive and may be contained in installations constructed of any common metals to withstand the pressures involved.

Manufacture. Ethane is produced commercially from the cracking of light petroleum fractions, and also by fractionation from natural gas.

Commercial Uses. Major uses of ethane include its employment as a fuel, in organic synthesis (it can be chlorinated to give ethyl chloride, for example, and can yield ethylene with a greater heat input than is required for obtaining ethylene by propane cracking), and as a refrigerant.

Physiological Effects. Inhalation of ethane in concentrations in air up to 5 per cent produces no definite symptoms, but inhalation of higher concentrations has an anesthetic effect. It can act as a simple asphyxiant by displacing the oxygen in the air. Contact between liquid ethane and the skin can cause freezing of the tissues, and should be avoided.

All the precautions required for the safe handling of any flammable compressed gas must be observed with ethane.

ETHANE CONTAINERS

Ethane is authorized by the ICC for shipment in cylinders.

Filling Limits. The maximum filling densities authorized for ethane in cylinders are: 35.8 per cent (per cent water capacity by weight), or 36.8 per cent in cylinders meeting ICC specifications 3A2000 or 3AA2000.

SHIPPING METHODS; REGULATIONS

Under the appropriate regulations and tariffs, ethane is authorized for shipment as follows:

By Rail: In cylinders (via freight or express).

By Highway: In cylinders on trucks.

By Water: In cylinders on cargo vessels only. In cylinders on barges of U. S. Coast Guard classes A, CA and CB only.

By Air: In cylinders aboard cargo aircraft only up to 300 lb (140 kg) maximum net weight per cylinder.

ETHANE CYLINDERS

Cylinders that comply with the following ICC specifications are authorized for ethane service: 3A1800, 3AA1800 and 3E1800. (ICC-3 cylinders may also be continued in service, but new construction is not authorized.)

Valve Outlet and Inlet Connections. Standard connection, U. S. and Canada—No. 350.

Cylinder Requalification. All types of cylinders authorized for ethane service must be requalified by periodic hydrostatic retest every 5 years under present regulations, except that no periodic retest is required for 3E cylinders.

Ethylene

$CH_2{:}CH_2$ (or $H_2C{:}CH_2$)
Synonym: Ethene (also, olefiant gas, bicarbuttetted hydrogen, elayl, or etherin)
ICC Classification: Flammable compressed gas; red label

PHYSICAL CONSTANTS

International symbol	C_2H_4
Molecular weight	28.05
Density at 32 F and 1 atm, lb/cu ft	0.0787
Specific gravity of gas at 32 F (air = 1)	0.978
Density, liquid, at boiling point, lb/cu ft	35.42
Boiling point at 1 atm	−154.8 F
Melting point at 1 atm	−272.9 F
Critical temperature	49 F
Critical pressure, psia	745
Latent heat of vaporization at boiling point, Btu/lb	208
Latent heat of fusion at melting point, Btu/lb	51.2
Gross heat of combustion, Btu/lb	21,625
Flammable limits in air, by volume	
lower	3–3.5%
upper	16–29%
Flammable limits in oxygen, by volume	
lower	2.9%
upper	79.9%
Autoignition temperature	
in air	914 F
in oxygen	905 F
Solubility in water at 32 F, by volume	26%

Properties. Ethylene is a colorless, highly flammable gas with a faint odor that is sweet and musty. It is nontoxic, being used as an anesthetic, and is hazardous only as a flammable substance or as a simple asphyxiant. Chemically, it reacts chiefly by addition to give saturated paraffins, or derivatives of paraffin hydrocarbons, and is widely used as a raw material in the synthetic organic chemical industry. It is slightly lighter than air, and is shipped as a gas at about 1250 psig at 70 F. Below 50 F at such charging pressure, it is a liquefied gas in the cylinder.

Materials of Construction. Any common commercially available metals may be used with ethylene because it is noncorrosive.

Manufacture. The most commonly used of a number of methods for producing ethylene commercially is high-temperature coil cracking of propane or of ethane and propane. Recovery of ethylene from the cracked gases is often accomplished by low-temperature, high-pressure straight fractionation. Another customary manufacturing method is by catalytic decomposition of ethyl alcohol.

Commercial Uses. Roughly half of the ethylene produced in the U. S. has been used to make ethyl alcohol, with another substantial portion going into the production of ethylene glycol. Other chemical raw materials made with ethylene include ethyl chloride, dichloroethane and vinyl chloride, ethyl ether, methyl acrylate and styrene.

Ethylene is also employed as an anesthetic (as noted), a refrigerant, and a fuel for metal cutting and welding, and also to accelerate plant growth and fruit ripening.

Physiological Effects. As an anesthetic drug, ethylene is a nontoxic gas found pleasant and nonirritating by patients. Prolonged inhalation of substantial concentrations results in unconsciousness; light and moderate anesthesia is attained, and deep anesthesia seldom occurs. Inhalation is fatal only if the gas acts as a simple asphyxiant, depriving the lungs of necessary oxygen.

No deleterious action by ethylene on circulatory, respiratory or other systems or organs has been observed, and the gas is not altered or combined in the body with any tissue. Exhalation eliminates the major portion of ethylene within minutes, although complete desaturation from body fat takes several hours. Minute traces can be detected in the blood a number of hours after anesthesia has ended.

Ethylene poses hazards to personnel through its high flammability, and the precautions necessary for the safe handling of any flammable gas must be observed in its use.

ETHYLENE CONTAINERS

Ethylene is authorized for shipment in cylinders under ICC regulations. It is also shipped in bulk quantities under special permits of the ICC.

Filling Limits. The maximum filling limits prescribed for ethylene in cylinders are as follows (per cent water capacity by weight): for cylinders of 1800 psig maximum service pressure, 31 per cent; for cylinders of 2000 psig maximum service pressure, 32.5 per cent; and for cylinders of 2400 psig maximum service pressure, 35.5 per cent. It is also shipped by highway in tube trailers. Under special permit of the ICC, ethylene is also shipped liquefied at low temperatures in insulated cargo tanks on trucks and truck trailers, and in insulated, low-temperature tank cars of the ICC-113A type. Bulk shipment of ethylene is also made over relatively short distances by pipeline.

SHIPPING METHODS; REGULATIONS

Under the appropriate regulations and tariffs, ethylene is authorized for shipment as follows:

By Rail: In cylinders (freight, express, or baggage), and by special ICC permit in 113A tank cars in liquid, low-temperature form.

By Highway: In cylinders on trucks, and in tube trailers; by special ICC permit, in insulated truck cargo tanks liquefied at low temperatures.

By Water: In cylinders via only cargo vessels. In cylinders on barges of U. S. Coast Guard classes A and C only.

By Air: Aboard cargo aircraft only in appropriate cylinders up to 300 lb (140 kg) maximum net weight per cylinder.

ETHYLENE CYLINDERS

Cylinders that meet the following ICC specifications are authorized for ethylene service: 3A1800, 3AA1800, 3E1800, 3A2000, 3AA2000, 3A2400, and 3AA2400 (cylinders manufactured under the now obsolete specification ICC-3 may be continued in service but new construction is not authorized).

Valve Outlet and Inlet Connections. Standard connection, U. S. and Canada—No. 350, No. 900.

Cylinder Requalification. Cylinders of types 3A and 3AA (as well as 3) used in ethylene service must be requalified by hydrostatic retest every 5 years under present regulations. For cylinders of type 3E, no periodic retest is required.

Medical Gas Cylinders. For use as an anesthetic, ethylene is shipped and stored in special medical gas cylinders. See Section B of Chapter 1 in Part III for descriptions of these cylinders and an explanation of their safe handling.

Fluorine

F_2 ICC Classification: Flammable compressed gas; red label

PHYSICAL CONSTANTS

International symbol	F_2
Molecular weight	38.0
Density of gas at 1 atm, lb/cu ft	
at 32 F	0.106
at 70 F	0.098
Specific volume of gas, at 70 F and 1 atm, cu ft/lb	10.2
Density of vapor, lb/cu ft	
at −306.6 F (liquid fluorine boiling point)	0.362
at −320.4 F (liquid nitrogen boiling point)	0.142
Vapor pressure, at −320.4 F, psia	5.6
Density of liquid, lb/cu ft	
at −306.6 F	94.1
at −320.4 F	97.9
Boiling point at 1 atm	−306.6 F
Melting point at 1 atm	−363.2 F
Triple point	−363.2 F at 0.0324 psia
Critical temperature	−200.4 F
Critical pressure, psia	808.5
Latent heat of vaporization at boiling point, Btu/lb	74.08
Latent heat of fusion (at freezing point), Btu/lb	5.779
Molar heat capacity, C_p, at 32 F, Btu/(lb mole)(R)	7.5183
Ratio of molar heat capacities, C_p/C_v	1.3583
Thermal conductivity of gas at 32 F and 1 atm, Btu/(hr)(sq ft)(°F/in.)	0.172
Viscosity, lb mass/(ft)(hr)	
of vapor at 32 F and 1 atm	0.0527
of liquid at −306.3 F	0.621
of liquid at −340.5 F	1.002
Surface tension, lb force/ft	
of liquid at −315.7 F	0.0010
of liquid at −340.5 F	0.0012
Weight per gallon of liquid, lb	
at −306.6 F	12.6
at −320.4 F	13.1

Properties. Fluorine is a highly toxic, pale yellow gas about 1.7 times as heavy as air at atmospheric temperature and pressure. When cooled below its low boiling point (−306.6 F), it is a liquid about 1.1 times as heavy as water.

Fluorine is the most powerful oxidizing agent known, reacting with practically all organic and inorganic substances. Exceptions are some of the inert gases, metal fluorides and a few completely fluorinated organic compounds in pure form. However, the latter may also react with fluorine if they are contaminated with a combustible material.

Heats of reaction with fluorine are always

high and most reactions take place with ignition.

Fluorine reacts slowly with many metals at room temperatures, however, and the reaction often results in formation of a metal fluoride film on the metal's surface; in the case of some metals, this film retards further action.

Fluorine readily displaces the other halogens from their compounds, but such reactions are not always feasible for preparing fluorides. It reacts with water to form a mixture containing principally oxygen and hydrogen fluoride plus small amounts of ozone, hydrogen peroxide and oxygen fluoride.

Materials of Construction. Nickel, iron, aluminum, magnesium, copper and certain of their alloys are quite satisfactory for handling fluorine at room temperature, for these are among the metals with which formation of a surface fluoride film retards further action. Listed on page 82 are various materials that have been used with satisfactory results in gaseous fluorine service at normal temperatures and liquid service at low temperatures.

Nickel and "Monel" are generally considered to be by far the best materials for fluorine service at high temperatures, but selection of suitable materials for service at elevated temperatures and pressures must be based on the conditions of the specific application.

Manufacture. Fluorine is produced commercially by electrolytic decomposition of an anhydrous hydrofluoric acid, potassium bifluoride (KF.HF) solution. The melt formed (approximately KF.2HF) is a solid at normal ambient temperatures and liquefies at approximately 160 F. The commercial electrolytic cell uses carbon anodes and a metal cathode with a method for separate collection of the gas released at each electrode. (These cells are heated when not operating or operating at low current rates to prevent their freezing, and are cooled when operated at normal rates to prevent overheating.) When voltage is applied to the cell, current passing through the melt decomposes the hydrofluoric acid with the release of fluorine at the anode and hydrogen at the cathode. Continuous addition of hydrofluoric acid replaces the acid decomposed.

The fluorine thus produced is purified and then either used directly in a production process, compressed for cylinder filling, or condensed to a liquid for charging into refrigerated containers.

Commercial Uses. Fluorine is used in producing uranium hexafluoride, sulfur hexafluorides, the halogen fluorides and other fluorine compounds that require the high reactivity of elemental fluorine for their preparation.

Some authorities see a large potential for liquid fluorine as an oxidizer for the fuel in rocket engine propellant combinations, on viewing its hypergolicity, high liquid density and high specific impulse as properties of particular value for this application.

Physiological Effects. Fluorine gas is a powerful caustic irritant and is highly toxic.[1] In one series of animal experiments, inhalation of acute exposures of 10,000 ppm for 5 minutes, 1000 ppm for 30 minutes, and 500 ppm for 1 hour produced 100 per cent mortality in rats, mice, guinea pigs and rabbits. Inhalation of 100 ppm for 7 hours produced wide variation in species mortality, ranging from 0 per cent in guinea pigs to 96 per cent in mice. Daily subacute exposures to a concentration of 2 ppm for periods of time varying from totals of 30 to 176 hours resulted in high mortality, rabbits appearing to be the most susceptible species and guinea pigs the least. Pulmonary irritation varying from severe at 16 ppm in some species to mild at 2 ppm represented the major pathological change, while similar subacute exposure at 0.5 ppm resulted in no significant pathology but some retention of fluorine in osseous tissues.

Contact between the skin and high concentrations of fluorine gas under pressure will produce burns comparable to thermal burns; contact with lower concentrations results in a chemical type of burn resembling that caused by hydrofluoric acid. For recommended first aid and medical treatment in the event of

Materials Giving Satisfactory Results in Fluorine Service*

Type of Equipment	Gaseous Service, Normal Temp.	Liquid Service, Low Temp.
Storage tanks	Stainless steel 304L Aluminum 6061 Mild steel (low pressure)	"Monel" Stainless steel 304L Aluminum 6061
Lines and fittings	Nickel "Monel" Copper Brass Stainless steel 304L Aluminum 2017, 2024, 5052, 6061 Mild steel (low pressure)	"Monel" Stainless steel 304L Copper Aluminum 2017, 2024, 2050
Valve bodies	Stainless steel 304 Bronze Brass	"Monel" Stainless steel 304 Bronze
Valve seats	Copper Aluminum 1100 Stainless steel 303 Brass "Monel"	Copper Aluminum 1100 "Monel"
Valve plugs	Stainless steel 304 "Monel"	Stainless steel 304 "Monel"
Valve packing	Tetrafluoroethylene polymer	Tetrafluoroethylene polymer
Valve bellows	Stainless steel 300 series "Monel" Bronze	Stainless steel 300 series "Monel" Bronze
Gaskets	Aluminum 1100 Lead Copper Tin Tetrafluoroethylene polymer Red rubber (5 psig) Neoprene (5 psig)	Aluminum 1100 Copper

* It is not necessarily implied that other materials not listed would not give adequate service.

hydrofluoric acid exposure, see Safety Data Sheet SD-25 of the Manufacturing Chemists' Association.[2]

It is unlikely that persons not injured or trapped would continue to inhale highly toxic concentrations of fluorine because of its strong odor and its irritation of eyes, nose and mucous membranes. Should exposure by inhalation occur, immediately remove the patient to fresh air and call a physician. Administer oxygen as necessary to help prevent pulmonary irritation.

Liquid fluorine will severely burn the skin and eyes.

Before introducing any application of fluorine, users should fully work out all details of first aid and treatment with the medical personnel who would be called to administer aid in case of accident.

Only trained and competent personnel should be permitted to handle fluorine. It is recommended that they work in pairs and within sight and sound of each other, but not in the same working area. Supervisory personnel should make frequent checks of the operation.

Precautions in Handling Fluorine. All precautions necessary for the safe handling of any flammable gas must be observed with fluorine, in addition to the precautions outlined below. Fluorine fires accidentally breaking out may be most simply extinguished by cutting off the fluorine supply at a primary point, then employing conventional fire-fighting methods. The dry types of extinguishers are recommended.

Essential additional precautions in the handling of fluorine cylinders are outlined in the subsequent section on cylinders.

Personal Protective Equipment. Clean neoprene gloves must be worn when handling equipment which contains or has recently contained fluorine. This precaution affords not only limited protection against fluorine contact but protection against contact with possible films of hydrofluoric acid that are formed by escaping fluorine and air moisture and that collect on valve handles and other surfaces. Neoprene coats and boots afford over-all body protection for short intervals of contact with low-pressure fluorine. All protective clothing must be designed and used so that it can be shed easily and quickly. Safety glasses must be worn at all times.

Face shields made preferably of transparent, highly fluorinated polymers like "Aclar" (registered trade mark) should be worn whenever operators must approach equipment containing fluorine under pressure. Face shields made of any conventional materials afford limited, though valuable, protection against air-diluted blasts of fluorine.

Leak Detection. All areas containing fluorine under pressure should be inspected for leaks at suitable intervals, and any leaks discovered should be repaired at once after fluorine has been removed from the system. Ammonia vapor expelled from a squeeze bottle of ammonium hydroxide at suspected points of leakage may be used to detect leaks. Filter paper moistened with potassium iodide provides a very sensitive means for detecting fluorine (effective down to about 25 ppm); in using it, hold the paper with metal tongs or forceps about 18 to 24 in. long.

Equipment Preparation and Decontamination. Equipment to be used for fluorine service should first be thoroughly cleaned, degreased and dried, then treated with increasing concentrations of fluorine gas so that any impurities will be burned out without the simultaneous ignition of the equipment. The passive metal fluoride film thus formed will inhibit further corrosion by fluorine.

Before opening or refilling equipment that has contained fluorine, thoroughly purge it with a dry inert gas (such as nitrogen) and evacuate it if possible. Minor quantities of fluorine to be vented and purged can be converted to harmless carbon fluoride gases by passage through a lump-charcoal-packed column. Large quantities to be purged require a purge system with a fluorine-hydrocarbon-air burner, scrubber and stack to prevent any undue exit hazards. Should a purged fluorine system require evacuation, a soda-lime tower followed by a drier should be included in the vacuum system to pick up trace amounts of fluorine in order to protect the vacuum pump.

Liquid Fluorine Spills. In the event of a large spillage of liquid fluorine, the contaminated area can be neutralized with sodium carbonate. The dry powder can be sprayed on the spill area from a fluidized system similar in principle to that of dry-chemical fire extinguishers. If major spillages occur in areas where the formation of hydrofluoric acid liquid and vapor pose no undue danger, water in the form of a fine mist or fog is recommended. The major portion of the fluorine will be converted to hot, light, gaseous products which rise vertically and diffuse quickly into the atmosphere.

FLUORINE CONTAINERS

Fluorine is authorized for shipment in cylinders as a compressed gas under ICC regulations, and as a liquefied, low-temperature gas in liquid-nitrogen refrigerated tanks mounted on truck trailers by special permit of the ICC.

Filling Limits. The maximum filling density authorized for fluorine in cylinders is 400 psig at 70 F, and cylinders must not contain over 6 lb of fluorine gas.

SHIPPING METHODS; REGULATIONS

Under the appropriate regulations and tariffs, fluorine is authorized for shipment as follows:

By Rail: In cylinders (via freight, and express to a maximum quantity of 6 lb in one outside container).

By Highway: In cylinders on trucks, and, under special ICC permit, in trailer-mounted tank transports.

By Water: In cylinders on cargo vessels only. On barges of U. S. Coast Guard classes A, CA, and CB only.

By Air: Not acceptable for shipment.

FLUORINE CYLINDERS

Cylinders that meet ICC specifications 3A1000, 3AA1000, and 3BN400 are authorized for fluorine service. The cylinders must not be equipped with safety relief devices, and must be fitted with valve protection caps. Commonly available sizes of cylinders are ½-lb, 4½-lb and 6-lb net weight.

Valve Outlet and Inlet Connections. Standard connection, U. S. and Canada—No. 670.

Cylinder Requalification. All cylinders authorized for fluorine must be requalified by hydrostatic retest every 5 years under current regulations.

Safe Handling and Storage of Cylinders. Personnel working with fluorine cylinders must be protected by use of a cylinder enclosure or barricade and remote-control valves, preferably ones operated by manual extension handles passing through the barricade. The main function of a barricade is to dissipate and prevent the breakthrough of any flame or flow of molten metal which, in case of equipment failure, could issue from any part of a system containing fluorine under pressure. Barricades of ¼-in. steel plate, brick or concrete provide satisfactory protection for fluorine in cylinder quantities. Adequate ventilation of enclosed working spaces is essential. Installation in a fume hood is recommended for laboratory use of fluorine cylinders.

Fluorine cylinders should be securely supported while in use to prevent movement or straining of connections.

Store full or empty fluorine cylinders in a well ventilated area, making sure that they are protected from excessive heat, located away from organic or flammable materials, and chained in place to prevent falling. Valve protection caps and valve outlet caps must be securely attached to cylinders not in use.

Always protect fluorine cylinders from mechanical shock or abuse and never heat them with a torch or heat lamp.

Additional precautions required for the safe handling of fluorine are explained in a preceding section.

TRAILER-MOUNTED FLUORINE TANKS

Fluorine is authorized for shipment as a liquid at low temperature and atmospheric pressure in tanks mounted on motor vehicle trailers under special permit of the ICC. These tanks commonly have a 5000-lb capacity.

Such trailer-mounted units consist of three concentric tanks. The liquid fluorine is contained in an inner baffled tank made of "Monel" metal or stainless steel. The inner tank is enclosed by a stainless steel tank filled with liquid nitrogen. The third, outer tank is made of carbon steel, and the annular space between it and the liquid nitrogen cooling jacket is filled with insulation and evacuated.

Vaporization of the liquid nitrogen (at −320.4 F and 1 atm) in the specially constructed units keeps the liquid fluorine below its boiling point (of −306.6 F at 1 atm), and radiation heat loss is minimized by the outer insulation. The ullage or vacant space in the liquid fluorine tank is brought to atmospheric pressure with helium to prevent subsequent in-leakage of moist air in case of valve or piping failure.

The liquid fluorine tank has two connections, one to a vapor-space line and the other to a dip line in the liquid. Both lines are double-valved. There is no rupture disc or safety valve for relieving excess pressure, but a pressure gage and high-pressure alarm are installed in the vapor line.

More than 300 gallons of liquid nitrogen are held by the inner cooling jacket, which has fill, vent and drain lines, each protected as required with a safety valve and rupture disc. Gages showing the level and pressure of the liquid nitrogen are on the control panel. The outer jacket of insulation is protected from excess pressure and has an electronic vacuum gage attached.

Safety equipment on the trailer body includes a water fire extinguisher, a dry-chemical fire extinguisher, and a special tool chest (containing hand tools, safety clothing, tubing and breathing apparatus for use by drivers only, in case of emergency on the road, and not for unloading). Full information on unloading the units is available from fluorine producers operating them.

STORAGE AND PIPING EQUIPMENT

Stationary installations for storing and piping fluorine must be designed with due regard for the reactivity of fluorine and must be made of materials proven in the type of fluorine service planned, such as high-temperature, high-pressure or cryogenic. Extreme care must be taken to keep all lines, fittings, tanks and other equipment clean. Pipe and fittings for service in lines not to be dismantled should be welded and, in general, the number of nonwelded lines should be kept to a minimum. Valves should have seatings of dissimilar metals in order to prevent galling, and packless stem seals should be used if possible.

REFERENCES

1. Voegtlin, C., and Hodge, H. C., *Pharmacology and Toxicology of Uranium Compounds*, National Nuclear Energy Series, Division VI, Vol. **1**, pp. 1021–1042.
2. "Hydrofluoric Acid" (Safety Data Sheet SD-25), Manufacturing Chemists' Association, 1825 Connecticut Ave., N. W., Washington, D. C., 20009.

Fluorocarbons

12, DICHLORODIFLUOROMETHANE: CCl_2F_2
13, MONOCHLOROTRIFLUOROMETHANE (Synonym—Chlorotrifluoromethane): $CClF_3$
22, MONOCHLORODIFLUOROMETHANE (Synonym—Chlorodifluoromethane): $CHClF_2$
115, MONOCHLOROPENTAFLUOROETHANE (Synonym—Chloropentafluoroethane): $CClF_2CF_3$
500, (12) DICHLORODIFLUOROMETHANE/(152a) DIFLUOROETHANE: CCl_2F_2/CH_3CHF_2 (73.8/26.2 per cent by weight)

ICC Classification: Nonflammable compressed gas; green label

(*Note:* The number preceding the name of each gas is its standard designation in the system developed for identifying refrigerant gases by the American Society of Heating, Refrigerating and Air-Conditioning Engineers (ASHRAE); the system is American Standard B79.1—1960.)

PHYSICAL PROPERTIES

	No. 12	No. 13	No. 22	No. 115	No. 500
International symbol	CCl_2F_2	—	$CHClF_2$	—	CCl_2F_2/CH_3CHF_2
Molecular weight	120.93	104.47	86.48	154.48	99.29
Vapor pressure, psia					
at −100 F	1.4	22.2	2.4	2.3	—
at −50 F	7.1	71.7	11.7	10.7	—
at 0 F	23.8	176.8	38.7	34.4	27.96
at 50 F	61.4	365.9	98.7	85.9	72.5
at 70 F	84.9	473.4	136.1	117.7	100.5
at 100 F	131.9	—	210.6	181.0	156.6
at 115 F	161.5	—	257.4	222.0	190
at 130 F	195.7	—	311.5	266.8	232.9
at 150 F	249.3	—	396.2	338.9	297.0
Density, saturated vapor, at boiling point, lb/cu ft	0.391	0.438	0.301	0.522	0.326
Density, liquid, lb/cu ft					
at −100 F	100.15	93.02	93.77	103.69	—
at −50 F	95.62	85.69	89.00	98.14	—
at 0 F	90.66	76.98	83.83	92.00	80.59
at 50 F	85.14	64.68	78.03	84.96	75.42
at 70 F	82.72	56.46	75.47	81.77	73.14
at 105 F	78.09	—	70.47	75.56	—
at 115 F	76.65	—	68.88	73.07	—
at 130 F	74.37	—	66.31	69.44	—
at 150 F	71.04	—	62.40	63.17	61.78

PHYSICAL PROPERTIES—*continued*

	No. 12	No. 13	No. 22	No. 115	No. 500
Boiling point at 1 atm	-21.62 F	−114.6 F	−41.36 F	−37.7 F	−28.0 F
Freezing point at 1 atm	−252 F	−294 F	−256 F	−159 F	−254 F
Critical temperature	233.6 F	83.9 F	204.8 F	175.9 F	221.1 F
Critical pressure, psia	596.9	561	721.9	453	631
Critical density, lb/cu ft	34.8	36.1	32.8	37.2	—
Critical volume, cu ft/lb	0.0287	0.0277	0.0305	0.0269	—
Latent heat of vaporization at boiling point, Btu/lb	71.04	63.85	100.45	54.20	—
Specific heat, liquid, Btu/(lb)(°F)					
at −22 F	0.213	0.247	0.260	0.233	—
at 77 F	0.232	—	0.300	0.282	—
Specific heat, vapor, at 77 F and 1 atm, Btu/(lb)(°F)	0.145	0.158	0.157	0.164	—
Specific heat ratio, C_p/C_v, at 77 F and 1 atm	1.137	1.145	1.184	1.091	—
Solubility in water at 77 F and 1 atm, lb/gal of water	0.0025	0.0008	0.024	0.0005	—
Weight per gallon, liquid, at 70 F, lb	11.05	7.54	10.08	10.92	—

Properties. The fluorocarbons constitute a large family of fluorinated hydrocarbon compounds that exhibit similar chemical properties and a wide range of physical characteristics. Their inert character and the range of their vapor pressures, boiling points and other physical properties makes them especially well suited for a large variety of differing conditions in two chief uses which have been developed for them: as refrigerants, and as propellants with many different kinds of pressure-packaged products.

Only widely used fluorocarbons which the ICC has classified as compressed gases and for which it has set regulations are included in the regular parts of this section. There are other fluorcarbons which do not fall within the definition of a compressed gas, and still more which are shipped under special permit of the ICC. The prefix, "mono-," given here in the names of several fluorocarbons, is increasingly omitted in modern terminology; it continues in use, though, in the ICC regulations.

The fluorocarbons are inert, nonflammable (in all concentrations in air under ordinary conditions), colorless and relatively nontoxic. Shipped as liquefied compressed gases under their own vapor pressures, they are colorless as liquids and they freeze to white solids. The fluorocarbons are odorless in concentrations of less than 20 per cent by volume in air, and have a faint and somewhat ethereal odor in higher concentrations. Chemically, they are analogs of hydrocarbons in which all or nearly all the hydrogen has been replaced by fluorine or fluorine and chlorine, accounting for their pronounced stability. They are more volatile and dense than corresponding hydrocarbons, and have lower refractive indices, lower solubilities and lower surface tension, but they have viscosities comparable to those of hydrocarbons. They also have relatively high dielectric strength.

The fluorocarbons show no appreciable decomposition at temperatures up to 400 F, or higher in many cases. In contact with a flame, these gases decompose to yield toxic products that are very irritating and hence are thought to give adequate warning of their presence in air even in very low concentrations. The fluorocarbons oxidize only with extreme difficulty and at very high temperatures. They are hydrolytically stable, unlike other halogen-containing compounds, but those containing hydrogen, like gas 22, are more susceptible to alkaline hydrolysis.

Fluorocarbons derived from the methane and ethane series are of the largest commercial significance today, but uses are developing for those of the cyclobutane and propane series.

Materials of Construction. The fluorcarbons are noncorrosive to all common metals except at very high temperatures. At elevated temperatures, the following metals resist fluorocarbon corrosion (and are named in decreasing order of their corrosive resistance): "Inconel," stainless steel, nickel, steel and bronze. Water or water vapor in fluorocarbon systems will corrode magnesium alloys or aluminum containing over 2 per cent magnesium (though such corrosion will be neither speeded nor slowed by the fluorocarbon's presence). In consequence, such metals are not recommended for use with fluorocarbon systems in which water may be present.

Systems using fluorocarbons as refrigerants (or similarly using other halogen-containing compounds) should be dry in order to prevent the possibility of faulty functioning of components like regulating valves, bellows, diaphragms, hermetically sealed coils or lubricating oils, due to the freezing of moisture in those components.

Some fluorocarbon compounds exert a high solvent action on natural rubber packing and gaskets, but certain newer types of synthetic rubber and plastics satisfactorily resist such action, even under very severe operating conditions.

Manufacture. In general, fluorocarbons are produced commercially by the reaction of hydrofluoric acid with chlorocarbons, or by the disproportionation of other fluorocarbons. Gas 12, for example, is made by the reaction of carbon tetrachloride and hydrogen fluoride in the presence of antimony chloride as a catalyst. Gas 13 is produced by the disproportionation of gas 12 in the vapor phase in the presence of aluminum chloride or bromine. Gas 22 is made by treating chloroform with anhydrous hydrogen fluoride in the presence of a small amount of antimony chloride at elevated temperatures and pressures. The gas 152a that is a component of the gas 500 mixture is manufactured by the addition of hydrogen fluoride to acetylene in liquid hydrogen fluoride with boron trifluoride as a catalyst.

Commercial Uses. The fluorocarbons are used most widely as refrigerants (and industrial coolants, as in shrink fitting of metal parts), and as aerosol propellants for products applied in spray or foam form. They are also used as polymer intermediates in the chemical industry, and as solvents, cleaning liquids and fire extinguishing agents. Other uses include their employment as dielectric fluids, as the vapor in wind-tunnel experiments in aerodynamics, and as power-transmitting fluids.

Special mixtures of two or more fluorocarbons are often used to provide desired properties in particular refrigerating and propellant applications.

Physiological Effects. The five fluorocarbons treated here are considered or proven to be relatively nontoxic when inhaled. They can, of course, act as simple asphyxiants by displacing needed oxygen in the air, and sites at which they are handled must have adequate ventilation. Gases 12 and 115 have been classified as "Group 6" compounds by the Underwriters' Laboratories, and, as such, are ones which result in no ill effects when inhaled in concentrations up to 20 per cent by volume for at least 2 hours. One other of the five fluorocarbons, gas 15, is thought to qualify for the Group 6 category; the remaining two, gases 22 and 500, have been classified as compounds of Group 5, which includes carbon dioxide. All of these gases decompose into toxic compounds in contact with a flame but, as previously noted, the decomposition products are irritating if inhaled and hence noticeable in very low concentrations.

Liquid fluorocarbons in contact with the skin can cause severe freezing or frostbite because of their low temperatures. Should liquid splash into the eyes, wash them thoroughly with water for at least 15 minutes and call an eye specialist at once.

Leak Detection. A large leak in a fluorocarbon system will be apparent through frosting around the leak. Smaller leaks may be detected with a soap water solution or a halide torch or leak detector.

FLUOROCARBON CONTAINERS

Cylinders, cargo tanks, portable tanks, single-unit tank cars and TMU tank cars are variously authorized for the shipment of the fluorocarbons, as explained below.

Filling Limits. The maximum filling densities authorized for fluorocarbons are as follows (per cent water capacity by weight):

In cylinders—gas 12, 119 per cent; gas 13, 100 per cent; gas 22, 105 per cent; gas 115, 110 per cent; gas 500, not liquid full at 130 F.

In single-unit tank cars and TMU tanks—gas 12, 119 per cent in TMU tanks, 125 per cent in single-unit tank cars; gas 22, 105 per cent in TMU tanks, 110 per cent in single-unit tank cars; gas 500, not liquid full at 105 F if the tank is insulated, or not liquid full at 130 F if the tank is uninsulated (also, the gas pressure at 105 F in a single-unit tank car tank that is insulated must not exceed three-fourths the prescribed retest pressure of the tank, or the gas pressure at 130 F in a TMU tank that is not insulated must not exceed three-fourths of the prescribed retest pressure of the tank).

In cargo tanks and portable tanks—gas 12, 119 per cent; gas 22, 105 per cent; gas 500 (authorized for cargo tank shipment only), not liquid full at 105 F if the tank is insulated or not liquid full at 115 F if the tank is not insulated (as determined by weight or liquid level gaging device).

SHIPPING METHODS; REGULATIONS

Under the appropriate regulations and tariffs, the fluorocarbon gases treated here are authorized for shipment as follows:

By Rail: In cylinders (via freight or express); also, for gases 12, 22 and 500, in single-unit tank cars and TMU tank cars.

By Highway: In cylinders on trucks; also, for gases 12 and 22, in cargo tanks and portable tanks; also, for gas 500, in cargo tanks.

By Water: For gases 12, 22 and 500, in cylinders, authorized TMU tanks, and portable tanks (meeting ICC-51 specifications and not exceeding 20,000 lb gross weight) aboard passenger and cargo vessels, and aboard ferry and railroad car ferry vessels (passenger or vehicle); also, in authorized tank cars aboard cargo vessels and railroad car ferry vessels (passenger or vehicle). Gases 13 and 115, in cylinders aboard passenger vessels, cargo vessels, cargo vessels and ferry and railroad car ferry vessels (passenger or vehicle). On barges (for any fluorocarbon, as a nonflammable gas) in cylinders on barges of U. S. Coast Guard classes A, BA, BB, CA and CB.

FLUOROCARBON CYLINDERS

Cylinders that meet the following ICC specifications are authorized for the fluorocarbon service indicated:

Gas 12—3A225, 3AA225, 3B225, 4A225, 4B225, 4BA225, 4B240ET, 4E225, 9, 41 and 3E1800. (Cylinders of types 9 and 41 must be shipped in strong outside containers.)

Gas 13—3A1800, 3AA1800, 3 and 3E1800.

Gas 22—3A240, 3AA240, 3B240, 4B240, 4BA240, 4B240ET, 4E240, 41 and 3E1800. (Cylinders of type 41 must be shipped in strong outside containers.)

Gas 115—3A225, 3AA225, 3B225, 4A225, 4B225, 4BA225 and 3E1800.

Gas 500—3A240, 3AA240, 3B240, 3E1800, 4A240, 4B240, 4BA240 and 9. (Cylinders of type 9 must be shipped in strong outside containers.)

Valve Outlet and Inlet Connection. Standard connection, U. S. and Canada—No. 620 (for gases 12, 13, 22 and 500; no standard connection is specified for gas 115).

Cylinder Requalification. All cylinders authorized for fluorocarbon service as indicated above must be requalified by hydrostatic retest every 5 years under present ICC regulations, except as follows:

(1) No periodic retest is required for 3E cylinders; and

(2) External visual inspection may be given in lieu of hydrostatic retest for cylinders that are used exclusively for fluorinated hydrocarbons and mixtures thereof which are commercially free from corroding components and that are of the following types (including cylinders of these types with higher service pressures): 3A480, 3AA480, 3A480X, 4B300 and 4BA300.

SINGLE-UNIT TANK CARS AND TMU TANK CARS

Gases 12, 22 and 500 are authorized by the ICC for shipment in single-unit tank cars that comply with specifications 105A300W. Common capacities for such tank cars range from 40,000 lb (about 4000 gal) to 120,000 lb.

Gases 12, 22 and 500 are also authorized for shipment in TMU tanks meeting ICC specifications 106A500X and 110A500W. Gases 22 and 500 may be transported in such tanks stenciled "Dispersant Gas" or "Refrigerant Gas."

CARGO TANKS AND PORTABLE TANKS

Gases 12, 22 and 500 are authorized for shipment in truck cargo tanks complying with specifications MC-330 and MC-331; many such tanks have capacities of some 3600 gal (36,000 lb). Gases 12 and 22 are also authorized for shipment in portable tanks meeting specification ICC-51. All three gases may be transported in cargo tanks and portable tanks stenciled "Dispersant Gas" or "Refrigerant Gas." The minimum design pressure specified for these tanks is 150 psig in gas 12 service and 250 psig in gas 22 or gas 500 service. TMU tanks that meet ICC specifications 106A500X and 110A500W are also authorized for motor vehicle shipment of gases 12, 22 and 500.

STORAGE AND HANDLING EQUIPMENT

Fluorocarbon storage tanks are commonly made of steel, and are built to the ASME code for unfired pressure vessels or the code of the authority having jurisdiction. Tanks and installations must, of course, comply with all state and local regulations. Recommended working pressures for installations are at least the vapor pressure of the compound at 135 F; as examples, 200 psig is a recommended pressure for systems handling gas 12, and 325 psig for systems handling gas 22.

For piping, copper with silver-soldered fittings is often used for lines of 1 in. or less, and ASA Schedule 40 steel with welded joints is used for larger lines. Screwed fittings should not be used. "Teflon" is among materials recommended for gaskets in flanged joints. Installations should be designed and built in consultation with experts thoroughly familiar with the fluorocarbons and their handling.

Helium

He ICC Classification: Nonflammable compressed gas; green label

PHYSICAL CONSTANTS

International symbol	He
Molecular weight (atomic weight)	4.003
Density at 70 F and 1 atm, lb/cu ft	0.01034
Density, gas, at boiling point, lb/cu ft	1.06
Density, liquid, at boiling point, lb/cu ft	7.798
Boiling point at 1 atm	−452.1 F
Freezing point at 1 atm	—*
Critical temperature	−450.3 F
Critical pressure, psia	33.2
Latent heat of vaporization at boiling point, Btu/lb	10.26
Specific heat, C_p, at 70 F and 1 atm, Btu/(lb)(°F)	1.25
Specific heat, C_v, at 70 F, Btu/(lb)(°F)	0.752
Ratio of specific heats, C_p/C_v, at 70 F	1.66
Liquid/gas ratio (liquid at boiling point, gas at 70 F and 1 atm), vol/vol	1/754.2

Properties. Helium is the second lightest element; only hydrogen is lighter. It is one-seventh as heavy as air. Helium is one of the rare gases of the atmosphere, in which it is present in a concentration of only 5 ppm. Natural gas containing up to 2 per cent helium has been found in the American Southwest; these fields, representing the major source of helium in the U. S., lie within 250 miles of Amarillo, Texas. Other helium-bearing natural gas fields have been discovered in Saskatchewan, Canada, and near the Black Sea.

Helium is chemically inert. It has no color, odor, or taste. Liquid helium is extremely important in cryogenic research since it is the only known substance to remain fluid at temperatures near absolute zero, and hence has a unique use as a refrigerant in cryogenics. It is also the only known nuclear reactor coolant that does not become radioactive. Helium is only slightly soluble in water, and will neither burn nor explode. It is shipped at high pressures—at or above 2200 psig at 70 F in cylinders and 2400 psig in bulk units. It is also shipped as a cryogenic liquid.

Materials of Construction. Because helium gas is inert, any commercially available metals may be used to contain it in installations designed to meet requirements for the pressures involved. Special materials such as certain stainless steels, and insulation are required for equipment handling liquid helium.

Manufacture. The principal source of helium is natural gas, from which it is recovered in essentially a stripping operation involving liquefaction and purification. Helium is also recovered from the atmosphere by fractionation, although infrequently on a commercial basis because of the small amounts contained in the atmosphere.

Commercial Uses. Helium is used as an inert gas shield in arc welding; as a lifting gas for lighter-than-air aircraft; as a gaseous cooling medium in nuclear reactors; to provide a protective atmosphere for growing germanium and silicon crystals for transitors; to provide a protective atmosphere in the production of such reactive metals as titanium and zirconium; to fill cold-weather fluorescent lamps;

* Helium will not solidify at 1 atm. The approximate pressure required for solidification near absolute zero (which is −459.69 F) is calculated to be 25 atm or 367.7 psia at −458 F.

to trace leaks in refrigeration and other closed systems; and to fill neutron and gas thermometers. It is used in cryogenic research, and, in mixtures with oxygen, it has medical applications. Radioactive mixtures of helium with krypton are available to users licensed by the Atomic Energy Commission.

Physiological Effects. Helium is nontoxic, and its gas poses hazards only as a simple asphyxiant that can deprive the lungs of needed oxygen. Liquid helium and cold helium gas given off by the liquid can damage living tissues with injuries like high-temperature burns on contact, and must be handled with all the precautions required for a liquefied gas at extremely low temperatures.

Special Handling Precautions for Liquid Helium. Users of liquid helium must also take special precautions in addition to those necessary for the safe handling of such inert liquefied gases as nitrogen and argon. Two properties of liquid helium make these special precautions imperative: its extreme cold solidifies all other gases; and its cold causes oxygen to condense on any uninsulated or inadequately insulated pipe through which it passes. In consequence, liquid helium must not be allowed to come in contact with air, and must be handled in closed systems having vacuum-jacketed lines. These systems must be pressurized internally at greater than atmospheric pressure, and must be equipped with pressure relief devices that prevent back leakage of air in order to prevent backflow of air into liquid helium equipment. Air backflow and plugging by solidified air constitutes a serious safety hazard. Similarly, if air enters and plugs the vent of a helium container, a serious hazard is created; therefore, the vents of liquid helium containers must be tested on delivery and periodically checked to make sure their vents remain clear. The use of open-neck Dewar flasks for liquid helium also increases the possibility of necktube plugging from transfer or gaging of contents. Users of liquid helium should obtain full information on safe handling precautions and equipment from their suppliers.

HELIUM CONTAINERS

Helium gas is shipped in cylinders, tank cars, special tanks on trucks and tank semitrailers. Liquid helium is shipped in special insulated cylinders and tank trucks, as a commodity not regulated by the ICC (when kept at pressures below 25 psig), or by special ICC permit. Atmospheric helium gas is shipped in cylinders and in liter flasks of pyrex or soft glass.

Filling Limit. The maximum filling limit at 70 F authorized by the ICC for the shipment of helium gas in cylinders and tube trailers is the service pressure of the container. The authorized ICC 107A tanks on tank cars may be charged with helium gas to a pressure 10 per cent in excess of the maximum gas pressure at 130 F of each tank.

SHIPPING METHODS; REGULATIONS

Under the appropriate regulations and tariffs, helium is authorized for shipment, as follows (in gaseous form, unless liquid is noted; liquid helium at pressure below 25 psig is not classified by the ICC as a "dangerous article," and no ICC regulations apply to its shipment at such pressures):

By Rail: In cylinders (by freight or express) and in tank cars.

By Highway: In cylinders on trucks, and in tube trailers. Liquid helium, in special containers on trucks and in insulated tank trucks or tank trailers either at pressures below 25 psig or by special ICC permit.

By Water: In cylinders on passenger and cargo vessels and on ferry vessels of all types; in authorized tank cars on cargo vessels and railroad car ferries (passenger or vehicle). In cylinders on barges of U. S. Coast Guard classes A, BA, BB, CA, and CB.

By Air: In cylinders on passenger aircraft up to 150 lb (70 kg), and on cargo aircraft up to 300 lb (140 kg). In containers meeting special requirements, pressurized liquid helium on cargo aircraft only up to 300 lb (140 kg), and low-pressure liquid helium on cargo aircraft only up to 660 lb (300 kg). (Quantities are maximum net weight per container.)

HELIUM CYLINDERS

Cylinders authorized by the ICC for the shipment of any nonliquefied compressed gas may be used to ship helium gas. (These are cylinders meeting ICC specifications 3A, 3AA, 3B, 3C, 3D, 3E, 4, 4B, 4BA, and 4C. In addition, cylinders meeting ICC specifications 3, 7, 25, 26, 33, and 38 may be continued in helium service, but new construction is not authorized.)

Valve Outlet and Inlet Connections: Standard connection, U. S. and Canada—No. 580, No. 930. Alternate standard connection, U. S. and Canada—No. 350 (obsolete, effective 1/1/72).

Cylinder Requalification. All cylinders authorized for helium gas service must be requalified by hydrostatic retest every 5 years except as follows: ICC-4 cylinders, every 10 years; and no retest is required for cylinders of types 3C, 3E, 4C, and 7.

LIQUID HELIUM CONTAINERS

A large part of the liquid helium used in the U. S. is shipped in special containers of relatively small capacity. One type, insulated by a liquid nitrogen shield between vacuum jackets, is usually made in 25-liter (6.6-gal) and 50-liter (13.2-gal) capacities. Containers of this type stand some 4 ft high and are about 20 to 25 in. in diameter. Those of a second type, built in 50-liter and 100-liter (26.4-gal) capacities, use a high-efficiency insulation. These containers stand over 5 ft high and are about 22 in. in diameter. Both types are vented through relief valves set at about 0.5 psig, and have helium evaporation rates of 1 to 2 per cent capacity per day. Helium shipped at such low pressures is not regulated by the ICC, but its shipment and storage may have to comply with local and state regulations. Containers of these types are commonly shipped by motor vehicle.

Some use has also been made of specially insulated and equipped containers which meet ICC cylinder specifications 4L for shipping liquid helium (either at pressures below 25 psig or by special ICC permit).

TANK CARS

Tank cars meeting ICC specifications 107A are authorized for the shipment of helium gas.

The specifications result in a car of special construction, containing in one design 30 nested cylinders, each about 18 in. in diameter and running the length of the car. The permanently mounted cluster of tubular cylinders is manifolded at the car end. At pressures of 2000 to 3500 psig, the car capacity is some 250,000 cu ft.

TUBE TRAILERS AND TANK TRAILERS

Bulk shipment of helium gas at high pressure is also made in tube trailers. The tube trailers consist of a nested cluster of long tubular cylinders manifolded at the end. Helium gas is shipped at pressures of some 2400 psig in these tanks and tank trailers, and their capacities range from about 10,000 to 128,000 cu ft.

Liquid helium is shipped at low pressure in motor vehicle tank semitrailers under special ICC permit. One model of such a tank trailer has a 5300 gal capacity; it can operate for 20 days without loss of helium, which gradually increases in pressure to less than 180 psig in the 20-day interval. Liquefied helium is also shipped at low pressure in tank trucks with capacities up to 14,000 gal (either at pressures below 25 psig or by ICC special permit).

STORAGE FACILITIES

Installations for the high-pressure storage of helium gas usually consist of manifolded cylinders each made from a 40-ft length of 10-in. heavy-wall pipe. Installations range in capacity up to 2.5 million cu ft.

REFERENCE

"Standard Density Data, Atmospheric Gases and Hydrogen" (Pamphlet P-6), Compressed Gas Association, Inc.

ADDITIONAL REFERENCE

"Helium Specification," Compressed Gas Association, Inc.

Hydrogen

H_2 ICC Classification: Flammable compressed gas; red label

PHYSICAL CONSTANTS

International symbol	H_2
Molecular weight	2.016
Specific gravity of gas at 32 F and 1 atm (air = 1)	0.06950
Specific volume at 70 F and 1 atm, cu ft/lb	192.0
Density of gas at 70 F and 1 atm, lb/cu ft	0.005209
Density of gas at boiling point and 1 atm, lb/cu ft	0.084
Density of liquid at boiling point and 1 atm, lb/cu ft	4.428
Liquid/gas ratio (liquid at boiling point, gas at 70 F and 1 atm), vol/vol	1/850.1
Boiling point at 1 atm	−423.0 F
Freezing point at 1 atm	−434.6 F
Critical temperature	−399.91 F
Critical pressure, psia	190.8
Triple point	−434.56 F at 1.0414 psia
Latent heat of vaporization at boiling point, Btu/lb	192.7
Specific heat, C_p, at 70 F, Btu/(lb)(°F)	3.416
Specific heat, C_v, at 70 F, Btu/(lb)(°F)	2.430
Ratio of specific heats, C_p/C_v, at 70 F	1.41
Heat of combustion, Btu/cu ft	
Gross	325
Net	275
Solubility in water at 60 F, vol/1 vol of water	0.019
Weight per gallon, liquid, at boiling point, lb	0.5920

Properties. Hydrogen is a colorless, odorless, tasteless, flammable and nontoxic gas at atmospheric temperatures and pressures. It is the lightest gas known, being only some seven-hundredths as heavy as air. Hydrogen is present in the atmosphere, occurring in concentrations of only about 0.01 per cent by volume at lower altitudes.

Hydrogen burns in air with a pale blue, almost invisible flame. Its ignition temperature will not vary greatly from 1050 F in mixtures with either air or oxygen at atmospheric pressure. Its flammable limits in dry air at atmospheric pressure are 4.1 to 74.2 per cent hydrogen by volume. In dry oxygen at atmospheric pressure, the flammable limits are 4.7 to 93.9 per cent hydrogen by volume. Its flammable limits in air or oxygen vary somewhat with initial pressure, temperature and water vapor content.

When cooled to its boiling point of −423 F, hydrogen becomes a transparent and odorless liquid only one-fourteenth as heavy as water. All gases except helium become solids at the temperature of liquid hydrogen. Because of its extremely low temperature, it can make ductile or pliable materials with which it comes in contact brittle and easily broken (an effect that must be considered whenever liquid hydrogen is handled). Liquid hydrogen has a relatively high thermal coefficient of expansion compared with other cryogenic liquids.

The hydrogen molecule exists in two forms: *ortho* and *para*, named according to their types of nuclear spin. (*Ortho*-hydrogen molecules have a parallel spin; *para*-hydrogen molecules, an anti-parallel spin.) There is no difference in the chemical properties of these forms, but there is a difference in physical properties. *Para*-hydrogen is the form preferred for rocket fuels. Hydrogen consists of about three parts *ortho* and one part *para* as a gas at room temperature. The equilibrium concentration of *para* increases with decreasing temperature until, as a liquid, the *para* concentration is nearly 100 per cent. If hydrogen should be cooled and liquefied rapidly, the relative three-to-one concentration of *ortho* to *para* would not immediately change. Conversion to the *para* form takes place at a relatively slow rate and is accompanied by the release of heat. For each pound of rapidly cooled liquid hydrogen that changes to the *para* form, enough heat is liberated to vaporize 1.5 lb of liquid hydrogen. However, if a catalyst is used in the liquefaction cycle, *para*-hydrogen can be produced directly without loss from self-generated heat.

Throttled expansion from high to low pressure at ordinary temperatures cools most common gases (such as oxygen, nitrogen and carbon dioxide). Hydrogen, though, is an exception, becoming heated to a slight extent under these conditions (increasing about 10 F in temperature, for example, when throttled from 2000 psig to atmospheric pressure).

Hydrogen diffuses rapidly through porous materials and through some metals at red heat. It may leak out of a system which is gas-tight for air or other common gases at equivalent pressures.

In its chemical properties, hydrogen is fundamentally a reducing agent and is frequently applied as such in organic chemical technology.

Materials of Construction. Hydrogen gas is noncorrosive, and may be contained at normal temperatures by any common metals used in installations designed to have sufficient strength for the working pressures involved. Metals used for liquid hydrogen equipment must have satisfactory properties at very low operating temperatures. Ordinary carbon steels lose their ductility at liquid hydrogen temperatures and are considered too brittle for this service. Suitable materials include austenitic chromium-nickel steels, copper, copper-silicon alloys, aluminum, "Monel," and some brasses and bronzes.

Manufacture. Hydrogen is produced industrially by several methods. A large part of the hydrogen produced in the U. S. is consumed at or within pipeline distance of the production site.

Hydrogen is frequently obtained as a by-product of cracking operations using petroleum liquids or vapors as feed stock.

Electrolysis of water, aqueous acid, or alkali solutions, is also used to produce hydrogen. Some plants have been built to make hydrogen as the primary desired product from the electrolysis of water. Considerable amounts of hydrogen are also obtained as by-products of other types of electrolytic operations designed to yield different chemicals as main products.

Passing steam over heated, spongy iron reduces the steam to hydrogen with accompanying formation of iron oxide. Several varieties of this "steam-iron" process are employed to make hydrogen commercially.

The reaction of steam with incandescent coke or coal (called the "water gas reaction") is also used as a source of hydrogen, with carbon monoxide as an additional product. In a catalytic version of the water gas process, excess steam breaks down to form more hydrogen while oxidizing the carbon monoxide to carbon dioxide.

Natural gas or light hydrocarbons may also be used as a raw material for a method of hydrogen production in which they react with steam in the presence of a catalyst. Hydrogen is also produced through the dissociation of ammonia.

Commercial Uses. Large quantities of hydrogen are consumed in chemical syntheses, primarily of ammonia and methanol.

In the organic chemical field, hydrogenation of edible oils in soy beans, fish, cotton seed and corn yields solids used as shortenings and other foods. Various alcohols are produced by the hydrogenation of the corresponding acids or aldehydes.

Hydrogen is used with oxygen in oxy-hydrogen welding and cutting, being employed largely in certain brazing operations, welding aluminum and magnesium (especially in thin sections), and welding lead. The oxy-hydrogen flame has a temperature of about 4000 F, and is well suited for such comparatively low-temperature welding and brazing. It is also used to some extent in cutting metals, particularly in underwater cutting because hydrogen can be safely compressed to the pressures necessary to overcome water pressures at the depths involved in salvage operations. The oxy-hydrogen flame is also applied in the working and fabrication of quartz and glass.

Atomic hydrogen welding, another important application of hydrogen, is particularly suitable for thin stock and can be used with practically all nonferrous metals and alloys as well as with ferrous alloys; jobs performed with this welding process range from the fabrication of nickel and monel tanks to the welding of aluminum aircraft parts and propellers and the repair of steel molds and dies. In the process, an arc with a temperature of about 11,000 F is maintained between two nonconsumable metal electrodes. Molecular hydrogen fed into the arc is transformed into atomic hydrogen, which transmits heat from the arc to the weld zone. At the relatively colder surface of the weld area, the atomic hydrogen recombines to molecular hydrogen with the release of heat. The hydrogen also shields the weld area from the air.

Hydrogen also serves as a non-oxidizing shield, alone or with other gases, in annealing, furnace brazing, producing parts from sintered carbides and other metals, and in the refining and heat treatment of such metals as tungsten and molybdenum.

Large electrical generators are sometimes run in a hydrogen atmosphere to reduce windage losses and remove heat.

Liquid hydrogen is assuming importance as a fuel for powering missiles and rockets. It is also employed in laboratory research on the properties of materials at cryogenic temperatures, among them the super-conductivity of metals (a state in which they have no electrical resistance).

Physiological Effects. Hydrogen is nontoxic but it can, of course, act as a simple asphyxiant by displacing the oxygen in the air. Unconsciousness from inhaling air which contains a sufficiently large amount of hydrogen can occur without any warning symptoms such as dizziness. Still lower concentrations than those which could lead to unconsciousness would be flammable, for the lower flammable limit of hydrogen in air is only some 4 per cent by volume. All the precautions necessary for the safe handling of any flammable gas must of course be observed with hydrogen.

Liquid hydrogen and also the cold gas evolving from the liquid can produce severe burns similar to thermal burns upon contact with the skin and other tissues, and eyes can be injured by exposure to the cold gas or splashed liquid too brief to affect the skin of the hands or face. Contact between unprotected parts of the body with uninsulated piping or vessels containing liquid hydrogen can cause the flesh to stick and to tear when an attempt is made to withdraw.

Safe Handling of Liquid Hydrogen. The potential hazards of liquid hydrogen stem mainly from three important properties: (1) its extremely low temperature; (2) its very large liquid-to-gas expansion ratio (with one volume of liquid giving rise to 850 volumes of gas at room temperature); and (3) its wide range of flammable limits after it has vaporized to a gas.

Only persons who have been thoroughly instructed in the hazards of liquid hydrogen and corresponding protective measures should be allowed to handle it. All precautions necessary for the safe handling of any gas liquefied at extremely low temperatures must be observed with liquid hydrogen. In addition, liquid hydrogen must be handled in closed systems so that air does not enter vents and lines and plug them by solidifying. If air, oxygen or another gas condenses and solidifies in the openings of a liquid hydrogen container, pressure may build up to the point of damaging or bursting the container. As noted previously, all gases except helium become solids at the low temperature of liquid hydrogen. Quantities of liquid hydrogen up to five liters may be handled for laboratory or test purposes in open-mouth Dewar vessels, but only if the vessels are stoppered down to the smallest opening needed for the work and if they have vents which permit the release of hydrogen vapor but exclude the entry of air. Glass is not recommended as a material for Dewar flasks or other vessels that are to be filled with liquid hydrogen because of the possibility that the liquid may burst a vessel into which it is being poured with explosive violence unless the vessel is completely prepared for the liquid.

Users of liquid hydrogen should obtain full information on safe handling, protective equipment and first aid from their suppliers. Measures to insure safe handling are also described in Pamphlet G-5 of Compressed Gas Association, Inc.[1]

HYDROGEN CONTAINERS

Hydrogen gas is authorized for shipment in cylinders and tank cars; it is also authorized for shipment in tube trailers under special ICC permit.

Liquid hydrogen is authorized for shipment in tank cars; it is also shipped in specially designed, insulated portable containers and tank trucks under special permit of the ICC.

Filling Limits. The maximum filling limits authorized for hydrogen are as follows (hydrogen gas, unless liquid hydrogen is noted):

In cylinders—the service pressure marked on the cylinder at 70 F and not in excess of 5/4 of the marked service pressure at 130 F (hydrogen gas is usually shipped in cylinders at pressures of around 2000 psig). For liquid hydrogen shipped under special permit in cylinders that meet ICC-4L specifications, the cylinder pressure must be limited by a pressure-controlling valve to one and one-fourth times the marked service pressure (or, for 4L cylinders insulated by a vacuum, at least 15 psi lower than one and one-fourth times the marked service pressure).

In tank cars—not more than seven-tenths of the marked test pressure at 130 F in uninsulated cars of the ICC-107A type. Liquid hydrogen in tank cars (meeting ICC specifications 113A60-W-2), to a maximum filling density of 6.6 per cent (per cent water capacity by weight), and each car must have a pressure controlling valve set at a maximum of 17 psig.

SHIPPING METHODS; REGULATIONS

Under the appropriate regulations and tariffs, hydrogen is authorized for shipment as follows (hydrogen gas, except where liquid hydrogen is noted):

By Rail: In cylinders (via freight, express or baggage) and in tank cars. Liquid hydrogen, in tank cars and in portable containers under ICC special permit.

By Highway: In cylinders on trucks; in tube trailers. Liquid hydrogen, in special portable containers on trucks and in tank trucks under ICC special permit.

By Water: In cylinders on cargo vessels only, and in ICC-authorized tank cars aboard trainships only. In cylinders on barges of U. S. Coast Guard classes A, BA, BB, CA and CB only. Liquid hydrogen is not authorized for shipment by water under present Coast Guard regulations, but it is shipped under special permit in tank barges.

By Air: In cylinders aboard cargo aircraft only up to 300 lb (140 kg) maximum net weight per cylinder. Liquid hydrogen is not accepted for air shipment under present regulations.

HYDROGEN CYLINDERS AND PORTABLE CONTAINERS

Cylinders built to comply with ICC specifications 3A or 3AA are the types chiefly used to ship hydrogen gas, but it is also authorized for shipment in any cylinders approved for nonliquefied gases (these include cylinders meeting ICC specifications 3B, 3C, 3D, 3E, 4, 4A, 4B, 4BA, and 4C; in addition, cylinders meeting ICC specifications 3, 7, 25, 26, 33, and 38 may be continued in gaseous hydrogen service, but new construction is not authorized).

Valve Outlet and Inlet Connections. Standard Connection, U. S. and Canada—No. 350.

Cylinder Requalification. All cylinders authorized for gaseous hydrogen service must be requalified by hydrostatic retest every 5 years under present regulations, with the following exceptions: ICC-4 cylinders, every 10 years; and no periodic retest is required for cylinders of types 3C, 3E, 4C and 7.

Portable Containers for Liquid Hydrogen. Portable containers which comply in some cases with ICC-4L specifications are used for the shipment of liquid hydrogen under special permit of the ICC. These containers employ a "super" insulating material to keep the liquid at its boiling point for weeks at a time, and also rely on venting a small portion of the vapor in the container to remove what heat is absorbed. The containers maintain the liquid under low positive pressure and are outfitted with complex venting, safety relief device, and liquid withdrawal equipment. One commonly used size stands about 5 ft high and is some 20 in. in diameter; it holds about 40 gal of liquid hydrogen. One of the largest such containers has a capacity of some 270 gal.

Estimating Quantity Left in Cylinders. The pressure in a cylinder charged with a nonliquefied gas such as hydrogen is related to both the amount of gas in the cylinder and the temperature of the gas. When the temperature of hydrogen in a cylinder is close to room or atmospheric temperature, the volume of hydrogen may be estimated by use of the chart shown in Fig. 1. For example, if the room temperature is 70 F and the cylinder pressure is 1200 psig, follow the diagonal line 70 F until it crosses the vertical line representing 1200 psig. From the point of intersection of these two lines trace horizontally across the chart to

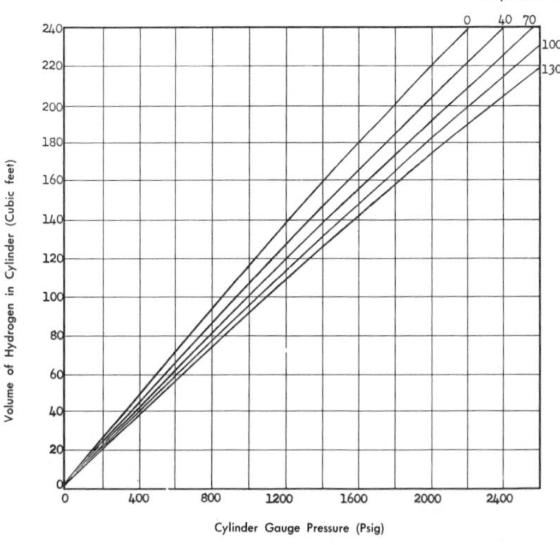

FIG. 1. Chart for estimating volume of hydrogen in a cylinder having a volumetric capacity of 2640 cu in. (Based on data in Bureau of Standards Miscellaneous Publication M191 of November 17, 1948).

read the figures at the left. This gives the volume of hydrogen in a cylinder with 2640 cu in. capacity of about 120 cu ft at 70 F and 1200 psig. To use the chart for hydrogen cylinders with volumes differing from 2640 cu in., multiply the hydrogen volume figure read off the left side of the chart by $V/2640$, where V is the volume in cubic inches of the given cylinder.

Special Precautions with Hydrogen Cylinders. Two special precautions must be observed with hydrogen gas in cylinders in addition to the various precautions necessary for the safe handling of any flammable compressed gas in cylinders.

It is vital to note that the first precaution differs from the usual safe procedure. Contrary to general practice with other gas cylinders, it is inadvisable to "crack" hydrogen cylinder valves before connecting them to a regulator or manifold since "self-ignition" of the issuing hydrogen may occur.

Second, leaking hydrogen cylinders should be guarded with special care. If hydrogen leaks from the cylinder valve even when the valve is closed, or if a leak occurs at the safety device, carefully remove the cylinder to an open space that is outdoors and well away from any possible source of ignition. Plainly tag the cylinder as having an unserviceable valve or safety device, and immediately notify the cylinder's supplier and ask for his instructions. *Extreme care is recommended in protecting access to the defective cylinder* because the leaking hydrogen may ignite in the absence of any normally apparent source of ignition and, if so, will burn with an almost invisible flame that can instantly injure anyone coming into contact with it.

Storing Hydrogen Cylinders. Hydrogen cylinders used with oxygen for welding and cutting and stored inside a building should be limited to a total capacity of 2000 cu ft exclusive of cylinders in use or attached for use. Quantities exceeding this total should be stored in a special building or separate room as recommended in Pamphlet No. 51 of the National Fire Protection Association,[2] or should be stored out-of-doors.

Storage of hydrogen gas not used in conjunction with oxygen should conform with the recommendations of CGA Pamphlet G-5.1.[3]

HYDROGEN TANK CARS

Gaseous hydrogen is authorized for rail shipment in tank cars that comply with ICC specifications 107A. These cars consist of a number of seamless containers permanently mounted on a special car frame. ICC regulations require that on such cars the safety relief device header outlet must be equipped with an approved ignition device which will instantly ignite any hydrogen discharged through the safety devices. Other requirements are also set forth in the regulations.

Liquefied hydrogen containing a minimum of 95 per cent *para*-hydrogen is authorized for shipment in tank cars meeting ICC specifications 113A60-W-2 which, as previously noted, must be outfitted with pressure controlling valves set at a pressure not exceeding 17 psig. A common size for these cars is 68-ft length with 28,300-gal capacity. The cars are so well insulated for internal cooling that they usually do not vent any hydrogen in being shipped distances up to 3000 miles; should the pressure approach the 17 psig relief setting, excess hydrogen is mixed with air in a special device to a concentration of less than half its lower flammable limit before being vented.

TUBE TRAILERS AND CARGO TANKS

Hydrogen gas is authorized for bulk highway shipment at high pressure (around 2000 psig) in tube trailers. These trailers are made up of a number of large tubes, constructed to ICC specifications 3A or 3AA and manifolded together to a common header. Valves must be installed on each tube and be closed while in transit. Such trailers are usually towed to the point of utilization and replace an empty unit which is then towed to the supplier's plant for refilling. Capacities of the tube trailers range up to 40,000 cu ft or more.

Liquefied hydrogen is shipped in insulated cargo tanks on trucks or truck trailers under special ICC permit. Some 8000 gal is a common capacity for the tanks, which consist of an inner shell surrounded by very efficient vacuum insulation and enclosed in an outer case of $\frac{3}{16}$ or $\frac{1}{2}$-in. steel. Though often equipped with an initial pressure relief device set at 17 psig, the tanks do not normally vent any hydrogen in transit, since they are ordinarily filled at a pressure of only some 2 psig at the plant and will rise in pressure to only about 6 to 8 psig during journeys of up to 3000 miles. Should the insulation deteriorate in transit, drivers are usually required to leave the highway and go to a safe area in which they reduce the pressure with a manual blowdown valve.

STORAGE AND HANDLING SYSTEMS

Hydrogen gas is most often stored at the user's location in cylinders or tube trailers, as noted earlier. Liquefied hydrogen is stored in systems that range in capacity to two million cu ft of gas or more and that frequently include a vaporizer to convert the liquid to gas at ordinary temperatures.

Systems for handling liquefied hydrogen, as well as piping and manifold systems for gaseous hydrogen, should be constructed only under the supervision of competent engineers who are thoroughly familiar with the problems incident to the piping of cryogenic fluids and flammable gases. Care should be taken to comply with the requirements of all state, municipal, and insurance authorities.

Recommended standards for gaseous hydrogen systems are given in the previously cited CGA Pamphlet G-5.1,[3] and recommended tentative standards for liquefied hydrogen systems are given in CGA Pamphlet G-5.2T.[4]

Standards for the storage of hydrogen cylinders used with oxygen for welding and cutting are presented, as stated before, in NFPA Pamphlet No. 51.[2]

Extensive information on equipment for handling liquid hydrogen appears in a handbook issued by the U. S. Department of Commerce.[5]

Electrical equipment in areas for the storage of large quantities of liquid hydrogen should be installed in accordance with Article 500 of NFPA Pamphlet No. 70.[6] Separate rooms or buildings devoted to the handling of liquid hydrogen in substantial quantities should conform with the recommendations of NFPA Pamphlet No. 68, "Guide for Explosion Venting."[7]

REFERENCES

1. "Hydrogen" (Pamphlet G-5), Compressed Gas Association, Inc.
2. "Standard for the Installation and Operation of Oxygen-Fuel Gas Systems for Welding and Cutting" (Pamphlet No. 51), National Fire Protection Association, 60 Batterymarch St., Boston, Mass., 02110.
3. "Standard for Gaseous Hydrogen Systems at Consumer Sites" (Pamphlet G-5.1), Compressed Gas Association, Inc.
4. "Standard for Liquefied Hydrogen Systems at Consumer Sites" (Pamphlet G-5.2), Compressed Gas Association, Inc.
5. "Handbook for Hydrogen Handling Equipment" (PB161835), U. S. Department of Commerce, Office of Technical Services, Washington 25, D. C. (Deals with liquid hydrogen only.)
6. "National Electrical Code" (Pamphlet No. 70), National Fire Protection Association.
7. "Guide for Explosion Venting" (Pamphlet No. 68), National Fire Protection Association.

ADDITIONAL REFERENCE

"Standard Density Data, Atmospheric Gases and Hydrogen" (Pamphlet P-6), Compressed Gas Association, Inc.

Hydrogen Chloride, Anhydrous

HCl

Synonym: Hydrochloric acid, anhydrous

ICC Classification: Nonflammable compressed gas; green label

PHYSICAL CONSTANTS

International symbol	HCl
Molecular weight	36.465
Vapor pressure, psig	
at 70 F	612
at 77 F	676
at 105 F	950*
at 115 F	1075*
Vapor density at 32 F and 1 atm, lb/cu ft	0.102
Specific gravity of gas at 32 F (air = 1)	1.268
Boiling point at 1 atm	−121 F
Melting point at 1 atm	−168 F
Critical temperature	124.7 F
Critical pressure, psia	1198
Latent heat of vaporization at −121 F, Btu/lb	190.5
Specific weight, lb/cu ft	
Liquid, at −50 F	67.80
Liquid, at +2 F	62.43
Vapor, at −50 F	0.7370
Vapor, at +2 F	1.7305

Properties. Anhydrous hydrogen chloride is a colorless gas which fumes strongly in moist air and has a highly irritating effect on body tissues. It has a sharp, suffocating odor, and is shipped in cylinders as a liquefied compressed gas under its vapor pressure of about 612 psig at 70 F.

Chemically, hydrogen chloride is relatively inactive and noncorrosive in the anhydrous state. However, it is readily absorbed by water to yield the highly corrosive hydrochloric (muriatic) acid. It also dissolves readily in alcohol and ether and reacts rapidly (violently, in some cases) with many organic substances. At high temperatures (3240 F and above), hydrogen chloride tends to dissociate into its constituent elements.

Materials of Construction. Piping, valves and other equipment used in direct contact with hydrogen chloride should be of stainless steel or of cast or mild steel. Carbon steel may be employed in some components, but only if their temperature is controlled to remain below about 265 F. Hydrogen chloride corrodes zinc, copper and copper alloys, and galvanized pipe or brass or bronze fittings must not be used with it.

Manufacture. Hydrogen chloride is produced as a by-product from the chlorination of benzene and other hydrocarbons, and by burning hydrogen, methane or water gas in a chlorine atmosphere.

Commercial Uses. An old and important industrial chemical, hydrogen chloride is widely used in the manufacture of rubber, pharmaceuticals and chemicals, as well as in gasoline refining, metals processing and wool reclaiming. Rubber hydrochloride, which results from the treatment of natural rubber with hydrogen chloride, can be cast in film

* Approximate values.

from solutions; such rubber hydrochloride films provide a strong, water-resistant packaging material for meats and other foods, paper products and textiles.

The chemicals industry uses hydrogen chloride to produce a large variety of chlorinated derivatives through both addition and subtraction reactions with organic compounds —as in the manufacture of ethyl chloride from ethylene in making vinyl plastics, of chloromethanes and monochlorobenzene, of alkyl chlorides, and of methyl chloride from methyl alcohol. Hydrogen chloride is utilized in the gasoline industry as a promoter for the aluminum chloride catalyst for converting *n*-butane to isobutane.

In the metals industry, uses of gaseous hydrogen chloride include application as a flux in bonding Babbitt metal to steel strip in manufacturing insert bearing blanks, and for treating steel strip at elevated temperatures to improve the bond in later hot galvanizing. The textiles industry uses hydrogen chloride to decompose vegetable fibers with which wool has been woven in reclaiming wool for fabrics. Cotton seeds are also delinted and disinfected by being tumbled in a current of gaseous hydrogen chloride.

Physiological Effects. Hydrogen chloride is toxic, causing severe irritation of the eyes and the skin on contact, and severe irritation of the upper respiratory tract on inhalation. Should contact occur, the eyes and eyelids must be washed immediately and thoroughly with large quantities of flowing water in order to avoid impairment of vision or even loss of sight. Prompt washing of the skin is also required after contact to prevent severe burns. Repeated exposure of the skin to concentrated anhydrous hydrogen chloride vapor may also result in burns or dermatitis.

Inhalation of excessive quantities produces coughing, burning of the throat and a choking sensation; occasionally, ulceration of the nose, throat and larynx, or edema of the lungs, has resulted. Prolonged inhalation causes death. The irritating character of the vapors provides warning of dangerous concentrations well before injury can result, unless personnel are trapped or disabled. Concentrations up to 5 ppm for an 8-hour working day are generally accepted as a safe threshold limit.[1] A concentration of 50 ppm in air cannot be tolerated for more than one hour, and concentrations of some 1500 to 2000 ppm are fatal for human beings in a few minutes.

Safety precautions include adequate ventilation of working areas, readily accessible safety shower and eye bath facilities, rubber or plastic aprons and gloves as protective clothing with wool or other acid-resistant outer garments, the availability of proper full-face gas masks, and thorough familiarity with safety equipment and measures on the part of personnel. Should eye contact, skin contact, or inhalation occur, immediately give first aid and call a physician.

Detailed information on safety precautions and first aid in the event of hydrogen chloride exposure is given in Safety Data Sheet SD-39 of the Manufacturing Chemists' Association.[1]

ANHYDROUS HYDROGEN CHLORIDE CONTAINERS

Anhydrous hydrogen chloride is shipped as a compressed liquefied gas under its own vapor pressure in cylinders and in tube trailers. It is also shipped in liquefied form in insulated tank cars and tank trucks by special permit of the ICC or the Board of Transport Commissioners for Canada.

Filling Limit. The maximum filling density authorized for hydrogen chloride in cylinders by the ICC is 65 per cent (per cent water capacity by weight).

SHIPPING METHODS; REGULATIONS

Under the appropriate regulations and tariffs, anhydrous hydrogen chloride is authorized for shipment as follows:

By Rail: In cylinders (freight or express), and in insulated tank cars (by special permit).

By Highway: In cylinders on trucks; also in

insulated tank trucks and tube trailers (by special permit).

By Water: In cylinders via cargo vessels, passenger vessels, and ferry and railroad car ferry vessels (passenger or vehicle). In cylinders on barges of U. S. Coast Guard classes A, BA, BB, CA and CB.

By Air: Aboard cargo aircraft only in appropriate cylinders up to 300 lb (140 kg) maximum net weight per cylinder.

ANHYDROUS HYDROGEN CHLORIDE CYLINDERS

Cylinders that meet the following ICC specifications are authorized for hydrogen chloride service: 3A1800, 3AA1800 and 3E1800 (cylinders manufactured under the now obsolete specification ICC-3 may be continued in service but new construction is not authorized). ICC-3A2015 cylinders are also often used in anhydrous hydrogen chloride service.

Valve Outlet and Inlet Connections. Standard connection, U. S. and Canada—No. 330.

Cylinder Requalification. Cylinders of types 3A and 3AA (as well as 3) used in hydrogen chloride service must be requalified by hydrostatic retest every 5 years under present regulations. For 3E cylinders, no periodic retest is required.

Corrosion Precaution. Hydrogen chloride cylinders should be used within 60 days to avoid valve corrosion. (This precaution should be taken in addition to the regular recommended practice of using cylinders in the order in which they are received.)

Cylinder Manifolding. Hydrogen chloride cylinders are manifolded by users, but care must be taken to have any special manifolding equipment meet all safety requirements.

Leak Detection. Leaks in anhydrous hydrogen chloride system connections may be detected by the white fumes which form when the gas comes into contact with the moisture of the atmosphere. Small leaks may be found with an open bottle of concentrated ammonium hydroxide solution (which forms dense white fumes in the presence of hydrogen chloride) or with wet blue litmus paper (which is turned pink by hydrogen chloride).

SPECIAL TANK TRUCKS AND TUBE TRAILERS

Under special permit of the ICC, hydrogen chloride is authorized for shipment in insulated tank trucks of special design. Constructed of stainless steel with urethane foam insulation, the tanks conform to ICC specification MC-330 and are loaded with liquid hydrogen chloride at −70 F and 50 psig. Unloading of the truck must be started within 3 days, unless special methods are used to keep truck pressure between 50 to 80 psig. Special-permit shipment of hydrogen chloride is also made in tube trailers carrying the liquefied gas at pressures of up to 1100 psig.

SPECIAL SINGLE-UNIT, INSULATED TANK CARS

Special permit of the ICC is also required for shipment of hydrogen chloride in single-unit tank cars of special design (in conformance with ICC specification 105A600-W, made of carbon steel with 10-in. insulation, and cooled by external coils). The cars are loaded with liquid hydrogen chloride at −50 F, and are designed for unloading without additional refrigeration within 25 to 35 days after filling.

STORAGE AND HANDLING EQUIPMENT

Hydrogen chloride is most often unloaded from special tank cars and tank trucks directly to process, without being stored by users. Unloading piping and valves must be of stainless steel or of cast or mild steel (Charpy tested to −50 F and capable of withstanding the high pressures of the unloading process). Galvanized or copper-alloyed metals such as brass or bronze are corroded by hydrogen chloride, and must not be used in pipe or fittings. Design and specification of components for unloading systems should be made with the help of suppliers or engineers thoroughly familiar with anydrous hydrogen chloride.

REFERENCE

1. "Hydrogen Chloride," Safety Data Sheet SD-39, Manufacturing Chemists' Association, 1825 Connecticut Ave., N. W., Washington, D. C., 20009.

Hydrogen Cyanide

HCN

Synonyms: Hydrocyanic acid, prussic acid, formonitrile

ICC Classification (as hydrocyanic acid, liquefied): Poison gas, class A; poison gas label

PHYSICAL CONSTANTS

International symbol	HCN
Molecular weight	27.03
Vapor pressure, psia	
at 14 F	3.7
at 32 F	5.7
at 50 F	8.5
at 68 F	12.5
at 86 F	17.7
at 104 F	24.8
Density, liquid, lb/cu ft	
at 32 F	44.64
at 70 F	42.85
at 105 F	41.23
at 115 F	40.75
at 130 F	40.05
Boiling point at 1 atm	78.3 F
Melting point at 1 atm	9 F
Critical temperature	362.3 F
Critical pressure, psia	782
Latent heat of vaporization at 77 F, Btu/lb	444.2
Latent heat of fusion at 9.3 F, Btu/lb	133.8
Specific heat, Btu/(lb)(°F)	
at −27.5 F	0.516
at 62.4 F	0.627
Heat of combustion, Btu/lb	10,555
Flammable limits in air, by volume	6–41%
Weight/gal, liquid, lb	
at 32 F	5.97
at 68 F	5.74

Properties. Hydrogen cyanide (hydrocyanic acid) is a colorless liquid with an atmospheric boiling point of 78.3 F. The vapor phase is also colorless and has a faint odor of bitter almonds. Hydrogen cyanide vapors are highly toxic, and hydrogen cyanide is designated a poison gas of the "extremely dangerous" class A variety by the ICC. Its gas is lighter than air, having a specific gravity of about 0.93 compared to air, and its liquid is lighter than water (specific gravity about 0.7). Hydrogen cyanide polymerizes spontaneously with explosive violence when not completely pure or stabilized; it is commonly stabilized by adding acids, such as 0.05 per cent phosphoric acid. Unstabilized hydrogen cyanide is not accepted for shipment under ICC regulations. Hydrogen cyanide is miscible in all proportions with water or alcohol, and miscible in ether, benzene, and most other organic solvents. It reacts with water, steam, acids or acid fumes to form very toxic fumes of cyanides.

Materials of Construction. Moderately low-carbon steels are among materials ordinarily used for hydrogen cyanide equipment. Certain stainless steels are employed in applications or equipment that must avoid contamination, or operate at high temperatures, or resist corrosion by other chemicals being handled with hydrogen cyanide.

Manufacture. Hydrogen cyanide is produced on a large scale by reacting ammonia and air with methane in the presence of a catalyst. Other methods by which it is made include treatment of a cyanide with dilute sulfuric acid; recovery from coke oven gas; decomposition of formamide; and synthesis from ammonia and hydrocarbons in an electrofluid reactor.

Commercial Uses. Hydrogen cyanide is used as an intermediate in producing acrylic plastics, including "Plexiglas," and in the manufacture of dyes, fumigants, rubber, cyanide salts, acrylonitrile, acrylates, adiponitrile and chelates.

Physiological Effects. Acute cyanide poisoning by hydrogen cyanide proves fatal with extreme rapidity, though breathing may sometimes continue for a few minutes. Less acute cases result in headache, dizziness, unsteadiness, and a sense of suffocation and nausea. Hydrogen cyanide can enter the system through absorption by the skin as well as by inhalation. Concentrations of from 100 ppm to 200 ppm in air can be lethal if inhaled for 30 to 60 minutes. Among persons who recover, disability rarely occurs. Hydrogen cyanide and other cyanides act as protoplasmic poisons, combining in the tissues with enzymes necessary for cellular oxidation. Suspension of tissue oxidation lasts only while the cyanide is present, with normal functions returning upon its removal if death has not occurred.

All the precautions necessary for the safe handling of any flammable gas, as well as all precautions required with a highly toxic gas, must be observed with hydrogen cyanide.

First Aid. Should a person be exposed to hydrogen cyanide, carry him to fresh air, have him lie down, and immediately call a physician and start first aid. Also, if clothing has been splashed by the liquid, remove contaminated garments but keep the patient warm.

If hydrogen cyanide gas has been inhaled, break an amyl nitrite pearl in a cloth and hold the cloth lightly under the patient's nose for 15 seconds, repeating this five times at intervals of about 15 seconds. Give artificial respiration if breathing has stopped.

If liquid hydrogen cyanide has been splashed on the skin or eyes, immediately flush the skin for at least 15 minutes. Get medical attention if cyanide has entered the eyes.

Detailed information on safety precautions and first aid with hydrogen cyanide is given in Safety Data Sheet SD-67 of the Manufacturing Chemists' Association.[1]

HYDROGEN CYANIDE CONTAINERS

Hydrogen cyanide is authorized for shipment in cylinders and tank cars. In addition, liquid hydrogen cyanide completely absorbed in inert material is authorized for shipment in wooden or fiberboard boxes which meet detailed ICC requirements.

Filling Limits. The maximum filling densities permitted for hydrogen cyanide are:

In cylinders—Not more than 0.6 lb of liquid for each 1-lb water capacity of the cylinder.

In tank cars—63 per cent (per cent water capacity by weight).

SHIPPING METHODS; REGULATIONS

Under the appropriate regulations and tariffs, hydrogen cyanide is authorized for shipment as follows (only if stabilized; unstabilized hydrogen cyanide is not accepted for shipment under ICC regulations):

By Rail: In cylinders (via freight only), and in single-unit tank cars.

By Highway: In cylinders on trucks.

By Water: Not accepted.

By Air: Not accepted. (Accepted only if in solution not exceeding 5 per cent, aboard cargo aircraft only in special containers to a maximum of 26.4 lb (12 kg) net weight per container.)

HYDROGEN CYANIDE CYLINDERS

Cylinders that meet the following ICC specifications are authorized for hydrogen cyanide service: 3A480, 3AA480 and 3A480X. In addition, these cylinders must be of not over 278-lb water capacity (nominal); valve protection caps must also be used, and be at least 3/16-in. thick, gas-tight, with 3/16-in. faced seat for gasket and with United States standard form thread. These caps must be capable of preventing injury or distortion of the valve when it is subjected to an impact caused by allowing the cylinder, prepared as if for shipment, to fall from an upright position with the side of the cap striking a solid steel object projecting not more than 6 in. above floor level. Additional requirements are that each filled cylinder must be tested for leakage before shipment and must show absolutely no leakage. This test must consist in passing over the closure of the cylinder, without the protection cap attached, a piece of Guignard's sodium picrate paper to detect any escape of hydrocyanic acid from the cylinder. Other equally efficient test methods may also be used instead of picrate paper.

Valve Outlet and Inlet Connections. Standard connection, U. S. and Canada—No. 160.

Cylinder Requalification. The 3A and 3AA cylinders authorized for hydrogen cyanide service must be requalified by hydrostatic retest every 5 years under present regulations.

SINGLE-UNIT TANK CARS

Bulk shipment of hydrogen cyanide is authorized by the ICC in single-unit tank cars meeting specifications 105A500W or 105A600W, with these further requirements: (1) the tank car must be equipped with safety valves of the type and size used on the specification 105A300W tank car and restenciled 105A300W; (2) the tank must be equipped with approved dome fittings and safety devices, and with cork insulation at least 4 in. thick; (3) the tank must be stenciled on both sides, in letters at least 2 in. high, "HYDROCYANIC ACID ONLY"; and (4) a written statement of procedure covering details of tank car appurtenances, dome fittings and safety devices, and marking, loading, handling, inspection and testing practices, must be filed with and approved by the Bureau of Explosives (63 Vesey St., New York City) before any car may be offered for transportation of hydrogen cyanide.

REFERENCE

1. "Hydrocyanic Acid" (SD-67), Manufacturing Chemists' Association, 1825 Connecticut Ave., N. W., Washington, D. C., 20009.

Hydrogen Sulfide

H_2S
Synonym: Sulfuretted hydrogen
ICC Classification: Flammable compressed gas; red label

PHYSICAL CONSTANTS

International symbol	H_2S
Molecular weight	34.08
Vapor pressure, psig	
at 70 F	252
at 80 F	285
at 105 F	425*
at 115 F	480*
at 130 F	570*
Specific gravity of gas (air = 1), at 59 F	1.189
Specific volume, at 1 atm and 70 F, cu ft/lb	11.23
Density, liquid, lb/cu ft	
at boiling point	61.96
at 70 F	48.33
at 105 F	45.34
at 115 F	44.29
at 130 F	42.99
Boiling point at 1 atm	−75.3 F
Melting point at 1 atm	−117.2 F
Triple point (H_2S liquid) (H_2S gas) (aqueous liquid)	160 F at 750 psia
Critical temperature	212.7 F
Critical pressure, psia	1306.8
Specific heat, vapor, at constant pressure, C_p, at 70 F and 1 atm, Btu/(lb)(°F)	0.2532
Specific heat, vapor, at constant volume, C_v, at 70 F and 1 atm, Btu/(lb)(°F)	0.1918
Specific heat ratio, C_p/C_v, at 70 F and 1 atm	1.32
Latent heat of vaporization at boiling point Btu/lb	234
Latent heat of fusion at melting point Btu/lb	30.0
Heat of combustion (H_2S to liquid water and SO_2 gas), Btu/lb	8340
Flammable limits, in air by volume	4.3–46%
Solubility in water, at 80 F and 1 atm, by weight	0.32%
Weight per gallon, liquid, at 70 F, lb	6.46

* Approximate values.

Properties. Hydrogen sulfide is a colorless, flammable gas with an offensive odor often referred to as "rotten eggs." It is instantly fatal if inhaled in very high concentrations, and irritating to the eyes and upper respiratory tract in low concentrations. Its burning in air forms sulfur dioxide and water. A reducing agent, hydrogen sulfide reacts readily with all the metals in the electromotive series down to and including silver. It is soluble in water, alcohol, petroleum solvents and crude petroleum.

Materials of Construction. Dry hydrogen sulfide is satisfactorily handled under pressure in steel or black iron piping. Aluminum and certain stainless steels are recommended for use with either wet or dry hydrogen sulfide. Brass valves, though tarnished by dry hydrogen sulfide, have been found to withstand years of service without appreciable malfunction.

Manufacture. Most hydrogen sulfide produced today is made as a by-product of other processes. The largest tonnage currently made originates as a by-product in the natural gas industry, which mainly converts the hydrogen sulfide derived to commercial sulfur. Most of the hydrogen sulfide sold commercially in high-pressure cylinders is a purified, liquefied by-product of the chemical industry.

Many methods have been used in the past to produce hydrogen sulfide, with the classical method for many years consisting of iron sulfide plus an acid (usually hydrochloric). It can be made by combining sulfur with hydrogen in a noncatalytic reactor at elevated pressure and temperature.

Commercial Uses. Hydrogen sulfide is used commercially to purify hydrochloric and sulfuric acid, to precipitate sulfides of metals, and to manufacture elementary sulfur. It is also employed as a reagent in analytical chemistry.

Physiological Effects. A toxic and asphyxiant gas, hydrogen sulfide in concentrations of from 20 to 150 ppm irritates the eyes, often causing severe pain and incapacitating workers for 10 days or more. Inhalation of slightly higher concentrations irritates the upper respiratory tract and, if prolonged, may result in pulmonary edema. Hydrogen sulfide acts on the nervous system if inhaled in concentrations of several hundred parts per million; inhalation of 500 ppm for 30 minutes produces headache, dizziness, excitement, staggering and disorders of the digestive tract, followed in some cases by bronchitis or bronchial pneumonia. Exposure to concentrations above 600 ppm can be fatal within 30 minutes through respiratory paralysis. As noted previously, exposure to very high concentrations is almost immediately fatal.

Relying on the strong, rotten egg-like odor of hydrogen sulfide to warn of its presence can prove very dangerous under certain conditions. High concentrations have a very sweetish odor and rapidly paralyze the sense of smell. Even low concentrations will exhaust the sense of smell, after prolonged exposure, to the point where exposed personnel may fail to detect the presence of the gas.

The maximum allowable concentration recommended by The American Standards Association for an 8-hour period is 20 ppm as given in American Standard Z37.2–1941, "Allowable Concentration of Hydrogen Sulfide."

Suggestions for Treatment. First aid measures and treatment of personnel affected by hydrogen sulfide must be conducted under the direction of competent medical authorities. A detailed description of safety precautions, protective equipment and first aid for hydrogen sulfide hazards is given in Safety Data Sheet SD-36 of the Manufacturing Chemists' Association.[1] Among suggested steps to take if personnel should be overcome by inhaling hydrogen sulfide, move the person to fresh air and cover him with a blanket to keep him warm. Recovery should be quick if he is breathing and conscious. If he is not breathing, call a physician and start artificial respiration at once. Continue artificial respiration until breathing has been restored and give the patient oxygen until completely recovered.

Leak Detection. Leaks can be investigated by coating suspected areas with soap suds and watching for bubbles, or by the use of a moist lead acetate paper.

All the precautions required for the safe handling of any flammable gas must be observed with hydrogen sulfide.

HYDROGEN SULFIDE CONTAINERS

Hydrogen sulfide is shipped as a liquefied compressed gas in cylinders, TMU tank cars, and TMU tanks on trucks.

Filling Limits. The maximum filling densities

authorized for hydrogen sulfide are (per cent water capacity by weight): in cylinders—62.5 per cent, in TMU tanks—68 per cent.

SHIPPING METHODS; REGULATIONS

Under the appropriate regulations and tariffs, hydrogen sulfide is authorized for shipment as follows:

By Rail: In cylinders (freight or express), and in TMU tank cars.

By Highway: In cylinders on trucks, and in TMU tanks on trucks.

By Water: In cylinders via cargo vessels only, and in authorized tank cars aboard trainships only. In cylinders on barges of U. S. Coast Guard classes A and C only.

By Air: Aboard cargo aircraft only, in appropriate cylinders up to 300 lb (140 kg) maximum net weight per cylinder.

HYDROGEN SULFIDE CYLINDERS

Cylinders that meet the following ICC specifications are authorized for hydrogen sulfide service: 3A480, 3AA480, 3B480, 4A480, 4B480, 4BA480, 26–480, and 3E1800.

Valve Outlet and Inlet Connections. Standard connections, U. S. and Canada—No. 330.

Cylinder Requalification. Under present regulations, cylinders of all types authorized for hydrogen sulfide service must be requalified by hydrostatic test every 5 years with the exception of type 3E (for which periodic hydrostatic retest is not required).

TMU TANKS

Hydrogen sulfide is authorized for shipment in TMU tanks meeting ICC specifications 106A800–X (with the further requirements that the tanks must not be equipped with safety relief devices of any description, and that valves must be protected by supplemental gas-tight closures approved by the Bureau of Explosives). Such tanks are authorized for shipment by tank car and by truck.

REFERENCE

1. "Hydrogen Sulfide," Safety Data Sheet SD-36, Manufacturing Chemists' Association, 1825 Connecticut Ave., N. W., Washington, D. C., 20009.

Isobutane

See **Liquefied Petroleum Gases**

Isobutene

See **Liquefied Petroleum Gases**

Krypton

See **Rare Gases of the Atmosphere**

Liquefied Petroleum Gases

BUTANE (Synonyms—Normal butane, *n*-butane, butyl hydride): C_4H_{10}

BUTYLENES (Synonym—butenes): C_4H_8

1-BUTENE (Synonyms—Ethylethylene, alpha-butene): $CH_2{:}CHCH_2CH_3$

CIS-2-BUTENE (Synonyms—Dimethylethylene, beta-butylene, "high-boiling" butene-2): $CH_3CH{:}CHCH_3$

TRANS-2-BUTENE (Synonyms—Dimethylethylene, beta-butylene, "low-boiling" butene-2): $CH_3CH{:}CHCH_3$

ISOBUTENE (Synonyms—2-methylpropene, isobutylene): $(CH_3)_2C{:}CH_2$

ISOBUTANE (Synonyms—2-methylpropane, trimethylmethane): C_4H_{10} [or $(CH_3)_2CHCH_3$]

PROPANE (Synonym—Dimethylmethane): C_3H_8

PROPYLENE (Synonym—Propene): C_3H_6 (or $CH_3CH{:}CH_2$)

ICC Classification: Flammable compressed gas; red label

PHYSICAL CONSTANTS
(for Research Grade Products)

(See following table for Butylenes)

	Butane	Iso-butane	Propane	Propylene
International Symbol	LP-Gas or LPG		LP-Gas or LPG	Pry
Molecular Weight	58.120	58.120	44.094	42.078
Density of liquid at saturation pressure (apparent value from weight in air of the air-saturated liquid), lb/cu ft				
at 60 F	36.39	35.05	31.57	32.49
at 70 F	35.95	34.82	31.2	32.07
at 105 F	34.38	33.26	29.33	29.89
at 115 F	34.01	32.82	28.70	29.20
at 130 F	33.38	32.01	27.77	28.08
Vapor pressure at 100 F, psia	51.6	72.2	190	226.4
Boiling point at 1 atm	31.10 F	10.89 F	−43.73 F	−53.86 F
Freezing point at 1 atm	−217.03 F	−255.28 F	—	—
Freezing point at saturation pressure (triple point)	—	—	−305.84 F	−301.45 F
Critical temperature	305.62 F	274.96 F	206.26 F	197.4 F
Critical pressure, psia	550.7	529.1	617.4	667
Critical volume, cu ft/lb	0.0702	0.0724	0.0728	0.0689
Specific volume of the real gas at 60 F and 1 atm, cu ft/lb	6.3120	6.3355	8.4515	8.8736

PHYSICAL CONSTANTS—*continued*

	Butane	Iso-butane	Propane	Propylene
Specific gravity of gas at 60 F and 1 atm (air = 1)	2.0757	2.06805	1.5503	1.4765
Specific gravity of liquid, 60 F/60 F, at saturation pressure (absolute value from weights in vacuum for the air-saturated liquid)	0.5844	0.5631	0.5077	0.5220
Specific heat of ideal gas at 60 F, C_p, Btu/(lb)(°F)	0.3908	0.3872	0.3885	0.3541
Specific heat of ideal gas at 60 F, C_v, Btu/(lb)(°F)	0.3566	0.3530	0.3434	0.3069
Ratio of specific heats, C_p/C_v, ideal gas at 60 F	1.096	1.097	1.131	1.154
Specific heat of liquid at 1 atm	0.5636	0.5695	0.5920	0.585
Latent heat of vaporization at boiling point and 1 atm, Btu/lb	165.65	157.51	183.05	188.18
Latent heat of fusion, Btu/lb	—	—	—	0.0673
Gross heat of combustion gas at 60 F and 1 atm, Btu/cu ft of the real gas	3374.4	3352.15	2563.3	2371.7
Liquid at 77 F and saturation pressure, Btu/lb	21122.	21072.	21484.	20943.*
Liquid at 77 F and saturation pressure, Btu/gal	100984.	96889.	88342.	90891.*
Net heat of combustion Gas at 60 F and 1 atm, Btu/cu ft of the real gas	3114.2	3092.9	2358.3	2218.3
Liquid at 77 F and saturation pressure, Btu/lb	19494.	19444.	19768.	19578.*
Liquid at 77 F and saturation pressure, Btu/gal	93201.	89404.	81286.	84971.*
Air required for combustion, cu ft of air per cu ft of the real gas, at 60 F and 1 atm	32.089	31.970	24.300	21.827
Air required for combustion, lb of air per lb	15.459	15.459	15.674	14.783
Flammable limits in air, % by volume				
Lower	1.8	1.8	2.1	2.0
Upper	8.4	8.4	9.5	10
Flash point	−101 F	−117 F	−156 F	−162 F

* At 60 F and 1 atm.

Physical Constants—Butylenes (for Research Grade Products)

	1-Butene	*cis*-2-Butene	*trans*-2-Butene	Isobutene
International symbol	—	—	—	—
Molecular weight	56.104	56.104	56.104	56.104
Vapor pressure at 100 F, psia	63.05	45.54	49.80	63.40
Boiling point at 1 atm	20.73 F	38.70 F	33.58 F	19.58 F
Freezing point at 1 atm	—	−218.04 F	−157.99 F	−220.63 F
Freezing point at saturation pressure (triple point)	−301.63 F	—	—	—
Critical temperature	295.6 F	311 F	311 F	292.51 F
Critical pressure, psia	583	600	600	579.8
Critical volume, cu ft/lb	0.0689	0.0503	0.0503	0.0513
Specific volume of the real gas at 60 F and 1 atm, cu ft/lb	6.5571	6.561	6.561	6.561
Specific gravity of the real gas at 60 F and 1 atm (air = 1)	1.9982	1.997	1.997	1.997
Density of liquid at saturation pressure (apparent value from weight in air of the air-saturated liquid), lb/cu ft				
at 60 F	37.43	39.04	37.97	37.37
at 70 F	36.82	38.50	37.47	36.90
at 105 F	35.26	37.01	35.89	35.25
at 115 F	34.69	36.55	35.43	34.76
at 130 F	33.82	35.86	34.71	34.01
Specific gravity of liquid, 60 F/60 F, at saturation pressure (absolute value from weights in vacuum for the air-saturated liquid)	0.6013	0.6271	0.6100	0.6004
Specific heat of ideal gas at 60 F, C_p, Btu/(lb)(°F)	0.3548	0.3269	0.3654	0.3701
Specific heat of ideal gas at 60 F, C_v, Btu/(lb)(°F)	0.3192	0.2915	0.3300	0.3347
Ratio of specific heats, C_p/C_v, ideal gas at 60 F	1.112	1.121	1.107	1.106
Specific heat of liquid at 1 atm	0.535	0.5271	0.5351	0.549
Latent heat of vaporization at boiling point and 1 atm, Btu/lb	167.94	178.91	174.39	169.48
Latent heat of fusion, Btu/lb	0.06501	0.12347	0.16483	0.10020
Gross heat of combustion				
Gas at 60 F and 1 atm, Btu/cu ft of the real gas	3177.9	3168.	3163.	3156.
Liquid at 77 F and saturation pressure, Btu/lb	20727.	20655.	20633.	20618.

Physical Constants: Butylenes—continued

	1-Butene	*cis*-2-Butene	*trans*-2-Butene	Isobutene
Gross heat of combustion				
Liquid at 77 F and saturation pressure, Btu/gal	103658.	107633.	104711.	102964.
Net heat of combustion				
Gas at 60 F and 1 atm, Btu/cu ft of the real gas	2970.3	2960.5	2957.	2949.
Liquid at 77 F and saturation pressure, Btu/lb	19364.	19291.	19267.	19254.
Liquid at 77 F and saturation pressure, Btu/gal	96838.	100528.	97781.	96154.
Air required for combustion, cu ft of air per cu ft of the real gas, at 60 F and 1 atm	29.538	29.52	29.52	29.52
Air required for combustion, lb of air per lb	14.78	14.78	14.78	14.78
Flammable limits in air, % by volume				
Lower (est. value)	1.6	(1.6)	(1.6)	(1.6)
Upper	9.3	—	—	—
Flash point	−112 F	−100 F	−100 F	−105 F

Properties. The liquefied petroleum gases are butane, isobutane, propane, propylene (propene), butylenes (butenes) and any mixtures of these hydrocarbons, in the generally accepted definition of the National Fire Protection Association. They are flammable, colorless, noncorrosive and nontoxic. Easily liquefied under pressure at atmospheric temperature, they are shipped and stored compactly as liquids and are used in gaseous form in their very large and diverse applications as fuels. Chemically, propane, isobutane and butane are among the lightest hydrocarbons in the paraffin series, occurring between ethane (natural gas) and pentane (the lightest natural gasoline fraction). Propylene, isobutene, 1-butene and 2-butene are among the lightest hydrocarbons in the monoolefin series, occurring between ethene and pentene. (The ending "-ane" indicates a member of the paraffin series, while "-ene" indicates a member of the monoolefin series.) The liquefied petroleum gases in the paraffin series are chemically inert and odorless, and the ICC and other regulating bodies require artificial odorization of propane and butane (except in technical uses where the odorant would harm further processing and the odorant warning action would not be important). Propylene and the butylenes have an unpleasant odor characteristic of petroleum refineries or coal gas.

The liquefied petroleum gases are supplied in various scientific and commercial grades. Their properties differ according to the grade being used. Properties are given for the research grades of these gases in the preceding tables of "Physical Constants," as noted in the tables. (Properties appearing in the tables, like those presented for gases throughout this second part of the *Handbook*, are drawn from authoritative scientific and industry sources, as stated in the introduction to Part II; in the case of these liquefied petroleum gases, large parts of the properties data were based on research studies of the American Petroleum Institute. The source of the liquid density properties data, here and in the sections on Butadiene, Ethane, and Methane, is the Phillips Petroleum Company Engineering Standards Data Book and Data Program 4027, selected best values.)

Properties of the various commercial grades of the liquefied petroleum gases accordingly differ to a greater or lesser extent from the properties for research grades that are shown in the "Physical Constants" tables. As an example, one producer gives average vapor pressure properties for commercial grades of propane and butane as follows:

	Commercial Propane	Commercial Butane
Vapor pressure, psig		
at 70 F	124	31
at 100 F	192	59
at 105 F	206	65
at 130 F	286	97

Moreover, these gases are often used in mixtures designed to have certain desired properties; in particular, propane and butane are frequently ordered as mixtures to meet certain boiling-point and other requirements of individual applications. Suppliers furnish physical constants data for the various grades and mixtures they make available.

All the liquefied petroleum gases are soluble to varying degrees in alcohol and ether. Propane and propylene are slightly soluble in water.

Materials of Construction. Any common, commercially available metals may be used with the liquefied petroleum gases because they are noncorrosive, though installations must of course be designed to withstand the pressures involved and must comply with all state and local regulations. Widely accepted recommendations on storage systems and safe usage are given in "Standard for the Storage and Handling of Liquefied Petroleum Gases," Pamphlet No. 58 of the National Fire Protection Association (which is also American Standard Z106.1).[1] Similar recommendations for larger storage systems appear in NFPA Pamphlet No. 59.[2]

Manufacture. Butane and propane (with other hydrocarbons in the paraffin series) are recovered from "wet" natural gas, from natural gas associated with or dissolved in crude oil, and from petroleum refinery gases. They may be separated from wet natural gas or crude oil through absorption in light "mineral seal" oil, adsorption on surfaces such as activated charcoal, or by refrigeration, followed in each case by fractionation. Propylene and other gases in the monoolefin series are recovered from petroleum gases by fractionation.

Commercial Uses. Propane and butane—known most extensively in commercial and popular terms as LP-Gas or LPG—have an extremely wide range of domestic, industrial, commercial, agricultural and internal combustion engine uses. It is estimated that the two gases, unmixed and in mixtures, have several thousand industrial applications and scores more in other fields. Their very broad application stems from their occurrence as hydrocarbons between natural gas and natural gasoline and their corresponding properties. Some of their major uses include:

Domestic, Industrial and Commercial Heating Uses. As appliance fuel for space heating, water heating, boiler heating, cooking, baking, air conditioning and refrigeration in rural or even urban areas beyond the reach of gas mains.

Public Utility and Industrial Uses. In bulk by utilities and industries (especially industries using kilns or furnaces which must be maintained continuously at given temperatures), as standby fuel to protect against failure or interruption of natural or artificial gas supply. By utilities to bridge peak load demands for natural or artificial gas, and for gas enrichment. By utilities serving rural communities from central plants.

Construction Industry Use. For space heating during the erection of buildings.

Industrial Processing Uses. As fuel for the entire range of industrial heating processes, especially those where the Btu value must be accurately controlled. Industrial heating process uses include heat treating, stress relieving, annealing, enamel baking and firing ceramic kilns and furnaces. For welding, brazing, metal cutting and soldering, and also as an atmosphere-producing gas for bright annealing.

Agricultural Uses. As fuel for such operations as poultry brooding, cotton and grain drying, tobacco curing, crop dehydration, weed burning and orchard heating.

Internal Combustion Engine Uses. As fuel for vehicles such as trucks, buses, taxicabs, and fork lift trucks, and for mobile farm machinery like tractors and harvesters. As fuel for stationary engines powering well pumps, electric generators, etc. Engines especially designed for LP-Gas are available, and gasoline engines may readily be converted for LP-Gas operation.

Uses of Isobutane and the Monoolefins. Isobutane and the gases in the monoolefin series are used less extensively than propane and butane. Isobutane and isobutene are used to improve the octane rating of gasoline, particularly gasoline of aviation grade, and they are also employed in manufacturing synthetic rubber. Propylene is used substantially in the chemical industry for synthesis in the production of a wide range of products. The butenes are used in preparing a large number of organic compounds and in manufacturing alkylate for increasing gasoline octane ratings.

Physiological Effects. The liquefied petroleum gases are nontoxic. Prolonged inhalation of high concentrations has an anesthetic effect; also, due to their ability to displace oxygen in the air, they can act as simple asphyxiants. Contact between the skin and these gases in liquid form can cause freezing of tissue and results in injury similar to a thermal burn. All precautions necessary for the safe handling of any flammable compressed gas must be observed with the liquefied petroleum gases.

Leak Detection. Never use flame to test for an LP-Gas leak; instead, apply soap water solution to areas suspected of leaking. Frost around valve stems, at piping joints or at other points indicates a liquid leak.

LIQUEFIED PETROLEUM GAS CONTAINERS

Propane and butane are authorized for shipment as liquefied compressed gases in cylinders, portable tanks and cargo tanks, and in insulated or uninsulated single-unit tank cars. ICC regulations also provide for their shipment in TMU (ton multi-unit) tanks and tank cars, but they are not generally shipped in TMU containers. The two gases are stored in large tanks above ground and below ground, and also in very large underground chambers usually developed in natural salt deposits.

In addition to such storage in liquefied form under their vapor pressures at normal atmospheric temperatures, refrigerated storage in liquefied form under atmospheric pressures is used for propane and butane. Refrigerated storage systems are closed and insulated, and in them the LP-Gas vapor is circulated through pumps and compressors to serve as the systems' refrigerant. Propane and butane are stored in pits in the earth capped by metal domes as well as in underground chambers; one of the largest storage pits of this kind, located in Utah, holds propane in the millions of gallons in refrigerated liquid form.

Isobutane and the monoolefins are authorized for shipment in single-unit tank cars and truck cargo tanks, and are usually shipped in bulk units because they are generally used in large quantities. They are also authorized for shipment in cylinders, portable tanks and TMU tanks.

Filling Limits. The maximum filling densities authorized by the ICC for liquefied petroleum gases are prescribed according to the specific gravity of the liquid material at

60 F in detailed tables that are part of the ICC regulations. Producers and suppliers who charge LP-Gas containers should consult these tables in the current regulations for the maximum densities to which to fill containers with the particular grades and mixtures they are handling. The lower and upper limits of the maximum filling densities authorized in the present regulations are as follows (per cent water capacity by weight):

In cylinders—from 26 per cent for 0.271–0.289 specific gravity, to 57 per cent for 0.627 to 0.634 specific gravity.

In single-unit tank cars and TMU tanks—from 45.500 per cent (insulated tanks, April through October) for 0.500 specific gravity, to 61.57 per cent (uninsulated tanks, November through March with no storage in transit) for 0.635 specific gravity. (In addition, filling must not exceed various specified limits of pressure and liquid content at temperatures of 105 F, 115 F or 130 F as given in ICC regulations.)

In cargo tanks and portable tanks—from 38 per cent (tanks of 1200-gal capacity or less) for 0.473 to 0.480 specific gravity, to 60 per cent (tanks of over 1200-gal capacity) for 0.627 and over specific gravity (except when using fixed length dip tube or other fixed maximum liquid level indicators). Moreover, the tank must not be liquid full at 105 F if insulated nor at 115 F if uninsulated, and the gage vapor pressure at 115 F must not exceed the tank's design pressure.

SHIPPING METHODS; REGULATIONS

Under the appropriate regulations and tariffs, liquefied petroleum gases are authorized for shipment as follows:

By Rail: In cylinders (via freight, express and as baggage), in insulated and uninsulated single-unit tank cars, and in TMU tank cars.

By Highway: In cylinders on trucks, in portable tanks, and in cargo tanks on trucks or on semi and full trailers.

By Water: In cylinders on passenger vessels and on ferry or railroad car ferry vessels (passenger or vehicle); in cylinders and portable tanks (meeting ICC-51 specifications and not over 20,000 lb gross weight, with vapor pressure at 115 F not exceeding the container service pressure) on cargo vessels; in authorized tank cars on trainships only, and in authorized tank trucks on trailerships and trainships only. In cylinders on barges of U. S. Coast Guard classes A, CA and CB only. In cargo tanks aboard tankships and tank barges (to maximum filling densities by specific gravity as prescribed in Coast Guard regulations).

By Air: In cylinders on cargo aircraft only up to 300 lb (140 kg) maximum net weight per cylinder.

By Pipeline: Propane and butane are also transported by pipeline from points of production to distant bulk storage facilities.

LIQUEFIED PETROLEUM GAS CYLINDERS

Cylinders authorized for liquefied petroleum gas service include those which comply with the following ICC specifications: 3A, 3AA, 3B, 3E, 4A, 4B, 4BA, 4B240ET, 4B240FLW, 4BW, 4E, 4, 9 and 41. (Cylinders meeting the following ICC specifications may be continued in use, but new construction is not authorized: 3, 4B240X, 25, 26 and 38. ICC regulations also authorize shipment in several special types of small containers.)

Valve Outlet and Inlet Connection. Standard connection, U. S. and Canada (for butane, isobutane, propane, and propylene or propene)—No. 510. Alternate standard connection, U. S. and Canada (for butane and propane)—No. 300. Alternate standard connection, U. S. and Canada (for propane)—No. 350 (obsolete effective 1/1/72).

Cylinder Requalification. All types of cylinders listed above must be requalified by periodic hydrostatic retest every 5 years under present ICC regulations except as follows:

(1) no periodic retest is required for 3E cylinders;

(2) 10 years is the required retest interval for type 4 cylinders; and

(3) external visual inspection may be used in lieu of hydrostatic retest for cylinders that are used exclusively for liquefied petroleum gas which is commercially free from corroding components and that are of the following types (including cylinders of these types with higher service pressures): 3A480, 3AA480, 3A480X, 3B, 4B, 4BA, 26–240 and 260–300.

SINGLE-UNIT TANK CARS AND TMU TANK CARS

Bulk shipment of liquefied petroleum gas is authorized for single-unit tank cars meeting ICC specifications of the 105A series, 105A tanks of given service pressures being specified for gases having vapor pressures of stipulated maximum values at 105 F or 115 F; single-unit tank cars in the 112A series and of the 114A340W type are also authorized under specific conditions. TMU tank car shipment is also authorized (though little used, as noted before) for LP-Gas with pressure not exceeding 375 psi at 130 F.

CARGO TANKS AND PORTABLE TANKS

Liquefied petroleum gas is authorized for shipment in truck or truck-trailer cargo tanks that comply with ICC specifications MC-330 or MC-331, and for shipment in portable tanks meeting specifications ICC-51. Various design pressures may be used for these tanks, but the gage vapor pressure at 115 F of a shipment must not exceed the tank's design pressure.

STORAGE CONTAINERS AND FACILITIES

Steel tanks and large underground chambers are used for the storage of liquefied petroleum gas. Steel tanks above ground or below ground range up to 70,000 gal or more in capacity, while below-ground caverns or pits have been found to offer safe and economical storage where a suitable geological formation exists at a site of large-volume handling, such as at pipeline or marine terminals and at gasoline plants or refineries. Widely recognized recommendations for LP-Gas installations are presented in the previously cited NFPA Pamphlets No. 58[1] and No. 59.[2]

REFERENCES

1. "Standard for the Storage and Handling of Liquefied Petroleum Gases" (Pamphlet No. 58), National Fire Protection Association, 60 Batterymarch St., Boston, Mass., 02110.
2. "Standard for the Storage and Handling of Liquefied Petroleum Gases at Utility Gas Plants" (Pamphlet No. 59), National Fire Protection Association.

Methane

CH_4

Synonyms: Marsh gas, methyl hydride

ICC Classification: Flammable compressed gas; red label

PHYSICAL CONSTANTS
(for Research Grade product)

International symbol	CH_4
Molecular weight	16.042
Boiling point at 1 atm	−258.68 F
Freezing point at saturation pressure (triple point)	−296.46 F
Critical temperature	−115.78 F
Critical pressure, psia	673.1
Critical volume, cu ft/lb	0.0991
Specific volume of gas at 60 F and 1 atm, cu ft/lb	23.6113
Specific gravity of gas at 60 F and 1 atm (air = 1)	0.55491
Specific heat of ideal gas at 60 F, C_p, Btu/(lb)(°F)	0.5271
Specific heat of ideal gas at 60 F, C_v, Btu/(lb)(°F)	0.4032
Ratio of specific heats, C_p/C_v, ideal gas at 60 F	1.307
Latent heat of vaporization at boiling point and 1 atm, Btu/lb	219.22
Latent heat of fusion, Btu/lb	0.05562
Gross heat of combustion at 60 F and 1 atm, Btu/cu ft of the real gas	1011.6
Net heat of combustion at 60 F and 1 atm, Btu/cu ft of the real gas	910.77
Air required for combustion at 60 F and 1 atm	
cu ft of air per cu ft of the real gas	9.563
lb of air per lb	17.233
Flammable limits in air, by volume	5.0–15%
Flash point	−306 F

Properties. Methane is a colorless, odorless, tasteless flammable gas. It is the major component of natural gas and one source of its commercial grade is actually Tennessee natural gas taken directly from the pipeline (which usually consists of at least 93 per cent methane). It is also a major constituent of coal gas and is present to an extent in air in coal mines. Chemically, methane is the first member of the paraffin (aliphatic or saturated) series of hydrocarbons. It is soluble in alcohol or ether, and slightly soluble in water. Methane is shipped as a nonliquefied compressed gas in cylinders at pressures of around 2000 psig at 70 F.

Materials of Construction. Methane is noncorrosive, and may be contained by any common, commercially available metals. Handling equipment must, however, be designed to withstand safely the pressures involved.

Manufacture. Methane is produced commercially from natural gas by absorption or adsorption methods of purification; supercooling and distillation are sometimes employed to secure methane of very high purity. Some California natural gas wells produce methane of high purity. It can also be obtained by cracking petroleum fractions.

Commercial Uses. Methane in natural gas of course serves very widely as a fuel. In the chemical industry, it is used heavily as a raw material for making important products that include acetylene, ammonia, ethanol and methanol; its chlorination also yields carbon tetrachloride, chloroform, methyl chloride and methylene chloride. The burning of high-purity methane is used to make carbon black of special quality for electronic devices.

Physiological Effects. Methane is generally considered nontoxic. Coal miners inhale concentrations of up to 9 per cent methane in air without apparent ill effects; inhalation of higher concentrations eventually causes a feeling of pressure on the forehead and eyes, but the sensation ends after returning to fresh air.

All the precautions necessary for the safe handling of any flammable compressed gas must be observed in working with methane.

METHANE CONTAINERS

Methane may be shipped in any cylinders of the types authorized for nonliquefied compressed gases. Bulk industrial users of methane usually receive it in natural gas by pipeline, and purify it if necessary for further processing. Transatlantic shipment of methane has been made in ships carrying it liquefied in insulated tanks at a temperature of some −260 F.

Filling Limits. The maximum filling densities permitted for methane in cylinders at 70 F are the authorized service pressures of the cylinders into which it is charged.

SHIPPING METHODS; REGULATIONS

Under the appropriate regulations and tariffs, methane is authorized for shipment as follows:

By Rail: In cylinders (via freight or express).

By Highway: In cylinders on trucks.

By Water: In cylinders aboard cargo vessels only. In cylinders on barges of U. S. Coast Guard classes A, CA and CB only. Transatlantic tankships have also transported methane liquefied at about −260 F, as noted above.

By Air: In cylinders aboard cargo aircraft only up to 300 lb (140 kg) maximum net weight per cylinder.

By Pipeline: As previously stated, pipeline transmission of methane in natural gas represents its chief means of transport in bulk.

METHANE CYLINDERS

Any cylinders authorized for the shipment of a nonliquefied compressed gas may be used in methane service under ICC regulations, but cylinders of the 3A and 3AA types are probably those most commonly used for methane. Authorized cylinders for methane service include those meeting ICC specifications 3A, 3AA, 3B, 3C, 3D, 3E, 4, 4A, 4BA and 4C; cylinders meeting specifications 3, 7, 25, 26, 33 and 38 may also be continued in methane service, but new construction is not authorized.

Valve Outlet and Inlet Connections. Standard connections, U. S. and Canada—No. 350.

Cylinder Requalification. All types of cylinders authorized for methane shipment must be requalified by periodic hydrostatic retest every 5 years under present regulations, with the following exceptions: ICC-4 cylinders, every 10 years; and no periodic retest is required for cylinders of types 3C, 3E, 4C and 7.

Methylamines (Anhydrous)

MONOMETHYLAMINE (Synonyms—Methylamine, Aminomethane): CH_3NH_2
DIMETHYLAMINE: $(CH_3)_2NH$
TRIMETHYLAMINE: $(CH_3)_3N$
ICC Classification: Flammable compressed gas; red label

PHYSICAL CONSTANTS

	Methylamines		
	Mono.	Di.	Tri.
International symbol	—	—	—
Molecular weight	31.06	45.08	59.11
Vapor pressure, psig			
at 68 F	28.8	11	13
at 105 F	67*	35*	35*
at 115 F	82*	43*	43*
at 130 F	109*	59.8	59*
at 140 F	130	68	65
Vapor density at 32 F and 1 atm, lb/cu ft	0.081	0.125	0.144
at 70 F	41.4	40.8	39.4
at 105 F	40.1	39.4	38.1
at 115 F	39.8	39.0	37.7
at 130 F	39.2	38.5	37.1
Boiling point at 1 atm	20.6 F	44.4 F	37.2 F
.Freezing point at 1 atm	−136.3 F	−134.0 F	−178.8 F
Critical temperature	314.4 F	328.3 F	320.2 F
Critical pressure, psia	1081.9	760.0	590.9
Latent heat of vaporization at boiling point, Btu/lb	357.5	252.8	166.9
Latent heat of fusion at freezing point, Btu/lb	84.9	56.7	47.6
Ignition temperature in air	806 F	756 F	374 F
Upper flammable limit (volume, % in air)	20.8	14.4	11.6
Lower flammable limit (volume, % in air)	5	2.8	2.0
Weight per gal, liquid, at 68 F, lb	5.55	5.48	5.30

* Approximate values.

Properties. The methylamines (monomethylamine, dimethylamine and trimethylamine) are colorless, flammable and toxic gases at room temperatures and pressures in their anhydrous form. They have a distinct and disagreeable fishy odor in concentrations up to 100 ppm. In higher concentrations they have an odor like ammonia, which they resemble and from which they are derived. They are easily liquefied, and are shipped in their anhydrous form as liquefied compressed gases. They are highly soluble in water and in alcohol, ether and various other organic solvents.

Vapors of the methylamines in air can burn within certain concentration ranges. Gaseous methylamines and their solutions are alkaline materials and in sufficient concentrations can irritate and burn the skin, eyes and respiratory system.

Chemically, the methylamines are slightly stronger than ammonia as bases. They hydrate in water solutions and neutralize acids to form methylammonium salts. They do not corrode iron and steel, but do attack copper and its alloys, zinc, and aluminum. The methylamines can form explosive compounds with mercury, and must never be brought into contact with mercury.

Methylamines are used and shipped both in the form of anhydrous gases and in aqueous solutions. Only the anhydrous form is treated here.

Materials of Construction. Iron, steel, stainless steels and Monel have proven satisfactory in methylamines service. Some plastics and elastomers also withstand their action. Copper, copper alloys (including brass and bronze), zinc (together with zinc alloys and galvanized surfaces), and aluminum are corroded by the methylamines and should not be used in direct contact with them. Mercury and the methylamines can explode on contact, and instruments containing mercury must never be used with the methylamines. Among gasket and packing materials satisfactory for use with them are compressed asbestos, polyethylene, Teflon and carbon steel or stainless steel wound asbestos.

Manufacture. The methylamines are produced by having methyl alcohol and ammonia interact over a catalyst at high temperature.

Commercial Uses. As sources of reactive organic nitrogen, the methylamines serve as intermediates in synthesizing pharmaceuticals, agricultural chemicals, dyes, rubber chemicals and explosives. Derivatives serve in agriculture as fungicides, insecticides and feed supplements. Derivatives have also been employed in producing antihistamines, tranquillizers and other drugs, in making dyestuffs, explosives and rocket fuel, and in curing resins. Rubber industry applications of derivatives include use as accelerators, vulcanizing agents and chain terminators in synthetic rubber production. Derivatives are also solvents for various organic plastics, resins, gums, dyes and pharmaceuticals.

Physiological Effects. The methylamines are toxic, and contact with them must be avoided. They are irritating to the nose, throat and eyes in low concentrations, and require suitable gas masks and eye protective devices for safe handling by exposed personnel. Severe exposure of the eyes may lead to loss of sight. Dermatitis results from contact of the methylamines with the skin. Inhalation of sufficiently high concentrations is followed by violent sneezing, a burning sensation of the throat with constriction of the larynx, and difficulty in breathing with congestion of the chest and inflammation of the eyes.

Detailed information on first aid and medical treatment for persons injured by the methylamines is given in Safety Data Sheet SD-57 of the Manufacturing Chemists' Association.[1]

CONTAINERS FOR ANHYDROUS METHYLAMINES

The anhydrous methylamines are shipped as compressed liquefied gases under their own vapor pressures in cylinders, tank cars and cargo tank trucks. They are also authorized by the ICC for shipment in TMU (ton multi-unit) tank cars and portable tanks.

Filling Limits. The maximum filling densities allowed under ICC regulations for cylinders, TMU tank car tanks (ICC 106A type) and truck cargo tanks and portable tanks are (per cent water capacity by weight): monomethylamine, 60 per cent; dimethylamine, 59 per cent; and trimethylamine, 57 per cent. Corresponding maximum filling densities for single-unit tank cars ICC 105A300-W with properly fitted loading and unloading valves) are: monomethylamine, 62 per cent; dimethylamine, 62 per cent; and trimethylamine, 59 per cent.

SHIPPING METHODS; REGULATIONS

Under the appropriate regulations and tarifts, the anhydrous methylamines are authorized for shipment as follows:

By Rail: In cylinders (via freight or express), and by insulated single-unit tank cars and TMU tank cars.

By Highway: In cylinders on trucks, in tank trucks, and in portable tanks on trucks.

By Water: In cylinders and portable tanks (not over 20,000 lb gross weight) via cargo vessels, and in railroad tank cars complying with ICC provisions on trainships only. Dimethylamine and trimethylamine are not permitted on passenger vessels, ferry vessels and railroad car ferry vessels. Monomethylamine may be shipped in cylinders on passenger vessels, ferries and railroad car ferries (including passenger or vehicle ferry vessels). The methylamines are also authorized for shipment in cylinders on barges of U. S. Coast Guard classes A and C only, and in cargo tanks aboard tankships and tank barges (with maximum filling densities by specific gravity as prescribed in Coast Guard regulations).

By Air: Aboard cargo aircraft only in appropriate cylinders up to 300 lb (140 kg) maximum net weight per cylinder.

CYLINDERS FOR ANHYDROUS METHYLAMINES

Cylinders meeting the following ICC specifications are authorized for methylamines service: 3A150, 3AA150, 3B150, 4B150, 4BA225, and 3E1800. Safety relief devices are not required on cylinders charged with the methylamines.

Valve Outlet and Inlet Connections: Standard connection, U. S. and Canada—No. 240.

Cylinder Requalification: 3A, 3AA and 3B cylinders used in methylamines service must be requalified by hydrostatic retest every 5 years under present regulations. 4B and 4BA cylinders used exclusively for anhydrous methylamines must similarly be requalified every 12 years. Periodic hydrostatic retest is not required for 3E cylinders.

CARGO TANKS AND PORTABLE TANKS

Anhydrous methylamines are authorized for shipment in motor vehicle cargo tanks complying with ICC specification MC-330 and MC-331, and in steel portable tanks built to ICC-51 specification. The minimum design pressure required for cargo and portable tanks in methylamines service is 150 psig.

SINGLE-UNIT TANK CARS AND TMU TANK CARS

Bulk quantities of the anhydrous methylamines are commonly shipped by rail in tank cars of ICC specification 105A300W (an insulated single-unit tank car with properly fitted loading and unloading valves). Authorized rail shipment of the methylamines may also be made in TMU tank cars of ICC specification 106A500X.

STORAGE AND HANDLING EQUIPMENT

Storage tanks for the methylamines should be made of steel and designed to comply with the Unfired Pressure Vessel Code of the ASME as well as with all state and local regulations. Safe working pressures vary with the vapor pressure-temperature relationship of the particular methylamine being stored, and with the high temperature ranges at the plant location. Important parts of well-designed tanks include proper dual pressure relief valves, a vapor absorbing system, liquid level and pressure gages, liquid and vapor transfer valves and an adequate electrical ground. Pipes, fittings, pumps, gages and other equipment should be of steel, iron or other material not subject to corrosion by the methylamines. It is best to have storage and handling installations designed with the help of engineers thoroughly familiar with the gases.

REFERENCE

1. *Methylamines*, Safety Data Sheet SD-57, Manufacturing Chemists' Association, 1825 Connecticut Ave., N. W., Washington, D. C., 20009.

Methyl Chloride

CH_3Cl
Synonym: Chloromethane
ICC Classification: Flammable compressed gas; red label

PHYSICAL CONSTANTS

International symbol	CH_3Cl
Molecular weight	50.491
Vapor pressure at 70 F, psig	58.71
Specific gravity of gas at 32 F and 1 atm (air = 1)	1.74
Density of liquid, lb/cu ft	
at 70 F	57.3
at 105 F	55.6
at 115 F	53.4
at 130 F	53.0
Specific gravity of liquid	
at 70 F	0.919
at −11.11 F	1.000
Boiling point at 1 atm	−10.8 F
Freezing point at 1 atm	−143.7 F
Critical temperature	289.4 F
Critical pressure, psia	968.7
Latent heat of vaporization at boiling point, Btu/lb	184.3
Specific heat of gas, C_p, at 77 F and 1 atm, Btu/(lb)(°F)	0.199
Specific heat of gas, C_v, at 77 F and 1 atm, Btu/(lb)(°F)	0.155
Ratio of specific heats, C_p/C_v, at 77 F and 1 atm	1.28
Specific heat of liquid, average 5 F to 86 F, Btu/(lb)(°F)	0.376
Flammable limits in air, by volume	8.1–17.2%
Autoignition temperature	1170 F
Solubility of gas in water at 1 atm, vol/1 vol of water	
at 32 F	3.4
at 68 F	2.2
at 86 F	1.7
at 104 F	1.3
Weight per gallon, liquid, at 70 F, lb	7.68

Properties. Methyl chloride is a colorless, flammable gas with a faintly sweet odor at room temperatures, and has anesthetic, narcotic and sickening effects when inhaled in sufficiently high concentrations. It is shipped as a transparent liquid under its vapor pressure of about 59 psig at 70 F.

Methyl chloride burns feebly in air but forms

mixtures with air that can be explosive within its flammability range.

Dry methyl chloride is highly stable at normal temperatures and in contact with air, but decomposes at temperatures above 700 F into end products which may be toxic (with traces of hydrochloric acid, phosgene, chlorine and carbon monoxide). Methyl chloride hydrolyzes slowly in the presence of moisture with the formation of corrosive hydrochloric acid. It is slightly soluble in water and highly soluble in alcohol, mineral oils, chloroform, and most organic liquids.

Materials of Construction. Dry methyl chloride may be contained in such common metals as steel, iron, copper and bronze, but it has a corrosive action on zinc, aluminum, die castings, and, it is thought, magnesium alloys. Methyl chloride must not be used with aluminum, for it forms spontaneously flammable methyl aluminum compounds upon contact with that metal. No reaction occurs, however, with the drying agent, activated alumina.

Gaskets made of natural rubber and many neoprene compositions should be avoided because methyl chloride dissolves many organic materials. Pressed-fiber gaskets, including those made of asbestos, may be used with methyl chloride. Polyvinyl alcohol is unaffected by methyl chloride and its use is also recommended. Medium-soft metal gaskets may be used for applications where alternating stresses like those resulting from large temperature changes do not lead to "ironing out" and consequent leakage.

Manufacture. Methyl chloride is made commercially in the U. S. mainly by two methods: the reaction of hydrogen chloride gas or hydrochloric acid with methyl alcohol (in the presence of a catalyst to accelerate the reaction); and the chlorination of methane.

In the first process, the products are gaseous methyl chloride with unreacted hydrogen chloride and methyl alcohol and several by-products. These are removed in a series of chemical purification steps and the methyl chloride gas is compressed and dried. A small amount of air remaining in the condensate is distilled off before the liquid is charged into the shipping container.

Natural gas is the source of the methane used in the second process, in which the methane is removed by fractional distillation and reacted with chlorine. Undesired reaction products (including methylene chloride, chloroform, carbon tetrachloride, hydrochloric acid, and some chlorinated hydrocarbons) are similarly removed from the methyl chloride in subsequent chemical purification steps.

Commercial Uses. Methyl chloride is used as a catalyst carrier in the low-temperature polymerization of such products as the silicones and Butyl and other types of synthetic rubber; as a refrigerant gas; as a methylating agent in organic synthesis of such compounds as methylcellulose, and also as a chlorinating agent; as an extractant for greases, waxes, essential oils and resins, and as a low-temperature solvent; and as a fluid for the thermometric and thermostatic equipment.

Physiological Effects. Methyl chloride is toxic, and areas in which it is handled must be adequately ventilated. It is an anesthetic about one-fourth as potent as chloroform, and also acts as a narcotic. Inhalation of it must be avoided, for inhalation in concentrations of several hundred ppm or more leads successively to dizziness, headache, an unsteady walk, weakness, nausea and vomiting, abdominal pain, tremors, extreme nervousness, mental confusion, convulsion, unconsciousness, and death. Apparent recovery from what seems a mild exposure through inhalation may be followed by serious and prolonged or even fatal after-effects within a few days or weeks. Repeated exposures are dangerous because methyl chloride is eliminated slowly from the body, in which it is converted into hydrochloric acid and methyl alcohol (wood alcohol). A concentration of 100 ppm is generally accepted as the safe upper limit for a daily 8-hour exposure.

Contact between the skin or the eyes and methyl chloride liquid (or vapor in a concentrated stream) must also be avoided, for such contact can result in frostbite of the tissues.

The physiological effects of methyl chloride in increasing concentrations are described in Bulletin 185 of the U. S. Public Health Service.[1]

All the precautions necessary for the safe handling of any flammable, toxic gas must be observed with methyl chloride.

Leak Detection. It is advisable to transfer the methyl chloride vapor or liquid from refrigerating units into gas-tight containers before testing the units for suspected leaks. The units may then be placed under carbon dioxide, air or nitrogen pressure and tested by the application of soapy water (or glycerine, in freezing weather) to the suspected points. Soapy water and glycerine are also recommended for testing possible leaks in cylinders and other containers. An open flame or a halide torch should not be used to detect leaks.

METHYL CHLORIDE CONTAINERS

Methyl chloride is authorized for shipment in cylinders, insulated single-unit tank cars, and TMU tanks and tank cars. It is also shipped in tank barges. Though authorized for shipment in portable tanks and cargo tanks on trucks, it is rarely transported in such tanks at the present time.

Filling Limits. The maximum filling densities authorized (per cent water capacity by weight): for methyl chloride under present regulations are as follows:

In cylinders—84 per cent.

In TMU tanks—84 per cent.

In single-unit tank cars (complying with ICC-105A300-W specifications)—86 per cent.

In cargo tanks and portable tanks—84 per cent (by weight; 88.5 per cent by volume).

SHIPPING METHODS; REGULATIONS

Under the appropriate regulations and tariffs, methyl chloride is authorized for shipment as follows:

By Rail: In cylinders (via freight, express or baggage), in TMU tank cars, and in single-unit tank cars.

By Highway: In cylinders on trucks, in TMU tanks on trucks, and in portable tanks and cargo tanks (though the latter are little used, as previously noted).

By Water: In cylinders on cargo vessels, passenger vessels, and ferry and railroad car ferry vessels (passenger or vehicle). On cargo vessels only in portable tanks (complying with ICC-51 specification) not over 20,000 lb gross weight. In authorized tank cars on trainships only, and in authorized tank trucks on trailerships and trainships only. In cylinders on barges of U. S. Coast Guard classes A, CA, and CB only. In cargo tanks on tankships and tank barges (to maximum filling densities by specific gravity as prescribed in Coast Guard regulations).

By Air: In cylinders aboard cargo aircraft only up to 300 lb (140 kg) maximum net weight per cylinder.

METHYL CHLORIDE CYLINDERS

Methyl chloride is authorized for shipment in cylinders meeting ICC specifications as follows: 3A225, 3AA225, 3B225, 4A225, 4B225, 4BA225, 4, 3E1800, and 4B240ET (cylinders which comply with ICC specifications 3, 25, 26-300 and 38 may also be continued in methyl chloride service, but new construction is not authorized).

Valve Outlet and Inlet Connections. Standard connection, U. S. and Canada—No. 620. Alternate standard connection, U. S. and Canada—No. 360.

Cylinder Requalification. Cylinders authorized for methyl chloride service must be requalified by hydrostatic retest every 5 years under present regulations, with the following exceptions: ICC-4 cylinders, hydrostatic retest every 10 years; no retest is required for 3E cylinders; and 4B, 4BA and 26-300 cylinders, retest every 12 years (if they are used exclusively for methyl chloride that is free from corroding components and if they are protected externally by suitable corrosion resistant coatings such as galvanizing, painting, etc.; as an alternative, these cylinders may also be requalified by being retested every 7 years to an internal hydrostatic pressure at least two times the marked service pressure and without deterimination of expansions).

SINGLE-UNIT TANK CARS AND TMU TANK CARS

Methyl chloride is authorized for shipment in insulated, single-unit tank cars which comply with ICC specification 105A300-W (provided that interior pipes of loading are equipped with excess-flow valves of approved design). It is also authorized for shipment in TMU tank cars meeting ICC specification 106A500-X.

TMU TANKS, CARGO TANKS AND PORTABLE TANKS ON TRUCKS

TMU tanks complying with specification 106A500-X are also authorized for the shipment of methyl chloride via motor vehicle. Portable tanks meeting ICC-51 specification and cargo tanks that comply with specifications MC-330 or MC-331 are also authorized, but currently are little used, as noted earlier. The minimum design pressure required for these portable and cargo tanks is 150 psig.

REFERENCE

1. "United States Public Health Service Bulletin No. 185," U. S. Public Health Service. Available from the Superintendent of Documents, U. S. Government Printing Office, Washington, D. C.

ADDITIONAL REFERENCE

"Methyl Chloride" (Safety Data Sheet SD-40), Manufacturing Chemists' Association, 1825 Connecticut Ave., N. W., Washington, D. C., 20009.

Methyl Mercaptan

CH_3SH
Synonym: Methanethiol
ICC Classification: Flammable compressed gas; red label

PHYSICAL CONSTANTS

International symbol	—
Molecular weight	48.10
Vapor pressure, psig	
at 70 F	11
at 105 F	39*
at 115 F	47*
at 130 F	61*
Specific gravity of gas at 59 F (air = 1)	1.66
Specific gravity of liquid at 32 F	0.896
Density, liquid, lb/cu ft	
at 70 F	54.08
at 105 F	52.46
at 115 F	51.97
at 130 F	51.28
Boiling point at 1 atm	42.7 F
Melting point at 1 atm	−185.8 F
Critical temperature	386.2 F
Critical pressure, psia	1049.6
Flash point (open cup)	Below 0 F
Average weight per gal, liquid, lbs	7.2
Coefficient of expansion, liquid, per °F	0.00088
Solubility in water (% by weight)	
at 59 F	2.4
at 77 F	1.3

Properties. Methyl mercaptan is a colorless, flammable gas with an extremely strong and disagreeable odor that nauseates some individuals. It is toxic, but its unpleasant odor provides ample warning of its presence. The odor also makes it necessary to evacuate premises in which leaks of any quantity have occurred. Methyl mercaptan is easily liquefied under pressure at room temperatures and is water white as a liquid.

Chemically, methyl mercaptan resembles methyl alcohol in many respects and provides a means of introducing the methylthio linkage into many compounds.

* Approximate values.

Materials of Construction. Stainless steel is a satisfactory material for use with methyl mercaptan, but can be costly for entire installations. Iron and steel with stainless steel trim are successfully employed in methyl mercaptan service if proper preparatory and maintenance steps are taken. However, stainless steel must be used for components with which intermittent exposure to air or moisture cannot be avoided and in which slight corrosion could interfere with proper functioning. Such components include unloading risers, flame arrestor cores, gages and instruments. Aluminum has been considered one of the least desirable materials for use with methyl mercaptan, for caution must be exercised in selecting a proper type of aluminum to handle it under pressure. Copper and copper-bearing alloys should not be used for fixed equipment under constant exposure to methyl mercaptan.

Before continuous exposure to liquid methyl mercaptan, all iron and steel equipment should be passivated with methyl mercaptan vapor or hydrogen sulfide to coat the surfaces with ferrous sulfide. Treated surfaces must subsequently be kept free from moisture and under inert atmosphere to prevent the protective sulfide film from deteriorating. Ferrous sulfide oxidizes with a red glow in the presence of air and ignites highly flammable methyl mercaptan vapor. When cleaning or disassembling treated equipment, large quantities of water should be used to reduce the hazards of fire or explosion.

Asbestos and "Teflon" are among various materials recommended by methyl mercaptan suppliers as satisfactory for gaskets and packing.

Manufacture. Methyl mercaptan is produced primarily by chemical synthesis of methyl alcohol and hydrogen sulfide.

Commercial Uses. One of the most important uses of methyl mercaptan is in the manufacture of methionine, an important amino acid used as an animal feed supplement, especially for poultry. It is also used in the synthesis of insecticides.

Physiological Effects. Methyl mercaptan, with the other lower alkyl mercaptans, has a fairly high degree of toxicity. It acts primarily on the central nervous system after inhalation, at first resulting in great stimulation and then depression. Little research has been done in mercaptan toxicity with humans because its strong odor has been thought to provide sufficient protection for personnel. However, inhalation of increasing methyl mercaptan concentrations by laboratory animals has led to increased respiration and restlessness, convulsions, muscular paralysis and finally death by respiratory paralysis. Fish show special sensitivity to methyl mercaptan, and have been fatally poisoned by as little as 0.5 to 3 ppm in water.

Liquid methyl mercaptan may severely irritate the skin.

Handling Precautions and Safety Equipment. Prolonged inhalation of methyl mercaptan vapors must be avoided and, although its odor will become extremely disagreeable before lethal concentrations are reached, the nose is temporarily desensitized to mercaptan after initial exposure. Air-line or oxygen-type gas masks must always be on hand in operating areas.

Low flash and fire points and high volatility of methyl mercaptan gas poses another principal hazard in its handling. Personnel must avoid spills and leaks. To spills, immediately apply liquid household bleach, calcium or sodium hypochlorite in 5 per cent aqueous solution. *Do not* use powdered bleach, which could cause fire or explosion.

Methyl mercaptan severely irritates the skin and the eyes, and safety showers and eyewashing facilities must be readily available for possible emergencies. Flush the skin or eyes thoroughly with water should contact occur, and get medical attention promptly if irritation continues after washing.

Other precautions that must be observed in handling methyl mercaptan include the following:

(1) Store cylinders away from open flames and away from locations in which the gas could flow downward to a flame or a spark from a motor.

(2) Require the wearing of rubber gloves, goggles and non-sparking shoes by personnel handling methyl mercaptan.

(3) Provide explosion-proof equipment or, if this is not possible, make sure that operating areas are well ventilated and that vapors can not accumulate and be ignited in them.

(4) Provide at least two men for work requiring the opening of any equipment, piping or vessels used in methyl mercaptan service so that one can summon aid if the other is overcome by fumes.

Further information about the toxic effects of mercaptan and their treatment is given in

Medical Bulletin 5, 78 (1941) of the Standard Oil Co. of New Jersey.[1]

Small leaks may exist in even the tightest processing system and result in an objectionable odor. This odor may be masked by suitable odor-masking chemicals. However, masking of odor must never be allowed to substitute for an essentially tight system.

METHYL MERCAPTAN CONTAINERS

Methyl mercaptan is shipped as a compressed liquefied gas under its own vapor pressure in cylinders, single-unit tank cars and tank trucks. It is also authorized for shipment in TMU (ton multi-unit) tank cars, and in TMU tanks and portable tanks on trucks.

Filling Limits. The maximum filling densities permitted for methyl mercaptan under ICC regulations are as follows (per cent of water capacity by weight):

For cylinders, TMU tanks, cargo tanks and portable tanks—80 per cent. (Cargo tanks and portable tanks are also authorized for a maximum filling density of 90 per cent by volume.)

For single-unit tank cars—82 per cent.

SHIPPING METHODS; REGULATIONS

Under the appropriate regulations and tariffs, methyl mercaptan is authorized for shipment as follows:

By Rail: In cylinders (via freight or express), and by single-unit tank cars and TMU tank cars.

By Highway: In cylinders on trucks, and in tank trucks, portable tanks and TMU tanks.

By Water: In cylinders and portable tanks (not over 20,000 lb gross weight) via cargo vessels only; in authorized tank cars aboard trainships only; and in authorized tank trucks on trailer ships and trainships only. In cylinders on barges of U. S. Coast Guard classes A and C only. In cargo tanks aboard tank ships and tank barges (to maximum filling densities by specific gravity as prescribed in Coast Guard regulations).

By Air: Aboard cargo aircraft only in appropriate cylinders up to 300 lb (140 kg) maximum net weight per cylinder.

METHYL MERCAPTAN CYLINDERS

Cylinders that meet the following ICC specifications are authorized for methyl mercaptan service: 3A240, 3AA240, 3B240, 4B240, 4BA240, 4B240ET, and 3E1800. Cylinders of the same types with higher service pressures are also authorized. Safety relief devices are not required on cylinders charged with methyl mercaptan.

Valve Outlet and Inlet Connections. Standard connection, U. S. and Canada—No. 330.

Cylinder Requalification. All cylinders authorized for methyl mercaptan service must be requalified by hydrostatic test every 5 years under present regulations, except that no periodic retest is required for 3E cylinders.

SINGLE-UNIT TANK CARS AND TMU TANK CARS

Methyl mercaptan is authorized for shipment in tank cars meeting ICC specification 105A300-W (an insulated, single-unit tank car with properly fitted valves). One supplier ships in 6000-gal cars of this type, and recommends unloading by pressurizing with dry nitrogen or natural gas with a maximum withdrawal rate of 30 GPM that results in complete unloading within 3 to 4 hours.

In addition rail shipment of methyl mercaptan is authorized for TMU tank cars complying with ICC specification 106A500-X; these TMU tanks must not be equipped with safety relief devices of any description.

CARGO TANKS AND PORTABLE TANKS

Methyl mercaptan is authorized for shipment in motor vehicle cargo tanks conforming with ICC specification MC-330 or MC-331, and in steel portable tanks meeting specification ICC-51.

In addition, TMU tanks complying with ICC-106A500-X specifications and not equipped with safety relief devices of any kind are authorized for transportation by motor vehicle.

STORAGE AND HANDLING EQUIPMENT

Welded low-carbon steel storage tanks complying with the unfired pressure vessel Code of the ASME have proven satisfactory for use with methyl mercaptan. A recommended design pressure is 85 psig at 150 F. Explosion-proof, stainless steel pumps and carbon steel or stainless steel valves and fittings are among recommended equipment. Equipment requires treating before use as previously described in the "Materials of construction" section. Storage and handling facilities should be designed with the help of professional personnel thoroughly familiar with methyl mercaptan.

REFERENCE

1. V. Cristensen, "A Case of Poisoning with Mercaptans," Standard Oil Co. (New Jersey), Medical Bulletin **5**, 78 (1941).

Monochlorodifluoromethane

See **Fluorocarbons.**

Monochloropentafluoroethane

See **Fluorocarbons.**

Monochlorotrifluoromethane

See **Fluorocarbons.**

Monomethylamine

See **Methylamines.**

Neon

See **Rare Gases of the Atmosphere.**

Nitrogen

N_2 ICC Classification: Nonflammable compressed gas; green label

PHYSICAL CONSTANTS

International symbol	N_2
Molecular weight	28.016
Specific gravity of gas at 70 F and 1 atm (air = 1)	0.9670
Specific volume of gas at 70 F and 1 atm, cu ft/lb	13.80
Density of gas at 70 F and 1 atm, lb/cu ft	0.07245
Density of gas at boiling point, lb/cu ft	0.2878
Density of liquid at boiling point and 1 atm, lb/cu ft	50.46
Liquid/gas ratio (liquid at boiling point, gas at 70 F and 1 atm), vol/vol	1/696.5
Boiling point at 1 atm	−320.36 F
Melting point at 1 atm	−345.7 F
Critical temperature	−232.87 F
Critical pressure, psia	492.2
Triple point	−346.027 F at 1.830 psia
Latent heat of vaporization at boiling point, Btu/lb	85.67
Latent heat of fusion at melting point, Btu/lb	11.0
Specific heat of gas, C_p, at 70 F and 1 atm, Btu/(lb)(°F)	0.2484
Specific heat of gas, C_v, at 70 F and 1 atm, Btu/(lb)(°F)	0.1774
Ratio of specific heats, C_p/C_v, at 70 F	1.400
Weight per gallon, liquid, at boiling point, lb	6.745

Properties. Nitrogen makes up the major portion of the atmosphere (78.03 per cent by volume, 75.5 per cent by weight). It is a colorless, odorless, flavorless, nontoxic and almost totally inert gas, and is colorless as a liquid. Nitrogen does not burn, and supports neither combustion nor respiration. It combines with some of the more active metals, such as lithium and magnesium, to form nitrides, and at high temperatures it will also combine with hydrogen, oxygen and other elements. It is consequently not recommended for use as a shielding inert atmosphere in welding, though it is employed to give inert protection against atmospheric contamination in many nonwelding applications. Nitrogen is only slightly soluble in water and most other liquids, and is a poor conductor of heat and electricity. As a liquid at cryogenic temperatures it is nonmagnetic, stable against mechanical shock, and free of toxic or irritant vapors. It is shipped as a nonliquefied gas at pressures of 2000 psig or above, and also as a cryogenic fluid at pressures below 200 psig.

Materials of Construction. Gaseous nitrogen is noncorrosive and inert, and may consequently be contained in systems constructed of any common metals and designed to withstand safely the pressures involved. At the temperature of liquid nitrogen, ordinary carbon steels and most alloy steels lose their ductility, and are considered unsatisfactory for liquid nitrogen service. Satisfactory materials for use with liquid nitrogen include 18-8 stainless steel and other austenitic nickel-chromium alloys, copper, "Monel," brass, and aluminum.

Manufacture. Nitrogen is produced commercially at air separation plants by liquefaction of atmospheric air and removal of the nitrogen from it by fractionation.

Commercial Uses. Nitrogen has many

commercial and technical applications. As a gas it is used in: agitation of color film solution in photographic processing; blanketing of oxygen-sensitive liquids, and of volatile liquid chemicals; the production of semiconductor electronic components, as a blanketing atmosphere; the blowing of foam-type plastics; the de-aeration of oxygen-sensitive liquids; the degassing of nonferrous metals; food processing and packing; inhibition of aerobic bacteria growth; magnesium reduction of aluminum scrap; and the propulsion of liquids through pipelines.

Gaseous nitrogen is also used in: pressurizing aircraft tires and emergency bottles to operate landing gear; purging, in the brazing of copper tubing for air conditioning and refrigeration systems; the purging and filling of electronic devices; the purging, filling and testing of high-voltage compression cables; the purging and testing of pipelines and related instruments; and the treatment of alkyd resins in the paint industry.

Liquid nitrogen also has a great many uses, among them: the cold-trapping of materials such as carbon dioxide from gas streams (and it is commonly employed in this way in systems which produce high vacuums); as a coolant for electronic equipment, for pulverizing plastics, and for simulating the conditions of outer space; for creating very high-pressure gaseous nitrogen (15,000 psig) through liquid nitrogen pumping; in food and chemical pulverization; for the freezing of expensive and highly perishable foods, such as shrimp; for the freezing of liquids in pipelines for emergency repairs; for low-temperature stabilization and hardening of metals; for low-temperature research; for low-temperature stress-relieving of aluminium alloys; for the preservation of whole blood, livestock sperm, and other biologicals; for refrigerating foods in long-distance hauling as well as local delivery; for refrigeration shielding of liquid hydrogen, helium and neon; for the removal of skin blemishes in dermatology; and for shrink fitting of metal parts.

Liquid nitrogen also has a number of classified applications in the missile and space programs of the U. S., in which it is used in large quantities.

Physiological Effects. Nitrogen is nontoxic and largely inert. It can act as a simple asphyxiant by displacing needed oxygen in the air. Inhalation of it in excessive concentrations can result in unconsciousness without any warning symptoms, such as dizziness.

Gaseous nitrogen must be handled with all the precautions necessary for safety with any nonflammable, nontoxic compressed gas.

All the precautions necessary for the safe handling of any gas liquefied at very low temperatures must be observed with liquid nitrogen. Severe burn-like injuries result from contact between the tissues and liquid nitrogen.

NITROGEN CONTAINERS

Nitrogen gas is authorized for shipment in cylinders, tank cars and tube trailers. Liquid nitrogen is shipped as a cryogenic fluid in insulated cylinders, insulated tank trucks and insulated tank cars.

Filling Limits. The maximum filling limits authorized for gaseous nitrogen are as follows:

In cylinders and tube trailers—the authorized service pressures marked on the cylinders or tube assemblies at 70 F and not in excess of 5/4ths of the marked service pressures at 130 F (also, in the case of cylinders of specifications 3A and 3AA *that meet special requirements*, up to 10 per cent in excess of their marked service pressures).

In tank cars—not more than seven-tenths of the marked test pressure at 130 F in uninsulated cars of the ICC-107A type.

The maximum filling limits authorized for liquid nitrogen are:

In cylinders that meet ICC-4L specifications—maximum filling density of 68 per cent (per cent of water capacity by weight); also, the cylinder pressure must be limited by a pressure-controlling valve to one and one-fourth times the marked service pressure (or, for 4L cylinders insulated by a vacuum, at least 15 psi lower than one and one-fourth times the marked service pressure).

Liquid nitrogen shipped in insulated truck cargo tanks or in other insulated containers at pressures below 25 psig is not subject to the regulations of the ICC.

SHIPPING METHODS; REGULATIONS

Under the appropriate regulations and tariffs, nitrogen is authorized for shipment as follows (nitrogen gas, except where liquid nitrogen is indicated):

By Rail: In cylinders and in tank cars. Liquid nitrogen, also in cylinders, and, in insulated tank cars not subject to ICC regulations.

By Highway: In cylinders on trucks, and in tube trailers. Liquid nitrogen, in cylinders, and in tank trucks not subject to ICC regulations.

By Water: In cylinders on cargo and passenger vessels, and on ferry and railroad car ferry vessels (passenger or vehicle). In authorized tank cars on cargo vessels only. In cylinders on barges of U. S. Coast Guard classes A, BA, BB, CA, and CB. Liquid nitrogen, in pressurized cylinders on cargo and passenger vessels and ferry and railroad car ferry vessels (passenger or vehicle).

By Air: In cylinders aboard passenger aircraft up to 150 lb (70 kg), and aboard cargo aircraft up to 300 lb (140 kg), maximum net weight per cylinder. Nonpressurized liquid nitrogen, aboard passenger and cargo aircraft up to about 12 gal (50 liters) maximum net contents per container; low-pressure liquid nitrogen, or pressurized liquid nitrogen, aboard cargo aircraft only up to 300 lb (140 kg) maximum net weight per container.

NITROGEN CYLINDERS

Cylinders which comply with ICC specifications 3A and 3AA are the types usually used to ship gaseous nitrogen, but it is authorized for shipment in any cylinders approved for nonliquefied compressed gas. (These include cylinders meeting ICC specifications 3A, 3AA, 3B, 3C, 3D, 3E, 4, 4A, 4B, 4BA, and 4C; in addition, continued use of cylinders complying with ICC specifications 3, 7, 25, 26, 33 and 38 is authorized, but new construction is not authorized.)

Liquid nitrogen is authorized for shipment in cylinders which meet ICC specifications 4L200; such cylinders must be equipped with pressure-controlling valves as previously stated.

Valve Outlet and Inlet Connections. Standard Connection, U. S. and Canada—No. 580. Alternate Standard Connection, U. S. and Canada—No. 590.

Cylinder Requalification. All cylinders authorized for gaseous nitrogen service must be requalified by hydrostatic retest every 5 years under present regulations, with the following exceptions: ICC-4 cylinders, every 10 years; and no periodic retest is required for cylinders of types 3C, 3E, 4C and 7.

Also, for cylinders of the 4L type authorized for liquid nitrogen service, no periodic retest is required for requalification.

Small Portable Containers for Liquid Nitrogen. Liquid nitrogen is shipped and stored in small portable containers which hold quantities ranging from 1 to some 25 gal. These long-necked, spherical containers encased in cylindrical shells are heavily insulated; they maintain the liquid at atmospheric pressure, and are consequently not subject to ICC regulations.

NITROGEN TANK CARS

Gaseous nitrogen is authorized for rail shipment in tank cars that comply with ICC specifications 107A. These cars consist of a number of seamless, tubular containers permanently mounted on a special frame. ICC regulations require that the pressure to which the containers are charged must not exceed seven-tenths of the marked test pressure at 130 F.

Liquid nitrogen is also shipped in special insulated tank cars with capacities equal to a million or more cubic feet of nitrogen gas at atmospheric pressure. These cars carry the liquid at pressures below 25 psig.

TUBE TRAILERS AND TANK TRUCKS AND TRAILERS

Gaseous nitrogen is shipped in tube trailers with capacities ranging up to more than 40,000 cu ft. The nested and manifolded tubes of these trailer are built to comply with ICC-3A or 3AA (or ICC-3AX or 3AAX, which are also authorized) cylinder specifications. The trailers commonly serve as the storage supply for the user, with the supplier replacing trailers as they are emptied.

Liquid nitrogen is shipped in bulk at pressures below 25 psig in special insulated tank trucks and truck trailers. These low-temperature cargo tanks hold the equivalent of 400,000 or more cu ft of nitrogen at atmospheric pressures.

STORAGE SYSTEMS

High-pressure tube trailers and tank cars often serve as the storage supply for gaseous nitrogen, while nitrogen is often stored in compact liquid form at the consumer's site. Liquid storage systems that frequently include vaporizing equipment for conversion to gas range in capacity from 25,000 to more than one million cu ft. Liquid storage systems should be designed and installed only under the direction of engineers thoroughly familiar with liquid nitrogen equipment, and in full compliance with all state and local requirements.

REFERENCE

"Standard Density Data, Atmospheric Gases and Hydrogen" (Pamphlet P-6), Compressed Gas Association, Inc.

Nitrous Oxide

N_2O
Synonyms: Nitrogen monoxide, dinitrogen monoxide, laughing gas
ICC Classification: Nonflammable compressed gas, green label

PHYSICAL CONSTANTS

International symbol	N_2O
Molecular weight	44.013
Vapor pressure, psia	
at −4 F	262
at 32 F	455
at 68 F	736
at 98 F	1055
Density, gas, at 1 atm, lb/cu ft	
at 32 F	0.1230
at 68 F	0.1146
Specific gravity, gas, at 32 F and 1 atm (air = 1)	1.529
Specific volume, gas, at 1 atm, cu ft/lb	
at 32 F	8.130
at 68 F	8.726
Density, saturated vapor, lb/cu ft	
at boiling point	0.194
at −4 F and 262 psia	2.997
at 68 F and 736 psia	10.051

Density, liquid	
at boiling point and 1 atm	76.6
at 70 F	48.3
Specific gravity, liquid, at 68 F and 736 psia	0.785
Boiling point at 1 atm	−127.3 F
Melting point at 1 atm	−131.5 F
Triple point	−131.5 F at 12.74 psia
Critical temperature	97.7 F
Critical pressure, psia	1054
Critical density, lb/cu ft	28.15
Latent heat of vaporization, Btu/lb	
at boiling point	161.8
at 32 F	107.5
at 68 F	78.7
Latent heat of fusion at triple point, Btu/lb	63.9
Specific heat, gas, at 1 atm, Btu/(lb)(°F)	
C_p, 77 F to 212 F	0.212
C_p, at 59 F	0.2004
C_v, at 59 F	0.1538
Ratio of specific heats, C_p/C_v, at 59 F and 1 atm	1.303
Solubility in water at 1 atm, vol/1 vol of water	
at 32 F	1.3
at 68 F	0.72
at 77 F	0.66
Solubility in alcohol at 68 F and 1 atm, vol/1 vol of alcohol	3.0
Weight/gal, liquid, lb	
at boiling point	10.23
at −4 F and 262 psia	8.35
at 68 F and 736 psia	6.54
Viscosity, gas, centipoises	
at 32 F	0.0135
at 80 F	0.0149
Thermal conductivity, gas, at 32 F (Btu)(ft)/(sq ft)(hr)(°F)	0.0083

Properties. Nitrous oxide at normal temperatures and pressures is a colorless, practically odorless and tasteless, nontoxic gas about 50 per cent heavier than air. It is shipped in liquefied form at its vapor pressure which at 70 F is about 745 psig. Nitrous oxide is nonflammable, but, being a mild oxidizing agent, will support combustion of flammable materials in a manner similar to oxygen but to a lesser extent than oxygen. Under ordinary conditions, nitrous oxide is stable and generally inert. Decomposition of the pure gas in the absence of catalysts is negligible at temperatures below 1200 F. Compared to air, nitrous oxide is relatively soluble in water, alcohol, and oils, and in many food products. Unlike some higher oxides of nitrogen, nitrous oxide dissolves in water without change in its acidity.

Materials of Construction. Nitrous oxide is noncorrosive and may therefore be used with any of the common, commercially available metals. Because of its oxidizing action, however, care must be taken to insure freedom from oil, grease, and other readily combustible materials in all equipment being prepared to handle nitrous oxide, particularly at high pressures.

Manufacture. The only practical commercial method yet developed for manufacturing nitrous oxide is by the thermal decomposition of ammonium nitrate, which yields nitrous oxide and water in the primary reaction. However, a number of side reactions also occur, depending upon the temperature of decomposition, and these are catalyzed by the presence of certain contaminants and metals or metallic compounds. The impurities formed, mostly the higher oxides of nitrogen, are highly toxic. Accordingly, after the water is condensed out, the gas is passed through a series of scrubbing towers to remove the impurities so that the final product contains only a small amount of nitrogen. In addition, the gas is usually passed through a bed of dessicant after compression in order to dry it.

Commercial Uses. The largest use of nitrous oxide probably still is a long established one, as an inhalant type of anesthetic or analgesic gas. Extensive use of nitrous oxide has more recently developed in the field of pressure packaging, in which it serves as a propellant for various aerosol products, particularly with foods such as whipped cream. Other applications include its employment as a leak-detecting agent, as an oxidizing agent in blow-torches, as both a refrigerant gas and a refrigerant liquid for immersion freezing of food products, and as a chemical reagent in the manufacture of various compounds (both organic, as with detonants, and inorganic, as in obtaining nitrites from alkali metals). Nitrous oxide has also served as an ingredient in rocket fuel formulations, and its use as part of the working fluid in hypersonic wind tunnels has recently been investigated.

Physiological Effects. Nitrous oxide is nontoxic and nonirritating as well as being chemically stable. When inhaled in high concentrations for a few seconds, it affects the central nervous system and may induce symptoms closely resembling alcoholic intoxication. Its colloquial name, "laughing gas," stems from the fact that some persons exhibit hilarity while in this condition.

For use as a general anesthetic in medicine, high concentrations of nitrous oxide are mixed with oxygen. Continued inhalation without an ample supply of oxygen results in simple asphyxia.

In view of its nontoxic nature, nitrous oxide has been used for a number of years for the food type of aerosols and in treating and preserving foodstuffs. It has been defined by the U. S. Food and Drug Administration as a food additive substance generally recognized as safe when used as a propellant for dairy and vegetable-fat toppings in pressurized containers.

Liquid nitrous oxide evaporates so rapidly that prolonged contact of the liquid with the skin may result in freezing or frostbite.

NITROUS OXIDE CONTAINERS

Cylinders, cargo tanks and portable tanks are authorized for the shipment of nitrous oxide. It is transported as a liquefied compressed gas under high pressure in cylinders, and at lower pressures and reduced temperatures in refrigerated cargo tanks and insulated portable tanks.

Filling Limits. The maximum filling densities authorized for nitrous oxide are as follows:

In cylinders—68 per cent (per cent water capacity by weight); or 75 per cent in cylinders made before Feb. 1, 1917, that have less than 12 lb water capacity and are known to have passed a pressure test of not less than 3500 psi.

In cargo tanks and portable tanks—95 per cent by volume (with the additional requirement that tanks be equipped with suitable pressure controlling devices, and that the vapor pressure at 115 F must not exceed the design pressure of the tank.

SHIPPING METHODS; REGULATIONS

Under the appropriate regulations and tariffs, nitrous oxide is authorized for shipment as follows:

By Rail: In cylinders (via freight, express, and baggage), and in portable tanks.

By Highway: In cylinders, cargo tanks, and portable tanks.

By Water: In cylinders aboard passenger vessels, cargo vessels, and ferry and railroad car ferry vessels (either passenger or vehicle for both types of ferries). Also in authorized tank motor vehicles aboard cargo vessels and ferry and railroad car ferry vessels (passenger or vehicle for both ferry types); and aboard cargo vessels and railroad car ferry vessels (passenger or ferry) in portable tanks meeting ICC-51 specifications (if shipped at stowage, the tanks must be not over 20,000 lb gross weight per tank, and must not be equipped with fixed length dip tube gaging devices). In cylinders on barges of U. S. Coast Guard classes A, BA, BB, CA, and CB only.

By Air: In cylinders aboard passenger aircraft up to 150 lb (70 kg), and in cylinders aboard cargo aircraft up to 300 lb (140 kg), maximum net weight per cylinder.

NITROUS OXIDE CYLINDERS

Cylinders that meet the following ICC specifications are authorized for nitrous oxide service: 3A1800, 3AA1800, 3E1800, and 3HT2000 (ICC-3 cylinders may also be continued in nitrous oxide service, but new construction is not authorized; also, 3HT cylinders are restricted to aircraft use only, and in shipment must be boxed in strong outside containers). The manifolding of cylinders transporting nitrous oxide is permitted under ICC regulations if each cylinder is individually equipped with an approved safety relief device, and if all cylinders are supported and held together as a unit by structurally adequate means.

Valve Outlet and Inlet Connection. Standard connection, U. S. and Canada—No. 320 (obsolete effective 11/1/66), No. 1320, No. 910.

Cylinder Requalification. Cylinders authorized for nitrous oxide service must be requalified by hydrostatic retest every 5 years under present regulations, except that periodic retest is required every 3 years for 3HT cylinders, and no periodic retest is required for cylinders of the 3E type. (3HT cylinders must also be withdrawn from service after a service life of 12 years or 4380 pressurizations.)

Safe Handling. All the precautions necessary for the safe handling of any compressed gas, and of any gas used medicinally, must be observed with nitrous oxide. In addition to these and to all applicable state and local regulations, the special rules below must be followed in handling nitrous oxide.

Never permit oil, grease or any other readily combustible substance to come in contact with cylinders or other equipment containing nitrous oxide. Oil and nitrous oxide may combine with explosive violence.

Remove any paper wrappings so that the cylinder label is clearly visible before placing cylinders in service.

Store nitrous oxide cylinders in an assigned, little-frequented location, making sure not to store them in the same room with cylinders containing reserve stocks of flammable gases. Never store medical gas cylinders of nitrous oxide in the hospital operating room.

Take care to avoid exhausting a nitrous oxide cylinder completely when using it with ether in anesthesia in order to prevent the possibility of having ether drawn back into the cylinder. Always protect nitrous oxide cylinders against feed back of any other gases or foreign material by suitable traps or check valves in lines to which the cylinders are connected.

Do not transfer nitrous oxide from one cylinder to another. Instead, always return the cylinders to charging plants for refilling under recognized safe practices.

Medical Gas Cylinders. See Section B of Chapter 1 in Part III for an identification and description of standard styles of medical gas cylinders used for nitrous oxide, together with an explanation of their safe handling.

CARGO TANKS AND PORTABLE TANKS

Nitrous oxide is authorized for shipment by the ICC in portable tanks that comply with specification ICC-51, and in motor vehicle cargo tanks meeting specifications MC-330 or

MC-331. (Cargo tanks meeting ICC specifications MC-320 may be continued in service if qualified by periodic retest as required, but new construction is not authorized.) The minimum design pressure stipulated for portable or cargo tanks is 200 psig except that it may be reduced to 100 psig (or the controlled pressure, whichever is greater), if the tanks are also designed to comply with the requirements for low temperature operation of the ASME Boiler and Pressure Vessel Code, Section VIII, Unfired Pressure Vessels.[1] The maximum service pressure authorized for portable or cargo tanks under current ICC regulations is 500 psig.

Portable tanks for nitrous oxide service must be lagged with a noncombustible insulating material thick enough so that conductance does not exceed 0.08 Btu/(sq ft)(hr)(°F temperature differential), determined at 60 F. Requirements for safety relief devices, excess flow valves, piping, valves, fittings and accessories must also be met by nitrous oxide portable tanks. One or more pressure-controlling devices may be installed to allow controlled escape of gas above a maximum operating temperature (such escape also exerts a self-refrigerating effect). Coils for refrigerating or heating or both may also be used, as may liquid level gaging devices (and additional gaging devices), but the coils and gaging devices must meet specified requirements if they are used. Portable tanks must be requalified by hydrostatic retest every 5 years under current regulations, and must comply with various other requirements concerning such matters as their certification, registration and repair. As with portable tanks for any compressed gas, those used for nitrous oxide must be of more than 1000 lb water capacity, and must be equipped with skids, mountings or other accessories to provide for moving the tanks with handling equipment.

For refrigerated cargo tanks in nitrous oxide service, similar regulations apply to insulation, safety relief and pressure control devices, refrigerating and heating coils, excess flow valves, piping, valves, fittings, accessories and certification and registration. For a cargo tank, it is further required that all inlet and outlet valves (except safety relief devices) must be marked to show whether they end in liquid or gas when the tank is at maximum filling density; it is also required that the tank must be fitted with a pressure gage having a shutoff valve between it and the tank (such a gage need be used only in the filling operation). The refrigerating unit may be mounted on the motor vehicle if desired. A manufacturer's data report on each cargo tank (as well as on each portable tank) must be kept in the files of the motor carrier operating the tank while the carrier has the tank in service. (Special inspection requirements for specified kinds of cargo tanks are also set forth in Section 77.824 of "Agent T. C. George's Tariff No. 15, Publishing Interstate Commerce Commission Regulations.")

STORAGE CONTAINERS AND EQUIPMENT

Industrial and medical consumers of nitrous oxide store the gas either in high-pressure systems (which often employ manifolded cylinders as the source of supply) or in bulk low-pressure storage containers in liquid form at reduced temperatures. Common operating conditions for low-pressure systems are in the ranges of 300 to 400 psig and 5 to 25 F. In addition to meeting all state and local requirements, high-pressure storage containers should comply with either the ASME Code ("Boiler and Pressure Vessel Code, Section VIII, Unfired Pressure Vessels"[1]) or ICC specifications and regulations, and low-pressure storage containers should comply with the ASME Code. Special construction is necessary for operating temperatures below −20 F. Piping for nitrous oxide systems should be of steel, stainless steel, wrought iron, or brass or copper pipe; or of seamless tubing made of copper, brass, or stainless steel.

Detailed recommendations for nitrous oxide storage installations are given in Pamphlet G-8.1 of Compressed Gas Association, Inc.[2] Installations should be designed and made by persons thoroughly familiar with nitrous oxide systems. Personnel who operate the installations must be adequately trained, and legible instructions should be maintained at operating locations.

Persons responsible for the use of nitrous oxide in hospitals or other facilities should see, "Safe Handling of Gases Used Medicinally," in this volume (Section B of Chapter 1 in Part III). Recommendations pertaining to nitrous oxide, particularly in hospitals, are also presented in two publications of the National Fire Protection Association, NFPA No. 565 and NFPA No. 56.[3,4]

REFERENCES

1. "ASME Boiler and Pressure Vessel Code, Section VIII, Unfired Pressure Vessels," American Society of Mechanical Engineers, 345 E. 47th St., New York, N. Y., 10017.
2. "Standard for the Installation of Nitrous Oxide at Consumer Sites" (Pamphlet G-8.1), Compressed Gas Association, Inc.
3. "Standard for Nonflammable Medical Gas Systems" (NFPA No. 565), National Fire Protection Association, 60 Batterymarch St., Boston, Mass. 02110.
4. "Recommended Safe Practice for Hospital Operating Rooms" (NFPA No. 56), National Fire Protection Association.

Oxygen

O_2 ICC Classification: Nonflammable compressed gas; green label

PHYSICAL CONSTANTS

International symbol	O_2
Molecular weight	32.000
Specific gravity of gas at 70 F and 1 atm (air = 1)	1.1053
Specific volume of gas at 70 F and 1 atm, cu ft/lb	12.08
Density of gas at 70 F and 1 atm, lb/cu ft	0.08281
Density of gas at boiling point, lb/cu ft	0.2959
Density of liquid at boiling point, lb/cu ft	71.27
Liquid/gas ratio (liquid at boiling point, gas at 70 F and 1 atm)	1/860.6
Boiling point at 1 atm	−297.4 F
Melting point at 1 atm	−361.1 F
Critical temperature	−181.1 F
Critical pressure, psia	736.9
Triple point	−361.89 F at 0.2321 psia
Latent heat of vaporization at boiling point, Btu/lb	91.7
Specific heat, C_p, at 70 F and 1 atm, Btu/(lb)(°F)	0.2193
Specific heat, C_v, at 70 F and 1 atm, Btu/(lb)(°F)	0.1566
Ratio of specific heats, C_p/C_v, at 70 F and 1 atm	1.400
Solubility of gas in water at 32 F, vol (gas)/vol (water)	1/32
Weight per gallon, liquid, at boiling point, lb	9.55

Properties. Oxygen, the colorless, odorless, tasteless elemental gas that supports life and makes combustion possible, constitutes about a fifth of the atmosphere (20.99 per cent by volume; by weight, almost a fourth—23.2 per cent). It is a transparent, pale blue liquid slightly heavier than water at temperatures ranging below some −300 F. All elements but the inert gases combined directly with oxygen, usually to form oxides. However, oxidation occurs over a wide range of temperatures, with elements like phosphorus and magnesium igniting spontaneously in air and the noble metals oxidizing only at very high temperatures. The common burning of hydrocarbons in an excess of air at temperatures below 3000 F to produce heat, power or light yields carbon dioxide, water, nitrogen and unreacted oxygen. Hydrogen and carbon monoxide are also produced at temperatures above 3000 F and with a deficiency of oxygen.

All materials that are flammable in air burn much more vigorously in oxygen. Some combustibles, such as oil or grease, burn with nearly explosive violence in oxygen if ignited. Pure oxygen itself is of course nonflammable.

Oxygen is shipped as a nonliquefied gas at pressures of 2000 psig or above, and also as a cryogenic fluid at pressures below 200 psig.

Materials of Construction. Oxygen is noncorrosive and can be contained in any common metals. However, care must be taken to remove all oil, grease and other combustible material from piping and containers before putting them into oxygen service. Cleaning methods employed by manufacturers of oxygen equipment are described in, "Equipment Cleaned for Oxygen Service," a pamphlet of Compressed Gas Association, Inc.[1] Oxygen-handling systems must also be designed to withstand safely the working pressures involved.

Manufacture. Almost all commercial oxygen made in North America today is produced at air separation plants by liquefaction of atmospheric air and removal of the oxygen from it by fractionation. Very small quantities are produced in parts of the U. S. where electric power is exceptionally inexpensive by the electrolysis of water, which of course also yields hydrogen.

Commercial Uses. Oxygen's major uses stem from its life-sustaining and combustion-supporting properties. It is used extensively in medicine for therapeutic purposes, for resuscitation in asphyxia, and with other gases in anesthesia. It is also used in high-altitude flying, deep-sea diving, and as both an inhalant and a power source in the U. S. space program. Industrial applications include its very wide utilization with acetylene, hydrogen and other fuel gases for such purposes as metal cutting, welding, hardening, scarfing, cleaning and dehydrating. Oxygen helps increase the capacity of steel and iron furnaces on a growing scale in the steel industry. One of its major uses is in the production of synthesis gas (a hydrogen-carbon monoxide mixture) from coal, natural gas or liquid fuels; synthesis gas is in turn used to make gasoline, methanol and ammonia. Oxygen is similarly employed in manufacturing some acetylene through partial oxidation of the hydrocarbons in methane. It is also used in the production of nitric acid, ethylene and other compounds in the chemical industry.

Physiological Effects. The inhalation of gaseous oxygen of course has a tonic effect on the human system rather than any toxic effect, and its tonic properties have led to the many therapeutic applications of oxygen. Specifically, the inhalation of oxygen in 100 per cent concentrations and at atmospheric pressure for 16 hours a day over many days has produced no observable harmful effects. However, exposures to oxygen at higher pressures for prolonged periods has been found to affect neuromuscular coordination and attentive powers.

All combustible materials—especially oil and greases—must be kept from contact with high oxygen concentrations, and all possible sources of ignition must be safely enclosed or kept completely away from either gaseous or liquid oxygen.

All the precautions necessary for the safe handling of any compressed gas must be observed with gaseous oxygen in addition. Liquid oxygen must also be handled with all the precautions required for safety with any cryogenic fluid. Contact between the skin and liquid oxygen, or uninsulated piping or vessels containing it, can cause severe burn-like injuries.

The safe handling of oxygen used medicinally, either alone or in mixtures with carbon dioxide or helium, is discussed in Section B of Chapter 1, Part III. Valve connection systems which help insure safety in medical applications of oxygen are described in Sections B and C of Chapter 4, Part III.

OXYGEN CONTAINERS

Gaseous oxygen is authorized for shipment in cylinders, tank cars and tube trailers. Liquid oxygen is shipped as a cryogenic fluid in insulated cylinders, insulated tank trucks and insulated tank cars.

Filling Limits. The maximum filling limits authorized for gaseous oxygen in shipment are as follows:

In cylinders and tube trailers—the authorized service pressures marked on the cylinders or tube assemblies at 70 F and not in excess of 5/4ths of the marked service pressures at 130 F (also, in the case of cylinders of specifications 3A and 3AA *that meet special requirements*, up to 10 per cent in excess of their marked service pressures).

In tank cars—not more than seven-tenths of the marked test pressure at 130 F in uninsulated cars of the ICC-107A type.

The maximum filling limits authorized for liquid oxygen are:

In cylinders that meet ICC-4L specifications —maximum filling density of 96 per cent (per cent water capacity by weight); also, the cylinder pressure must be limited by a pressure-controlling valve to one and one-fourth times the marked service pressure (or, for 4L cylinders insulated by a vacuum, at least 15 psi lower than one and one-fourth times the marked service pressure).

Liquid oxygen shipped in insulated truck cargo tanks or in other insulated containers at pressures below 25 psig is not subject to the regulations of the ICC.

SHIPPING METHODS; REGULATIONS

Under the appropriate regulations and tariffs, oxygen is authorized for shipment as follows (gaseous oxygen, except where liquid oxygen is noted):

By Rail: In cylinders and tank cars, for both gaseous and liquid oxygen (special insulated tank cars for liquid oxygen are not subject to ICC regulations).

By Highway: In cylinders on trucks, and in tube trailers. Liquid oxygen, in cylinders, and in tank trucks or trailers not subject to ICC regulations.

By Water: In cylinders on cargo and passenger vessels, and on ferry and railroad car ferry vessels (passenger or vehicle). In authorized tank cars on cargo vessels only. In cylinders on barges of U. S. Coast Guard classes A, BA, BB, CA and CB. Pressurized liquid oxygen, in cylinders on cargo and passenger vessels, and on ferry and railroad car ferry vessels (passenger or vehicle).

By Air: In cylinders aboard passenger aircraft up to 150 lb (70 kg), and aboard cargo aircraft up to 300 lb (140 kg), maximum net weight per cylinder. Liquid oxygen, either pressurized or nonpressurized, is not accepted for air shipment under present regulations.

OXYGEN CYLINDERS

Cylinders meeting ICC specifications 3A or 3AA are the types usually used to ship gaseous oxygen, but oxygen is authorized for shipment in any cylinders designated for nonliquefied compressed gases. (These include cylinders which comply with ICC specifications 3A, 3AA, 3B, 3C, 3D, 3E, 4, 4A, 4B, 4BA, and 4C; in addition, cylinders meeting ICC specifications 3, 7, 25, 26, 33, and 38 may be continued in gaseous oxygen service, but new construction is not authorized.)

Liquid oxygen is authorized for shipment in cylinders which meet ICC specifications 4L200; such cylinders must be equipped with pressure-controlling valves as stated earlier.

Small cylinders of special sizes for oxygen used in medicine (alone or in mixtures with helium or carbon dioxide) are described in Section B of Chapter 1 in Part III; the section also explains the applications and safe handling of oxygen used medicinally. Cylinder valve connections for medicinal oxygen are discussed in Sections B and C of Chapter 4, Part III.

Valve Outlet and Inlet Connections. Standard Connections, U. S. and Canada—No. 540, No. 870. Standard Connections, U. S. and Canada, for oxygen-carbon dioxide mixtures (carbon dioxide not over 7 per cent)—No. 280,

No. 880. Standard Connections, U. S. and Canada, for oxygen-helium mixtures (helium not over 80 per cent)—No. 280, No. 890.

Cylinder Requalification. All cylinders authorized for gaseous oxygen service must be requalified by hydrostatic retest every 5 years under present regulations, with the following exceptions: ICC-4 cylinders, every 10 years; and no periodic retest is required for cylinders of types 3C, 3E, 4C and 7.

Cylinders of the 4L type authorized for liquid oxygen service are also exempt from periodic retest requirements at present.

Small Portable Containers for Liquid Oxygen. Liquid oxygen is shipped and stored in small portable containers which hold quantities ranging from 1 to about 25 gal. These long-necked spherical containers encased in cylindrical steel shells are heavily insulated; they maintain the liquid at atmospheric pressure, and are consequently not subject to ICC regulations.

OXYGEN TANK CARS

Gaseous oxygen is authorized for rail shipment in tank cars meeting ICC specifications 107A. These cars consist of a number of seamless, tubular containers permanently mounted on a special frame. ICC regulations require that the pressure to which the car is charged must not exceed seven-tenths of the marked test pressure at 130 F in uninsulated 107A cars, as previously noted.

Liquid oxygen is also shipped in special insulated tank cars with capacities equal to a million or more cubic feet of oxygen gas at atmospheric pressure. The cars operate at pressures below 25 psig.

TUBE TRAILERS AND TANK TRUCKS AND TRAILERS

Gaseous oxygen is shipped in tube trailers with capacities ranging up to more than 40,000 cu ft of gas at atmospheric pressure. The nested and manifolded tubes of these trailers are built to comply with ICC-3A or 3AA cylinder specifications (or ICC-3AX or 3AAX specifications, which are also authorized). The trailers commonly serve as the storage supply for the user, with empty trailers being replaced periodically by the supplier.

Liquid oxygen is shipped in bulk at pressures below 25 psig in special insulated cargo tanks on trucks and truck trailers. These low-temperature tanks hold the equivalent of 400,000 cu ft or more of oxygen at atmospheric pressure.

STORAGE SYSTEMS

High-pressure cylinders, tube trailers and tank cars often serve as the storage supply for gaseous oxygen, while oxygen is also frequently stored in compact liquid form at the user's site. An essential factor that must be taken into account in designing an oxygen storage system is the power of oxygen to promote combustion. Standards to insure safety with oxygen systems are set forth in pamphlets of the National Fire Protection Association[2,3,4] and the Compressed Gas Association, Inc.[4]

REFERENCES

1. "Equipment Cleaned for Oxygen Service" (Pamphlet G-4.1), Compressed Gas Association, Inc.
2. "Standard for the Installation and Operation of Oxygen-Fuel Gas Systems for Welding and Cutting" (NFPA Pamphlet No. 51), National Fire Protection Association, 60 Batterymarch St., Boston, Mass, 02110.
3. "Nonflammable Medical Gas Systems" (NFPA Pamphlet No. 565), National Fire Protection Association.
4. "Standard for Bulk Oxygen Systems at Consumer Sites" (CGA Pamphlet G-4.2), Compressed Gas Association, Inc; (NFPA Pamphlet No. 566), National Fire Protection Association.

ADDITIONAL REFERENCES

"Oxygen Specification," Compressed Gas Association, Inc.

"Standard Density Data, Atmospheric Gases and Hydrogen" (Pamphlet P-6), Compressed Gas Association, Inc.

Phosgene

$COCl_2$

Synonyms: Carbonyl chloride, carbon oxychloride, chloroformyl chloride

ICC Classification: Poison gas, class A; poison gas label

PHYSICAL CONSTANTS

International symbol	$COCl_2$
Molecular weight	98.924
Vapor pressure at 68 F, psia	23.44
Boiling point at 1 atm	46.7 F
Freezing point at 1 atm	−195 F to − 198 F
Critical temperature	360 F
Critical pressure, psia	823
Critical density, lb/cu ft	32.5
Specific volume at 70 F, cu ft/lb	3.9
Specific gravity of gas at (68) F (air = 1)	3.5
Specific gravity of liquid at 68 F	1.388
Density of liquid at 68 F, lb/cu ft	86.50
Latent heat of vaporization at 46.4 F, Btu/lb	106.2
Heat capacity of liquid at boiling point, Btu/(lb)(°F)	0.244
Coefficient of expansion at 32 F per °F	0.002207
Weight per gallon, liquid, at 68 F, lb	11.58

Properties. Phosgene is a nonflammable colorless gas more than three times as heavy as air, and is designated a poison of the class A or "extremely dangerous" group by the ICC. Phosgene under pressure is a colorless to light yellow liquid. It has its own characteristic odor which is often stifling or suffocating and strong, but sometimes not unpleasant, depending on the concentration; it has been said to resemble sour green corn or moldy hay in odor when greatly diluted with air. Phosgene vapors strongly irritate the eyes. The vapors do not ordinarily persist in dangerous concentrations where there is adequate ventilation because of phosgene's low boiling point.

Completely dry, pure phosgene is stable at ordinary temperatures. It dissociates into its component parts, carbon monoxide and chlorine, at elevated temperatures, to an extent ranging from 0.45 per cent dissociation at 214 F to 100 per cent at 1472 F.

Phosgene is slightly soluble in water and is slowly hydrolized by water to form corrosive hydrochloric acid and carbon dioxide. It is imperative to prevent moisture from entering any closed phosgene container because the formation of these compounds could build up sufficient pressure to rupture the container.

For other solvents, phosgene dissolves as follows (parts per 100 parts of solvent by weight, at 1 atm and 68 F, or as indicated): carbon tetrachloride, 28; chloroform, 59; glacial acetic acid, 62; Russian mineral oil, 35; chlorinated paraffin, 81; ethyl acetate, 98; benzene, 99; toluene (at 63 F), 244; xylene (at 54 F), 457; and chlorobenzene (at 54 F), 422.

Field neutralization of phosgene, in emergencies or after possible gas warfare use, is accomplished with alkali or alkali solutions.

Materials of Construction. Anhydrous phosgene in the liquid state is compatible with a variety of common metals, including aluminum (of 99.5 per cent purity), copper, pure iron or cast iron, steel (including cast steel and chrome-nickel steels), lead (up to 250 F), nickel, and silver; it is also compatible with platinum and platinum alloys in instruments. Nonmetallic materials with which liquid anhydrous phosgene is also compatible include acid-resistant linings (ceramic plates and carbon blocks), enamel on cast iron or glass-lined steel, Jena special glass (as well as "Pyrex" and "Kimax"), porcelain, quartzware, granite or basalt natural stone, stoneware and "Teflon."

In the presence of moisture, phosgene is not compatible with copper, steel, nor pure or cast iron. Detailed data on the corrosion resistance of various materials to phosgene under a range of conditions are given in a 1960 survey of the Shell Development Company.[1]

Steel piping with seamless fittings is recommended for handling phosgene as a general rule, and pipe no smaller than $\frac{3}{4}$-in. nominal size should be used to insure rigidity and minimize possible leaks. For pipe size up to 4-in., schedule 80 seamless (or alloy steel to ASTM A333 GR3) piping is recommended; 6-in.-diameter schedule 40 seamless may be used as a larger pipe size. Screwed or flanged joints should be kept to the minimum, and cast iron or malleable iron fittings and valves should not be used; nonarmored porcelain valves must not be used, regardless of the pressure with either liquid or gas phosgene. Only O.S.Y. or rising stem valves are recommended, to reduce the possibility of accidents; nonindicating valves should not be used. "Monel" is the material generally used in manually operated valves for disc, seat and stem.

A pipe joint compound composed of litharge and glycerin is recommended, as are bonded asbestos fiber, chemical lead 2 to 4 per cent antimony, or "Teflon" envelope for flat gaskets, depending on the temperature. Detailed recommendations on these and other materials for various purposes in phosgene service may be obtained from phosgene manufacturers.

Manufacture. Phosgene is produced commercially by passing chlorine and carbon monoxide (in excess) over activated carbon as a catalyst under carefully controlled conditions.

Commercial Uses. Phosgene is used mainly as an intermediate in the manufacture of many types of compounds (including: barbiturates; chloroformates and thiochloroformates; carbamoyl chlorides, acid chlorides and acid anhydrides; carbamates; carbonates and pyrocarbonates; urethanes; ureas, azo-urea dyes, triphenylmethane dyes and substituted benzo phenones; isocyanates and isothiocyanates; carbazates and carbohydrazides; malonates; carbodiimides; and oxazolidinediones). It is also used in bleaching sand for glass manufacture, and as a chlorinating agent.

Physiological Effects. Phosgene is a lung irritant and also attacks other parts of the respiratory system. Low concentrations in air cause watering of the eyes and coughing which may result in a thin, frothy expectoration. High concentrations cause greater distress. Phosgene is more than ten times as toxic as chlorine and about 680 ppm (0.5 mg per liter) is lethal within 10 minutes. Exposure to concentrations of 3 to 5 ppm can cause an irritation of the eyes and throat, with coughing; 25 ppm represents a dangerous exposure if prolonged for 30 to 60 minutes; 50 ppm proves rapidly fatal even after short exposure; and about 120 ppm (0.5 mg per liter) is lethal within 10 minutes. The maximum allowable concentration for an 8-hour exposure is generally accepted to be 0.1 ppm (0.4 mg per cubic meter of air).

One serious difficulty with the treatment of persons exposed to phosgene is that symptoms may not appear until hours after the exposure. The delayed action of phosgene can be particularly injurious if the victim performs violent exercise after having been exposed.

All persons who have been gassed with phosgene must be examined by a physician as soon as possible, because serious symptoms may develop at a later stage.

First Aid for Phosgene Exposure. First summon a physician to examine any person exposed to phosgene. Then take the following steps:

(1) Remove contaminated clothing if impregnated with phosgene.

(2) Do not permit the patient to walk.

(3) Keep the patient at rest. An occasional change of position from lying to sitting may be beneficial. Keep the patient calm and have him try to suppress desires to cough.

(4) Warm the patient with blankets or hot water bottles and give drinks of hot sweetened tea or coffee to which a lump of butter has been added.

(5) Any difficulty of breathing or cyanosis should be relieved by administering a 93 per cent oxygen-7 per cent carbon dioxide mixture or oxygen through a mask.

(6) If and only if breathing has ceased, apply artificial respiration by the Schaefer prone-pressure method (which will help to empty the lungs of fluid), and at the same time have an assistant administer gases as in (5) above.

Customary Steps in Medical Treatment. Physicians usually observe the following practices in the treatment of rarely occurring cases of phosgene gassing:

(1) Absolute rest is essential, and the patient should be transported to the hospital in an ambulance.

(2) The patient should be watched very closely for 24 hours, because serious lung edema can develop suddenly in patients who are apparently normal.

(3) Continuous administration of pure oxygen by means of a mask is of the utmost importance, and may be needed for several days.

(4) Cardiac weakness is often apparent and coramine (1 cc), or camphor in oil if coramine is not available, should be given every 4 hours.

(5) Venesection should be performed only if there is definite evidence of embarrassment of the right heart, and not for cyanosis alone.

(6) If the pulse is rapid and feeble, the heart should be fully digitalized.

PHOSGENE CONTAINERS

Phosgene is authorized for shipment by the ICC in cylinders and in TMU tanks, which usually have net capacities of 150 lb and 2000 lb, respectively.

Filling Limits. The maximum filling densities permitted for phosgene are as follows:

In cylinders—125 per cent (per cent of water capacity by weight) plus the requirement that the cylinder must not contain more than 150 lb of phosgene.

In TMU tanks—not liquid full at 130 F.

SHIPPING METHODS; REGULATIONS

Under the appropriate regulations and tariffs, phosgene is authorized for shipment as follows:

By Rail: In cylinders (via freight only), and in TMU tank cars.

By Highway: In cylinders on trucks, and in TMU tanks on trucks and on full or semi-trailers (provided that tanks are securely chocked or clamped to prevent shifting, and that adequate facilities are available for handling tanks where transfer in transit is necessary).

By Water: In cylinders and TMU tank cars authorized by the ICC on cargo vessels only. In authorized cylinders on barges of U. S. Coast Guard classes A, BA, BB, CA and CB.

By Air: Not accepted for shipment.

PHOSGENE CYLINDERS

Cylinders that meet ICC specifications 3D and that are of not over 125-lb water capacity (nominal) are authorized for the shipment of phosgene. (Cylinders meeting specifications ICC-33 may also be continued in phosgene service, but new construction is not authorized.) Gas-tight valve protection caps must be affixed to cylinders for shipment unless the cylinders are packed in wooden boxes as prescribed; if gaskets are used between the caps and cylinder necks, they must be renewed for each shipment even though they may appear to be in good condition.

Each filled cylinder must show absolutely no leakage in an immersion test made before shipment. Cylinders must be tested without their valve protection caps. For the test, the cylinder and valve must be kept submerged in a bath of water heated to approximately 150 F for at least 30 minutes, and frequent examinations must be made during that time to note any escape of gas. The cylinder valve must not be loosened after the test and before shipment.

All the precautions necessary for the safe handling, shipping and storage of any compressed gas cylinders must of course be observed with phosgene cylinders.

Valve Outlet and Inlet Connections. Standard connections, U. S. and Canada—No. 640; Small Valve Series—No. 160.

Cylinder Requalification. Both the 3D and 33 types of cylinders authorized for phosgene service must be requalified by hydrostatic retest every 5 years under present regulations.

Leak Detection. Suspected leaks should be investigated only by personnel who are wearing gas masks of an approved type (these are gas masks approved for protection against phosgene by the U. S. Bureau of Mines; the Bureau recommends either an acid-gas-and-organic-vapor mask or a universal mask for concentrations of less than 2 per cent, and masks with a self-contained breathing apparatus for larger concentrations or sites otherwise deficient in oxygen). Phosgene users should consult U. S. Bureau of Mines publications for recommendations of necessary masks.[2]

Personnel wearing masks can easily detect phosgene leaks with ammonia vapor devices, as phosgene produces white fumes in the presence of ammonia. In case of leakage around the valve stem, an operator should tighten down on the valve packing nut only with the special wrench supplied with the cylinder for this purpose.

Warming Cylinders to Help Remove Contents. Phosgene cylinders may be heated by warm air or warm water to facilitate removal of their contents. Never use steam, boiling water or direct flame for this purpose. Never under any circumstances allow the outside of a cylinder to reach temperatures above 125 F.

TMU TANKS

TMU tanks that meet ICC specifications 106A500X are authorized for phosgene service. (Cylinders meeting specifications 106A500 may also be continued in phosgene service, but new construction is not authorized.) These tanks may be shipped either by rail or by motor vehicle. The tanks must be equipped with gas-tight valve protection caps approved by the Bureau of Explosives, and they must not be equipped with safety relief devices of any type. TMU tanks equipped in this way are authorized only for phosgene.

REFERENCES

1. "Corrosion Data Survey" (P-3), 1960 edition, Shell Development Co.
2. Circular No. 7885 (pp. 5, 6, 8, 9); Supplemental List 1C 7885, pp. 2, 3, U. S. Bureau of Mines.

Propane

See **Liquefied Petroleum Gases**

Propene and Propylene

See **Liquefied Petroleum Gases**

Rare Gases of the Atmosphere

ARGON (see separate section)
HELIUM (see separate section)
KRYPTON: Kr
NEON: Ne
XENON: Xe
ICC Classification (neon only): Nonflammable compressed gas; green label

PHYSICAL CONSTANTS

	Neon	Krypton	Xenon
International symbol	Ne	Kr	Xe
Molecular weight	20.183	83.70	131.3
Density at 70 F and 1 atm, lb/cu ft	0.05215	0.2172	0.3416
Specific volume, at 70 F and 1 atm, cu ft/lb	19.18	4.604	2.927
Specific gravity (air = 1) at 70 F and 1 atm	0.6959	2.894	4.560
Density, saturated vapor, at 1 atm, lb/cu ft	0.5862	—	—
Density, gas, at boiling point, lb/cu ft	0.6068	—	—
Density, liquid, at boiling point, lb/cu ft	75.35	150.6	190.8
Liquid/gas ratio (liquid at boiling point, gas at 70 F and 1 atm), vol/vol	1/1445	1/693.4	1/558.5
Boiling point, at 1 atm	−410.9 F	−244.0 F	−162.6 F
Melting point, at 1 atm	−415.6 F	−272 F	−220 F
Triple point temperature	−415.48 F	−252.78 F	−169.24 F
Triple point pressure, psia	6.284	10.617	11.838
Critical temperature	−379.75 F	−82.79 F	+61.9 F
Critical pressure, psia	394.74	796.25	855.32
Latent heat of vaporization at boiling point, Btu/lb	37.44	46.40	41.4
Latent heat of fusion, at freezing point, Btu/lb	7.1	—	—
Specific heat at constant pressure, C_p, at 70 F	0.25	0.06	0.04
Specific heat at constant volume, C_v, at 70 F	0.153	0.036	0.024
Specific heat ratio, at 70 F	1.642	1.689	1.666

Properties. Krypton, neon and xenon are rare atmospheric gases. Each is odorless, colorless, tasteless, nontoxic, monatomic and chemically inert. All three together constitute less than 0.002 per cent of the atmosphere, with approximate concentrations in the atmosphere of 18 ppm for neon, 1.1 ppm for krypton, and 0.09 ppm for xenon. Few users of the three gases need them in bulk quantities and the three are shipped most often in single cylinders and liter flasks.

Two of the other rare gases of the atmosphere, argon and helium, are treated in separate sections, for they are far more plentiful and have more varied uses. Radon, the sixth rare gas, is not treated in the Handbook because it has little or no practical application at present. It is radioactive, and is the heaviest gas known (density at 70 F and 1 atm, 0.61 lb/cu ft).

Among the rare gases, neon, krypton and xenon in particular ionize at lower voltages than other gases, and the brilliant, distinctive light they emit while conducting electricity in the ionized state accounts for one of their primary uses. Their characteristic colors as ionized conductors are: neon, red; krypton, yellow-green; and xenon, blue to green (and similarly, argon, red or blue; and helium, yellow).

Materials of Construction. These three inert gases may be used with containers and equipment made of any common metals. Installations must, of course, be designed to meet all requirements for the pressures involved.

Manufacture. Neon, krypton and xenon are produced commercially at air separation plants in two stages—an initial stage of partial separation by liquefaction and fractional distillation, and a final purification stage requiring complex processing.

Commercial Uses. Neon, krypton and xenon are used principally to fill lamp bulbs and tubes. The electronics industry uses them singly or in mixtures in many types of gas-filled electron tubes (among them, voltage regulator tubes, starter tubes, phototubes, counter tubes, T. R. Tubes, xenon thryatron tubes, half-wave xenon rectifier tubes and Geiger-Muller tubes). Large quantities of neon (as well as of atmospheric helium and specially purified argon) are employed as fill gases in illuminated signs, small quantities of krypton and xenon being used for special effects. In the lamp industry the three gases serve as fill gas in specialty lamps, neon glow lamps, 100-watt fluorescent lamps, ultra violet sterilizing lamps and very high output lamps. The three gases have additional applications in the atomic energy field as fill gas for ionization chambers, bubble chambers, gaseous scintillation counters and other detection and measurement devices.

Neon, krypton and xenon are produced on a comparatively small scale. One estimate placed the total U. S. production of the three in 1958 at about 5000 cu ft (most of it neon), as compared to some 500 million cu ft of argon in 1959 and over 640 million cu ft of helium in 1960.[1]

Physiological Effects. Neon, krypton and xenon are nontoxic. As gases, they can act as simple asphyxiants by displacing air, and they cannot be detected by odor or color. Liquefied neon has come into use, and it must be handled with all the precautions required for safety with a low-temperature liquefied gas (see Part III, Chapter 1, section on liquefied atmospheric gases).

NEON, KRYPTON AND XENON CONTAINERS

Seldom used in bulk quantities, neon, krypton and xenon are shipped in individual cylinders or in liter quantities in glass flasks. Liquefied neon is also shipped in special insulated containers.

Filling Limit. The maximum authorized filling limit for neon, krypton and xenon in approved types of cylinders is the marked service pressure of the cylinder at 70 F (or, in the case of only 3A and 3AA cylinders meeting additional specified requirements, 10 per cent in excess of the marked service pressure).

SHIPPING METHODS; REGULATIONS

Under the appropriate regulations and tariffs, neon, krypton and xenon are authorized for shipment as follows:

By Rail: In cylinders (freight or express).

By Highway: In cylinders on trucks.

By Water: In cylinders via cargo and passenger vessels, and in ferry and railroad car ferry vessels (either passenger or vehicle). In cylinders on barges of U. S. Coast Guard classes A, BA, BB and C.

By Air: For gaseous neon and krypton, aboard passenger aircraft in appropriate cylinders up to 150 lb (70 kg) maximum net weight per cylinder, and aboard cargo aircraft in appropriate cylinders up to 300 lb (140 kg) maximum net weight per cylinder. Liquefied neon, either pressurized or low-pressure, is authorized for shipment aboard cargo aircraft only in containers meeting specific requirements up to 300 lb (140 kg) maximum net weight per container. Xenon is shipped by air as a "Nonflammable compressed gas, n. o. s. (not otherwise specified)."

NEON, KRYPTON AND XENON CYLINDERS

Neon, krypton and xenon are authorized for shipment in cylinders of any type approved by the ICC for nonliquefied compressed gases (these are cylinders meeting ICC specifications 3A, 3AA, 3B, 3C, 3D, 3E, 4, 4A, 4BA, and 4C; also, cylinders meeting ICC specifications 3, 7, 25, 26, 33, and 38 may be continued in service with these gases, but new construction is not authorized). Krypton and xenon are shipped by rail, highway or water as "Nonflammable compressed gas, n. o. s. (not otherwise specified)."

Valve Outlet and Inlet Connections. Standard connection, U. S. and Canada—No. 580; small valve series, No. 120.

Cylinder Requalification. Under present regulations, cylinders of all types authorized for service with neon, krypton and xenon must be requalified by hydrostatic test every 5 years with the exception of types 3C, 3E, 4C, and 7 (for which periodic hydrostatic retest is not required).

REFERENCE

1. G. A. Cook, ed., *Argon, Helium, and the Rare Gases*, 2 vols., pp. 435–437, New York, John Wiley & Sons—Interscience, 1961.

ADDITIONAL REFERENCE

"Standard Density Data, Atmospheric Gases and Hydrogen" (Pamphlet P-6), Compressed Gas Association, Inc.

Sulfur Dioxide

SO_2
Synonym: Sulfurous acid anhydride
ICC Classification: Nonflammable compressed gas; green label

PHYSICAL CONSTANTS

International symbol	SO_2
Molecular weight	64.06
Vapor pressure at 70 F, psig	34.4
Density of gas at 32 F and 1 atm, lb/cu ft	0.1827
Specific gravity of gas at 32 F and 1 atm (air = 1)	2.2636
Specific gravity of liquid at 32 F	1.436
Density of liquid, lb/cu ft	
at 70 F	86.06
at 105 F	82.55
at 115 F	81.50
at 130 F	79.94
Boiling point at 1 atm	14.0 F
Freezing point at 1 atm	−103.9 F
Critical temperature	314.82 F
Critical pressure, psia	1141.5
Latent heat of vaporization at 70 F, Btu/lb	155.5
Solubility in water, by weight	
at 32 F	18.59%
at 68 F	10.14%
at 86 F	7.30%
at 104 F	5.13%
Weight per gallon, liquid, at 70 F, lb	11.53

Properties. Sulfur dioxide is a nonflammable, colorless gas that has a characteristic, pungent odor and is highly irritating at room temperatures and atmospheric pressures. At temperatures below 14 F, or under moderate pressures, sulfur dioxide is a colorless liquid. It is more than twice as heavy as air in gaseous form and roughly one and a half times the weight of water as a liquid, and consists by weight of 50.05 per cent sulfur and 49.95 per cent oxygen. It is shipped as a liquefied compressed gas under its vapor pressure of some 35 psig at 70 F.

Chemically, sulfur dioxide is an outstanding oxidizing and reducing agent. Dry sulfur dioxide is not corrosive to ordinary metals. However, zinc will react with sulfur dioxide containing minute quantities of moisture, and most common metals will be corroded by sulfur dioxide holding sufficient amounts of moisture. Sulfur dioxide dissolved in water will form sulfurous acid, which is unstable toward heat, and decreasing proportions of sulfur dioxide go into solution in water as temperature increases.

Table 1 gives the vapor pressure, volume, density, and latent heat properties of sulfur dioxide in containers over a range of temperatures.

Materials of Construction. Steel and other common metals except zinc have been found to give satisfactory service with dry sulfur dioxide. Among materials suitable with moisture-bearing sulfur dioxide are certain stainless steels (such as type 316) and lead.

Manufacture. In North America sulfur

TABLE 1. VAPOR PRESSURE, VOLUME, DENSITY, AND LATENT HEAT OF SULFUR DIOXIDE IN CONTAINERS AT VARIOUS TEMPERATURES[1]

Temperature		Vapor Pressure		Volume (cu ft/lb)		Density (lb/cu ft)		Latent Heat (Btu/lb)
(°F)	(°C)	(psia)	(psig)	Liquid	Vapor	Liquid	Vapor	
−40	−40.0	3.12	*23.6″	.01044	22.2	95.79	.04505	178.4
−20	−28.9	5.88	*18.0″	.01062	12.5	94.16	.08000	174.4
0	−17.8	10.26	*9.07″	.01082	7.35	94.42	.13605	170.3
10	−12.2	13.3	*2.85″	.01092	5.77	91.58	.17331	168.3
20	− 6.7	16.9	2.2	.01103	4.59	90.66	.21786	166.3
30	− 1.1	21.3	6.6	.01114	3.70	89.77	.27027	164.2
40	4.4	26.6	11.9	.01125	3.02	88.89	.33113	162.2
50	10.0	32.9	18.2	.01137	2.48	87.95	.40323	160.0
60	15.6	40.3	25.6	.01149	2.05	87.03	.48780	157.8
70	21.1	49.1	34.4	.01162	1.70	86.06	.58824	155.5
80	26.7	59.3	44.6	.01175	1.42	85.11	.70423	153.1
90	32.2	71.0	56.3	.01189	1.20	84.10	.83333	150.7
100	37.8	84.1	69.4	.01204	1.02	83.06	.98039	148.2
110	43.3	99.1	84.4	.01219	.868	82.03	1.15207	145.7
120	48.9	116.3	101.6	.01235	.746	80.97	1.34048	143.0
130	54.4	135.8	121.1	.01251	.646	79.94	1.54799	140.0
140	60.0	157.7	143.0	.01269	.554	78.80	1.80505	137.1

* Indicates inches of mercury below atmospheric pressure.

dioxide is produced by the combustion of sulfur in burners of special design, by burning pyrites, as a by-product of smelter operations and as a by-product of chemical operations. Sulfur dioxide can be purified by passing the gas into water which dissolves it and certain impurities. This liquor is then heated to drive off the sulfur dioxide, the liberated gas being dried and liquefied.

The gas thus produced is sold in two grades: the commercial grade and the refrigeration grade. The refrigeration grade is purified to contain not more than .005 per cent of moisture, its quality being controlled by constant sampling and anslysis to meet the exacting specifications of the refrigeration industry. The commercial grade contains not more than .010 per cent of moisture but is not considered suitable for refrigerating machines.

Commercial Uses. Sulfur dioxide's useful properties as a refrigerant, fumigant, preservative, bleach, antichlor, etc. are utilized in a diverse group of industries. Small quantities of sulfur dioxide are used by the refrigerating and air conditioning industries. The petroleum industry consumes sulfur dioxide in the Edeleanu Process for the refining of kerosene and light lubricating oils. Another principal application of sulfur dioxide is found in the manufacture of sulfite pulp for paper and artificial silk.

Other uses of sulfur dioxide include its utilization in the multiple role of preservative, bleach and fumigant as it is used to preserve fruits, to bleach fruits, sugar and grains and to fumigate vermin-infested grains. Sulfur dioxide in the textile industry is employed as an antichlor and sour in bleaching and is used in the preparation of sodium hydrosulfite for dyeing and printing. As an antichlor it is also used in water treatment to remove objectionable odors remaining after purification. Liquid sulfur dioxide is used in the preparation of chlorine dioxide for pulp bleaching and other uses. Other uses include its employment in tanning leather, metal refining, fumigating ships and as a catalyst and reagent in the manufacture of various resins and plastics.

Large quantities of sulfur dioxide are used in the southwestern United States in a process which involves the addition of sulfur dioxide to irrigating water to increase crop yields in alkaline soils.

Physiological Effects. Sulfur dioxide is an

extremely irritating gas which is practically irrespirable to those unaccustomed to it. It readily elicits respiratory reflexes. It affects the upper respiratory tract, but with deeper breathing affects the lower system also. Four parts per million can be readily detected by odor but as the nose becomes accustomed to it, the amount necessary to produce a reflex respiratory defense response increases. In higher concentrations the severely irritating effect of the gas makes it inconceivable that a person would remain in a sulfur dioxide contaminated atmosphere unless he were trapped and unable to flee. Since the gas serves in this way as its own warning agent, cases of severe exposure to sulfur dioxide are not numerous.

Exposure to sulfur dioxide gas in low concentrations produces an irritating effect on the mucous membranes of the eyes, nose, throat and lungs due to the formation of sulfurous acid as the gas comes in contact with the moisture on these surfaces. This irritation is accompanied by a desire to cough which must be controlled to minimize injury. Exposure to higher concentrations produces a suffocating effect due to the closing of the glottis to shut out the gas. Table 2 shows the physiological response to various concentrations of sulfur dioxide. There is no evidence to indicate that exposure to sulfur dioxide in allowable concentrations produces a cumulative effect.

TABLE 2. PHYSIOLOGICAL RESPONSE TO VARIOUS CONCENTRATIONS OF SULFUR DIOXIDE

	Parts of Sulfur Dioxide per Million Parts of Air
Least detectable odor*†	3 to 5
Least amount causing immediate irritation to the eyes†	20
Least amount causing immediate irritation to the throat†	8 to 12
Least amount causing coughing†	20
Maximum concentration allowable for prolonged exposure†	10
Maximum concentration allowable for short exposure (½ to 1 hr)†‡	50 to 100
Dangerous for even short exposure‡	400 to 500

* U. S. Dept. of Interior, Bureau of Mines, Bulletin 98, 1915.

† A Fieldner and S. Katz, *Eng. and Mining J.*, 1919, 107, page 693.

‡ Lehman, *Arch. f. Hyg.*, **18**, 180.

Exposure to escaping sulfur dioxide liquid will result in a freezing action of the skin in addition to any effects of inhalation of gas. This freezing effect is the natural result of the escape of any liquefied refrigerant in the lower boiling point range.

Provisions for an Emergency. Since sulfur dioxide represents a panic hazard, and in a sufficiently high concentration may cause injury or even death from suffocation, provisions should be made in advance by the sulfur dioxide user for action in an emergency. All employees handling sulfur dioxide should be impressed with the potential danger which it represents and should be trained in its safe handling. In addition they should be provided with personal protective equipment for use in an emergency and should be drilled until they are familiar with its use. This protective equipment should include a gas mask of a type approved by the Bureau of Mines for sulfur dioxide service. Care should be taken to assure that masks are kept in proper working order and that they are stored so as to be readily available in case of need. Cannister type masks are unsafe for high concentrations of sulfur dioxide and employees should be warned of possible failure of this type of mask in the event of a really serious leak. Self-contained breathing apparatus or a mask with a long air hose and outside source of air may be required under extreme conditions. Other protective equipment provided should include goggles or large lens spectacles to eliminate the possibility of liquid sulfur dioxide coming in contact with the eyes and causing possible injury.

If sulfur dioxide should be released, the irritating effect of the gas will force personnel to leave the area before they have long been exposed to dangerous concentrations. To facilitate their rapid evacuation there should be sufficient well-marked, easily accessible exists. If despite all precautions a person should be trapped in a contaminated atmosphere, he should breath as little as possible and open his eyes only when necessary. Since sulfur dioxide gas is heavier than air a trapped person should seek a high position to take advantage of lower gas concentrations at that level.

Since sulfur dioxide neither burns nor supports combustion, there is no danger of fire or explosion due to igniting gas or liquid. Should fire break out due to some other cause in an area containing sulfur dioxide, every effort should be made to remove the containers from the area to prevent overheating which would lead to melting of the fuse plugs and release of the sulfur dioxide. If they cannot be removed, firemen should be informed of their location.

Whenever it is necessary to enter a storage tank or other vessel containing sulfur dioxide or one that has previously contained liquid sulfur dioxide, a self-contained respirator, gas mask or air pack should be worn; and the person entering the tank should be connected to an attendant by a rope even though the tank has been thoroughly purged. It is further advisable that the person entering the tank first apply a protective cream to those parts of the body which tend to perspire to avoid irritation of the skin and that gas-tight goggles or face mask be worn to avoid irritation of the eyes.

First Aid Suggestions. Any person who has been burned or overcome by sulfur dioxide should be placed under a physician's care immediately. Prior to the physician's arrival the following first aid suggestions are presented, based upon what is believed to be common practice in industry. Their adoption in a particular case should of course be subject to prior endorsement by the user's medical adviser.

(a) Remove patient burned or overcome by gas to an area free from fumes, preferably a warm room (about 70 F).

(b) Place patient in a reclining position with head and shoulders slightly elevated. Keep the patient warm, providing blankets or other covers if necessary. Keep the patient quiet and urge him to resist the desire to cough if possible. Rest is essential.

For asphyxiation. Where breathing is weak, administer mixtures of carbon dioxide and oxygen or carbon dioxide and air, containing not more than 5 per cent of carbon dioxide. Give the mixture continuously until all sounds of "gurgling" have ceased or until the lungs have cleared. (Apparatus required for this treatment is on the market.)

If breathing has ceased, start artificial respiration immediately, using slow, even motions, not to exceed 14 movements per minute. If possible assist respiration with an inhalator. Continue artificial respiration until ordered to stop by a physician or until the patient is breathing normally.

For the eyes. If liquid sulfur dioxide enters the eye, wash immediately with large quantities of water (keeping the lids open) for at least 15 min. Instill a drop of mineral oil or castor oil and send patient to an eye specialist as soon as possible.

For the skin. Immediately remove any clothing splashed with sulfur dioxide and wash the affected skin areas with large quantities of water. Cover burn areas with sterile gauze and keep patient warm. Treat for shock. Burns more serious than first degree should be dressed and treated by a physician.

Sulfur Dioxide Leaks. A sulfur dioxide leak indicates its presence by the characteristic, pungent odor of sulfur dioxide gas. The location of even the smallest leak may be readily determined by means of ammonia vapor dispensed from an aspirator or squeeze bottle in the region where a leak is suspected, or by the use of an ammonia swab, prepared by securing a small piece of cloth or sponge to a wire and soaking it in a strong solution of aqua ammonia. When the ammonia vapor or swab is passed over points of suspected leaks, dense white fumes form near the leak where the sulfur dioxide and ammonia come in contact. Leaks may also be less satisfactorily detected by applying oil or a soap solution to joints and noting where bubbles of escaping gas appear.

Measures in case of leakage. Where quantities of sulfur dioxide are being handled, possible hazards due to ruptured lines, broken gage glasses, leaking joints, etc., must be considered. When a leak does occur, only an authorized employee should attempt to stop it; and, if there is any question as to the seriousness of the leak, a suitable gas mask should be worn. (See above, Provisions for Emergency.) In general, where serious leaks develop, the person responsible, even when equipped with a suitable mask, should remain in the contaminated area only long enough to close the necessary valves and to make emergency adjustments.

Leaks which might develop are ordinarily not serious and can be readily controlled. Where leaks do occur, the supply of sulfur dioxide should be shut off by closing the appropriate valve. Leaks at unions or other fittings may often be eliminated by tightening the connection. If corrosion is indicated, care must be taken to empty the lines before working on them, as a broken fitting might lead to a serious loss of sulfur dioxide before the supply valves could be shut off.

Handling of leaking containers. Although serious leaks in shipping containers rarely occur, careless handling will sometimes result in this condition. Cylinder and TMU tank valves are made of brass and, if struck by a heavy object, might be broken. Leaks may also occur from carelessness in heating cylinders or drums, resulting in melting of a fuse plug and discharge of the contents. Care in handling will eliminate these dangers. Occasionally, leaks may develop in the valve packing, but these can ordinarily be checked by tightening the packing nut.

In the event of a leaking container, it should, wherever possible, be moved to an open area where the hazard due to escaping sulfur dioxide will be minimized. If a container is discharging too freely to permit movement, it should, if possible, be arranged in such a position that the leak will be at the top, thus discharging gaseous sulfur dioxide and not liquid. If large quantities of gas can be withdrawn rapidly into equipment, or satisfactorily vented, the evaporation will often lower the pressure in the container to such a point that it can be moved to the open without difficulty, or possibly the leak can be repaired by persons provided with suitable masks. If the leak still prevents removal of the container to an open area, gaseous or liquid sulfur dioxide can be vented into a solution of lime, caustic soda, or other alkaline material. One pound of sulfur dioxide is equivalent to about 2 lb of lime or 1½ lb of caustic soda.

If the above information proves insufficient under unusual circumstances, call the shipper at once.

SULFUR DIOXIDE CONTAINERS

Sulfur dioxide is authorized for shipment in cylinders, single-unit tank cars, TMU tank cars, truck cargo tanks, and portable tanks and TMU tanks on trucks.

Withdrawing Sulfur Dioxide from Shipping and Storage Containers. In order to withdraw sulfur dioxide liquid or vapor from shipping or storage containers, it is often necessary to warm the contents or to employ other special methods in ways which have proven safe and practical. These withdrawal methods are described in CGA Pamphlet G-3.[2]

Filling Limits. The maximum filling limits for all types of containers authorized for sulfur dioxide shipment under ICC regulations is the maximum filling density of 125 per cent (per cent water capacity by weight). This is roughly equivalent to the filling density of 87.5 per cent by volume that is authorized as an alternative for truck cargo tanks and portable tanks.

SHIPPING METHODS; REGULATIONS

Under the appropriate regulations and tariffs, sulfur dioxide is authorized for shipment as follows:

By Rail: In cylinders (via freight, express, or baggage), in single-unit tank cars, and in TMU tank cars.

By Highway: In cylinders on trucks, in cargo tanks, and in portable tanks and TMU tanks.

By Water: In cylinders on cargo and passenger vessels, and on ferry and railroad car ferry vessels (passenger or vehicle). In authorized truck cargo tanks on cargo vessels only, and on ferry and railroad car ferry vessels (passenger or vehicle). In authorized single-unit and TMU tank cars, and in portable tanks (meeting ICC-51 specifications, not over 20,000 lb gross weight, and not equipped with fixed-length dip-tube gaging devices), on cargo vessels only and on railroad car ferry vessels (passenger or vehicle). In cylinders on barges of U. S. Coast Guard classes A, BA, BB, CA, and CB.

By Air: In cylinders aboard cargo aircraft only up to 300 lb (140 kg) maximum net weight per cylinder.

SULFUR DIOXIDE CYLINDERS

Sulfur dioxide is authorized for shipment in cylinders which comply with the following ICC specifications: 3A225, 3AA225, 3B225, 4A225, 4B225, 4BA225, 4B240ET, 4, and 3E1800 (cylinders that meet ICC specifications 3, 25, 26 and 38 may also be continued in sulfur dioxide service, but new construction is not authorized).

Valve Outlet and Inlet Connections. Standard Connection, U. S. and Canada—No. 620. Alternate Standard Connection, U. S. and Canada—No. 360.

Cylinder Requalification. All types of cylinders authorized for sulfur dioxide service must be requalified by hydrostatic retest every 5 years under present regulations, with the following exceptions: ICC-4 cylinders, every 10 years; and no periodic retest is required for cylinders of type 3E.

SINGLE-UNIT TANK CARS AND TMU TANK CARS

Insulated single-unit tank cars complying with ICC specifications 105A200-W are authorized for rail shipment of sulfur dioxide in bulk. Rail shipment is also authorized in TMU tank cars meeting ICC specifications 106A500-X or 110A500-W.

CARGO TANKS, PORTABLE TANKS, AND TMU TANKS ON TRUCKS

Cargo tanks meeting specifications MC-330 or MC-331, and portable tanks complying with specifications ICC-51, are authorized for motor vehicle shipment of sulfur dioxide. Also authorized for shipment on trucks are TMU tanks of specifications 106A500-X or 110A500-W. The minimum design pressure for cargo or portable tanks not over 1200 gal. water capacity must be 150 psig plus a corrosion allowance, and other stipulated requirements must be met. For tanks exceeding 1200 gal. water capacity, the minimum design pressure must be 125 psig plus a corrosion allowance. Also authorized is an optional ICC-51 portable tank of 1000 to 2000 lb water capacity and with fusible plugs at each end; the minimum design pressure for such tanks must be 225 psig, and these tanks must be filled by weight.

STORAGE AND STORAGE TANKS

Sulfur dioxide must be stored as well as handled with all the precautions necessary for safety with any compressed gas. Cylinders and other portable containers in particular should not be stored in sub-surface locations, since sulfur dioxide gas, being heavier than air, might accumulate in the bottom of the storeroom and not be carried away by drafts from openings located above the surface level.

TMU tank containers should be stored on their sides and blocked to prevent rolling. A convenient horizontal storage rack may be made by supporting the drums at each end on a railroad rail or I-beam. Valve protecting plates or hoods should be in place at all times that containers are not in use.

In general, the storing of sulfur dioxide in cargo and portable tanks involves the same factors as does its storage in cylinders or multi-unit tank car tanks.

Sulfur dioxide consumers who require bulk storage equipment should secure competent engineering assistance in its design, construction and installation. Sulfur dioxide suppliers will often be able to provide valuable advice in connection with sulfur dioxide storage.

Sulfur dioxide bulk storage containers should be designed and constructed in accordance with recognized engineering practice as exemplified by Section VIII of the Boiler and Pressure Vessel Code of the American Society of Mechanical Engineers. Since it is possible that the design of the vessel, its capacity and its location may be influenced by state or municipal regulations and insurance restrictions, a thorough investigation of pertinent restrictions should be made prior to beginning construction of the vessel.

After the storage vessel has been installed, competent technical personnel should be trained in the operation and maintenance of this equipment. Personnel should be impressed with the great importance of preventing overfilling tanks with sulfur dioxide with the consequent risk of the vessel becoming liquid-full and failing due to hydrostatic pressure. The amount of liquid sulfur dioxide that may be stored safely in a vessel may be determined in two ways:

(a) By weight at 125 per cent of the weight of water that the container will hold at 60 F.

(b) By volume in accordance with the provisions of Table 3.

Each storage container, except containers mounted on scales, should be equipped with a liquid level gaging device to permit ready determination of the amount of liquid in the tank at any time. Liquid level gaging devices should conform to the following recommendations:

Table 3. Maximum Safe Volume of Liquid Sulfur Dioxide in a Storage Tank at Various Temperatures

Temperature of Liquid Sulfur Dioxide in Tank (°F)	Maximum Safe Volume Liquid Sulfur Dioxide in % of Full Volume at 125% Filling Density
30	86
40	87
50	88
60	89
70	90
80	91
90	92
100	93

Liquid level indicating devices should be designed and installed so as to permit reading the liquid level within plus or minus 1 per cent of the capacity of the container from full tank level down to at least 20 per cent below full tank. Readings below this level are usually desirable for other purposes but are not necessary for avoidance of overfilling.

Gage glasses may be used as liquid level devices if certain precautions are observed. They should be protected by solid transparent shields and guards and should be provided with excess flow valves or with weighted shut-off cocks that must be opened to take a reading and will close automatically. Each gage glass should be provided with a drain valve and should be drained after each reading to a point where it is no more than 85 per cent full. This should be done to prevent bursting that could be caused by thermal expansion. Gage glasses should be located so that the glass and its content do not differ greatly in temperature from the contents of the container.

All gaging devices should be arranged so that the maximum liquid level to which the container may be filled safely is readily determinable. For this purpose the information presented in Table 3 should be duplicated on a metal plate mounted at the operating position of the gaging device.

Gaging devices that require bleeding of sulfur dioxide to the atmosphere, such as rotary tube, fixed tube and slip tube, should be designed so that the bleed valve maximum opening is not larger than a No. 54 drill size, unless provided with excess flow valve.

Gaging devices should be designed for a working pressure of at least 200 psig.

REFERENCES

1. Rynning, D. F., and Hurd, C. O., "Thermodynamic Properties of Sulfur Dioxide," *Trans. Am. Inst. Chem. Eng.*, **41**, No. 3 (June 25, 1945).
2. "Sulfur Dioxide" (Pamphlet G-3), Compressed Gas Association, Inc.

Sulfur Hexafluoride

SF_6 ICC Classification: Nonflammable compressed gas; green label

PHYSICAL CONSTANTS

International symbol	SF_6
Molecular weight	146.06
Vapor pressure, psia	
at 32 F	186
at 68 F	315
at 70 F	320
at 104 F	487
at 113 F	540
at 130 F	780
Density of gas at 70 F and 1 atm, lb/cu ft	0.387
Specific gravity of gas at 68 F and 1 atm (air = 1)	5.106
Specific gravity of liquid at 130 F	0.728
Density of liquid under its own approx vapor pressure, lb/cu ft	
at 70 F	73.9
at 105 F	84.9
Sublimation temperature at 1 atm	−83 F
Freezing point at 32.5 psia	−59.4 F
Critical temperature	114.2 F
Critical pressure, psia	546.59
Critical density, lb/cu ft	45.5
Latent heat of vaporization at 32 F and 32.5 psia, Btu/lb	37.3
Latent heat of fusion at −59.4 F and 32.5 psia, Btu/lb	17.1
Specific heat of gas at 70 F and 1 atm, C_p, Btu/(lb)(°F)	0.16
Weight per gallon, liquid, at 68 F and 315 psia, lb	11.4

Properties. Sulfur hexafluoride is a colorless, odorless, nontoxic, nonflammable gas that has high dielectric strength and serves widely as an insulating gas in electrical equipment. At atmospheric pressures it sublimes directly from the solid to the gas phase, and does not have a stable liquid phase unless under pressures of more than some 32 psia. It is shipped as a liquefied compressed gas at its vapor pressure of about 300 psig at 70 F. One of the most chemically inert gases known, it is completely stable in the presence of most materials to temperatures of about 400 F, and has shown no breakdown or reaction in quartz at 900 F. Sulfur hexafluoride is slightly soluble in water and oil. No change in pH occurs when distilled water is saturated with sulfur hexafluoride.

Materials of Construction. Sulfur hexafluoride is noncorrosive to all metals. It may be partially degraded if subjected to an electrical discharge, and some of the breakdown products may be hydrolyzed in the presence of moisture to give acidic and possibly corrosive secondary products. Sulfur hexaluoride decomposes very slightly in the presence of certain metals at temperatures in excess of 400 F, and this effect is most pronounced with silicon and carbon steels. Such breakdown, presumably catalyzed by the metals, is of the order of only several tenths of 1 per cent over a year's time. Among metals with which decomposition at elevated temperatures does not occur are aluminum, copper, brass and silver.

Most common gasket materials, including

asbestos, neoprene and natural rubber, are suitable for sulfur hexafluoride service.

Manufacture. Sulfur hexafluoride is made commercially by the direct fluorination of molten sulfur. Some lower fluorides formed in the process are scrubbed out with various caustic solutions, and the commercial product is more than 99 per cent pure. Common impurities include small amounts of carbon tetrafluoride, nitrogen and water vapor.

Commercial Uses. Sulfur hexafluoride is employed extensively as a gaseous dielectric in various kinds of electrical power equipment, such as switch gears, transformers and circuit breakers. It has also been used as a dielectric at microwave frequencies, and as an insulating medium for the power supplies of high-voltage machines.

Physiological Effects. Sulfur hexafluoride is completely nontoxic, and in fact has been used medically with humans in cases involving pneumoperitoneum, or the introduction of gas into the abdominal cavity. It can of course act as a simple asphyxiant by displacing oxygen in the air. Lower fluorides of sulfur, some of which are toxic, may be produced if sulfur hexafluoride is subjected to electrical discharge, and inhalation of the gas after electrical discharge must be guarded against.

Leak Detection. Standard fluorcarbon detecting devices can be employed to find sulfur hexafluoride leaks, and can identify apertures leaking quantities as small as a half-ounce per year.

SULFUR HEXAFLUORIDE CONTAINERS

Sulfur hexafluoride is authorized for shipment in cylinders under ICC regulations.

Filling Limit. The maximum filling density permitted for sulfur hexafluoride in cylinders is 110 per cent (per cent water capacity by weight).

SHIPPING METHODS; REGULATIONS

Under the appropriate regulations and tariffs, sulfur hexafluoride is authorized for shipment as follows:

By Rail: In cylinders.

By Highway: In cylinders on trucks.

By Water: In cylinders on cargo vessels, passenger vessels, ferry vessels and railroad car ferry vessels (passenger or vehicle for both types of ferry vessels. In cylinders on barges of U. S. Coast Guard classes A, BA, BB, CA and CB.

By Air: In cylinders aboard passenger aircraft up to 150 lb (70 kg), and aboard cargo aircraft up to 300 lb (140 kg), maximum net weight per cylinder.

SULFUR HEXAFLUORIDE CYLINDERS

Cylinders that meet the following ICC specifications are authorized for sulfur hexafluoride service: 3A1000, 3AA1000 and 3E1800 (ICC-3 cylinders may also be continued in use, but new construction is not authorized).

Valve Outlet and Inlet Connection. Standard connection, U. S. and Canada—No. 590.

Cylinder Requalification. Cylinders authorized for sulfur hexafluoride service must be requalified by hydrostatic retest every 5 years under present regulations with the exception of 3E cylinders, for which no periodic retest is required.

Trimethylamine

See **Methylamines**

Vinyl Chloride

CH_2CHCl (or C_2H_3Cl)
Synonyms: Chloroethylene, chloroethene
ICC Classification: Flammable compressed gas; red label

PHYSICAL CONSTANTS

International symbol	C_2H_3Cl
Molecular weight	62.50
Vapor pressure at 68 F, psig	35
Specific gravity of gas at 59 F (air = 1)	2.15
Specific gravity of liquid, at 68 F	0.9121
Density, liquid, lb/cu ft	
at 70 F	56.71
at 105 F	54.38
at 115 F	53.69
at 130 F	52.61
Boiling point at 1 atm	7 F
Melting point at 1 atm	−245 F
Critical temperature	317 F
Critical pressure, psia	775
Flammable limits in air, by volume	4–22%
Flash point, open cup	−108 F

Properties. Vinyl chloride is a colorless, flammable gas with a sweet ethereal odor. It is shipped as a liquefied compressed gas, and the colorless or water-white liquid is so highly volatile that prolonged contact of it with the skin results in freezing or frostbite through evaporation. In addition, vinyl chloride irritates the skin and the eyes on contact, while inhalation of concentrations of more than 500 ppm produces mild anesthesia.

Dry vinyl chloride does not corrode metals at normal temperatures and pressures, but vinyl chloride accelerates the corrosion of iron and steel in the presence of moisture at elevated temperatures.

The most important of the vinyl monomers, vinyl chloride polymerizes readily when exposed to air, sunlight, heat or oxygen, although it is otherwise quite stable chemically. The addition of small amounts of phenol or of hydroquinone inhibits polymerization.

Materials of Construction. Steel is recommended for use with vinyl chloride, though vinyl chloride and moisture speed its corrosion at elevated temperatures. Copper and copper alloys must not be employed in contact with vinyl chloride. Acetylene present as an impurity in vinyl chloride can form an explosive acetylide when exposed to copper.

Manufacture. Vinyl chloride is produced by the reaction of acetylene and hydrogen chloride, either as liquids or gases, with a copper chloride catalyst in the liquid process and a mercury chloride catalyst in the gas process. It is also made by the cracking of ethylene dichloride (a product of ethylene and chlorine) at high temperatures.

Commercial Uses. Vinyl chloride is used heavily as a monomer raw material for the polymerization of vinyl resins in manufacturing vinyl plastics. It is also used for making polyvinyl chloride and copolymers, as well as in organic synthesis.

Physiological Effects. Vinyl chloride acts as a general anesthetic in concentrations of well over 500 ppm. A 500 ppm concentration is generally accepted as the maximum allowable for safety to health through an 8-hour day. Some authorities believe that its odor does not provide adequate warning of its presence in concentrations sufficient to produce dizziness and unconsciousness, and urge special caution against leaks and poor ventilation.

Vinyl chloride can irritate or damage the eyes on contact, and personnel must wear chemical safety goggles when there is any chance of having the liquid or saturated vapor come in contact with the eyes. Liquid vinyl chloride also irritates the skin and can freeze the skin on protracted contact. Should contact occur, the skin must be washed with large amounts of running water. The eyes must similarly be washed for at least 15 minutes if vinyl chloride gets into them, and, except for cases of only very minor irritation, should be treated immediately by an eye specialist.

A detailed discussion of safety precautions and first aid measures to employ with vinyl chloride appears in Safety Data Sheet SD-56 of the Manufacturing Chemists' Association.[1]

Precautions required for the safe handling of any flammable gas must be observed with vinyl chloride.

VINYL CHLORIDE SHIPPING CONTAINERS

Vinyl chloride is shipped as a liquefied compressed gas in cylinders, single-unit tank cars and TMU (ton multi-unit) tank cars, and in tank trucks and TMU tanks on trucks. For all these types of containers, it is required that all parts of valves and safety relief devices in contact with the contents of the container must be of a metal or other material (suitably treated if necessary) which will not cause formation of any acetylides.

Filling Limits. The maximum filling densities authorized by the ICC for vinyl chloride are as follows (per cent water capacity by weight):

In cylinders—84 per cent.

In single-unit tank cars—of the type ICC-105A200W—87 per cent; of the type ICC-112A400W—86 per cent;

In TMU tanks of the 106A500-X type and in MC-330 or MC-331 cargo tanks—84 per cent.

SHIPPING METHODS; REGULATIONS

Under the appropriate regulations and tariffs, vinyl chloride is authorized for shipment as follows:

By Rail: In cylinders (freight or express), and in single-unit tank cars and TMU tank cars.

By Highway: In cylinders on trucks, and in tank trucks and TMU tanks on trucks.

By Water: In cylinders via cargo vessels only, in authorized tank cars aboard trainships only, and in authorized containers on trucks aboard trailerships and trainships only. In cylinders on barges of U. S. Coast Guard classes A and C only. In cargo tanks aboard barges (with special warning signs and manning standards as required) and aboard tankships (to maximum filling densities by specific gravity as specified in Coast Guard regulations).

By Air: Only inhibited vinyl chloride, aboard cargo aircraft in appropriate cylinders up to 300 lb (140 kg) maximum net weight per cylinder.

VINYL CHLORIDE CYLINDERS

Cylinders that meet the following ICC specifications are authorized for vinyl chloride service: 3A150, 3AA150, 3E1800, 4B150 (without brazed seams), and 4BA225 (without brazed seams). (Cylinders meeting specification ICC-25 may be continued in service, but new construction is not authorized.)

Valve Outlet and Inlet Connections. Standard connection, U. S. and Canada—No. 290.

Cylinder Requalification. Under present regulations, cylinders of all types authorized for vinyl chloride service must be requalified by hydrostatic test every 5 years with the exception of type 3E (for which periodic hydrostatic retest is not required).

SINGLE-UNIT TANK CARS AND TMU TANK CARS

Vinyl chloride is authorized for shipment in single-unit tank cars meeting ICC specifications 105A200W or 112A400W (provided that they have properly fitted loading and unloading valves; also, for cars of the 105A200W type, openings in tank heads to facilitate application of nickel linings are authorized). Rail shipment is also authorized for TMU tank cars of ICC specifications 106A500X.

CARGO TANKS AND TMU TANKS ON TRUCKS

Shipment of vinyl chloride is authorized by the ICC in cargo tanks on trucks meeting ICC specifications MC-330 or MC-331 and having a minimum design pressure of 150 psig. Shipment of TMU tanks of ICC specification 106A500X on trucks is also authorized.

STORAGE AND HANDLING EQUIPMENT

Steel is recommended for all piping, storage tanks and equipment used with vinyl chloride. Valves in particular must not contain copper or copper alloys. Adequate electrical grounding and diking or ditching in tank areas to control the liquid in the event of vessel rupture are among recommended precautions. Installations must be designed to comply with requirements for unfired pressure vessels and all state and local regulations.

REFERENCE

1. "Vinyl Chloride," Chemical Safety Data Sheet SD-56, Manufacturing Chemists' Association, 1825 Connecticut Ave., N. W., Washington, D. C., 20009.

Vinyl Methyl Ether

CH_2:$CHOCH_3$ (or C_3H_6O)
Synonym: Methyl vinyl ether
ICC Classification (in inhibited form): Flammable compressed gas; red label

PHYSICAL CONSTANTS

International symbol	—
Molecular weight	58.08
Vapor pressure at 70 F, psig	8.80
Specific gravity of liquid at 68 F	0.75
Density, liquid, lb/cu ft	
at 70 F	46.66
at 105 F	45.05
at 115 F	44.60
at 130 F	43.91
Boiling point at 1 atm	42 F
Freezing point at 1 atm	−187 F
Flammable limits in air at 77 F, by volume	2.6–39%
Flash point (open cup)	−69 F
Solubility in water by weight at 68 F	1%
Weight per gallon, liquid, at 77 F, lbs	6.2

Properties. Vinyl methyl ether is a colorless, flammable gas that has a sweet, pleasant odor and is heavier than air. It will react violently with itself in polymerization unless inhibited by low temperature or pressure, or both, or the addition of an inhibiting agent (among inhibitors or stabilizers used are triethanolamine, dioctylamine and solid potassium hydroxide pellets). Colorless or water-white as a liquid, it is shipped as a liquefied compressed gas at about 9 psig at 70 F.

Materials of Construction. Steel, stainless steel, aluminum, "Hastelloys" and nickel and its alloys are among metals used with vinyl methyl ether, which is noncorrosive. Copper and copper alloys must not be used with vinyl methyl ether in which acetylene is present as an impurity (because acetylene can form explosive acetylides on exposure to copper and its alloys). Vinyl methyl ether is unsafe with any materials not properly cleaned to remove acidic salts from their surface.

Manufacture. Vinyl methyl ether is produced by the combination of acetylene and methyl alcohol in the presence of a catalyst. It is also made commercially by combining acetaldehyde with methanol and cracking the acetal formed to give the ether.

Commercial Uses. Vinyl methyl ether is used in producing copolymers that are ingredients in coatings and lacquers, as a modifier for alkyl and polystyrene resins, as a plasticizer for nitrocellulose and other plastics, and as a raw material in the manufacture of plastics and synthetic resins.

Physiological Effects. Little has been established concerning the physiological effects of vinyl methyl ether. Inhalation and skin contact are believed to pose minimal hazards. Avoidance of prolonged inhalation of the gas in areas where its odor can be detected is advised, for inhalation is thought to produce mild anesthesia. Contact of vinyl methyl ether with the eye produces no permanent damage even if untreated, according to one authority. Sufficiently long contact by the highly volatile liquid vinyl methyl ether can cause freezing or frostbite of the skin.

Precautions necessary for the safe handling of any flammable gas must be observed with vinyl methyl ether. Should it ignite, recommended fire extinguishing agents include dry chemicals, carbon dioxide and water spray. Should fire occur in or near a processing or storage area, do not put water into a closed system. Vinyl methyl ether must also be stored under an inert atmosphere. It can form peroxides in the presence of air or oxygen; samples containing peroxides must not be distilled to dryness.

VINYL METHYL ETHER CONTAINERS

Inhibited vinyl methyl ether is shipped as a compressed liquefied gas in cylinders, in single-unit tank cars, in TMU (ton multi-unit) tank cars, and in TMU tanks on trucks. For all these types of containers, it is required that all parts of valves and safety relief devices in contact with the contents of the container must be of a metal or other material (suitably treated if necessary) which will not cause formation of any acetylides.

Vinyl methyl ether without an inhibitor is also shipped in cylinders, TMU tanks and single-unit tank cars by special permit of the ICC.

Filling Limit. The maximum filling density authorized for inhibited vinyl methyl ether in cylinders, single-unit tank cars and TMU tanks is 68 per cent (per cent water capacity by weight).

SHIPPING METHODS; REGULATIONS

Under the appropriate regulations and tariffs, inhibited vinyl methyl ether is authorized for shipment as follows:

By Rail: In cylinders (freight, or express in one outside container to a maximum of 20 lb), and in single-unit tank cars and TMU tank cars.

By Highway: In cylinders on trucks, and in TMU tanks on trucks.

By Water: In cylinders via cargo vessels only, and in authorized tank cars via trainships only. In cylinders on barges of U. S. Coast Guard classes A and C only.

By Air: Aboard cargo aircraft only in appropriate cylinders up to 22 lb (10 kg) maximum net weight per cylinder.

VINYL METHYL ETHER CYLINDERS

Cylinders that meet the following ICC specifications are authorized for vinyl methyl ether service: 3A150, 3AA150, 3B150, 3E1800, 4B150 (without brazed seams) and 4BA225 (without brazed seams). (Cylinders meeting specification ICC-25 may be continued in service, but new construction is not authorized.)

Valve Outlet and Inlet Connections. Standard connection, U. S. and Canada—No. 290.

Cylinder Requalification. Under present regulations, cylinders of all types authorized for vinyl methyl ether service must be requalified by hydrostatic test every 5 years with the exception of type 3E (for which periodic hydrostatic retest is not required).

SINGLE-UNIT TANK CARS AND TMU TANK CARS

Inhibited vinyl methyl ether is authorized for shipment in single-unit tank cars meeting ICC specifications 105A100W (provided that they have properly fitted loading and unloading valves). Single-unit tank cars meeting ICC specifications 105A100 may also be continued in service with inhibited vinyl methyl ether, but new construction is not authorized. Rail shipment is also authorized for TMU tank cars of ICC specifications 106A500X.

TMU TANKS ON TRUCKS

Shipment of TMU tanks of ICC specifications 106A500X on trucks is also authorized.

STORAGE AND HANDLING EQUIPMENT

Mild steel is among the materials recommended for vinyl methyl ether storage tanks. Storage tanks must be designed for the vapor pressure developed at the highest storage temperature expected. One satisfactory design might consist of a mild steel tank built for a 50 to 100 psig working pressure and fitted with adequate relief valves and cooling coils to prevent excessive loss through the relief valves during summers. All acidic materials and water must be removed from equipment handling uninhibited vinyl methyl ether, particularly during distillation, to prevent accidental polymerization. Vinyl methyl ether can be handled at atmospheric pressure as a gas or, if kept cooled to 32 F to 36 F, as a liquid.

Xenon

See **Rare Gases of the Atmosphere**

PART III

Standards for Compressed Gas Handling and Containers

Part III of the "Compressed Gas Handbook" presents the major standards that have been developed for the many users of compressed gases, as well as those for the producers and shippers of compressed gases and gas equipment. These standards represent requirements or recommendations for all persons and organizations working with compressed gases. Standards are covered in the six chapters of Part III, as follows:

Chapter 1 outlines the general requirements for the safe handling of compressed gases in its opening section. It then gives fundamental information about the safe handling of gases used for medical purposes, and has a closing section on safety standards for medical vacuum systems in hospitals.

Chapter 2 thoroughly explains the minimum requirements for safety devices for relieving possible excess pressure in compressed gas containers, dealing separately with safety relief devices for cylinders, for cargo and portable tanks, and for storage tanks.

Chapter 3 covers standards relating primarily to compressed gas cylinders—their marking and labeling, their required retesting and inspection for requalification, their repair, and the disposition of cylinders no longer fit for service.

Chapter 4 defines the major systems of standards for compressed gas cylinder valve connections. These systems have been adopted to help prevent accidental connections between a gas and equipment not designed for that gas, and thus protect personnel and equipment. The standards apply to the valve connections of conventional compressed gas cylinders, and to connections for special cylinders and equipment used in medicine.

Chapter 5 describes the major types of containers used for shipping and storing compressed gases, and the standards to which these containers must conform.

Chapter 6 interprets and applies safety standards in a detailed and practical explanation of a frequent operation in industry: receiving and unloading bulk shipments of liquefied compressed gas.

The standards set forth in Part III have been developed by Compressed Gas Association, Inc. The Association issues standards and recommendations on a number of more highly specialized matters as well. Some of these are referred to in notes in the chapters. A full list of the current CGA pamphlets is given in appendix B.

CHAPTER 1

Safe Handling of Compressed Gases

The very properties that make compressed gases useful in almost every area of modern life can also make them dangerous when mishandled. Years of experience with compressed gases have led to practices and equipment which, if employed, result in complete safety. This chapter presents the basic standards developed by Compressed Gas Association, Inc., for the safe handling of compressed gases. The standards as phrased by the CGA reflect many years of very broad experience in dealing with the gases. The chapter's three sections are:

Section A. General Requirements for Safe Handling.

Section B. Safe Handling Requirements for Gases Used Medicinally.

Section C. Medical Vacuum Systems in Hospitals.

Essential knowledge for anyone involved with compressed gases in quantities of conventional cylinder size and larger is given in Section A. Section B covers essential information for all persons working with compressed gases in hospitals, clinics and emergency treatment facilities for medical purposes. Of more specialized significance, Section C gives the CGA standards for medical vacuum systems used in hospitals.

SECTION A

General Requirements for Safe Handling

1. INTRODUCTION

1.1. Compressed gas containers, when constructed according to the proper Interstate Commerce Commission (ICC) specifications, or the specifications of the Board of Transport Commissioners for Canada (BTC) and maintained in accordance with ICC or BTC regulations, may be considered safe for the purposes for which they are intended. Accidents occurring during the transportation, handling, use and storage of these containers can almost invariably be traced to failure to follow ICC or BTC requirements or to abuse or mishandling of the containers.

1.2. The following rules compiled by Compressed Gas Association, Inc., are primarily for the guidance of users of compressed gases in cylinders and are based upon accident prevention experience within these industries. (General precautions are also included for tank car handling and reference is made to cargo tank handling.) It should not be assumed that every acceptable safety precaution is contained herein, or that abnormal or unusual circumstances may not warrant or require further or additional procedure.

1.3. The information contained in this statement was obtained from sources believed to be reliable and is based upon the experience of members of Compressed Gas Association, Inc. However, by the issuance of this statement the association and its members, jointly and severally, make no guarantee of results and assume no liability in connection with the information herein contained, or for the safety suggestions herein made.

2. REGULATIONS APPLYING TO COMPRESSED GASES

2.1. The transportation of compressed gases is regulated by the United States Government under the provisions of an Act of Congress dated March 4, 1921, known as "The Transportation of Explosives Act." This act is administered by the Interstate Commerce Commission for railway and for highway transport. The transportation of compressed gases by water comes under the jurisdiction of the United States Coast Guard. Transportation of compressed gases by air is regulated by the Federal Aviation Agency.

2.2. The storage and use of compressed gases are regulated by many state or municipal authorities.

2.3. The Board of Transport Commissioners for Canada, an agency of the Canadian federal government, has authority over the transportation of hazardous commodities by rail, similar to that of the ICC in the United States. All such rail transportation requirements in Canada are identical with those prescribed by the ICC. Regulations for water transportation in Canada come within the jurisdiction of the Department of Transport, while highway transportation, where regulated, comes within the domain of provincial or local authorities.

3. SAFE HANDLING RULES FOR CYLINDERS OF COMPRESSED GASES*

3.1. General

3.1.1. Only cylinders meeting ICC regulations should be used for the transportation of compressed gases.

3.1.2. Cylinders must not be charged except by the owner or with the owner's consent, and then only in accordance with the regulations of the Interstate Commerce Commission.

3.1.3. The practice of transferring compressed gases from large to small cylinders by anyone other than the manufacturer or distributor is not recommended, except where consent by the owner has been given, and safe procedures for these operations are followed.

3.1.4. Compressed gas containers must not contain gases capable of combining chemically, nor should the gas service be changed without first removing the original content, and cleaning or purging to remove residues, if necessary.

3.1.5. It is illegal to remove or to change the prescribed numbers or marks stamped into cylinders without authority from the Bureau of Explosives, New York, N. Y.

3.1.6. If a cylinder leaks** and the leak cannot be remedied by simply tightening a valve gland or packing nut, close the valve and attach a tag stating that the cylinder is unserviceable. Remove the leaking cylinder out of doors to a well ventilated location. If the gas is flammable or toxic, place an appropriate sign at the cylinder, warning against these hazards. Notify the gas supplier and follow his instructions as to the return of the cylinder.

3.1.7. It is illegal to ship a leaking cylinder by common or contract carrier whether charged or partially charged. It is illegal to ship compressed gas in cylinders that have been ex-

* Rules pertaining to the storage and handling of cylinders apply with equal force to the storage and handling of spheres and drums where the alternate use of these containers is authorized by ICC regulations.

** See 3.7 for safe handling of type ICC-4L cylinders for which venting is normal.

posed to fire. Consult your supplier for advice under these circumstances.

3.1.8. Each cylinder must bear the proper ICC label required for the compressed gas contained, except under certain specified conditions set forth in ICC regulations.

3.1.9. Where the user is responsible for the handling of the cylinder and connecting it for use, such cylinders should carry a legible label or stencil identifying the content. See Chapter 3, Sections A and B.

3.1.10. Do not deface or remove any markings, labels, decals, tags and stencil marks used for identification of content attached by the supplier.

3.1.11. Before returning empty cylinders, close the valve and see that cylinder valve protective caps and outlet caps or plugs, if used, are replaced. Cover label with empty label meeting ICC requirements, or if cylinder is provided with combination shipping and caution tag, remove lower portion.

3.1.12. Cylinders containing compressed gases should not be subjected to a temperature above 125 F. A flame should never be permitted to come in contact with any part of a compressed gas cylinder.

3.1.13. Cylinders should not be subjected to artificially created low temperatures without the approval of the supplier. Many steels undergo decreased ductility at low temperatures.

3.1.14. Never tamper with the safety relief devices in valves or cylinders.

3.1.15. Never attempt to repair or to alter cylinders, valves, or safety relief devices.

3.1.16. Never use cylinders as rollers, supports, or for any purpose other than to contain the content as received.

3.1.17. Keep cylinder valve closed at all times, except when the cylinder is in active use.

3.1.18. Notify owner of cylinder if any condition has occurred which might permit any foreign substance to enter the cylinder or valve, giving details and cylinder serial number.

3.1.19. Do not place cylinders where they might become part of an electric circuit. When the cylinders are used in conjunction with electric welding, precautions must be taken against accidentally grounding compressed gas cylinders and allowing them to be burned by electric welding arc.

3.1.20. Do not repaint cylinders unless authorized by the owner.

3.1.21. When in doubt about the proper handling of a compressed gas cylinder or its content, consult the manufacturer or supplier of the gas.

3.2. Moving Cylinders

3.2.1. Where removable caps are provided for valve protection, such caps should be kept on cylinders at all times except when cylinders are in use.

3.2.2. Do not lift cylinders by the cap.

3.2.3. Never drop cylinders nor permit them to strike against each other or against other surfaces violently.

3.2.4. Never handle a cylinder with a lifting magnet. Slings, ropes or chains should not be used unless provisions have been made on the cylinder for appropriate lifting attachments, such as lugs. A crane may be used when a safe cradle or platform is provided to hold the cylinders.

3.2.5. Avoid dragging or sliding cylinders. It is safer to move cylinders even short distances by using a suitable truck.

3.2.6. Use suitable hand truck, fork truck, roll platform or similar device with cylinder firmly secured for transporting and unloading.

3.3. Storing Cylinders

3.3.1. Cylinders should be stored in accordance with all local, state and municipal regulations and in accordance with appropriate standards of Compressed Gas Association, Inc., and the National Fire Protection Association (see Bibliography).

3.3.2. Cylinder storage areas should be prominently posted with the names of the gases to be stored.

3.3.3. Where gases of different types are stored at the same location, cylinders should be grouped by types of gas, and the groups arranged to take into account the gases contained, e.g.; flammable gases should not be stored near oxidizing gases.

3.3.4. Charged and empty cylinders should be stored separately with the storage layout so planned that cylinders comprising old stock can be removed first with a minimum handling of other cylinders.

3.3.5. Storage rooms should be dry, cool and well ventilated. Where practical, storage rooms should be fire-resistant. Storage in subsurface locations should be avoided. Cylinders should not be stored at temperatures above 125 F, nor near radiators or other sources of heat.

3.3.6. Do not store cylinders near highly flammable substances such as oil, gasoline or waste.

3.3.7. Cylinders should not be exposed to continuous dampness and should not be stored near salt or other corrosive chemicals or fumes. Rusting will damage the cylinders and may cause the valve protective caps to stick.

3.3.8. Protect cylinders from any object that will produce a cut or other abrasion in the surface of the metal. Do not store cylinders near elevators or gangways, or in locations where heavy moving objects may strike or fall on them. Where caps are provided for valve protection, such caps should be kept on cylinders in storage.

3.3.9. Cylinders may be stored in the open but should be protected from the ground beneath to prevent rusting. Cylinders may be stored in the sun except in localities where extreme temperatures prevail, or in the case of certain gases where the supplier's recommendation for shading shall be observed. If ice or snow accumulate on a cylinder, thaw at room temperature, or with water at a temperature not exceeding 125 F.

3.3.10. Cylinders should be protected against tampering by unauthorized individuals.

3.4. Withdrawing Cylinder Content

3.4.1. Compressed gases should be handled only by experienced and properly instructed persons.

3.4.2. The user responsible for the handling of the cylinder and connecting it for use should check the identity of the gas by reading the label or other markings on the cylinder before using. If cylinder content is not identified by marking, return cylinder to the supplier without using.

3.4.3. Removable type valve protective caps should remain in place until ready to withdraw content, or to connect to a manifold.

3.4.4. Before using cylinder, be sure it is properly supported to prevent it from being knocked over.

3.4.5. Suitable pressure regulating devices must be used in all cases where gas is admitted to systems having pressure rating limitations lower than the cylinder pressure.

3.4.6. Never force connections that do not fit. Threads on regulator connections or other auxiliary equipment must be the same as those on cylinder valve outlet. Detailed, dimensioned drawings of standard cylinder valve outlet and inlet connections are published in the "American and Canadian Standard Compressed Gas Cylinder Valve Outlet & Inlet Connections" (ASA-B57.1 and CSA-B96).*

3.4.7. Where compressed gas cylinders are connected to a manifold, such a manifold and its related equipment, such as regulators, must be of proper design.

3.4.8. Regulators, gages, hoses and other appliances provided for use with a particular gas or group of gases must not be used on cylinders containing gases having different chemical properties unless information obtained from the supplier indicates that this can be done safely.

3.4.9. Open cylinder valve slowly. Point the valve opening away from yourself and other persons. Never use wrenches or tools except those provided or approved by the gas manufacturer. Avoid the use of a wrench on valves equipped with handwheels. Never hammer the valve wheel in attempting to open or close the valve. For valves that are hard to open, or

* See Chapter 4.

frozen because of corrosion, contact the supplier for instructions.

3.4.10. Never use compressed gas to dust off clothing, as this may cause serious injury to the eyes or body, or create a fire hazard.

3.4.11. Never use compressed gases where the cylinder is apt to be contaminated by the feedback of process materials, unless protected by suitable traps or check valves.

3.4.12. Connections to piping, regulators, and other appliances should always be kept tight to prevent leakage. Where hose is used, it should be kept in good condition.

3.4.13. Before a regulator is removed from a cylinder, close the cylinder valve and release all pressure from the regulator.

3.5. Flammable Gases

3.5.1. Do not store cylinders near flammable solvents, combustible waste material and similar substances, or near unprotected electrical connections, gas flames or other sources of ignition.

3.5.2. Never use a flame to detect flammable gas leaks. Use soapy water.

3.5.3. Do not store reserve stocks of cylinders containing flammable gases with cylinders containing oxygen. They should be segregated. Inside of buildings, stored oxygen and fuel gas cylinders should be separated by a minimum of 20 ft, or there should be a fire-resistive partition between the oxygen and fuel gas cylinders. This is in accordance with NFPA Standard No. 51, "Gas Systems for Welding and Cutting."*

3.6. Poison Gases

3.6.1. Personnel handling and using poison gases should have available for immediate use gas masks or self-contained breathing apparatus of a design approved by U. S. Bureau of Mines for the particular service desired. Such equipment should be located convenient to the place of work, but kept out of the area most likely to be contaminated.

3.6.2. Poison gases should be used only in forced ventilation areas or, preferably, in hoods with forced ventilation or out-of-doors. Poison gases emitted from equipment in high concentration should be discharged into appropriate scrubbing equipment which will remove it from effluent streams.

3.6.3. Before using, read all label information and data sheets associated with the use of the particular poison gas.

3.6.4. Use poison gases in cylinder sizes that will ensure complete usage of the cylinder content in a reasonable amount of time.

3.6.5. The Interstate Commerce Commission requires that containers charged with the following materials when offered for transportation bear the poison gas label and be subject to all other regulations prescribed by the ICC for such materials:

Bromoacetone
Chlorpicrin mixtures with nonliquefied, nonflammable compressed gas or with methyl chloride
Cyanogen
Cyanogen chloride (containing less than 0.9 per cent water)
Diphosgene
Ethyldichloroarsine
Hexaethyl tetraphosphate and compressed gas mixture
Hydrocyanic acid
Lewisite
Methyldichloroarsine
Mustard gas
Nitric oxide
Nitrogen dioxide, nitrogen peroxide, nitrogen tetroxide, nitrogen tetroxide-nitric oxide mixtures containing up to 33.2 per cent weight nitric oxide
Organic phosphates not otherwise specified mixed with compressed gas
Parathion and compressed gas mixtures
Phenylcarbylamine chloride
Phosgene (diphosgene)
Poisonous liquid or gas, not otherwise specified
Police grenades, poison gas, class A
Tetraethyl dithio pyrophosphate and compressed gas mixture

* Available from the National Fire Protection Association, 60 Batterymarch St., Boston, Mass. 02110.

Tetraethyl pyrophosphate and compressed gas mixture

3.6.6. Because of the hazardous nature of poison gases, persons handling such gases are advised to contact the supplier for more complete information.

3.7. Pressurized Liquid Oxygen, Nitrogen and Argon*

3.7.1. ICC specification cylinders containing pressurized liquid oxygen, nitrogen or argon must be transported, stored, and used in an upright position. These materials are maintained at extremely low temperatures, and cylinders must be kept upright to permit venting of vapor periodically to maintain safe internal pressures.

3.7.2. Persons handling these pressurized liquids are advised to contact the supplier for more complete handling information.

3.7.3. Avoid having any of the liquid atmospheric gases splash or spill on the skin or into the eyes. Stand clear and proceed very slowly when charging a warm container or when inserting objects into the liquid, for boiling and splashing always occur under these conditions. Make sure that there is a hose or a large open container of water nearby when handling these liquids, so that the water can be used to wash off any area of the body accidentally splashed with liquid. Always use tongs to withdraw objects immersed in liquid and handle them carefully. (*Note:* The vapor mist that appears upon exposure to air of a liquefied atmospheric gas is condensed moisture; the gas itself is invisible.)

3.7.4. When handling these liquids, wear protective clothing as follows: a face shield or safety goggles (which *must* be goggles with side shields); asbestos gloves that fit loosely enough to be thrown off quickly should liquid spill or splash into them; and high-top shoes with cuffless trousers worn outside the shoes. It is also advisable to wear quickly removed outer garments, such as shop coats or aprons, that cover all skin areas remaining exposed.

3.7.5. Always handle liquefied atmospheric gases in well ventilated areas, and never dispose of them in confined areas or closed places that other persons might enter. Excessive concentrations of oxygen constitute fire hazards, and excessive concentrations of the other atmospheric gases constitute asphyxiation hazards.

3.7.6. Use only containers designed for low-temperature liquids, and fill even these specially designed containers as slowly as possible. Use only the stopper supplied with the specific container, for all low-temperature containers are protected by vents or other means for permitting the escape of vapors. Fill containers only with the liquids that they are designed to hold. Take special care to put liquid oxygen only in oxygen containers, for oxygen mixed with other liquefied atmospheric gases is hazardous. Use only proper transfer equipment, such as a filling funnel for small containers and a discharge tube for all containers larger than 50 liters as well as for all small containers not safe or convenient to tilt.

3.7.7. Never allow liquid oxygen to come into contact with oil, grease or carbonaceous materials of any kind, because liquid oxygen can promote intense combustion. Immediately remove any clothing that becomes saturated with liquid oxygen.

3.7.8. Install equipment or piping for liquefied atmospheric gases only with the advice of persons thoroughly experienced in low-temperature work. Follow all procedures prescribed by the manufacturers of low-temperature equipment, and properly train and supervise all personnel working with these liquids. Make sure that only authorized personnel have access to liquid storage areas.

* Paragraphs 3.7.3 through 3.7.8 are not part of the official CGA standards. They are suggestions intended only for preliminary information and were compiled from industry sources.

4. GENERAL PRECAUTIONS FOR TANK CARS

4.1. Tank cars containing compressed gases must not be shipped unless they are charged by or with the consent of the owner thereof.

4.2. Care must be taken that compressed gases are charged only in tank cars designed and marked for the particular gas to be charged. Before a tank car can be charged with a compressed gas other than that for which commodity use has been approved by the Tank Car Committee of the Association of American Railroads, approval must be obtained from this committee by the car owner or party authorized by the owner.

4.3. Tank cars must not contain gases capable of combining chemically, nor should the gas service be changed without first removing the original content, and cleaning or purging to remove residues, if necessary.

4.4. Approval must be obtained from the Tank Car Committee of the Association of American Railroads by the car owner or party authorized by the car owner for any alterations or welded repairs to existing tanks.

4.5. In addition to car markings required by ICC Regulations, individual tanks should carry a stencil identifying the content.

4.6. Charged tank cars must carry the ICC "Dangerous" placard.

4.7. Follow ICC Regulations regarding placard requirements and the condition of empty tank cars before returning to the shipper. Before releasing empty tank cars, the dome cover should be tightly closed and the ICC "Dangerous" placard reversed, displaying the ICC "Dangerous-Empty" placard.

4.8. A defective or leaking tank car must not be shipped.

4.9. Safe handling of types 106A and 110A tanks ("TMU" or "ton multi-unit" containers) should follow the recommendations as specified for cylinders.

4.10. Numbers or marks stamped into compressed gas tank cars must not be changed without written authority from the Bureau of Explosives.

4.11. Never tamper with the safety relief devices in tank car tanks or valves. In the event of a leak in the tank car or fittings which cannot be repaired by simple adjustments, telephone the supplier for instructions.

4.12. Shipper's detailed instructions and diagrams for unloading should always be followed and all caution markings on both sides of the tank or dome must be read and observed. Angle valves should be opened slowly to avoid closing of check valves in the education pipe. The use of a hammer on valve or cover plate to release a check valve should be avoided.

4.13. Cars must be unloaded on a properly protected private track of the consumer, or under alternate conditions as set forth in ICC regulations.

4.14. When possible, railway sidings on which compressed gas tank cars are placed for unloading should be level and be devoted solely to this purpose.

4.15. Unloading operations should be carefully supervised and should be performed only by reliable persons properly instructed and made responsible for careful compliance with all safety regulations. Operators should be provided with proper personal protective equipment.

4.16. When a tank car is spotted on a siding for unloading, brakes must be set and wheels blocked. Caution signs must be so placed on the track or car as to give necessary warning to persons approaching car from open end or ends of siding and must be left up until after car is unloaded and disconnected. Signs must be of metal, at least 12 by 15 in. in size and bear the words "Stop—Tank Car Connected," or "Stop—Men at Work," the word "Stop" being in letters at least 4 in. high, and the other words in letters at least 2 in. high. The letters must be white on blue background.

4.17. Derails should be placed at one or both ends of the unloading track approximately one car length from the car being unloaded, unless the car is protected by a closed and locked switch or gate.

4.18. Cars should be electrically grounded before unloading if content is flammable.

4.19. For further information on tank car unloading, refer to Chapter 6, "How to Receive and Unload Liquefied Compressed Gases."

5. CARGO TANK MOTOR VEHICLES

5.1. Cargo tanks mounted on motor vehicles are normally not handled by the gas user. However, in cases where a cargo tank is handled by the user, he should consult the gas supplier for instructions on safe handling procedures.

SECTION B

Safe Handling Requirements for Gases Used Medicinally*

1. INTRODUCTION

1.1. This section relates to compressed medical gases and their containers. Its purpose is to promote safe practices, not only for the protection of those persons handling medical gases, but also for the general public as well, in order that these gases may be made available under proper and safe conditions.

1.2. This section has been prepared by experienced technical personnel in Compressed Gas Association, Inc., for the guidance of persons in the medical and dental professions, their co-workers and all other personnel engaged in and responsible for the handling of compressed medical gases. It is recommended that responsibility for the handling of compressed medical gases be delegated to reliable, well trained persons who are fully informed of the potential hazards to themselves and to others, inherent in the handling of these products. In order to familiarize such personnel with compressed medical gases and their handling it is recommended that they study and follow the practices and principles herein outlined.

1.3. In order to present adequately these safe practice recommendations, particularly for the guidance of those whose knowledge of this subject may be limited, there are included herein certain related descriptive data regarding properties of the medical gases as well as information on compressed gas containers and regulations governing these commodities.

1.4. This section does not purport to give any guidance for the administration of medical gases which is quite properly the responsibility of the medical and dental professions.

1.5. The information presented herein is obtained from sources believed to be reliable and based upon the experience of a number of members of Compressed Gas Association, Inc. However, by the issuance of this standard the Compressed Gas Association, Inc., and its members, jointly or severally, make no guarantee of results and assume no liability in connection with the information herein contained, or the safety suggestions herein made. Moreover, it should not be assumed that every acceptable safety procedure is contained herein or that abnormal or unusual circumstances may not warrant or require further or additional procedure. These suggestions should not be confused either with state, municipal or insurance requirements, or with national safety codes.

2. MEDICAL GASES AND THEIR PROPERTIES

2.1. General

2.1.1. Medical gases are prepared under carefully controlled conditions to meet the purity specifications prescribed in the *Pharmacopeia of the United States or the National Formulary*. They are normally shipped under pressure in metal containers in accordance with the regulations of the Interstate Commerce Commission.

2.1.2. In cylinders containing *liquefied* com-

* For safety standards concerning medical gas container connections, see Chapter 4, Section B, "The Pin-Index Safety System," and Section C, "The Diameter-Index Safety System."

pressed gas and vapor in equilibrium, the pressure in the container is determined almost solely by the vapor pressure of the contained liquid at the equilibrium temperature and bears no relation to the amount of liquid which remains in the cylinder. In cylinders containing low-temperature liquefied gases, such as liquid oxygen, the pressure in the cylinder is equal to or less than the predetermined pressure setting of the control valve on the cylinder, and this pressure bears no relation to the amount of liquid which remains in the cylinder. Hence, the pressure in a cylinder containing a *liquefied* compressed gas, such as nitrous oxide or carbon dioxide, or a low temperature *liquefied* compressed gas, such as liquid oxygen, will remain approximately constant as gas is slowly withdrawn from the cylinder until all of the liquid has been exhausted, at which time the pressure drops in direct proportion to the rate at which the gas is withdrawn. The pressure in a cylinder charged with a *liquefied* compressed gas is therefore not an indication of the amount of gas remaining in the cylinder. The quantity of gas in the cylinder can be determined only by weight.

2.1.3. In cylinders charged with a *nonliquefied* compressed gas, the pressure in the container is related both to temperature and the amount of gas in the container. For *nonliquefied* compressed gases such as oxygen, helium, carbon dioxide-oxygen mixtures, or helium-oxygen mixtures, cylinder content may be determined by pressure, i.e., at a given temperature when the pressure is reduced to half the original pressure the cylinder will be approximately half full. Cylinder content may also be determined by weight using appropriate weight-volume conversion factors.

2.1.4. Some of the properties and pertinent facts about currently used medical gases and their mixtures are given in 2.2. through 2.9. Further information about these gases appears in Part II of this volume.

2.2. Carbon Dioxide

Chemical formula	CO_2
Color	Colorless
Odor	Odorless
Taste	Slightly acid
Physical state in full cylinder	Liquefied gas below 88 F Nonliquefied gas at 88 F and higher
Number of gal in 1 oz at 1 atm and 70 F	4.08
Number of liters in 1 oz at 1 atm and 70 F	15.44
Specific gravity of gas compared to air	1.529
Combustion characteristics	Nonflammable Does not support combustion
Normal cylinder filling limit	68 per cent of the weight of water that the cylinder will hold
Pressure in normally charged cylinder	Pressures in cylinders of carbon dioxide charged to a filling density of 68 per cent will vary with temperature approximately as shown in the following table. In any cylinder of carbon dioxide at temperatures below 88 F the pressure will remain constant, varying only with temperature as shown in the table, as long as liquid and vapor are both present. At temperatures of 88 F or higher the pressure at a given temperature will decrease proportionately as the cylinder contents are withdrawn

Temperature (F)	Approximate Gage Pressure in psig in a Cylinder of Carbon Dioxide at 68% Filling Density
50	638
60	733
70	838
80	955
88	1057
90	1215
100	1455
110	1725
120	1930
130	2265

2.3. Cyclopropane

Chemical formula	C_3H_6
Color	Colorless
Odor	Characteristic odor resembling petroleum benzine
Taste	Pungent
Physical state in full cylinder	Liquefied gas
Number of gal in 1 oz at 1 atm and 70 F	4.29
Number of liters in 1 oz at 1 atm and 70 F	16.24
Specific gravity of gas compared to air	1.481
Combustion characteristics	Flammable
Ignition limits in air	2.40-10.3 per cent by volume
Ignition limits in oxygen	2.48-60.0 per cent by volume
Normal cylinder filling limit	55 per cent of the weight of water that the cylinder will hold
Pressure in normally charged cylinder	Pressure in cylinders of cyclopropane will vary with temperature approximately as shown in the following table as long as liquid and vapor are both present

Temperature (F)	Approximate Gage Pressure in psig in a Cylinder of Cyclopropane 55% Filling Density
50	53
60	65
70	79
80	86
90	116
100	134
110	154
120	174
130	199

2.4. Ethylene

Chemical formula	C_2H_4
Color	Colorless
Odor	Slightly sweet
Taste	Slightly sweet
Physical state in full cylinder	Liquefied gas below 50 F Nonliquefied gas at 50 F and higher
Number of gal in 1 oz at 1 atm and 70 F	6.40
Number of liters in 1 oz at 1 atm and 70 F	24.23
Specific gravity of gas compared to air	0.9749
Combustion characteristics	Flammable
Ignition limits in air	3.05-28.6 per cent by volume
Ignition limits in oxygen	2.90-79.9 per cent by volume
Normal cylinder filling limit	31.0-35.5 per cent of the weight of water that the cylinder will hold, depending upon the type of cylinder
Pressure in normally charged cylinder	Pressures in cylinders of ethylene charged to a filling density as prescribed above will be approximately 1250 pounds per square inch gage (psig) at 70 F. At temperatures of 50 F or higher the pressure at a given temperature will decrease proportionately as the cylinder contents are withdrawn

2.5. Helium

Chemical formula	He
Color	Colorless
Odor	Odorless
Taste	Tasteless
Physical state in the cylinder	Nonliquefied gas
Number of gal in 1 oz at 1 atm and 70 F	45.2
Number of liters in 1 oz at 1 atm and 70 F	171.1
Specific gravity of gas compared to air	0.1380
Combustion characteristics	Nonflammable Does not support combustion
Normal cylinder filling limit	1650-2000 psig at 70 F depending upon the type of cylinder
Pressure in normally charged cylinder	Pressures in cylinders of helium will vary as described in 2.1.3. At any given temperature the pressure will decrease proportionately as the cylinder contents are withdrawn

2.6. Nitrous Oxide

Chemical formula	N_2O
Color	Colorless
Odor	Odorless
Taste	Tasteless
Physical state in full cylinder	Liquefied gas below 98 F Nonliquefied gas at 98 F and higher

2.6. Nitrous Oxide—*continued*

Number of gal in 1 oz at 1 atm and 70 F	4.08
Number of liters in 1 oz at 1 atm and 70 F	15.44
Specific gravity of gas compared to air	1.5297
Combustion characteristics	Nonflammable Supports combustion
Normal cylinder filling limit	68 per cent of the weight of water that the cylinder will hold
Pressure in normally charged cylinder	Pressures in cylinders of nitrous oxide charged to a filling density of 68 per cent will vary with temperature approximately as shown in the following table. In any cylinder of nitrous oxide at temperatures below 98 F the pressure will remain constant, varying only with temperature as shown in the table as long as liquid and vapor are both present. At temperatures of 98 F or higher the pressure at a given temperature will decrease proportionately as the cylinder contents are withdrawn

Temperature (F)	Approximate Gage Pressure in psig in a Cylinder of Nitrous Oxide at 68% Filling Density
50	575
60	655
70	745
80	830
90	940
98	1040

2.7. Oxygen

Chemical formula	O_2
Color	Colorless
Odor	Odorless
Taste	Tasteless
Physical state in the cylinder	Nonliquefied gas in conventional compressed gas cylinders such as ICC-3A or 3AA. Pressurized liquid gas in ICC-4L cylinders
Number of gal in 1 oz at 1 atm and 70 F	5.65
Number of liters in 1 oz at 1 atm and 70 F	21.39
Specific gravity of gas compared to air	1.1053
Combustion characteristics	Non-flammable Supports combustion
Normal cylinder filling limit	*Nonliquefied gas:* 1800-2400 psig at 70 F, depending upon the type of cylinder

2.7. Oxygen—*continued*

Pressure in normally charged cylinder	*Nonliquefied gas:* Pressures in cylinders of oxygen will vary as described in 2.1.3. At any given temperature the pressure will decrease proportionately as the cylinder content is withdrawn *Pressurized Liquid Gas:* 75-235 psig in ICC-4L cylinders. Pressure in cylinder during normal operation is 75 psig. During nonusage, pressure will increase very slowly, over a period of 3-5 days, to 235 psig, after which the cylinder will vent gas at a rate of 2-5 cubic feet per hour (cfh)

2.8. Helium-Oxygen Mixtures

2.8.1. Mixtures of helium and oxygen are supplied in a number of different compositions, the most usual mixture containing 80 per cent helium and 20 per cent oxygen. These mixtures exist in the cylinder as nonliquefied gas of homogeneous composition. They are ordinarily charged in cylinders to a pressure of 1650 to 2000 pounds per square inch gage at 70 F, depending upon the type of cylinder. At any given temperature the pressure will decrease proportionately as the cylinder contents are withdrawn.

2.8.2. One ounce of a mixture of 80 per cent helium and 20 per cent oxygen by volume is equivalent to about 19 gallons at normal atmospheric pressure and temperature.

2.9. Carbon Dioxide-Oxygen Mixtures

2.9.1. Mixtures of carbon dioxide and oxygen are generally supplied in a number of different compositions ranging from 5 per cent carbon dioxide =95 per cent to 30 per cent carbon dioxide =70 per cent oxygen. These mixtures exist in the cylinder as nonliquefied gas of homogeneous composition. They are ordinarily charged in cylinders to a pressure of 1500 to 2200 psig at 70 F depending upon the type of cylinder. At any given temperature the pressure will decrease proportionately as the cylinder contents are withdrawn.

2.9.2. One ounce of the mixtures within the range of compositions given above is equivalent to about 5.4 gallons at normal atmospheric pressure and temperature. The factor varies with the composition of the mixture from about 5.55 to 5.07 gallons per ounce.

3. RULES AND REGULATIONS PERTAINING TO MEDICAL GASES

3.1. ICC Regulations and Specifications

3.1.1. General.

3.1.1.1. The gases described in the foregoing pages are classified by the Interstate Commerce Commission as compressed gases, and as such their transportation is regulated by the United States Government under the provisions of an Act of the Congress, dated March 4, 1921. Under this act, cylinders in which medical gases are shipped in commerce that is subject to the jurisdiction of the Commission must comply with ICC specifications. These specifications require, among other things, that the steel used in cylinders must meet certain chemical and physical requirements and that cylinder must pass a hydrostatic pressure test. In Canada, cylinders in which medical gases are shipped must comply with the Specifications of the Board of Transport Commissioners for Canada.

3.1.2. Cylinder Filling Limits.

3.1.2.1. Because of the characteristic of any gas in a closed container to increase in pressure with rising temperature, the possibility always exists that a cylinder charged with gas at a safe pressure at normal temperatures would reach a dangerously high pressure at elevated temperatures. This is equally true whether the contents of the cylinder are in the gaseous or liquid state. In the latter case, the liquid may expand to such a degree that excess hydrostatic pressure develops within the cylinder. To prevent these conditions from occurring with normal usage, the ICC has drawn up regulations which limit the amount of gas that may be charged into a cylinder.

3.1.2.2. In addition to certain other restrictions the charging of cylinders is in general limited to such that the pressure in the cylinder at 70 F does not exceed the service pressure for which the cylinder is designed. However, certain types of cylinders are permitted to be charged with some nonliquefied nonflammable gases which include oxygen, helium, helium-oxygen mixtures, and carbon dioxide-oxygen mixtures, to a pressure at 70 F which is 10 per cent in excess of their marked service pressure.

3.1.2.3. The charging of cylinders with medical gases other than carbon dioxide and nitrous oxide is also limited by the requirement that the pressure in the cylinder at 130 F must not exceed $1\frac{1}{4}$ times the maximum permitted filling pressure at 70 F. The pressure in pressurized liquid oxygen cylinders is limited to $1\frac{1}{4}$ times the service pressure.

3.1.2.4. The charging of cylinders with liquefied gases is further limited by setting a maximum permitted filling density for each gas, the term "filling density" being defined as "the per cent ratio of the weight of gas in a container to the weight of water that the container will hold at 60 F." Therefore, to determine the maximum amount of liquefied gas in pounds that may be charged into a container, multiply the water capacity in pounds by the maximum permitted filling density prescribed for the gas. The capacities of medical gas cylinders are also given in Tables 1 and 2, which appear near the end of this section.

3.1.3. Retesting Cylinders.

3.1.3.1. ICC regulations provide that cylinders, with a few exceptions, must be subjected to a test by interior hydrostatic pressure at least once in 5 years. These retests must be made at a minimum pressure which is specified in ICC Regulations for each type of cylinder; for example, ICC-3A cylinders must be retested at a minimum pressure of 5/3 times the marked service pressure. Cylinders which have been in a fire must be removed from service until they have been subjected to the prescribed procedures which, in most cases, include reheat-treating and retesting.

3.1.3.2. If a cylinder leaks, shows evidence of damage that may weaken it appreciably, or shows a permanent expansion which exceeds 10 per cent of the total expansion in the retest, it must be condemned. Cylinders condemned because of excessive permanent expansion may be reheat-treated and retested.

3.1.3.3. Records must be kept giving data showing the results of the tests made on all cylinders, and each cylinder passing the test must be plainly and permanently stamped with the month and year of the test, for example, 4-66 for April, 1966. Dates of previous tests must not be obliterated.

3.1.4. Marking and Labelling Cylinders.

3.1.4.1. Definite markings are prescribed by the ICC Regulations. For example, on ICC-3A cylinders the following markings are required to be stamped plainly and permanently on the shoulder, top head, or neck in letters or figures at least $\frac{1}{4}$ in. high, if space permits:

(a) The ICC specification number, followed by the service pressure, for example, ICC-3A2015.

(b) A serial number, except in the case of certain small sizes of cylinders which may be identified by lot number, and an identifying symbol (letters) to be located below the ICC mark. The symbol and numbers must be those of purchaser, user, or maker and the symbol must be registered with the Bureau of Explosives.

(c) The inspector's official mark near the serial number and the date of test so placed that dates of subsequent tests can be easily added.

(d) Spun and plugged cylinders must be specifically marked as such.

3.1.4.2. The markings on cylinders must not be changed except as specifically provided in ICC regulations. Attention is called to the fact that changing of serial numbers or ownership marks is prohibited unless a detailed report is filed with the Bureau of Explosives. Marking on cylinders must be kept in readable condition.

3.1.4.3. In addition to the above specified markings on cylinders, ICC regulations require, with certain exceptions generally not applicable to medical gases, that each package containing compressed gas must bear an appropriate identifying caution label. Cylinders containing nonflammable gases require an ICC green label, and those containing flammable gases require the ICC red label. When empty cylinders bearing these caution labels are to be shipped as such, the labels must be removed, obliterated, or completely covered by an ICC "empty" label.

3.1.4.4. ICC regulations permit the use of a combination label-tag, one side of which contains the prescribed wording of the caution label, while the other side is used as a shipping tag with space for the names and addresses of both the shipper and consignee. Medical gas manufacturers in general use such a combination label-tag on large cylinders attached to the cylinder cap and this tag is perforated so that when the cylinder is empty part of the tag may be torn off at the perforation thus effectively obliterating the caution label. At the same time, that part of the tag remaining attached to the cylinder contains the return address of the supplier of the gas.

3.1.4.5. Small cylinders of medical gases are usually packed in outside containers such as cartons or crates. In this case the individual cylinders do not need to bear the ICC caution label but such appropriate label must be attached to the outside package.

3.1.4.6. Outside packages through which the specification markings on inside containers are not visible must also be labeled or marked to the effect that the inside packages comply with prescribed specifications.

3.1.4.7. Each cylinder should carry a legible label or stencil identifying the content as prescribed in Chapter 3, Section A, "Marking Portable Compressed Gas Containers to Identify the material contained."

3.1.5. Cylinder Recharging.

3.1.5.1. ICC regulations prohibit the shipment of cylinders unless they were charged by or with the consent of the owner. If this consent should be granted, the recharging of cylinders must comply in every respect with regulations that govern the charging of cylinders at a manufacturer's plant.

3.1.5.2. For information on the hazards involved in recharging cylinders by inexperienced operators see 4.5, entitled "Transfilling Cylinders."

3.2. Federal Food, Drug and Cosmetic Act

3.2.1. The shipment of medical gases in interstate commerce is also regulated by the Federal Food, Drug and Cosmetic Act. Under the provisions of this Act the medical gases must conform to the standards of the *Pharmacopeia of the United States or National Formulary* and must be appropriately labeled.

3.2.2. The *Pharmacopeia* or *National Formulary* contains a section or monograph on each of the following gases—carbon dioxide, cyclopropane, ethylene, helium, nitrogen, nitrous oxide and oxygen. Definite standards and methods of testing are prescribed in each monograph to assure a product of appropriate quality, purity and potency.

3.2.3. Although mixtures of medical gases, such as carbon dioxide-oxygen or helium-oxygen, are not specifically covered in the *Pharmacopeia* or *Formulary*, the act provides that the individual components must be clearly identified on the label and must meet their respective requirements.

3.2.4. The act further provides that the gases must be packaged and labeled as stated in the *Pharmacopeia* or *National Formulatory*; also that each individual package or cylinder must bear a label containing the name and address of the manufacturer, packer, or distributor as well as an accurate statement of the quantity of the contents. General usage in the medical gas industry is to express the contents of full cylinders in terms of gallons or liters measured at 70 F and normal atmospheric pressure.

3.2.5. The cylinder labels of some medical gas suppliers bear a cautionary statement to the effect that the gas is to be administered only by or under the supervision of a physician, dentist, or others trained and experienced in the use of the gas. The *Pharmacopeia* and the *National Formulary* require that flammable gases be labeled as such with a statement that their mixtures with air or oxygen are explosive.

3.2.6. For the purposes of enforcing the act, inspectors of the Federal Food and Drug Administration are empowered to make factory or warehouse inspections and to enter and inspect any vehicles used for transportation of these commodities in interstate commerce.

3.3. U. S. Department of Commerce Color Code

3.3.1. A color code to aid in the identification of small medical gas cylinders has been adopted by the medical gas industry, The American Society of Anesthesiologists, and the American Hospital Association. This code has been published by the U. S. Department of Commerce as Simplified Practice Recommendation R176-41 of the Bureau of Standards. Copies may be obtained from the Superintendent of Documents, Washington, D. C. The code recommends that anesthetic gas cylinders approximately 4½ in. in diameter by 26 in. long, and smaller, for use on anesthesia machines be marked with the following colors or color combinations:

Kind of gas	Color
Oxygen	Green
Carbon dioxide	Gray
Nitrous oxide	Light blue
Cyclopropane*	Orange
Helium	Brown
Ethylene	Red
Carbon dioxide and oxygen	Gray and green
Helium and oxygen	Brown and green

3.3.2. These color markings shall be applied to the shoulders of the containers (except chromium-plated cylinders for cyclopropane)* so as to be clearly visible to the anesthetist. Where the marking is to consist of two colors, the pattern shall be such as to permit a sufficient amount of both colors to be seen together.

3.3.3. The label affixed to each cylinder, carrying the name of the gas and other information required by regulations, and tags, if used, shall also bear the same color or colors as the shoulder.

3.3.4. While the above code for color identification of small medical gas cylinders (sizes used on anesthesia machines) has been adopted as an industry standard, attention is called to the fact that this color code should be used only as a guide. The primary identification for gas cylinder content is the label (see 3.1.4.7).

4. RECOMMENDED SAFE PRACTICES FOR HANDLING MEDICAL GASES**

4.1. General Rules

4.1.1. Never permit oil, grease, or other readily combustible substance to come in contact with cylinders, valves, regulators, gages, hoses and fittings. Oil and certain gases such as oxygen or nitrous oxide may combine with explosive violence.

4.1.2. Never lubricate valves, regulators, gages, or fittings with oil or any other combustible substance.

4.1.3. Do not handle cylinders or apparatus with oily hands or gloves.

4.1.4. Connections to piping, regulators, and other appliances should always be kept tight to prevent leakage. Where hose is used it should be kept in good condition.

4.1.5. Never use an open flame to detect gas leaks. Soapy water is generally used.

* Since paint does not adhere well to polished metal surfaces, chromium-plated cylinders for cyclopropane shall bear orange labels. If tags are also used, they shall be orange.

** General precautions for handling compressed gases are given in Section A of this chapter.

4.1.6. Prevent sparks or flame from any source from coming in contact with cylinders and equipment.

4.1.7. Never interchange regulators or other appliances used with one gas with similar equipment intended for use with other gases.

4.1.8. Fully open the cylinder valve when cylinder is in use.

4.1.9. Never attempt to mix gases in cylinders. (Mixtures should be obtained already prepared, from recognized suppliers.)

4.1.10. Before placing cylinders in service any paper wrappings should be removed so that the cylinder label is clearly visible.

4.1.11. User should identify gas content by label or stencil on the cylinder before using. If the cylinder is not identified to show the gas contained, return the cylinder to the supplier without using.

4.1.12. Do not deface or remove any markings which are used for identification of content of cylinder. This applies to labels, decals, tags and stencilled marks.

4.1.13. When returning empty cylinders, close valve before shipment and see that cylinder valve protective caps and outlet caps or plugs, if used, are replaced before shipping. Cover ICC green or red label with ICC empty label, or if cylinder is provided with combination shipping and caution tag, remove lower portion.

4.1.14. No part of any cylinder containing a compressed gas should ever be subjected to a temperature above 125 F. A flame should never be permitted to come in contact with any part of a compressed gas cylinder.

4.1.15. Never tamper with the safety relief devices in valves or cylinders.

4.1.16. Never attempt to repair or to alter cylinders.

4.1.17. Never use cylinders for any purpose other than to contain gas.

4.1.18. Cylinder valves should be closed at all times except when gas is actually being used.

4.1.19. Notify supplier of cylinder if any condition has occurred which might permit any foreign substance to enter cylinder or valve, giving details and cylinder number.

4.1.20. Do not place cylinders where they might become part of an electric circuit.

4.1.21. Cylinders should be repainted only by the supplier.

4.1.22. Compressed gases should be handled only by experienced and properly instructed persons.

4.2. Moving Cylinders

4.2.1. Where caps are provided for valve protection, such caps should be kept on cylinders when cylinders are moved.

4.2.2. Never drop cylinders nor permit them to strike each other violently.

4.2.3. Avoid dragging or sliding cylinders. It is safer to move large cylinders even short distances by using a suitable truck, making sure that the cylinder retaining chain or strap is fastened in place.

4.3. Storing Cylinders

Wherever cylinders of medical gases are stocked (by hospitals, doctors, distributors, etc.) the question of storage is of great importance. Many cities have local ordinances regulating the storage of medical gases. Persons storing medical gases should be familiar with these ordinances and fully comply with them. In the storage of medical gas cylinders it is recommended that the rules listed below also be followed:

4.3.1. Cylinders should be stored in a definitely assigned location.

4.3.2. Full and empty cylinders should be stored separately with the storage layout so planned that cylinders comprising old stock can be removed first with a minimum of handling of other cylinders.

4.3.3. Storage rooms should be dry, cool and well ventilated. Where practical, storage rooms should be fire-resistive. Storage in subsurface locations should be avoided. Storage conditions should comply with local and state regulations.

4.3.4. Cylinder should be protected against excessive rise of temperature. Do not store cylinders near radiators or other sources of radiant heat. Do not store cylinders near highly flammable substances such as oil, gasoline, waste, etc. Keep sparks and flame away from cylinders.

4.3.5. Do not store reserve stocks of cylinders containing flammable gases in the same room with those containing oxygen or nitrous oxide. (It is good practice to include cylinders containing carbon dioxide in the storage room with those containing flammable gases, since carbon dioxide gas is in itself a fire extinguisher.)

4.3.6. Cylinders should never be stored in the operating room.

4.3.7. Small cylinders may best be stored in bins, grouped as to the various gases or mixtures of gases.

4.3.8. Large cylinders should be placed against a wall to offer some protection against being knocked over. They should not be placed along an aisle used for trucking traffic. The best practice is to provide means for a chain fastening of large cylinders to the wall.

4.3.9. ICC-4L cylinders must be stored in an upright position.

4.3.10. Be careful to protect cylinders from any object that will produce a cut or other abrasion in the surface of the metal. Do not store cylinders in locations where heavy moving objects may strike or fall on them. Where caps are provided for valve protection, such caps should be kept on cylinders in storage.

4.3.11. Cylinders may be stored in the open but in such cases should be protected against extremes of weather and from the ground beneath to prevent rusting. During winter, cylinders stored in the open should be protected against accumulations of ice or snow. In summer, cylinders stored in the open should be screened against continuous direct rays of sun in those localities where extreme temperatures prevail.

4.3.12. Cylinders should not be exposed to continuous dampness and should not be stored near corrosive chemicals or fumes. Rusting will damage the cylinders and may cause the valve protection caps to stick.

4.3.13. Never store cylinders where oil, grease, or other readily combustible substance may come in contact with them. Oil and certain gases such as oxygen or nitrous oxide may combine with explosive violence.

4.3.14. Cylinders should be protected against tampering by unauthorized individuals.

4.3.15. Valves should be kept closed on empty cylinders at all times.

4.4. Withdrawing Cylinder Content

4.4.1. Never attempt to use contents of a cylinder without a suitable pressure regulating device. Pressure regulators are preferred for reducing pressure from cylinders. If needle valves are used, particular attention shall be given to preventing excessive pressure from building up beyond the needle valve.

4.4.2. Do not remove valve protection cap until ready to withdraw contents or to connect to a manifold.

4.4.3. Where compressed gas cylinders are connected to a manifold, such a manifold must be of proper design and equipped with one or more pressure regulators where necessary.

4.4.4. After removing valve protection cap, slightly open valve an instant to clear opening of possible dust and dirt.

4.4.5. When opening valve, point the outlet away from you. Never use wrenches or tools except those provided or approved by the gas supplier. Never hammer the valve wheel in attempting to open or to close the valve.

4.4.6. Regulators, pressure gages and manifolds provided for use with a particular gas or group of gases must not be used with cylinders containing other gases.

4.4.7. Never use medical gases where the cylinder is liable to become contaminated by the feed-back of other gases or foreign material unless protected by suitable traps or check valves.

4.4.8. It is important to make sure that the threads on regulator-to-cylinder valve connections or the pin indexing devices on yoke-to-cylinder valve connected are properly mated. Never force connections that do not fit.

4.4.9. After attaching regulator and before cylinder valve is opened see that the regulator is turned to the off position in the case of regulators equipped with a pressure adjusting screw. This is accomplished by turning the screw counter-clockwise until it is free.

4.4.10. Never permit gas to enter the regulating device suddenly. Open the cylinder valve slowly.

4.4.11. Before regulating device and cylinder are disconnected close the cylinder valve and release all pressure from the device.

4.4.12. Cylinder valves should be closed at all times except when the gas is actually being used.

4.5. Transfilling Cylinders

4.5.1. There are serious hazards involved in transferring compressed gas from one cylinder to another. Compressed Gas Association, Inc., recognizes these hazards and in past communications with hospital executives and professional anesthetists urged that the practice of transferring medical gases from one cylinder to another be discontinued. This recommendation has also been made by the National Fire Protection Association, which in its Pamphlet No. 56 entitled *Code for the Use of Flammable Anesthetics*, states, "transfer of gas from one cylinder to another on the hospital site or by hospital personnel should be prohibited."

4.5.2. In the interests of public safety, Compressed Gas Association, Inc., urges that cylinders be returned to charging plants for recharging under recognized safe practices, and calls attention to the following:

4.5.2.1. The hazard of overcharging small cylinders is always present when the charging is done by inexperienced operators who lack adequate knowledge of proper filling conditions and properties of the gas being handled. Charging conditions and densities vary for cylinders of different manufacturers and owners, even though their sizes appear to be the same. Overcharging may result in cylinder rupture and damage. All operations involved in the transfer of high pressure gases from one container to another require experienced supervision and equipment maintenance of a high degree in order to avoid personal injury and property damage.

4.5.2.2. Unless proper precautions are taken, a dangerous mixture of gases may occur when charging one cylinder from another. Manufacturers report that each year there are returned to them supposedly empty cylinders which actually contain ether or a gas other than that originally shipped. Some of these contaminating gases are flammable. Intermixture of flammable and oxidizing gases may cause a serious explosion. To avoid this, manufacturers have established definite procedures for detecting contaminating gases, and have provided special equipment for the thorough cleaning, when necessary, and preparing of all medical gas cylinders before they are recharged.

4.5.2.3. Cylinders which have been used for one type of gas may inadvertently or with intent be recharged by inexperienced or improperly trained operators with a gas other than that originally or last contained in the cylinder. Such practice will definitely cause contamination and may introduce a serious explosion hazard as well.

4.5.2.4. The importance of purity of medical gases cannot be overemphasized. This is recognized by the fact that the sale and distribution in interstate commerce of gases which are adulterated or may be injurious to health are prohibited by the Federal Food, Drug and Cosmetic Act. Medical gas manufacturers are required to supply compressed gases labeled in accordance with the requirements of this federal act, and to furnish such gases in full compliance with standards of purity described by the *Pharmacopeia of the United States* and the *National Formulary*. Many states have similar laws affecting local distribution. The transfer of medical gases from one cylinder to another by inexperienced persons may adversely affect purity.

4.5.2.5. Safety relief devices, valves and parts must be inspected at frequent intervals ot insure safe operation, and repairs or replacements made when defects are found. Manufacturers regularly engaged in the production of gases are best equipped to perform this essential maintenance work.

4.6. Use of Ether with Medical Gases

4.6.1. While ether is not a product of the medical gas industry, it is used so much in common with medical gases that recognition of its hazards is important. Ether is a highly volatile liquid, giving off, even at comparatively low temperatures, vapors which form with air, or oxygen, flammable and explosive mixtures. The vapors are heavier than air and may travel a considerable distance to a source of ignition and flash back. Spontaneously explosive peroxides sometimes form on long standing or exposure in bottles to sunlight. In order to prevent these peroxides from forming in an anesthesia machine, the ether vaporizer wicks should be rinsed and dried after each day's use and the residual ether in the vaporizer jar should be discarded and the jar thoroughly cleaned before replacing on the machine. In storing containers of ether they should be safeguarded against mechanical injury and kept in unheated compartments away from sunlight and any source of ignition. When used in connection with nitrous oxide or oxygen in anesthesia, care must be taken to avoid exhausting the gas cylinders completely to prevent the possibility of ether being drawn back into the cylinder.

5. MEDICAL GASES EQUIPMENT

5.1. Standard Sizes and Capacities of Medical Gas Cylinders

5.1.1. Medical gases are supplied in standard cylinder styles having the dimensions and capacities as shown in the accompanying Tables 1 and 2.

5.2. Valve Outlet Connections

5.2.1. Style E and smaller cylinders are equipped with flush type valves which are used with yoke connections. For details refer to references in 5.2.2. Style F and larger cylinders

TABLE 1. MEDICAL GAS CYLINDERS: CAPACITIES IN GALLONS AND POUNDS

Style Cylinder and Approximate Dimensions***		Carbon Dioxide	Cyclopropane	Ethylene	Helium	Nitrous Oxide†	Oxygen*	Helium-Oxygen Mixtures	Carbon Dioxide-Oxygen Mixtures
A (3″ o.d. x 7″)	gal lb-oz	50 0–12.5	40 0–9.4	40 0–6.25	15 0–0.33		20 0–3.75	15 **	20 **
B (3½″ o.d. x 13″)	gal lb-oz	100 1–9.0	100 1–7.5	100 0–15.75	28 0–0.63	125 1–15	40 0–7.25	29 **	40 **
D (4¼″ o.d. x 17″)	gal lb-oz	250 3–14.5	230 3–5.5	200 1–15.5	80 0–1.8	260 4–0	95 1–1.0	82 **	95 **
E (4¼″ o.d. x 26″)	gal lb-oz	420 6–9.0		330 3–4.0	131 0–2.9	430 6–10	165 1–13.25	134 **	165 **
F (5½″ o.d. x 51″)	gal lb-oz	1280 20–0.0		1100 10–12.0	425 0–9.4	1400 21–8	550 6–2.0	435 **	550 **
M (7⅛″ o.d. x 43″)	gal lb-oz	2000 31–4		1640 15–14.0	605 0–13.75	2090 32–1	800 8–14.0	620 **	800 **
G (8½″ o.d. x 51″)	gal lb-oz	3200 50–0.0		2800 27–8.0	1100 1–8.0	3655 56–0	1400 15–8.5	1126 **	1400 **

* Oxygen is also supplied in styles H or K cylinders, 9″ o.d. × 51″ containing 220 to 244 cu ft of gas.
** Varies with composition of the mixture.
*** The approximate dimensions given are for the cylinder less valve.
† (1) These figures represent maximum filling limitations based on 68% filling density and minimum cylinder volumes as follows: B—84 cu in., D—168 cu in., E—275 cu in., F—900 cu in., M—1320 cu in., G—2300 cu in., and H—2640 cu in.
(2) H cylinders of 2640 cu in. minimum volume may be charged with 64 lb 6 oz (4200 gal) of nitrous oxide.

TABLE 2. MEDICAL GAS CYLINDERS: PRESSURES AND CAPACITIES IN CUBIC FEET AND LITERS*

Cylinder Style and Approximate Dimensions			Carbon Dioxide	Cyclo-propane	Ethylene	Helium	Nitrous Oxide	Oxygen	Helium-Oxygen Mixtures	Carbon Dioxide-Oxygen Mixtures
	Cylinder Pressure at 70 F (psig)		840	80	1,250	1,650–2,000[1]	745	1,800–2,400[1]	1,650–2,000[1,2]	1,500–2,200[1,2]
A	Contents Weight (lb)		0.8	0.6	0.4	0.03		0.23		
(3″ o.d. x 7″)	Gas Volume at 70 F	cu ft	6.6	5.3	5.3	2.0		2.7	2.0	2.7
	and 14.7 psia	liters	184.0	150.0	150.0	56.6		76.5	56.6	76.5
B	Contents Weight (lb)		1.6	1.5	1.0	0.04	2.0	0.5		
($3\frac{1}{2}$″ o.d. x 13″)	Gas Volume at 70 F	cu ft	13.3	13.3	13.3	3.7	16.6	5.3	4.0	5.3
	and 14.7 psia	liters	377.0	377.0	377.0	105.0	470.0	150.0	113.0	150.0
D	Contents Weight (lb)		4.0	3.3	2.0	0.1	4.0	1.0		
($4\frac{1}{4}$″ o.d. x 17″)	Gas volume at 70 F	cu ft	33.0	30.0	26.6	10.6	34.5	12.6	11.0	12.6
	and 14.7 psia	liters	934.0	848.0	752.0	300.0	975.0	356.0	310.0	356.0
E	Contents Weight (lb)		6.6		3.3	0.2	6.6	2.0		
($4\frac{1}{4}$″ o.d. x 26″)	Gas Volume at 70 F	cu ft	56.0		44.0	17.0	57.0	22.0	18.0	22.0
	and 14.7 psia	liters	1585.0		1245.0	480.0	1610.0	622.0	510.0	622.0
F	Contents Weight (lb)		20.0		11.0	0.6	21.0	6.0		
($5\frac{1}{2}$″ o.d. x 51″)	Gas Volume at 70 F	cu ft	170.0		146.0	56.0	186.0	73.0	58.0	73.0
	and 14.7 psia	liters	4800.0		4130.0	1585.0	5260.0	2062.0	1640.0	2062.0
M	Contents Weight (lb)		31.0		16.0	0.9	32.0	9.0		
($7\frac{1}{8}$″ o.d. x 43″)	Gas Volume at 70 F	cu ft	266.0		218.0	80.3	278.0	106.0	82.0	106.0
	and 14.7 psia	liters	7400.0		6160.0	2280.0	7850.0	3000.0	2320.0	3000.0
G	Contents Weight (lb)		50.0		28.0	1.5	56.0	16.0		
($8\frac{1}{2}$″ o.d. x 51″)	Gas Volume at 70 F	cu ft	425.0		372.0	146.0	485.0	186.0	150.0	186.0
	and 14.7 psia	liters	12000.0		10500.0	4130.0	13750.0	5260.0	4250.0	5260.0
H and K	Contents Weight (lb)						64.0	16–22		
(9″ o.d. x 51″)	Gas Volume at 70 F	cu ft					557.0	197–265		
	and 14.7 psia	liters					15800.0	5570–7500		

[1] Varies with cylinder size.
[2] Varies with per cent composition of mixtures.
* Values are for full cylinders and are approximate.

are equipped with valves having threaded outlet connections.

5.2.2. Detailed, dimensioned drawings of standard and alternate standard compressed gas cylinder valve outlets have been prepared by Compressed Gas Association, Inc., and recognized by the American Standards Association and the Canadian Standards Association as American and Canadian Standards, respectively. Further information about these standards appears in the following Chapter 4.

5.2.2.1. Chapter 4, Section B, gives detailed information on the pin-index safety system for flush-type cylinder valves. This system has been devised to prevent the interchangeability of medical gas cylinders equipped with flush-type valves as employed in Style E and smaller cylinders. The system consists of a combination of 2 pins projecting from the yoke assembly of the apparatus and so positioned as to fit into matching holes in the cylinder valve. All medical gas cylinders having flush-type valves and shipped by member companies of Compressed Gas Association, Inc., are now drilled in conformance to this standard. The Association has strongly recommended that all new gas apparatus now being manufactured incorporate the system and that older, existing equipment be modified in accordance with it. All users of medical gas administering apparatus are urged to take full advantage of the safety features gained by adoption of the pin-index safety system.

5.2.3. The use of standard valve outlet connections should be encouraged to the exclusion of adapters.

5.2.4. As a further step in its continuing efforts to minimize the possibility of accidental substitution of the wrong gases by users of medical gas administering equipment, Compressed Gas Association, Inc., has developed a standard that would apply to removable threaded connections from the low-pressure regulator outlet to the patient. All users of medical gas administering apparatus are urged to take full advantage of the safety features of this standard which is known as the diameter-index safety system.

5.2.5. The diameter-index safety system was developed by CGA to meet the need for a standard to provide noninterchangeable connections where removable exposed threaded connections are employed in conjunction with individual gas lines of medical gas administering equipment, at pressures of 200 psig or less, such as outlets from medical gas regulators and connectors for anesthesia, resuscitation and therapy apparatus. Removable threaded connections are those which are commonly and readily engaged or disengaged in routine use and service.

5.2.6. The long established 9/16 in. 18 thread connection has been retained as the standard for oxygen. For all other medical gas and air suction equipment, noninterchangeable indexing is achieved by a series of increasing and decreasing diameters in the component parts of the connections. Further information can be found in Chapter 4, Section C, "The Diameter-Index Safety System."

5.3. Regulators

5.3.1. The purpose of a regulator is to reduce the pressure of the gas as it issues from the cylinder to a usable pressure, and to control the flow of the gas. Other terms are often used erroneously for the regulator; including such names as "gage" and "valve." A "valve" is defined as a movable mechanism which opens and closes a passage. This term usually refers to the part on top of the cylinder to which the regulator is connected.

5.3.2. A "gage" is an instrument of measure, and most regulators are equipped with 2 gages. The gage nearest the cylinder measures the pressure of the gas in the cylinder in pounds per square inch. The other gage may register the reduced or working pressure in pounds per square inch or it may measure the rate of discharge or flow of gas from the regulator in liters or gallons per minute.

5.3.3. There are several methods of controlling the flow of gas from the regulators, the most common being by means of an adjusting screw on the bonnet of the regulator.

5.4. Piping and Manifold Systems

5.4.1. Piping and manifold systems for medical gases should be constructed only under the supervision of a competent engineer who is thoroughly familiar with the problems incident to piping compressed gases. Consultation with your gas supplier before installation of piping and manifolds may often be helpful. Standards for piping and manifold systems are published in NFPA Pamphlet No. 565—"Standards for Nonflammable Medical Gas Systems."*

SECTION C.

Medical Vacuum Systems in Hospitals

1. GENERAL

1.1. Scope

This section applies to piped vacuum systems for hospitals that are used for drainage of the patient or for medical laboratory use only. It does not apply to systems used for vacuum cleaning or as a vacuum condensate return system. Separate central vacuum systems that dispose drainage directly into the sanitary sewer are limited to dental suites (not dental operating or oral surgery) and are not covered by this standard.

At the time the *Handbook* goes to press, this standard is undergoing substantial revisions based on experience developed since the standard was first issued.

1.2. Definition

A piped vacuum distribution consists of a central vacuum supply system with control equipment, and a system of piping extending to the points in the hospital where suction may be required, and suitable station outlet valves at each use-point with suction trap bottles which are equipped with an overflow shutoff device to prevent carryover of fluids into the piping system.

2. SOURCE OF VACUUM

2.1. Central Vacuum Systems

2.1.1. The central vacuum source shall consist of two or more vacuum pumps, duplexed, which alternately or simultaneously on demand supply the vacuum system. Each is capable of maintaining adequate vacuum at peak calculated demand.

2.1.2. Means are provided to automatically alternate the pumps so that usage is divided evenly.

2.1.3. Means are provided to automatically activate the additional pump unit(s) should the first pump unit of the duplex system be incapable of maintaining adequate vacuum.

2.1.4. Each vacuum pump is provided with suitable motor starting devices with overload protection. In addition, suitable disconnect switches shall be installed in the electrical circuit ahead of the motor starting devices. The electrical control circuit shall be so arranged that shutting off or failure of one vacuum pump will not affect the operation of other pumps.

2.1.5. Electrical wiring to pump motors, controls and alarms shall conform to the provisions of the National Electrical Code.

2.1.6. The electrical system for the vacuum pumps shall be connected to both the normal and emergency electrical power systems.

2.1.7. Receivers of a suitable capacity shall be installed with each vacuum system. The receivers may be of the vertical or horizontal type.

2.1.8. A suitable drain shall be installed in

* Obtainable for a nominal charge from the National Fire Protection Association, 60 Batterymarch St., Boston, Mass. 02110.

the bottom of the receiver so that condensate which might accumulate may be drained from the receiver.

2.1.9. The exhaust from the vacuum pumps shall be piped to the outside with the end turned down and screened against insect life. The exhaust terminal shall be a minimum distance of 10 ft from any door, window, air intake, or other opening in any building, and a minimum distance of 20 ft above the ground.

2.1.10. The vacuum pump shall be installed in a light, clean location with ample accessibility. The pump shall be bolted to the floor or suitable pedestal.

3. VACUUM REQUIREMENTS

3.1. The vacuum pump together with the pipe sizing of the system shall be capable of maintaining a vacuum of 12 in. of mercury (Hg) at the outlet station farthest away from the pump, when the calculated demand for the hospital (as figured from Table 1), is drawn on the system.

4. WARNING SYSTEMS

4.1. Emergency Alarm System

4.1.1. An emergency alarm signal shall be provided to indicate by audible and visual signals when the vacuum in the hospital line drops below 8 in. Hg (mercury) vacuum.

4.1.2. Alarm signals shall be located to assure continuous responsible observation. Each signal shall be appropriately labeled.

Note: The emergency signal is usually in the office or principal working area of the individual responsible for the maintenance of the vacuum system. To assure continuous surveillance, an auxiliary visual signal should be located at the telephone switchboard.

4.1.3. The alarm signal system shall be connected to both the normal and emergency electrical power systems.

5. PIPELINE SYSTEMS

5.1. Pipelines

5.1.1. Central vacuum systems shall not be connected to a sanitary waste system for disposal of patient drainage.

5.1.2. All pipelines shall be seamless Type K or Type L copper tubing. Copper tubing should preferably be hard temper for exposed locations and soft temper for underground or concealed locations. Pipe sizes shall be in conformity with good engineering practice for proper delivery of maximum volumes specified. Vacuum piping shall not be supported by

TABLE 1. OUTLET RATING CHART FOR HOSPITAL AND CLINICAL CENTRAL VACUUM SYSTEMS

Location of Outlet	Allowance per Outlet (cfm at 1 atm)	Simultaneous Use Factor (%)	Actual Cubic Feet per Minute (1 atm/ Outlet)
Operating rooms, major	1	100	1.0
Operating rooms, minor dental, eye, ear, nose, throat, cystoscopy, diapsy, fracture	1	50	.5
Delivery room	1	100	1.0
Recovery, post anesthesia	$\frac{1}{2}$	100	.5
Intensive care unit	$\frac{1}{2}$	100	.5
Emergency room	1	100	1.0
Patient rooms, nurseries, wards and general medical	$\frac{1}{2}$	20	.1
Patient rooms and wards, t.b. hospitals and polio wards	$\frac{1}{2}$	40	.2
Treatment and examining room	$\frac{1}{2}$	20	.1
Dental clinic	$\frac{1}{2}$	50	.25
Autopsy, laboratory and inhalation therapy	$\frac{1}{2}$	20	.1

other piping but shall be supported with pipe hooks, metal pipe straps, bands or hangers suitable for the size of pipe and of proper strength and quality at proper intervals, so that piping cannot be moved accidently from the installed position as follows:

$\frac{1}{2}$-in. pipe or tubing	6 ft
$\frac{3}{4}$-in. or 1-in. pipe or tubing	8 ft
$1\frac{1}{4}$-in. or larger (horizontal)	10 ft
$1\frac{1}{4}$-in. or larger (vertical)	every floor level

5.1.3. All fittings used for connecting copper tubing shall be wrought copper, brass, or bronze fittings made especially for solder or brazed connections, except as provided in 5.1.4.

5.1.4. Approved tubing connection fittings may be used on distribution lines when pipe sizes are $\frac{1}{2}$ in. nominal or less if the fitting is accessible from the room. Such fittings may also be used in connecting copper tubing of $\frac{3}{4}$ in. nominal or less to shutoff valves described in 5.2, provided the fittings are readily accessible.

5.1.5. Buried piping shall be adequately protected against frost, corrosion and physical damage. Ducts or casings may be used. If piping is run outdoors and subject to freezing temperatures, precautions shall be taken to prevent freezing of condensate.

5.1.6. Vacuum risers may be installed in pipe shafts if suitable protection against physical damage or corrosion is provided.

5.1.7. Pipelines shall be suitably protected against physical damage from the movements of portable equipment such as carts, stretchers and trucks.

5.1.8. Vacuum pipelines shall be readily identified by appropriate labeling with the word "VACUUM." Such labeling shall be by means of metal tags, stenciling, stamping, or with adhesive markers, in a manner that is not readily removable. Labeling shall appear on the pipe at intervals of not more than 20 ft and at least once in each room and each story traversed by the pipeline.

5.1.9. Antivibration couplings shall be installed between the vacuum pump and the pipeline and between the vacuum pump and the exhaust pipe.

5.1.10. Piping systems for vacuum shall not be used as a grounding electrode.

5.2. Shutoff Valves

5.2.1. All shutoff valves accessible to other than authorized personnel shall be installed in valve boxes with frangible windows large enough to permit manual operation of valves and labeled in substance as follows:

> CAUTION—Vacuum Valves
> Do not close except in emergency
> This valve controls vacuum supply to . . .

5.2.2. The main vacuum supply line shall be provided with a shutoff valve so located as to be accessible in an emergency.

5.2.3. Each riser supplied from the main line shall be provided with a shutoff valve adjacent to the riser connection.

5.2.4. Patient outlet stations shall not be supplied directly from a riser unless a manual shutoff valve located in the same story is installed between the riser and the outlet station.

5.2.5. Each pump shall be provided with a shutoff valve to isolate it from the vacuum system for routine maintenance and repair.

5.3. Station Outlets

5.3.1. Each station outlet for vacuum shall be equipped with either a valve with threaded connection or a quick coupler valve of the noninterchangeable type and shall be legibly labeled "SUCTION" or "VACUUM."

5.3.2. Valves with threaded connections shall be of the noninterchangeable type and shall conform to the diameter-index safety system as described in Chapter 4, Section C, "The Diameter-Index Safety System."

5.3.3. A secondary check valve is not required in a vacuum station valve.

5.3.4. Station outlets in patient rooms shall be located approximately 5 ft above the floor or suitably located to avoid physical damage to the valve or control equipment.

5.3.5. All permanently installed vacuum gages and manometers for the vacuum system shall be those manufactured expressly for vacuum and labeled: "VACUUM."

5.3.6. Wall outlet assemblies furnished for installation by manufacturers shall be legibly marked "VACUUM" or "SUCTION" so that in its state of disassembly for hookup to the piping system proper identity remains existent.

6. INSTALLATION OF PIPING SYSTEMS

6.1. General

6.1.1. Before installation, all piping, valves and fittings—except those supplied especially prepared for vacuum service by the manufacturer and received sealed on the job—shall be thoroughly cleaned of oil, grease and other readily oxidizable materials by washing in a hot solution of sodium carbonate or trisodium phosphate (proportion of one pound to three gallons of water). *The use of organic solvents, for example, carbon tetrachloride, is prohibited.* Scrubbing shall be employed where necessary to insure complete cleaning. After washing, the material shall be rinsed thoroughly in clean hot water. After cleaning, particular care shall be exercised in the storage and handling of all pipe and fittings. Pipe and fittings shall be temporarily capped or plugged to prevent recontamination before final assembly. Tools used in cutting or reaming shall be kept free from oil or grease. Where such contamination has occurred, the items affected shall be rewashed and rinsed.

6.1.2. All joints in the piping, except those at valves or at equipment requiring pipe thread connections shall be made with silver brazing alloy or similar high melting point (at least 1000 F) brazing metal. Particular care shall be exercised in applying the flux to avoid leaving any excess inside the completed joints. The outside of the tube and fittings shall be cleaned by washing with hot water after assembly.

6.1.3. Pipe thread joints used in shutoff valves, including station outlet valves, shall be installed by tinning the male thread with soft solder. Litharge and glycerin or an approved oxygen luting compound are acceptable.

Note: The purpose of the above specifications is that oxygen, nitrous oxide and compressed air lines normally are installed at the same time as vacuum systems, and carelessness in installing the vacuum lines might result in damage or contamination to these lines.

6.2. Testing

6.2.1. After installation of the piping, but before attaching the vacuum line to the vacuum pump and before installation of the station outlet valves, the line shall be blown clear by means of oil-free nitrogen or air.

6.2.2. After installation of station outlet valves, but before attaching the vacuum line to the vacuum pump, each section of the piping system shall be subjected to a test pressure of 150 psig, by means of oil-free nitrogen or air. This test pressure shall be maintained until each joint has been examined for leakage by means of soapy water. All leaks shall be repaired and the section retested.

6.2.3. A 24 hr standing pressure test with oil-free nitrogen or air at 150 psig shall be made to check the completeness of previous joint tests. After completion of the final standing pressure test, the system shall be attached to the vacuum pump and put into operation.

REFERENCE

"Tentative Standard for the Use of Inhalant and Resuscitative Equipment and Agents" (Pamphlet P-2.2T), Compressed Gas Association, Inc.

CHAPTER 2

Safety Relief Device Standards

Almost all compressed gas containers are fitted with devices that let the gas escape should surrounding conditions (such as heat) cause the enclosed gas to increase in pressure to a dangerous degree. The standards that have been developed by Compressed Gas Association, Inc., for these safety relief devices are given in this chapter. Its sections present the CGA safety relief device standards as follows:

Section A. For Cylinders—Safety Relief Device Standards, Part 1;

Section B. For Cargo and Portable Tanks—Safety Relief Device Standards, Part 2;

Section C. For Storage Tanks—Safety Relief Device Standards, Part 3;

Section D. Recommended Practice for the Manufacture of Fusible Plugs.

SECTION A

For Cylinders—Safety Relief Device Standards, Part 1

This part of the safety relief device standards represents the minimum requirements recommended by Compressed Gas Association, Inc., for safety relief devices considered to be appropriate and adequate for use on cylinders having capacities of 1000 lb water, or less and ICC-3AX and 3AAX cylinders having capacities of over 1000 lb water, and which comply with the specifications and charging and maintenance regulations of the Interstate Commerce Commission (ICC) or the corresponding specifications and regulations of the Board of Transport Commissioners for Canada (BTC).

It is recognized that there are cylinders that conform to the specification requirements of the ICC or the BTC, which are used in services that are beyond the jurisdiction of either of these authorities. In such cases it is recommended that state, provincial, local or other authorities having jurisdiction over these cylinders be guided by Part 1 of these standards in determining adequate safety relief device requirements, provided that the cylinders are charged with gas and maintained in accordance with ICC and BTC regulations.

It is further recognized that there may be cylinders which are used in services beyond the jurisdiction of the ICC or BTC and which do not conform to the specification requirements of either authority. It is recommended that the authorities having jurisdiction over such cylinders be guided by Part 1 of these standards in determining safety relief device requirements, provided that such cylinders are considered by the authority to have a construction at least equal to the equivalent ICC specification requirements and further, provided that the cylinder shall be charged with gas and maintained in accordance with ICC or BTC requirements.

For cylinders that come within the jurisdiction of state and local regulatory authorities, the user should check for compliance with all local regulations. A number of states and cities have pressure vessel laws and regulations which include requirements for safety relief devices. The CGA Standards are prepared specifically for compressed gas containers and the safety relief devices may not be acceptable unless special permission is obtained from the authority having jurisdiction.

For cylinders that come within the jurisdiction of the ICC or BTC, safety relief devices must be approved as to type, size, quantity and location by the Bureau of Explosives, 63 Vesey St., New York, N. Y., 10007.

1. DEFINITIONS

For the purposes of these standards the following terms are defined:

1.1 A "safety relief device" is a device intended to prevent rupture of a cylinder under certain conditions of exposures. (The term as used herein shall include the approach channel, the operating parts, and the discharge channel.)

1.2. An "approach channel" is the passage or passages through which gas must pass from the cylinder to reach the operating parts of the safety relief device.

1.3. A "discharge channel" is the passage or passages beyond the operating parts through which gas must pass to reach the atmosphere exclusive of any piping attached to the outlet of the device.

1.4. A "safety relief device channel" is the channel through which gas released by operation of the device must pass from the cylinder to the atmosphere, exclusive of any piping attached to the inlet or outlet of the device.

1.5. The "operating part" of a safety relief device is the part that normally closes the safety discharge channel but when moved from this position as a result of the action of heat or pressure, or a combination of the two, permits escape of gas from the cylinder.

1.6. A "frangible disc" is an operating part in the form of a disc, usually of metal and which is so held as to close the safety relief device channel under normal conditions. The disc is intended to burst at a predetermined pressure to permit the escape of gas. (Such discs are generally of flat, preformed, reinforced, or grooved types.)

1.7. The "pressure opening" is the orifice against which the frangible disc functions.

1.8. The "rated bursting pressure" of a frangible disc is the maximum pressure for which the disc is designed to burst when in contact with the pressure opening for which it was designed when tested as required in 5.3.1.

1.9. A "fusible plug" is an operating part in the form of a plug of suitable low melting material, usually a metal alloy, which closes the safety relief device channel under normal conditions and is intended to yield or melt at a predetermined temperature to permit the escape of gas.

1.10. The "yield temperature" of a fusible plug is the temperature at which the fusible metal or alloy will yield when tested as required in 5.2.3.

1.11. A "reinforced fusible plug" is a fusible plug consisting of a core of suitable material having a comparatively high yield temperature surrounded by a low melting point fusible metal of the required yield temperature.

1.12. A "combination frangible disc-fusible plug" is a frangible disc in combination with a low melting point fusible metal, intended to prevent its bursting at its predetermined bursting pressure unless the temperature also is high enough to cause yielding or melting of the fusible metal.

1.13. A "safety relief valve" is a safety relief device containing an operating part that is held normally in a position closing the safety relief device channel by spring force and is intended to open and to close at predetermined pressures.

1.14. A "combination safety relief valve and fusible plug" is a safety relief device utilizing a safety relief valve in combination with a fusible plug. This combination device may be an integral unit or separate units and is intended to open and to close at predetermined pressures or to open at a predetermined temperature.

1.15. The "set pressure" of a safety relief valve is the pressure marked on the valve and at which it is set to start-to-discharge (see 3.3.2).

1.16. The "start-to-discharge pressure" of a safety relief valve is the pressure at which the first bubble appears through a water seal of not over 4 in. on the outlet of the safety relief valve (see 5.5.1).

1.17. The "flow capacity" of a safety relief device is the capacity in (cfm) of free air discharged at the required flow rating pressure.

1.18. "Flow rating pressure" is the pressure at which a safety relief device is rated for capacity.

1.19. A "nonliquefied compressed gas" is a gas, other than a gas in solution which under the charging pressure, is entirely gaseous at a temperature of 70 F.

1.20. A "liquefied compressed gas" is a gas which, under the charging pressure, is partially liquid at a temperature of 70 F. A flammable compressed gas which is normally nonliquefied at 70 F but which is partially liquid under the charging pressure and temperature, shall follow the requirements for liquefied compressed gases.

1.21. A "compressed gas in solution" (acetylene) is a nonliquefied gas which is dissolved in a solvent.

1.22. A "pressurized liquid compressed gas" is a compressed gas other than a compressed gas in solution, which cannot be liquefied at a temperature of 70 F, and which is maintained in the liquid state at a pressure not less than 40 psia by maintaining the gas at a temperature less than 70 F.

1.23. The "test pressure of the cylinder" is the minimum pressure at which it must be tested as prescribed in the specifications for compressed gas cylinders by the ICC or BTC.

1.24. "Free air" or "free gas" is air or gas measured at a pressure of 14.7 psia and a temperature of 60 F.

1.25. "ICC regulations" as used in these standards refers to "Interstate Commerce Commissions Regulations for Transportation of Explosives and Other Dangerous Articles by Land and Water in Rail Freight, Express and Baggage Services and by Motor Vehicle (Highway) and Water, including Specifications for Shipping Containers," Code of Federal Regulations, Title 49, Parts 71 to 78.

1.26. "BTC regulations" as used in these standards refers to Board of Transport Commissioners for Canada, "Regulations for the Transportation of Dangerous Commodities by Rail."

2. TYPES OF SAFETY RELIEF DEVICES

Types of safety relief devices as covered by this Part are designated as follows:

2.1. Type CG-1: Frangible disc.

2.2. Type CG-2: Fusible plug or reinforced fusible plug utilizing a fusible alloy with yield temperature not over 170 F, nor less than 157 F (165 F nominal).

2.3. Type CG-3: Fusible plug or reinforced fusible plug utilizing a fusible alloy with yield temperature not over 220 F, nor less than 208 F (212 F nominal).

2.4. Type CG-4: Combination frangible disc-fusible plug, utilizing a fusible alloy with yield temperature not over 170 F, nor less than 157 F (165 F nominal).

1.5. Type CG-5: Combination frangible disc-fusible plug, utilizing a fusible alloy with yield temperature not over 220 F, nor less than 208 F (212 F nominal).

2.6. Type CG-7: Safety relief valve.

2.7. Type CG-8: Combination safety relief valve and fusible plug.

3. APPLICATION REQUIREMENTS FOR SAFETY RELIEF DEVICES

3.1. General

3.1.1. Compressed gas cylinders, which under the regulations of the Interstate Commerce Commission must be equipped with safety relief devices, shall be considered acceptable when equipped with devices of proper construction, location and discharge capacity under the conditions prescribed in Table 1 (see page 203). Safety relief devices are prohibited on compressed gas cylinders authorized by ICC or BTC for shipment of the following Poisons, Class A:

- Bromacetone, liquid
- Chlorpicrin and methyl chloride mixtures
- Cyanogen chloride containing less than 0.9 per cent water
- Cyanogen gas
- Ethyldichlorarsine
- Hexaethyl tetraphosphate and compressed gas mixture
- Hydrocyanic acid, liquefied
- Hydrocyanic acid (prussic) liquid
- Lewisite
- Methyldichlorarsine
- Mustard gas
- Nitrogen dioxide, liquid
- Nitrogen peroxide (tetroxide)
- Parathion and compressed gas mixture
- Phenylcarbylamine chloride
- Phosgene (diphosgene)
- Poisonous liquid or gas, n.o.s.
- Tetraethyl dithiopyrophosphate and compressed gas mixture
- Tetraethyl pyrophosphate and compressed gas mixture

3.1.2. The design, material and location of safety relief devices shall have been proved to be suitable for the intended service. Consideration shall be given in the design and application of safety devices to the effect of the resultant thrust when the device functions.

3.1.3. Only replacement parts or assemblies provided by the manufacturer shall be used unless the advisability of interchange is proved by adequate tests.

3.1.4. When cylinders are required to be equipped with safety relief devices at both ends, they shall have the specified discharge capacity at each end.

3.1.5. When cylinders are not required to be equipped with safety relief devices at both ends, the discharge capacity of the individual devices may be combined to meet the minimum total discharge capacity requirement.

3.2. Frangible Discs

When a frangible disc is used with a compressed gas cylinder, the rated bursting pressure of the disc (when tested within the temperature range of 60 F to 160 F in accordance with Section 5.3) shall not exceed the minimum required test pressure of the cylinder with which the device is used, except as follows: For ICC-3E or BTC-3E cylinders the rated bursting pressure of the device shall not exceed 4500 pounds per square inch gage (psig).

3.3. Safety Relief Valves

3.3.1. When a safety relief valve is used on a compressed gas cylinder, the flow rating pressure shall not exceed the minimum required test pressure of the cylinder on which the safety relief valve is installed and the reseating pressure shall not be less than the pressure in a normally charged cylinder at 130 F.

3.3.2. The set pressure shall not be less than 75 per cent nor more than 100 per cent of the minimum required test pressure of the cylinder in which the safety relief valve is installed.

3.3.3. Safety relief valves should have direct communication with the vapor space of the cylinder when in normal use.

3.4 Piping of Safety Relief Devices

3.4.1. When fittings and piping are used on either the upstream or downstream side or both of a safety relief device or devices, the passages shall be so designed that the flow capacity of the safety relief device will not be reduced below the capacity required for the container on which the safety relief device assembly is installed, nor to the extent that the operation of the device could be impaired. Fittings, piping and method of attachment shall be designed to withstand normal handling and the pressures developed when the device or devices function.

3.4.2. No shutoff valve shall be installed between the safety relief devices and the cylinder.

4. DESIGN AND CONSTRUCTION REQUIREMENTS FOR SAFETY RELIEF DEVICES

4.1. The material, design and construction of a safety relief device shall be such that there will be no significant change in the functioning of the device and no serious corrosion or deterioration of the materials within the period between renewals, due to service conditions. The chemical and physical properties of the materials shall be uniform and suitable for the requirements of the part manufactured therefrom.

4.4. In combination frangible disc-fusible plug devices the fusible metal shall be on the discharge side of the frangible disc. The fusible metal shall not be used in lieu of a gasket to seal the disc against leakage around the edges. Gaskets if used shall be of material that will not deteriorate rapidly at the maximum temperature range specified for the fusible metal.

4.3. Methods of manufacture, inspection and tests shall conform to best practices in order to attain satisfactory performance of the safety relief devices (see references).

4.4. The flow capacity of each design and modification thereof of all types of safety relief devices shall be determined by actual flow test. Methods of making flow tests are given in 5.6.

4.5. For noninsulated cylinders for non-liquefied gas, the minimum required flow capacity of safety relief devices, except safety relief valves, shall be calculated by the following formula:

$$Qa = 0.154\text{Wc}$$

where

Qa = flow capacity at 100 psia test pressure in cubic feet per minute of free air.
Wc = water capacity of the cylinder in pounds, but not less than 25 pounds.

Note: The above formula expresses flow capacity requirements equal to 70 per cent of that which will discharge through a perfect orifice having a 0.00012 square inch area for each pound of water capacity of the cylinder.

4.6. For non-insulated cylinders for liquefied gas, the minimum required flow capacity of safety relief devices, except safety relief valves, shall be two times that required by the formula in 4.5.

4.7. For non-insulated cylinders for non-liquefied gas, the minimum required flow capacity of safety relief valves shall be calculated by the following formula:

$$Qa = 0.00154\text{PWc}$$

where

Qa = flow capacity in cubic feet per minute of free air.
P = flow rating pressure in pounds per square inch absolute.
Wc = water capacity of the cylinder in pounds, but not less than 12.5 pounds.

4.8. For non-insulated cylinders for liquefied gas, the minimum required flow capacity of safety relief valves shall be 2 times that required by the formula in 4.7.

4.9. For specification ICC-4L insulated cylinders containing pressurized liquid argon, nitrogen or oxygen, the minimum required flow capacity of safety relief devices shall be

½ that required by the formula in 4.5 or 4.7, whichever is applicable, provided the total thermal conductivity of insulation on the cylinder does not exceed 0.20 Btu per hr per sq ft per degree F at 100 F, and the insulation is enclosed in a steel jacket not less than 0.060 in. thick.

4.10. For acetylene cylinders, fire test will be used in determining safety relief device requirements (see Note M, Table 1).

5. TESTS

5.1. Test of Fusible Alloy

5.1.1. If a laboratory control test of fusible alloy is made, the following test is recommended:

5.1.1.1. Select at random two sticks of the fusible alloy from each batch.

5.1.1.2. A sample for test shall consist of a piece 2 in. long by approximately ¼ in. diameter cut from each stick. Each sample shall be suspended horizontally on suitable supports spaced 1 in. apart and presenting knife edges to the sample so that the ends of the sample will overhang the knife edges ½ in. The supported samples shall be immersed in a glycerine bath not closer than ¼ in. to the bottom of the container. This bath shall be suspended in and controlled by an outer glycerine bath.

5.1.1.3. Two samples from the same stick shall be tested at one time. A thermometer (bulb immersion) shall be inserted into the bath between and closely adjacent to the samples so that the bulb will be completely immersed at the same level as the samples. The bath temperature shall be raised at a rate not in excess of 5 degrees F per minute.

5.1.1.4. The yield temperature shall be taken as that temperature at which the second of the four ends of the samples loses its rigidity and drops.

Note: The test outlined above has been found to be a satisfactory laboratory control test for determining yield temperature for fusible alloys. While this test is recommended, equivalent tests using samples of different dimensions and different rates of heat rise are permissible, provided they yield reproducible results.

5.2. Tests of New or Reconditioned Fusible Plugs or Reinforced Fusible Plugs

5.2.1. Two representative samples shall be selected at random from each lot and subjected to the tests prescribed in 5.2.2. and 5.2.3. If both samples should fail to meet the requirements of 5.2.2. and 5.2.3, the lot shall be rejected. If one sample fails to meet the requirements of 5.2.2 and 5.2.3, four additional samples may be selected at random from the same lot and subjected to these tests. If any of these four additional samples fail to meet the requirements of 5.2.2. and 5.2.3, the lot shall be rejected. The production of either new or reconditioned fusible plugs, including reinforced fusible plugs, by a manufacturer on any one day for any one range of minimum to maximum specified yield temperature, but in no case greater than 3000, shall constitute a lot.

5.2.2. A test to determine resistance to extrusion of the fusible alloy and leaks at a temperature of 130 F or less in a fusible plug shall be made as follows: The finished fusible plug shall be subjected to a controlled temperature of not less than 130 F for 24 hours with a gas pressure of 500 psig on the end normally exposed to the contents of the cylinder. In order to pass this test no leakage nor visible extrusion of material shall be evident upon examination of the end exposed to atmospheric pressure.

5.2.3. A test for determining temperature of a fusible plug shall be made as follows:

5.2.3.1. Subject these plugs to an air pressure of not less than 3 psig applied to the end normally exposed to the content of the cylinder. While subjected to this pressure the plugs shall be immersed in a water bath or a glycerine-water bath, at a temperature in the 5 F range immediately below the specified minimum yield temperature, and held in that temperature range for a period of 10 minutes.

The temperature of the bath shall then be raised at a rate not in excess of 5 F per minute during which the pressure may be increased to not more than 50 psig. When the temperature of the bath reaches the point where metal is exuded or spewed out sufficiently to produce leakage of air, the temperature of the bath should be recorded as the yield temperature of the plugs. It shall be within the temperature limits specified in Section 2 for that type of fusible plug.

5.2.3.2. As an alternate method, these plugs, after passing the test at a temperature of not less than 5 F below the specified minimum yield temperature may at once be immersed in another bath held at a temperature not exceeding the specified maximum yield temperature. If air leakage occurs within 5 minutes at that temperature the requirements have been met.

5.2.3.3. Variation in temperature within the liquid bath in which the plug is immersed for either test in 5.2.3.1 or 5.2.3.2 shall be kept to a minimum by stirring while making these tests.

5.2.4. Fusible plugs to be used in chlorine service and manufactured and tested in accordance with The Chlorine Institute, Inc.* Drawings 110 and 112, latest issue, need not be tested under the foregoing provisions of 5.2.

5.3. Tests of Frangible Disc Safety Relief Devices

5.3.1. The production of frangible discs shall be segregated into lots of not more than 3000 discs with appropriate control exercised to assure uniformity of production. Representative samples shall be selected at random for testing to verify rated bursting pressure. The number of samples selected shall be appropriate for the manufacturing procedures followed, but at least 2 samples shall be tested from each lot. Samples shall be mounted in a proper holder with a pressure opening having dimensions identical with that in the device in which it is to be used and submitted to a bursting test at a temperature not lower than 60 F nor higher than 160 F. The test pressure may be raised rapidly to 85 per cent of the rated bursting pressure, held there for at least 30 seconds, and thereafter shall be raised at a rate not in excess of 100 psig per min, until the disc bursts. The actual bursting pressure shall not be in excess of the rated bursting pressure and not less than 90 per cent of the rated bursting pressure. If the actual bursting pressure is not within the limits prescribed above, entire lot of frangible discs shall be rejected; except that if the manufacturer so desires, he may subject four more discs selected at random from the same lot to similar tests. If all four additional discs meet the requirement, the lot may be used; otherwise, the entire lot shall be rejected. Any elevated temperature determinations may be arrived at by tests conducted at room temperature, provided that the relation of bursting pressure to different temperatures is established by test for the type of material used.

5.3.2. The production of frangible disc holders (that part containing the pressure opening) of 3000 or less shall be considered a lot. Two representative holders selected at random from the lot shall be assembled with proper frangible discs from acceptable lot as tested in 5.3.1 and subjected to the bursting pressure test of 5.3.1. The actual bursting pressure shall not be in excess of the rated bursting pressure of the disc nor less than 85 per cent of the rating. If the actual bursting pressure at a temperature not less than 60 F nor more than 160 F is not within the above limits, the entire lot of frangible disc holders shall be rejected; except if the manufacturer so desires, he may subject four more holders selected as above from the same lot to similar tests. If all four holders meet the requirement, the lot may be used; otherwise, the entire lot shall be rejected. Any elevated temperature determinations may be arrived at by tests conducted at room temperature, provided that the relation of bursting pressure to different temperatures is established by test for the type of material used.

* The Chlorine Institute, 342 Madison Ave., New York, N. Y. 10017.

5.3.3. Testing of the completed safety relief device for detail requirements specified in 5.3.1 and 5.3.2 in lieu of individual tests will be considered as complying with requirements of both 5.3.1 and 5.3.2.

5.4. Tests of Combination Frangible Disc-fusible Plug Safety Relief Devices

5.4.1. The production of combination frangible disc-fusible plug devices of any one rated bursting pressure and any one yield temperature on any one day shall be considered a lot. Two representative assembled devices shall be selected at random and submitted to a performance test conducted as follows:

5.4.1.1. Each assembled device shall be subjected to a pressure of 70 to 75 per cent rated maximum bursting pressure of the frangible disc used, and while under this pressure, shall be immersed in a liquid bath held at a temperature not less than 5 F below the minimum specified yield temperature of the fusible metal for at least 10 minutes. The fusible metal shall not show evidence of melting. The temperature of the bath shall then be raised at a rate not in excess of 5 F per minute without material change in pressure. When the maximum specified yield temperature of the fusible metal is reached the fusible metal must have melted. There shall be no leakage.

5.4.1.2. The frangible disc shall then be tested in accordance with the requirements of 5.3.1. The device may be removed from the bath for this test.

5.4.1.3. As an alternate to tests in 5.4.1.1 and 5.4.1.2, the frangible disc and fusible metal may be tested separately to requirements 5.2.3 and 5.3.1, provided the design of the device is such as to allow for the separation of the parts and the separate tests.

5.4.1.4. If either of the devices fail to meet the requirements given in 5.4.1.1; 5.4.1.2; or 5.4.1.3 the entire lot shall be rejected; except if the manufacturer so desires, he may subject four more such devices selected at random to similar tests. If all four additional devices meet the requirements, the lot may be used.

5.5 Pressure Tests of Safety Relief Valves

5.5.1. Each safety valve shall be subjected to an air or gas pressure test to determine the following:

5.5.1.1. That the start to discharge pressure at which the first bubble appears through a water seal of not over 4 inches on the outlet of the safety relief valve is not less than 75 per cent nor more than 100 per cent of the flow rated pressure for which the safety relief valve is marked.

5.5.1.2. That after the start to discharge pressure test, the resealing pressure or the pressure at which leakage ceases through the water seal on the outlet of the safety relief valve, is not less than 70 per cent of the flow rated pressure.

5.6. Flow Capacity Tests

5.6.1. The flow capacity of each design and modification thereof of all types of safety relief devices shall be determined by actual flow test. Three samples of each size of each device representative of standard production shall be tested. Each device shall be completely assembled from the inlet of the approach channel to the exit of the discharge channel in the manner normally assembled for use. Each device shall be caused to operate by either pressure or by temperature or by a combination of such effects but not exceeding the maximum temperature and pressure for which it was designed.

5.6.1.1. After pressure testing, without cleaning, removal of parts, or reconditioning, each safety relief device shall be subjected to an actual flow test wherein the amount of air or gas released by the device is measured. The average capacity of the 3 devices tested shall be recorded as the flow-rated capacity. The design of the devices shall be considered acceptable, provided the capacities of the three devices shall fall within 10 per cent of the highest capacity recorded.

5.6.2. Acceptable methods of flow testing.

5.6.2.1. Safety relief devices may be tested for flow capacities by testing with equipment conforming to the American Gas Association Gas Measurement Committee Report No. 3, "Orifice Metering of Natural Gas" as reprinted with revisions, 1956.

5.6.2.2. Air or gas shall be supplied to the safety relief device through a supply pipe provided with a pressure gage and a thermometer for indciating or recording the pressure and temperature of the supply. Observations shall be made and recorded after steady flow conditions have been established. Test conditions need not be the same as the conditions under which the device is expected to function in service, but the following limits must be met. The inlet pressure of the air or gas supplied to the safety relief device shall be not less than 100 psig, except that the flow test of a safety relief valve shall be made at the flow rating pressure.

5.6.2.3. Where the testing method prescribed in 5.6.2.1 is employed, such test may be made by the manufacturer of the safety relief device, provided that the request form for safety relief device approval is sent in duplicate to the Bureau of Explosives, showing the results of these tests and that copies thereof are maintained by the manufacturer (for approval request form, see Appendix A of these standards).

5.6.2.4. Where any other method of testing is used, the Bureau of Explosives shall be furnished with a certification of the accuracy of the rest results by a competent disinterested agency, which is acceptable to the Bureau.

5.7. Rejected safety relief devices or components may be reworked provided they are subjected to such additional tests as are required to insure compliance with all the requirements of these standards.

6. IDENTIFICATION

It is the purpose of Section 6 of these Standards to recommend certain safeguards or guides so that safety relief device performance may not be jeopardized by improper service practices. The aim in general, is to make it possible to identify the manufacturer of the device and to have the main replaceable parts so identified or coded that it may be readily determined, usually by reference to manufacturer's published data, whether parts are intended to function together, what operating pressure range or temperature range they will provide for, and whether they have adequate flow capacity for the cylinder with which they are to be employed. In particular, it is pointed out that frangible discs can be applied only against pressure openings for which they were specifically designed. Some manufacturers employ sharp pressure opening contours while others employ rounded or other shaped contours. Because of these contour variations, an interchange of discs will give widely different bursting pressures, even though the same diameter may be employed for the bursting pressure opening. In addition, variation in diameter for the pressure opening will give still wider variation in bursting pressure if discs are interchanged improperly.

6.1. Suitable marking shall be provided so that the manufacturer of the safety relief device may be determined.

6.2. Where a knowledge of the date of manufacture is necessary for proper maintenance of a safety relief device the month and year of manufacture shall be marked on the device.

6.3. When frangible discs and pressure opening parts are designed to be replaced as individual piece parts, they shall be marked to indicate the rated bursting pressure (with the proper mating part), the flow capacity, and the manufacturer.

6.4. Suggested Methods of Marking

6.4.1. Stamp the manufacturer's name or trademark and maximum service pressure or identifying part number on the part containing the pressure opening.

6.4.2. Ink or otherwise mark the number on the frangible disc or apply other code mark to facilitate determination of bursting pressure range and proper mating part.

6.5. When frangible discs and pressure opening parts are combined in factory assembled safety relief devices designed to be replaced as a unit (CG-1, CG-4 or CG-5) the assembly shall be marked externally to indicate pressure rating, flow capacity, manufacturer, and yield temperature rating, if any.

6.6. Fusible metal safety relief devices (CG-2 or CG-3) shall be marked externally to indicate yield temperature rating, flow capacity, and manufacturer.

6.7. Safety relief valves shall be marked to indicate:

(1) Manufacturer.

(2) The pressure for which the valve is "set to start to discharge."

(3) The flow rating pressure in pounds per square inch gage (psig) at which the flow capacity of the valve is determined.

(4) The flow capacity in cubic feet per minute of free air.

6.8. All markings required in 6.2 through 6.7, inclusive, may be coded. Coded designations shall be determinable from the manufacturer.

7. MAINTENANCE REQUIREMENTS FOR SAFETY RELIEF DEVICES

7.1. General Practices

7.1.1. As a precaution to keep cylinder safety relief devices in reliable operating condition, care shall be taken in the handling or storing of compressed gas cylinders to avoid damage. Care shall also be exercised to avoid plugging by paint or other dirt accumulation or safety relief device channels or other parts which could interfere with the functioning of the device. *Only qualified personnel shall be allowed to service safety relief devices.*

7.2. Routine Checks When Filling Cylinders

7.2.2. Each time a compressed gas cylinder is received at a point for refilling, all safety relief devices shall be examined externally for corrosion, damage, plugging of external safety relief device channels, and mechanical defects such as leakage or extrusion of fusible metal. If there is any doubt regarding the suitability of the safety relief device for service, the cylinder shall not be filled until it is equipped with a suitable device.

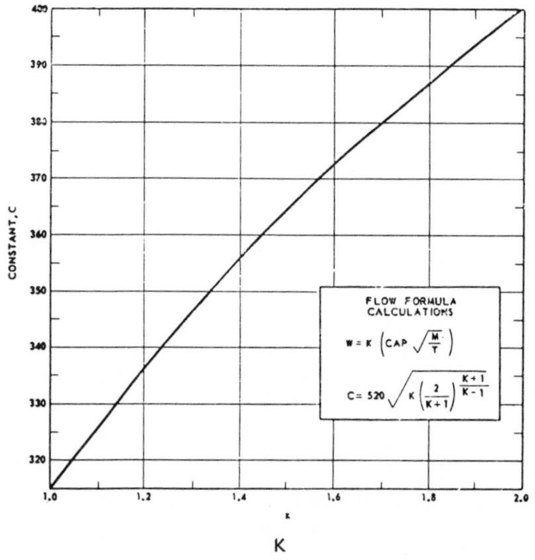

k	Constant C	k	Constant C	k	Constant C
1.00	315	1.26	343	1.52	366
1.02	318	1.28	345	1.54	368
1.04	320	1.30	347	1.56	369
1.06	322	1.32	349	1.58	371
1.08	324	1.34	351	1.60	372
1.10	327	1.36	352	1.62	374
1.12	329	1.38	354	1.64	376
1.14	331	1.40	356	1.66	377
1.16	333	1.42	358	1.68	379
1.18	335	1.44	359	1.70	380
1.20	337	1.46	361	2.00	400
1.22	339	1.48	363	2.20	412
1.24	341	1.50	364		

FIG. 1. Constant C for Gas or Vapor Related to Ratio of Specific Heats ($k = C_p/C_v$) at Standard Conditions. (Data from Figure UA-230, ASME Boiler and Pressure Vessel Code, Section VIII, Unfired Pressure Vessels.)

TABLE 1. RECOMMENDED SAFETY RELIEF DEVICES

The types of safety relief devices listed in the table below and indicated by the existence of a letter symbol or symbols are considered by Compressed Gas Association, Inc., to be suitable for application on cylinders for various compressed gases. When ICC or BTC regulations require safety relief devices on compressed gas cylinders, such devices must be approved as to type, size, quantity and location, by the Bureau of Explosives.

The table does not include all types of safety relief devices which might be acceptable for the different gases. Only those types of safety relief devices are shown which already have been approved by the Bureau of Explosives for the particular gas service and with which Compressed Gas Association Inc., concurs. Failure to show other types of safety relief devices for various gases in this table does not necessarily reflect unfavorably upon the use of such safety relief devices but merely indicates the absence of applications for approval of other types of safety relief devices at the time these standards were promulgated.

Approval requests for types of safety relief devices other than those listed below must be sent to the Bureau of Explosives and be accompanied by test data as shown on form suggested in Appendix A, of these standards.

TYPES OF SAFETY RELIEF DEVICES

Name of Gas	CG–1 Fran-gible Disc	CG–2 165 F Fusible	CG–3 212 F Fusible	CG–4 Fran-gible Disc/w 165 F Fusible	CG–5 Fran-gible Disc/w 212 F Fusible	CG–7 Safety Relief Valve	CG–8 Safety Relief Valve and Fusible
Acetylene (gas in solution)			KM				
Liquefied, Flammable							
Butane, normal	C	F		CF		A	J
Butadiene, inhibited	C	F		CF		A	J
Butene=1 (Pure)	C	F		CF		A	J
Cyclopropane	C	F		CF		A	J
Difluoreothane		F				A	J
Difluoromonochloroethane		F				A	J
Dimethylamine, anhydrous	Safety relief device not required by ICC or BTC regulations.						
Dimethyl ether		F				A	
Ethane	C			CF		A	J
Hydrogen, liquefied	N						
Hydrogen sulfide		BL		BCL			
Isobutane	C	F		CF		A	J
Liquefied petroleum gas	C	F		CF		A	J
Methyl chloride		F				A	J
Methyl mercaptan	Safety relief device not required by ICC or BTC regulations.						
Monomethylamine, anhydrous	Safety relief device not required by ICC or BTC regulations.						
Propane	C	F		CF		A	J
Propylene	C	F		CF		A	J
Trimethylamine, anhydrous	Safety relief device not required by ICC or BTC regulations.						
Vinyl chloride, inhibited		F				A	J
Vinyl methyl ether, inhibited		F				A	J
Liquefied, Nonflammable							
Ammonia, anhydrous		GF					
Carbon dioxide	A	E		D			
Carbon dioxide—Nitrous oxide mixture	A						
Chlorine		P					
Dichlorodifluoromethane		F				A	J
Dichlorodifluoromethane—difluoroethane mixture		F				A	J

Table 1 (*Continued*)

Name of Gas	CG–1 Frangible Disc	CG–2 165 F Fusible	CG–3 212 F Fusible	CG–4 Frangible Disc/w 165 F Fusible	CG–5 Frangible Disc/w 212 F Fusible	CG–7 Safety Relief Valve	CG–8 Safety Relief Valve and Fusible
Liquefied, Nonflammable—*cont.*							
Hydrogen bromide				F			
Hydrogen chloride				F	F		
Monochlorodifluoromethane		F				A	J
Monochlorotetrafluoroethane		F				A	J
Monochlorotrifluoromethane	A						
Nitrosyl chloride		F					
Nitrous oxide	A	E		D			
Sulfur dioxide		F				AH	
Sulfur hexafluoride	A			B	B		
Tetrafluoroethylene, inhibited	A						
Trifluorochloroethylene, inhibited		F					
Non liquefied, flammable							
Carbon Monoxide				BL	BL		
Ethylene	T	E		D			
Fluorine	Safety relief devices are prohibited by ICC or BTC regulations.						
Hydrogen	R			T	T		
Methane	R	T	T	T	T		
Non liquefied, Non flammable							
Air, compressed	B	S	S	S	S		
Argon	B	S	S	S	S		
Boron trifluoride				S	S		
Carbon dioxide—Oxygen mixture	B	S	S	S	S		
Helium	B	S	S	S	S		
Helium-oxygen mixture	B	S	S	S	S		
Krypton	B	S	S	S	S		
Neon	B	S	S	S	S		
Nitrogen	B	S	S	S	S		
Oxygen	B	S	S	S	S		
Xenon	B	S	S	S	S		
Poisons, Class A (See 3.1.1)	Safety relief devices are prohibited by ICC or BTC regulations.						
Pressurized liquid							
Argon	N						
Nitrogen	N						
Oxygen	N						

Key to Symbols Used in Table 1

A This device is required in one end of the cylinder only regardless of the length.
B When cylinders are over 65 in. long, exclusive of neck, this device is required in both ends.
C This device is permitted only in cylinders having a minimum required test pressure of 3000 psig or higher, and is required in one end only. The bursting pressure of the disc shall be at least 75 per cent of the minimum required test pressure of the cylinder.
D This device is permitted only in cylinders which are in direct medical service. It is not to be used in cylinders which are in transfer service even though the gas itself is intended for medical purposes. When cylinders are over 65 in. long, exclusive of neck, the device is required in both ends.
E This device is permitted only in cylinders which are in direct medical service. It is not to be used in cylinders which are in transfer service even though the gas itself is intended for medical purposes. When used in cylinders not over 30 in. long, exclusive of neck, it is required in one end of the cylinder only.
F When cylinders are over 30 in. long, exclusive of neck, this device is required at both ends.
G ICC and BTC regulations do not require cylinders containing less than 165 lbs of this gas to be equipped with a safety relief device.

H This device is permitted on cylinders having a water capacity not exceeding 10 lb.

J Type CG-8 combination safety relief valve and fusible plug may be used subject to the following:

(a) If 100 per cent of the safety relief device capacity requirement is supplied by a safety relief valve, the supplementary fusible plug may have a yield temperature above 170 F. Each safety relief device is required in one end of the cylinder only, regardless of length, and may be in the same or opposite ends of the cylinder.

(b) If at least 70 per cent of the safety relief device capacity requirement is supplied by a safety relief valve, the balance of the capacity requirement must be supplied by a type CG-2 fusible plug having a yield temperature of 157 to 170 F. Each safety relief device is required in one end of the cylinder only regardless of length and may be in the same or opposite ends of the cylinder.

(c) If less than 70 per cent of the safety relief device capacity requirement is supplied by a safety relief valve, the balance of the capacity requirement must be supplied by a type CG-2 fusible plug having a yield temperature range of 157 to 170 F and must be provided in each end of cylinders over 30 in. long, exclusive of neck. For cylinders 30 in. long or less, the above combination device is required in one end of the cylinder only.

(d) When the flow capacities of a safety relief valve and a type CG-2 fusible plug are to be combined to provide the required safety device flow capacity, the minimum capacity to be provided by the fusible plug is calculated as follows: Using the total water capacity of the container, calculate the flow capacity for a safety relief valve by using the formula in 4.7 or 4.8. Divide the rated flow capacity of the safety relief valve by the calculated flow capacity required to obtain the percentage supplied by the safety relief valve. The remaining percentage for the fusible plug is multiplied by the total water capacity to obtain the water capacity figure for use in calculating the minimum flow capacity to be supplied by the fusible plug using the formula in 4.5. or 4.6.

Example: A liquefied gas; a 1000 lb water capacity container; a safety relief valve with rated flow capacity of 1220 cfm at 494.7 psia.

Using formula in 4.8: $Qa = 2 \times 0.00154\ PWc = 2 \times 0.00154 \times 1000 \times 494.7 = 1524$ cfm

$$\frac{1220}{1524} = 80\% \text{ by safety relief valve}$$

Using formula in 4.6: $20\% \times 1000 = 200$ lb water capacity

$Qa = 2 \times 0.154 \times 200 = 61.6$ cfm minimum capacity at 100 psia to be provided by fusible plug.

K This schedule does not apply to acetylene cylinders for use in lighthouse service only and coming within the jurisdiction of the U. S. Coast Guard or the Canadian Department of Transport.

L Where safety relief devices are incorporated in cylinder valves used with these gases, such devices shall be an integral part of the valve or so designed that there is no projection beyond the valve body, if practical. If a projection is used, it shall be round.

M The number and location of safety relief devices for cylinders of any particular size shall be proved adequate as a result of the fire test, and any change in style of cylinder, a filler, or quantity of devices can only be approved if found adequate upon reapplication of the fire test.
Details of the fire test are:

(a) One fully charged cylinder will be connected to a remotely located pressure gage. With the valve opened, the cylinder shall be placed horizontally on suitable supports so that a fire can be built under and around the sides but not to the ends of the cylinder.

(b) The fire shall be of such proportion that the pressure within the cylinder will rise at least 25 psig before any safety relief device functions. One or more devices shall function in 10 minutes. If no device functions in 10 minutes, the test may be repeated.

(c) The fire shall be continued at approximately the same intensity for an additional 10 minutes following the first safety relief device release.

(d) Safety relief devices are adequate and acceptable if the cylinder does not fail; but retest is permitted when failure of the cylinder occurs without propulsion or fragmentation.

N This device is required in one end of the cylinder only, regardless of length. A pressure controlling valve as required in §73.304 (f) of ICC regulations must also be used. This valve must be so sized and set as to limit the pressure in the cylinder to 1¼ times its marked service pressure. The insulation jacket shall be provided with a pressure-actuated device which will function at a pressure of not more than 25 psig, and provide a discharge area of .00012 sq in. per lb water capacity of cylinder.

P When cylinders are over 55 in. long, exclusive of neck, this device is required in both ends, except for cylinders purchased after October 1, 1944, which must contain no aperture other than that provided in the neck of the cylinder for attachment of a valve equipped with an approved safety relief device. (Chlorine cylinders do not generally exceed 55 in. in length, since ICC regulations, §73.308(a) Note 6, require that cylinders purchased after November 1, 1935 must not contain over 150 lb chlorine.)

R This device is permitted only in cylinders over 65 in. long, exclusive of neck. It is required in both ends of cylinders. Each device shall be arranged to discharge upwards and unobstructed to the open air in such a manner as to prevent any impingement of escaping gas upon the containers.

S This device is permitted only in cylinders having a length not exceeding 65 in., exclusive of neck. It is required in one end only.

T This device is required in only one end of cylinders having a length not exceeding 65 in., exclusive of neck. For cylinders over 65 in. long this device is required in both ends, and each device shall be arranged to discharge upwards and unobstructed to the open air in such a manner as to prevent any impingement of escaping gas upon the containers.

TABLE 2. TEMPERATURE CORRECTION FACTORS TO 60F*

Degrees F	Factor	Degrees F	Factor	Degrees F	Factor
1	1.0621	51	1.0088	101	.9628
2	1.0609	52	1.0078	102	.9619
3	1.0598	53	1.0068	103	.9610
4	1.0586	54	1.0058	104	.9602
5	1.0575	55	1.0048	105	.9594
6	1.0564	56	1.0039	106	.9585
7	1.0552	57	1.0029	107	.9577
8	1.0541	58	1.0019	108	.9568
9	1.0530	59	1.0010	109	.9560
10	1.0518	60	1.0000	110	.9551
11	1.0507	61	.9990	111	.9543
12	1.0496	62	.9981	112	.9535
13	1.0485	63	.9971	113	.9526
14	1.0474	64	.9962	114	.9518
15	1.0463	65	.9952	115	.9510
16	1.0452	66	.9943	116	.9501
17	1.0441	67	.9933	117	.9493
18	1.0430	68	.9924	118	.9485
19	1.0419	69	.9915	119	.9477
20	1.0408	70	.9905	120	.9469
21	1.0398	71	.9896	121	.9460
22	1.0387	72	.9887	122	.9452
23	1.0376	73	.9877	123	.9444
24	1.0365	74	.9868	124	.9436
25	1.0355	75	.9859	125	.9428
26	1.0344	76	.9850	126	.9420
27	1.0333	77	.9840	127	.9412
28	1.0323	78	.9831	128	.9404
29	1.0312	79	.9822	129	.9396
30	1.0302	80	.9813	130	.9388
31	1.0291	81	.9804	131	.9380
32	1.0281	82	.9795	132	.9372
33	1.0270	83	.9786	133	.9364
34	1.0260	84	.9777	134	.9356
35	1.0249	85	.9768	135	.9349
36	1.0239	86	.9759	136	.9341
37	1.0229	87	.9750	137	.9333
38	1.0218	88	.9741	138	.9325
39	1.0208	89	.9732	139	.9317
40	1.0198	90	.9723	140	.9309
41	1.0188	91	.9715	141	.9302
42	1.0178	92	.9706	142	.9294
43	1.0168	93	.9697	143	.9286
44	1.0158	94	.9688	144	.9279
45	1.0147	95	.9680	145	.9271
46	1.0137	96	.9671	146	.9263
47	1.0127	97	.9662	147	.9256
48	1.0117	98	.9653	148	.9248
49	1.0108	99	.9645	149	.9240
50	1.0098	100	.9636	150	.9233

* From AGA Gas Measurement Report No. 3, "Orifice Metering of Natural Gas."

TABLE 3. BASIC ORIFICE FACTORS—FLANGE TAPS FOR FLOW PER MINUTE

Base Temperature 60 F
Base Pressure = 14.7 psia

Flow Temperature 60 F
Specific Gravity = 1.0

Orifice Diameter-Inches	Pipe Sizes — Extra Heavy, Schedule 80 Nominal and Published Inside Diameters (Inches)				
	2 1.939	3 2.900	4 3.826	6 5.761	8 7.981
.250	.2118	.2118*	.2115*	———	———
.375	.4740	.4730	.4726*	———	———
.500	.8431	.8386	.8372	.8364*	———
.625	1.3252	1.3114	1.3075	1.3049	———
.750	1.9270	1.8950	1.8858	1.8792	———
.875	2.6593	2.5902	2.5733	2.5605	2.5552
1.000	3.5412	3.4007	3.3700	3.3493	3.3398
1.125	4.6033	4.3325	4.2782	4.2453	4.2315
1.250	5.8930	5.3938	5.3005	5.2492	5.2297
1.375	7.4762	6.5967	6.4408	6.3617	6.3343
1.500		7.9560	7.7045	7.5838	7.5463
1.625		9.4942	9.0982	8.9172	8.8658
1.750		11.2407	10.631	10.363	10.293
1.875		13.2313	12.313	11.924	11.830
2.000		15.5108	14.157	13.602	13.475
2.125		18.1867	16.183	15.401	15.231
2.250			18.412	17.325	17.098
2.375			20.868	19.377	19.078
2.500			23.588	21.563	21.172
2.625			26.583	23.892	23.382
2.750			29.952	26.368	25.708
2.875				29.000	28.155
3.000				31.797	30.725
3.125				34.773	33.420
3.250				37.942	36.243
3.375				41.318	39.200
3.500				44.918	42.295
3.625				48.762	45.530
3.750				52.868	48.913
3.875				57.262	52.448
4.000				61.970	56.142
4.250				72.580	64.038
4.500					72.675
4.750					82.135
5.000					92.530
5.250					103.940
5.500					116.533
5.750					130.500
6.000					

*These orifices have diameter ratios lower than the minimum value for which the formulas used were derived and this size of plate should not be used unless it is understood that the accuracy of measurement will be relatively low.

(Data were taken from Gas Measurement Committee Report No. 3, "Orifice Metering of Natural Gas." (1956 Reprint), American Gas Association, and converted to calculations in cubic feet per minute.)

APPENDIX A

NOTE: This form is not suitable for acetylene cylinders. For further information contact the Compressed Gas Association,

COMPRESSED GAS ASSOCIATION, INC.

REQUEST TO BUREAU OF EXPLOSIVES FOR SAFETY RELIEF DEVICE APPROVAL

Date____________________

Manufacturer____

Address____

Catalog or Model No.__________________________________

Dwg. No.____________ Date of Dwg. and Latest Revision__________________

Safety Relief Device Type CG—__________ (See Table 1 of these Standards)

Set Pressure______________psig. Flow Rating Pressure______________psia.

Yield Temperature____________ F. Rated Bursting Pressure____________psig.

Chemical Name of Gas________________Liquefied ( ) Non Liquefied ( )

Commercial Name of Gas__________________________________

Percentage of Components for Mixed Gases________________________

Specification and Service Pressure of ICC Cylinder(s) to be Used______________

Maximum Container Size for Which Approval is Requested________________
(pounds water capacity)

Minimum Required Flow-CFM of Air (See Pars. 4.5 to 4.9)________________

Actual Flow-CFM of Air at 60 F and Base Pressure of 14.7 psia.______________
(Item 16 of Test Data)

Test Conducted By: ____________________Title________________

Company____

Approval Requested By:____________________Title________________

Company____

Approved by Bureau of Explosives: Date____________________________

Signature:________________________

NOTE: For safety relief devices on insulation jacket of ICC-4L cylinders, indicate:

Description of Device ____________ Discharge Area________ Set Pressure________psig

(For Test Data, see next page)

APPENDIX A (Continued)

TEST DATA

This Form is suitable for Test Data using Orifice Meters.

Test Medium - Air__________or Name of Gas_______________Specific Gravity____________

Molecular Weight______________Ratio of Specific Heats (k)______________

ITEM	SAMPLES	1	2	3
1.	Start to Discharge Pressure—psig.			
2.	Resealing Pressure—psig.			
3.	Frangible Disc—Bursting Pressure—psig.			
4.	Fusible Plug—Yield Temperature—degrees F.			
5.	Flow Rating Pressure—psia. (psig + 14.7)			
6.	Orifice Diameter—Inches			
7.	Meter Pipe—Inside Diameter—Inches			
8.	Orifice Factor (For Flow in CFM.) See Table 3			
9.	Constant (Item 8 x $\sqrt{\text{Item 5}}$)			
10.	Differential Pressure— $\sqrt{\text{Inches Water}}$			
11.	Flow Temperature—degrees F.			
12.	Temperature Correction Factor. See Table 2.			
13.	Supercompressibility Factor (Air = 1.0)			
14.	Gas Constant Ratio*			
15.	Flow (Items 9 x 10 x 12 x 13 x 14)			
16.	AVERAGE FLOW AT 60 F and 14.7 psia			

* Gas constant ratio for air = 1.0; for other than air = 356/Gas Constant (C). See Figure 1.

SECTION B

For Cargo and Portable Tanks—Safety Relief Device Standards, Part 2

This part of the Safety Relief Device Standards represents the minimum requirements recommended by Compressed Gas Association, Inc., for safety relief devices considered to be appropriate and adequate for use on cargo tanks and portable tanks for compressed gases; except tank car tanks, having water capacity exceeding 1000 lb and which comply with the specifications and charging and maintenance regulations of the Interstate Commerce Commission (ICC) or the corresponding portable tank specifications and regulations of the Board of Transport Commissioners for Canada (BTC).

It is recognized that there are cargo and portable tanks that conform with the specification requirements of the ICC and portable tanks that conform with the specification requirements of the BTC which are used in services that are beyond the jurisdiction of either of these authorities. In such cases it is recommended that state, provincial, local or other authorities having jurisdiction over these containers be guided by Part 2 of these standards in determining adequate safety relief device requirements, provided that the cargo and portable tanks are charged with gas and maintained in accordance with the ICC or BTC regulations that apply.

It is further recognized that there may be cargo and portable tanks which are used in services beyond the jurisdiction of the ICC or the BTC and which do not conform to the specification requirements of either authority. It is recommended that the authorities having jurisdiction over such cargo and portable tanks be guided by Part 2 of these standards in determining safety relief device requirments, provided that such cargo and portable tanks are considered by the authority as having a construction at least equal to the equivalent ICC or BTC specification requirements and further provided that the cargo and portable tanks shall be charged with gas and maintained in accordance with the ICC or BTC requirements that apply.

1. DEFINITIONS

For the purposes of these standards the following terms are defined:

1.1. The term "cargo tank" means any container designed to be permanently attached to any motor vehicle or other highway vehicle and in which is to be transported any compressed gas. The term "cargo tank" shall not be construed to include any tank used solely for the purpose of supplying fuel for the propulsion of the vehicle or containers fabricated under specifications for cylinders.

1.2. The term "portable tank" means any container designed primarily to be temporarily attached to a motor vehicle, other vehicle, railroad car other than tank car, or marine vessel, and equipped with skids, mountings or accessories to facilitate handling of the container by mechanical means, in which any compressed gas is to be transported. The term "portable tank" shall not be construed to include any cargo tank, any tank car tank, or any tank of the ICC-106A and ICC-110A-W type.

1.3. A "safety relief device" is a device intended to prevent rupture of a container under certain conditions of exposure.

1.4. A "safety relief valve" is a safety relief device containing an operating part that is held normally in a position closing the safety relief device channel by spring force and is intended to open and to close at predetermined pressures.

1.5. The "set pressure" of a safety relief valve is the pressure marked on the valve and at which the valve is set to start-to-discharge.

1.6. The "start-to-discharge pressure" of a safety relief valve is the pressure at which the first bubble appears through a water seal of not over 4 in. on the outlet of the valve.

Note: When the nature of the service requires the use of a metal-to-metal seat safety relief valve, with or without secondary sealing means, the start-to-discharge pressure may be considered the pressure at which an audible discharge occurs.

1.7. The "resealing pressure" of a safety relief valve is the pressure at which leakage ceases through a water seal of not over 4 in. on the outlet of the valve.

1.8. The "flow capacity" of a safety relief device is the capacity in cubic feet per minute of free air discharged at the required flow rating pressure.

1.9. The "flow rating pressure" is the pressure at which a safety relief device is rated for capacity.

1.10. "Free air" or "free gas" is air or gas measured at a pressure of 14.7 psia and a temperature of 60 F.

1.11. A "frangible disc" (refer to Part 1 of these standards) is a safety relief device in the form of a disc, usually of metal, which is so held as to close the safety relief device channel under normal conditions. The disc is intended to burst at a predetermined pressure to permit the escape of gas.

1.12. A "fusible plug" (refer to Part 1 of these standards) is a safety relief device in the form of a plug of suitable low-melting material, usually a metal alloy, which closes the safety relief device channel under normal conditions and is intended to yield or melt at a predetermined temperature to permit the escape of gas.

1.13. The term "design pressure" is identical to "maximum allowable working pressure" as used in the "Code" for the original design and construction of the container, except that for containers constructed in accord with Paragraph U-68 or U-69 of Section VIII of the ASME Boiler and Pressure Vessel Code, 1949 Edition, the maximum allowable working pressure for the purposes of these Standards is considered to be 125 per cent of the design pressure as provided for in §73.315 of ICC or BTC regulations.

1.14. The "Code" as used in these standards is defined as (1) Paragraph U-68, U-69, U-200 or U-201 of Section VIII of the Boiler and Pressure Vessel Code of the American Society of Mechanical Engineers, 1949 Edition; or (2) Section VIII of the Boiler and Pressure Vessel Code of the American Society of Mechanical Engineers, 1950, 1952, 1956, 1959 and 1962 editions; or (3) The Code for Unfired Pressure Vessels for Petroleum Liquids and Gases of the American Petroleum Institute and the American Society of Mechanical Engineers (API-ASME)*, 1951 Edition.[1]

1.15. "ICC regulations" as used in these standards refers to "Interstate Commerce Commission Regulations for Transportation of Explosives and Other Dangerous Articles by Land and Water in Rail Freight, Express and Baggage Services and by Motor Vehicle (Highway) and Water, including Specifications for Shipping Containers," Code of Federal Regulations, Title 49, Parts 71 to 78.[2]

1.16. "BTC regulations" as used in these standards refer to Board of Transport Commissioners for Canada, "Regulations for The Transportation of Dangerous Commodities by Rail."[3]

2. TYPES OF SAFETY RELIEF DEVICES

Types of safety relief devices covered by this Part are as follows:

2.1. Safety Relief Valve.

2.2. Fusible plug utilizing a fusible alloy with yield temperature not over 170 F, nor less than 157 F [165 F nominal] (see Part 1 of these standards, Type CG-2).

2.3. Frangible Disc (see part 1 of these Standards, Type CG-1).

* The API-ASME Code, as a joint publication and interpretation service, was discontinued as of December 31, 1956.

3. APPLICATION REQUIREMENTS FOR SAFETY RELIEF DEVICES

3.1. General

3.1.1. Each container shall be provided with one or more safety relief devices which, unless otherwise specified, shall be safety relief valves of the spring-loaded type.

3.1.2. Safety relief valves shall be set to start-to-discharge at a pressure not in excess of 110 per cent of the design pressure of the container nor less than the design pressure of the container except as follows:

3.1.2.1. If an over-designed container is used, the set pressure of the safety relief valve may be between the minimum required design pressure for the lading and 110 per cent of the design pressure of the container used.

3.1.2.2. For sulfur dioxide containers, a minimum set pressure of 120 and 110 psig is permitted for the 150 and 125 psig design pressure containers, respectively (see 4.4.2).

3.1.2.3. For carbon dioxide (refrigerated), nitrous oxide (refrigerated), and pressurized liquid argon, nitrogen and oxygen, there shall be no minimum set pressure.

3.1.2.4. For butadiene, inhibited, and liquefied petroleum gas containers, a minimum set pressure of 90 per cent of the minimum design pressure permitted for these ladings may be used (see 4.4.2).

3.1.2.5. For containers constructed in accord with Paragraph U-68 or U-69 of the Code, 1949 edition, the set pressure marked on the safety relief valve may be 125 per cent of the original design pressure of the container.

3.1.3. The design, material and location of safety relief devices shall have been proved to be suitable for the intended service.

3.1.4. Only replacement parts or assemblies provided by the manufacturer of the device shall be used unless the suitability of interchange is proved by adequate tests.

3.1.5. Safety relief valves shall have direct communication with the vapor space of the container.

3.1.6. Any portion of liquid piping or hose which at any time may be closed at each end must be provided with a safety relief device to prevent excessive pressure.

3.1.7. The following additional restrictions apply to safety relief devices on containers for carbon dioxide or nitrous oxide which are shipped in refrigerated and insulated containers:

3.1.7.1. The maximum operating pressure in the container may be regulated by the use of one or more pressure controlling devices, which devices shall not be in lieu of the safety relief valve required in paragraph 3.1.1.

3.1.7.2. All safety relief devices shall be so installed and located that the cooling effect of the contents will not prevent the effective operation of the device.

3.1.7.3. In addition to the safety relief valves required by 3.1.1, each container for carbon dioxide may be equipped with one or more frangible disc safety relief devices of suitable design set to function at a pressure not exceeding 2 times the design pressure of the container.

3.1.8. Subject to conditions of §73.315(a) (1) of ICC regulations for methyl chloride and sulfur dioxide optional portable tanks of 225 psig minimum design pressure, one or more fusible plugs approved by the Bureau of Explosives may be used in lieu of safety relief valves of the spring-loaded type. If the container is over 30 in. long, a safety relief device having the total required flow capacity must be at both ends.

3.1.9. When storage containers for liquefied petroleum gas are permitted to be shipped in accordance with §73.315(j) of ICC regulations, they must be equipped with safety relief devices in compliance with the requirements for safety relief devices on aboveground containers as specified in the current edition of National Fire Protection Association Pamphlet No. 58 "Standard for the Storage and Handling of Liquefied Petroleum Gases."[4]

3.1.10. When containers are filled by pumping equipment which has a discharge capacity in excess of the capacity of the container safety relief devices, and which is capable of producing pressures in excess of the design pressure of the container, precautions should be taken to prevent the development of pressures in the container in excess of 120 per cent of its design pressure. This may be done by providing additional capacity of the safety relief valves on the container, by including a bypass on the pump discharge, or by any other suitable method.

3.1.11. The following additional requirements apply to safety relief devices on containers for liquefied hydrogen and pressurized liquid argon, nitrogen and oxygen.

3.1.11.1. The liquid container shall be protected by one or more safety relief valves and one or more frangible discs.

3.1.11.2. The minimum capacity of the safety relief valves shall be as required by 4.3.3 and 4.3.4 or as provided in 3.1.11.3.

3.1.11.3. The frangible disc shall have a flow capacity at least equal to that required by 4.3.3 and 4.3.4 with the insulation space saturated with the gaseous lading at atmospheric pressure, and shall function at a pressure not exceeding 150 per cent of the design pressure (plus 15 psig if vacuum insulation is used), or the test pressure, whichever is less. When the frangible disc is designed to function at a pressure not exceeding 120 per cent of the design pressure, the minimum capacity for the safety relief valves on vacuum insulated containers shall be sized to provide adequate venting capacity for loss of a vacuum with insulation saturated with gaseous lading at atmospheric pressure.

3.2. Piping of Safety Relief Devices

3.2.1. When fittings and pipings are used on either the upstream or downstream side or both of a safety relief device or devices, the passages shall be so designed that the flow capacity for the safety relief device will not be reduced to the extent that the operation of the device could be impaired nor below the capacity required for the container on which the safety device assembly is installed.

3.2.2. Safety relief devices shall be arranged to discharge unobstructed to the open air in such a manner as to prevent any impingement of escaping gas upon the container. Safety relief devices shall be arranged to discharge upward except this is not required for carbon dioxide, nitrous oxide and pressurized liquid argon, nitrogen and oxygen.

3.2.3. No shutoff valves shall be installed between the safety relief devices and the container except, in cases where two or more safety relief devices are installed on the same container, a shutoff valve may be used where the arrangement of the shutoff valve or valves is such as always to insure full required capacity flow through at least one safety relief device.

4. DESIGN AND CONSTRUCTION REQUIREMENTS FOR SAFETY RELIEF DEVICES

4.1. The material, design and construction of a safety relief device shall be such that there shall be no significant change in the functioning of the device and no serious corrosion or deterioration of the materials within the period between renewals, due to service conditions. The chemical and physical properties of the materials shall be uniform and suitable for the requirements of the part manufactured therefrom.

4.2. Methods of manufacture, inspection and tests shall conform to best practices in order to attain satisfactory performances of the safety relief devices (see References).

4.3. Safety relief devices shall have a total flow capacity as calculated by the formulas in 4.3.2. and 4.3.3. These formulas are based on the principle of preventing the pressure in the container from exceeding a maximum of 120 per cent of the design pressure of the container.

4.3.1. The flow capacity of safety relief devices of each design and modification thereof shall be determined by an actual flow test. Methods of making flow tests are given in 5.2.

4.3.2. For uninsulated containers for liquefied gases the minimum required flow capacity of the safety relief device shall be calculated using the formula:

$$Q_a = G_u A^{0.82}$$

where

Q_a = flow capacity in cubic feet per minute of free air.

G_u = gas factor for uninsulated containers obtained from 4.3.5, Table 4, for the kind of gas involved.

$^*A$ = Total outside surface area of the container in square feet.

Note: Graph of A vs. $A^{0.82}$ is shown in Fig. 2.

TABLE 4. VALUES OF G_i AND G_u FOR COMMONLY USED SAFETY RELIEF VALVE SET PRESSURES*

Gas	Set Pressure psig	Value of G_i	Value of G_u
Anhydrous ammonia	265	2.80	22.1
Anhydrous dimethylamine	150	3.76	31.0
Anhydrous monomethylamine	150	3.88	28.6
Anhydrous trimethylamine	150	5.33	41.8
Argon, pressurized liquid	100	10.2	—
	200	12.2	—
Butadiene, inhibited	100	4.17	35.8
Carbon dioxide (refrigerated)	100	5.40	37.6
	325	6.46	47.4
Chlorine	225	6.74	54.3
Dichlorodifluoromethane	150	7.79	62.9
Dichlorodifluoromethane-dichlorotetrafluoroethane mixture	150	7.79	62.9
Dichlorodifluoromethane-monofluorotrichloromethane	150	7.79	62.9
Hydrogen, liquefied	100	11.9	—
Liquefied petroleum gas	250	6.56	53.6
Methyl chloride	150	4.96	40.4
Methyl mercaptan	100	6.05	51.2
Monochlorodifluoromethane	250	8.14	65.8
Nitrogen, pressurized liquid	100	10.2	—
	200	12.2	—
Nitrous oxide (refrigerated)	100	5.36	37.2
	350	6.20	46.0
Oxygen, pressurized liquid	100	10.2	—
	200	12.2	—
Sulfur dioxide	150	4.84	40.0
Vinyl chloride, inhibited	150	5.61	46.8

**Note:* When lower set pressures than those shown are used, the values of G_i and G_u are on the safe side and may be used as shown or calculated as covered below. For higher set pressures than shown, values of G_i and G_u must be calculated from the following formulas:

$$G_u = \frac{633{,}000}{LC} \sqrt{\frac{ZT}{M}} \quad \text{and} \quad G_i = \frac{73.2 \times (1200 - t)}{LC} \sqrt{\frac{ZT}{M}}$$

where

L = Latent heat at flowing conditions in Btu per pound.

C = Constant for gas or vapor related to ratio of specific heats ($k = C_p/C_v$) at standard conditions from Figure 1 (in Part 1 of these Standards).

Z = Compressibility factor at flowing conditions.

T = Temperature in degrees R (Rankin) of gas at pressure at flowing conditions ($t + 460$).

M = Molecular weight of gas.

t = Temperature in degrees F (Fahrenheit) of gas at pressure at flowing conditions.

When compressibility factor Z is not known, 1.0 is a safe value of Z to use. When gas constant C is not known, 315 is a safe value of C to use. For complete details concerning the basis and origin of these formulas, see "How to Size Safety Relief Devices," by F. J. Heller.[5]

4.3.3. For insulated containers for liquefied compressed gases, other than pressurized liquid oxygen, nitrogen and argon, which meet the requirements of 4.3.3.1 and 4.3.3.2, the minimum required flow capacity of the safety relief device shall be calculated using the formula:

$$Q_a = G_i U A^{0.82}$$

where

Q_a = flow capacity in cubic feet per minute of free air.

G_i = gas factor for insulated containers obtained from Table 4 for the kind of gas involved.

**U = total thermal conductivity of the container insulating material at a mean temperature of 100 F.

*A = total outside surface area of the container in square feet.

Note: Graph of A vs. $A^{0.82}$ is shown in Fig. 2.

4.3.3.1. A portion of the insulation shall remain in place in a fire*** of 30 minutes duration. If the insulation remains completely in place, the U value shall be as defined in 4.3.3. If the insulation remains partly in place, the U value shall be increased proportionately.

4.3.3.2. The total thermal conductivity of the insulation shall not exceed .20 Btu/hr/sq ft/degree F at 100 F, except for liquefied hydrogen.

4.3.3.3. For liquefied hydrogen, the total thermal conductivity U shall be chosen on the basis that the insulation space is saturated with gaseous hydrogen at atmospheric pressure.

* When the surface area is not stamped on the name plate or when the marking is not legible, the area can be calculated by using one of the following formulas:

(1) Cylindrical container with hemispherical heads: Area = over-all length × outside diameter × 3.1416.

(2) Cylindrical container with semiellipsoidal heads: Area = (over-all length + .3 outside diameter) × outside diameter × 3.1416.

(3) Spherical container: Area = (outside diameter)² × 3.1416.

** Total thermal conductivity = thermal conductivity in Btu per hour per square foot per inch of thickness per degree Fahrenheit divided by thickness of the insulation in inches.

*** For the purpose of this standard, a fire produces a heat input of 34,500 Btu/hr/sq ft to a bare tank.

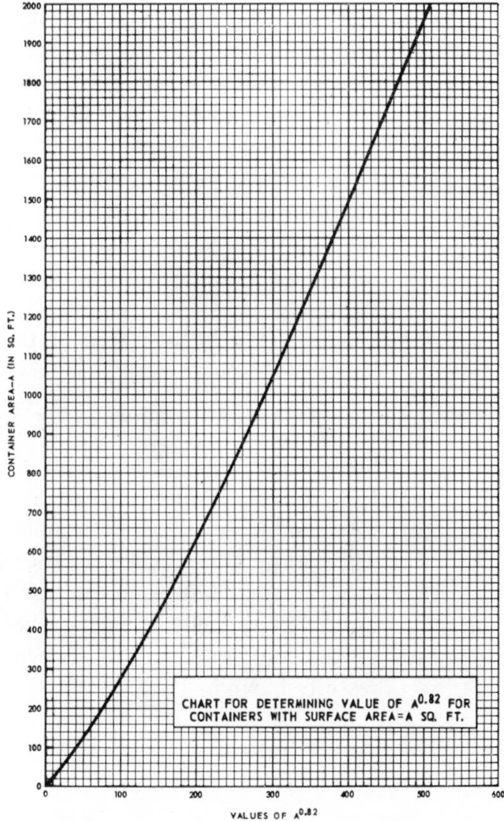

FIG. 2

4.3.4. For insulated containers for pressurized liquid oxygen, nitrogen and argon, the minimum required flow capacity of the safety relief device shall be calculated using the following formula; provided the flow capacity of the safety relief device shall not be less than .004 cu ft of free air per minute per pound of water capacity of the container at a set pressure of 25 psig. (Requirements for the insulation and jacketing material are specified in the container specifications.)

$$Q_a = G_i U A^{0.82}$$

where

Q_a = flow capacity in cubic feet per minute of free air.

G_i = Gas factor for insulated containers obtained from 4.3.5, Table 4 for pressurized liquid nitrogen. Value of G_i for nitrogen shall be used for pressurized liquid oxygen and argon.

$**U$ = Total thermal conductivity of the container insulating material saturated with the gaseous nitrogen at atmospheric pressure at a mean temperature of 100 F.

$*A$ = Total outside surface area of the container in square feet.

Note: Graph of A vs. $A^{0.82}$ is shown in Fig. 2.

4.3.5. Values are given in Table 4 for G_i (for insulated containers) and G_u (for uninsulated containers) for use in formulas $Q_a = G_i U A^{0.82}$ and $Q_a = G_u A^{0.82}$. These values for G_i and G_u are to be used in determining the required flow capacity of safety relief valves for a given commodity at 120 per cent of the set pressure shown in Table 4.

4.4 Safety Relief Valves:

4.4.1. Safety relief valves shall be of the spring-loaded type. The inlet connection shall not be less than $\frac{3}{4}$ in. nominal pipe size with physical dimensions for the wall thickness not less than those of Schedule 80 pipe (extra-heavy).

4.4.2 The minimum design pressures of containers required by the ICC and BTC for the various compressed gases are shown in Table 5.

4.4.3. Safety relief valves shall be designed so that the possibility of tampering will be minimized. If the pressure setting or adjustment is external, safety relief valves shall be provided with suitable means for sealing the adjustment.

4.5. If the design of a safety relief valve is such that liquid can collect on the discharge side, the valve shall be equipped with a drain at the lowest point where liquid can collect.

5. TESTS OF SAFETY RELIEF DEVICES

5.1. Pressure Tests of Safety Relief Valves

5.1.1. Each safety relief valve shall be subject to an air or gas pressure test to determine the following:

5.1.1.1. That the start-to-discharge pressure setting is not less than 100 per cent nor more than 110 per cent of the set pressure marked on the valve.

Note: In setting the valve care must be taken that evidence of start-to-discharge is due to opening of the valve and not due to a defect.

5.1.1.2. That after the start-to-discharge pressure test, the resealing pressure is not less than 90 per cent of the start-to-discharge pressure.

5.2. Flow Capacity Tests of Safety Relief Valves

5.2.1. The flow capacity of each design and modification thereof of a spring-loaded safety relief valve shall be determined by actual flow test at the flow rating pressure of 120 per cent of the set pressure marked on the valve. Three samples of each size of each valve representative of standard production shall be tested.

5.2.1.1. After pressure testing and without cleaning, removal of parts, reconditioning or adjusting, each sample valve shall be subjected to an actual flow test at 120 per cent of the set pressure and the rate of flow of air or gas released by the valve shall be measured. When gases other than air are used for flow test determinations, the results shall be converted to equivalent air flow figures. The average rate of air through the three sample valves shall be recorded as the rated flow capa-

* When the surface area is not stamped on the name plate or when the marking is not legible, the area can be calculated by using one of the following formulas:

(1) Cylindrical container with hemispherical heads: Area = over-all length × outside diameter × 3.1416.

(2) Cylindrical container with semiellipsoidal heads: Area = (over-all length + .3 outside diameter) × outside diameter × 3.1416.

(3) Spherical container: Area = (outside diameter)^2 × 3.1416.

** Total thermal conductivity = thermal conductivity in Btu per hour per square foot per inch of thickness per degree Fahrenheit divided by thickness of the insulation in inches.

TABLE 5.

Gas	Minimum Design Pressure of Container (psig)
Anhydrous ammonia	265
Anhydrous dimethylamine	150
Anhydrous monomethylamine	150
Anhydrous trimethylamine	150
Argon, pressurized liquid	25 psig or the controlled pressure, whichever is greater.
Butadiene, inhibited	100
Carbon dioxide (refrigerated)	100 psig or the controlled pressure, whichever is greater.
Chlorine	225
Dichlorodifluoromethane	150
Dichlorodifluoromethane-dichloro-tetrafluoroethane mixture	150
Dichlorodifluoromethane-monofluoro-trichloromethane mixture	150
Liquefied petroleum gas	Equivalent to the vapor pressure of the lading at 115 F. (See §73.315(b)(1) of ICC Regulations)
Methyl chloride	150*
Methyl mercaptan	100
Monochlorodifluoromethane	250
Nitrogen, pressurized liquid	25 psig or the controlled pressure, whichever is greater.
Nitrous oxide (refrigerated)	100 psig or the controlled pressure, whichever is greater.
Oxygen, pressurized liquid	25 psig or the controlled pressure, whichever is greater.
Sulfur dioxide:	
up to 1200 gal water capacity tank	150*
over 1200 gal water capacity tank	125*
Vinyl chloride, inhibited	150

* See 3.1.8 for optional portable tank.

city. The design of the valves shall be considered acceptable, provided the capacities of the three valves shall be within 10 per cent of the highest capacity recorded.

5.2.2. Acceptable Methods of Flow Testing

5.2.2.1. Safety relief valves may be tested for flow capacities by testing with equipment conforming to the American Gas Association "Gas Measure Committee Report No. 3, Orifice Metering of Natural Gas," as reprinted with revisions, 1956,[6] or ASME Power Test Code PTC 25-1958.[7]

5.2.2.2. Air or gas shall be supplied to the safety relief valve through a supply pipe provided with a pressure gage and a thermometer for indicating or recording the pressure and temperature of the supply. Observations shall be made and recorded after steady flow conditions have been established.

5.2.2.3. When the testing method prescribed in 5.2.2.1 is employed, such test may be made by the manufacturer of the safety relief valve, or a competent testing facility, provided that a record of the test is maintained by the manufacturer and a copy thereof certifying the results of test is filed with the Bureau of Explosives, if required (for sample of test data, see Appendix).

5.2.2.4. When any other method of testing is used, the Bureau of Explosives shall be furnished with a certification of the accuracy of the test results by a competent disinterested agency.

5.3. Tests of Safety Relief Devices other than Spring-Loaded Safety Relief Valves

5.3.1. When safety relief devices other than spring-loaded safety relief valves are used to satisfy the requirements of these standards, the flow capacity at a pressure of 120 per cent of the design pressure of the container may be determined by calculation if the capacity at some other pressure has been determined by actual flow tests. All other test requirements of Part 1 of these standards shall apply.

5.4. Rejected material may be reworked, provided the material is subject to such additional tests as are required to insure compliance with all requirements of these standards.

6. IDENTIFICATION

6.1. Safety relief valves shall be marked to indicate:

(a) Manufacturer's name or trademark and catalog number.

(b) The month and year of manufacture.

(c) The set pressure in psig.

(d) The flow capacity in cubic feet per minute of free air.

6.2. For safety relief devices other than spring-loaded safety relief valves, the requirements of Part 1 of these standards shall apply.

7. MAINTENANCE REQUIREMENTS FOR SAFETY RELIEF DEVICES

7.1. General Practices

7.1.1. Care should be exercized to avoid damage to safety relief devices. Care shall also be exercised to avoid plugging by paint or other dirt accumulation of safety relief device channels or other parts which could interfere with the functioning of the device.

7.1.2. Only qualified personnel shall be allowed to service safety relief devices. Any servicing or repairs which require resetting of safety relief valves shall be done only by or with the consent of the valve manufacturer.

7.2. Routine Checks When Filling Containers

7.2.1. Safety relief devices periodically shall be examined externally for corrosion, damage, plugging of external safety relief device channels, and mechanical defects such as leakage or extrusion of fusible metals. Valves equipped with secondary resilient seals shall have the seals inspected periodically. If there is any doubt regarding the suitability of the safety relief device for service, the container shall not be filled until it is equipped with a suitable safety relief device.

APPENDIX

COMPRESSED GAS ASSOCIATION, INC.

SAFETY RELIEF VALVE TEST DATA

(For Cargo and Portable Tanks)

Manufacturer____________________ Date____________

Address____________________ Cat. No.____________

Dwg. No.________ Date________ Inlet Con.________ In.________

Set Pressure____________ Psig. Name of Gas____________

Test Medium____________ Specific Gravity____________

Molecular Weight____________ Ratio of Specific Heats (k)____________

ITEM	SAMPLES	1	2	3
1.	Start to Dicharge Pressure—psig.			
2.	Resealing Pressure—psig.			
3.	Flow Rating Pressure—psia. (psig + 14.7)			
4.	Orifice Diameter—Inches			
5.	Meter Pipe—Inside Diameter—Inches			
6.	Orifice Factor (See Table 3 of Part 1 of these Standards, p. 207)			
7.	Constant (Item 6 x $\sqrt{\text{Item 3}}$)			
8.	Differential Pressure— $\sqrt{\text{Inches Water}}$			
9.	Flow Temperature—degrees F.			
10.	Temperature Correction Factor (See Table 2 of Part 1 of these Standards, p. 206)			
11.	Supercompressibility Factor (Air=1.0)			
12.	Constant (C) (See Figure 1 of Part 1 of these Standards, p. 202)			
13.	Flow (Items 7 x 8 x 10 x 11 x 12)			
14.	Average Flow at 60 F and 14.7 psia	CFM of AIR		

Tests Conducted by____________ Date________ File No.________

Company____________________

Submitted by____________ Title________ Date________

REFERENCES

1. (1) Paragraph U-68, U-69, U-200 or U-201, Section VIII, "Boiler and Pressure Vessel Code of the American Society of Mechanical Engineers," 1949 Edition; or (2) Section VIII, "Boiler and Pressure Vessel Code of the American Society of Mechanical Engineers," 1950, 1952, 1956, 1959 and 1962 Editions; or (3) "The Code for Unfired Pressure Vessels for Petroleum Liquids and Gases of the American Petroleum Institute and the American Society of Mechanical Engineers," 1951 Edition (the API-ASME Code, as a joint publication and interpretation service, was discontinued as of

Dec. 31, 1956). Available from American Society of Mechanical Engineers, 345 E. 47th St., New York, N. Y., 10017.
2. "Interstate Commerce Commission Regulations for Transportation of Explosives and Other Dangerous Articles by Land and Water in Rail Freight, Express and Baggage Services and by Motor Vehicle (Highway) and Water, including Specifications for Shipping Containers," Code of Federal Regulations, Title 49, Parts 71-78. Available from Bureau of Explosives, 63 Vesey St., New York, N. Y., 10007.
3. Board of Transport Commissioners for Canada, "Regulations for The Transportation of Dangerous Commodities by Rail." Available from Supervisor of Government Publications, Department of Public Printing and Stationery, Ottawa, Ontario, Canada.
4. "Standard for the Storage and Handling of Liquefied Petroleum Gases" (NFPA No. 58; ASA Z106.1) (see current edition); National Fire Protection Association, 60 Batterymarch St., Boston, Mass., 02110.
5. Heller, F. J., "How to Size Safety Relief Devices," 1954. Phillips Petroleum Company, Bartlesville, Okla.
6. "Gas Measurement Committee Report No. 3, Orifice Metering of Natural Gas" (reprinted with revisions, 1956). American Gas Association, 420 Lexington Ave., New York, N. Y., 10017.
7. "Safety and Relief Valves, Power Test Codes PTC 25-1958." American Society of Mechanical Engineers.

ADDITIONAL REFERENCE

See also Sections A, C, and D of this Chapter, and Chapter 4, Section A, "American-Canadian Standard Cylinder Valve Outlet and Inlet Connections."

SECTION C

For Storage Tanks—Safety Relief Device Standards, Part 3

This part of the Safety Relief Device Standards represents the minimum requirements recommended by Compressed Gas Association, Inc., for safety relief devices considered to be appropriate and adequate for use on compressed gas storage containers constructed in accordance with the American Society of Mechanical Engineers (ASME), the American Petroleum Institute (API)-ASME Codes,[1] or the equivalent. While these standards are written specifically to cover "Compressed Gas" as defined in Paragraph 1.2, the procedures are applicable to storage containers with a design pressure exceeding 15 psig.

For safety relief devices on storage vessels with design pressures of 15 psig or less, a standard is being developed at the time the *Handbook* goes to press. It will be designated Part 4 of these standards.

These standards do not apply to compressed gas manufacturing plants, nor are they intended to conflict with or supercede the following existing standards: current edition of National Fire Protection Association Pamphlet No. 58, "Standard for the Storage and Handling of Liquefied Petroleum Gases"; current editions of the American Standard Safety Requirements for the Storage and Handling of Anhydrous Ammonia, K61.1.[2]

A number of states and cities have pressure vessel laws and regulations which include requirements for safety relief devices. The CGA standards are prepared specifically for compressed gas containers and the relief devices may not be acceptable unless special permission is obtained from the authority having jurisdiction.

1. DEFINITIONS

For the purpose of these standards the following terms are defined:

1.1. The term "storage container" means any container designed to be permanently mounted on a stationary foundation and in which is to be stored any compressed gas.

1.2. A "compressed gas" is defined as any material or mixture having in the container either an absolute pressure exceeding 40 psi at 70 F, or an absolute pressure exceeding 104 psi at 130 F, or both; or any liquid flammable material having a Reid vapor pressure exceeding 40 psia at 100 F (as determined by the American Society for Testing and Materials Method of Test for Vapor Pressure of Petroleum Products D-323).

1.3. A "safety relief device" is a device intended to prevent rupture of a container under certain conditions of exposure.

1.4. A "safety relief valve" is a safety relief device containing an operating part that is held normally in a position closing the safety relief device channel by spring force and is intended to open and to close at predetermined pressures.

1.5. The "set pressure" of a safety relief valve is the pressure marked on the valve and at which the valve is set to start-to-discharge.

1.6. The "start-to-discharge pressure" of a safety relief valve is the pressure at which the first bubble appears through a water seal of not over 4 in. on the outlet of the valve.

Note: When the nature of the service requires the use of a metal-to-metal seat safety relief valve, with or without sealing means, the start-to-dishcarge pressure may be considered the pressure at which an audible discharge occurs.

1.7. The "flow capacity" of a safety relief device is the capacity in cubic feet per minute of free air discharge when tested at the required flow rating pressure.

1.8. The "flow rating pressure" is the pressure at which a safety relief valve is rated for capacity.

1.9. "Free air" or "free gas" is air or gas measured at a pressure of 14.7 psia and a temperature of 60 F.

1.10. A "frangible disc" (refer to Part 1 of these standards) is a safety relief device in the form of a disc, usually of metal, which is so held as to close the safety relief device channel under normal conditions. The disc is intended to burst at a predetermined pressure to permit the escape of gas.

1.11. A "fusible plug" (refer to Part 1 of these standards) is a safety relief device in the form of a plug of suitable low melting material, usually a metal alloy, which closes the safety relief device channel under normal conditions and is intended to yield or melt at a predetermined temperature to permit the escape of gas.

1.12. The term "design pressure" is identical to "maximum allowable working pressure" as used in the "code" for the original design and construction of the container, except that for containers constructed in accord with Paragraph U-68 and U-69 of Section VIII of the Boiler and Pressure Vessel Code of the ASME, 1949 edition, the maximum allowable working pressure is 125 per cent of the design pressure.

1.13. The "Code" as used in these standards is defined as (1) Paragraph U-68, U-69, U-200 or U-201 of Section VIII of the Boiler and Pressure Vessels Code of the American Society of Mechanical Engineers, 1949 edition; or (2) Section VIII of the Boiler and Pressure Vessel Code of the American Society of Mechanical Engineers, 1950, 1952, 1956, 1962 and 1965 editions; or (3) the Code for Unfired Pressure Vessels for Petroleum Liquids and Gases of the American Petroleum Institute and the American Society of Mechanical Engineers (API-ASME), 1951 edition.

1.14. A "Nonliquefied compressed gas" is a gas, other than a gas in solution which, under the charging pressure, is entirely gaseous at a temperature of 70 F.

1.15. A "liquefied compressed gas" is a gas which, under the charging pressure, is partially liquid at a temperature of 70 F. A flammable compressed gas which is normally nonliquefied at 70 F but which is partially liquid under the charging pressure and temperature shall follow the requirements for liquefied compressed gases.

1.16. A "pressurized liquid compressed gas" is a compressed gas other than a compressed gas in solution, which cannot be liquefied at a temperature of 70 F, and which is maintained in the liquid state at a pressure not less than 40 psia by maintaining the gas at a temperature less than 70 F.

2. TYPES OF SAFETY RELIEF DEVICES

Types of safety relief devices covered by this part are as follows:

2.1. Safety relief valve.

2.2. Combination safety relief valve and frangible disc.

2.3. Fusible plug utilizing a fusible alloy with yield temperature not over 170 F, nor less than 157 F (165 F nominal) (see Part 1 of these standards, Type CG-2).

2.4. Frangible disc (see Part 1 of these standards, Type CG-1).

3. APPLICATION REQUIREMENTS FOR SAFETY RELIEF DEVICES

3.1. Each container shall be provided with one or more safety relief devices which unless otherwise specified, shall be safety relief valves of the spring-loaded type.

3.1.1. Safety relief valves shall be set to start-to-discharge at a pressure not in excess of 110 per cent of the design pressure of the container nor less than the design pressure of the container, except as follows:

3.1.1.1. If an overdesigned container is used, the set pressure of the safety relief valve may be between the minimum required design pressure for the lading and 110 per cent of the design pressure of the container used.

3.1.1.2. For carbon dioxide (refrigerated), nitrous oxide (refrigerated), and pressurized liquid oxygen, nitrogen and argon, there shall be no minimum set pressure.

3.1.1.3. For butadiene, inhibited and liquefied petroleum gas containers, a minimum set pressure of 90 per cent of the minimum design pressure permitted for these ladings may be used.

3.1.1.4. For containers constructed in accord with Paragraph U-68 or U-69 of the code, 1949 edition, the set pressure marked on the safety relief valve may be 125 per cent of the original design pressure of the container.

3.1.2. Frangible discs may be used in lieu of safety relief valves on containers containing substances that may render a safety valve inoperative, or where a loss of valuable material by leakage should be avoided, or contamination of the atmosphere by leakage of noxious gases must be avoided.

3.1.3. For methyl chloride and sulfur dioxide storage containers of 1000 to 2000 lb water capacity, fusible plugs may be used in lieu of safety relief valves of the spring-loaded type. Each end shall have the total specified relieving capacity.

3.2. The design, material and location of safety relief devices shall be suitable for the intended service.

3.3. Safety relief devices shall have direct communication with the vapor space of the container.

3.4. Safety devices shall be so installed and located that the cooling effect of the contents will not prevent the effective operation of devices.

3.5. When fittings and piping are used on either the upstream or downstream side or both of a safety relief device or devices, the passages shall be so designed that the flow capacity of the safety relief device will not be reduced to the extent that the operation of the device could be impaired, nor below the capacity required for the container on which the safety relief device assembly is installed.

3.6. Safety relief devices shall be arranged to discharge upward and unobstructed to the open air in such a manner as to prevent any impingement of escaping gas upon the container. Safety devices shall be arranged to discharge upward, except this is not required for carbon dioxide, nitrous oxide and pressurized liquid argon, nitrogen and oxygen.

3.7. No shutoff valves shall be installed between the safety relief devices and the container, except in cases where two or more safety relief devices are installed on the same container, a shutoff valve may be used where the arrangements of the shutoff valve or valves is such as always to insure full required flow capacity through at least one safety relief device.

3.8. When storage containers are filled by pumping equipment which has a discharge capacity in excess of the capacity of the container safety relief devices, and which is capable of producing pressures in excess of the design pressure of the container, precautions should be taken to prevent the development of pressure in the container in excess of 120 per cent of its design pressure. This may be done by providing additional capacity of the safety relief valves on the container, by including a bypass on the pump discharge, or by any other suitable method.

3.9. The following additional requirements apply to safety relief devices on containers for liquefied hydrogen and pressurized liquid argon, nitrogen and oxygen.

3.9.1. The liquid container shall be protected by one or more safety relief valves and one or more frangible discs.

3.9.2. The minimum capacity of the safety relief valves shall be sized to provide adequate venting capacity with insulation saturated with gaseous lading or air at atmospheric pressure, whichever provides the greater thermal conductance.

3.9.3. The frangible disc shall have a flow capacity at 120 per cent of the design pressure at least equal to that required by 4.3.3 or 4.3.5 (whichever is applicable) with the insulation space saturated with gaseous lading or air at atmospheric pressure, whichever provides the greater thermal conductance, and shall function at a pressure not exceeding 150 per cent of the design pressure (plus 15 psig if vacuum insulation is used) or the test pressure, whichever is less.

4. DESIGN AND CONSTRUCTION REQUIREMENTS FOR SAFETY RELIEF DEVICES

4.1. The material, design and construction of a safety relief device shall be such that there shall be no significant change in the functioning of the device and no serious corrosion or deterioration of the materials within the period between renewals, due to service conditions. The chemical and physical properties of the materials shall be uniform and suitable for the requirements of the part manufactured therefrom.

4.2. Methods of manufacture, inspection and test shall conform to best practices in order to attain satisfactory performance of the safety relief devices (see References).

4.3. Safety relief devices shall have a total flow capacity as calculated by the formulas in 4.3.2, 4.3.3 and 4.3.5, except as provided in 4.3.4. These formulas are based on the principle of preventing the pressure in the container from exceeding a maximum of 120 per cent of the design pressure of the container.

4.3.1. The flow capacity of safety relief devices of each design and modification thereof shall be determined by an actual flow test. Methods of making flow tests are given in 5.2.

4.3.2. For uninsulated containers for nonliquefied gases the minimum required flow capacity of the safety relief device shall be calculated using the formula:

$$Q_a = 0.029 W_c$$

where

Q_a = flow capacity in cubic feet per minute of free air.

W_c = water capacity of the container in pounds.

4.3.3. For uninsulated containers not meeting the requirements of 4.3.4 and for insulated containers not meeting the requirements of 4.3.5 for liquefied compressed gases, the minimum flow capacity of the safety relief devices shall be calculated using the formula:

$$Q_a = G_u A^{0.82}$$

where

Q_a = flow capacity in cubic feet per minute of free air.

G_u = gas factor for uninsulated containers obtained from Table 4 (in Section B of this chapter), for the kind of gas involved.

*A = total outside surface area of the container in square feet.

Note: Graph of A vs. $A^{0.82}$ is shown in Fig. 2 (in Section B of this chapter).

4.3.4. For containers complying with any one of the following, the flow capacity of the safety relief device may be reduced to 30 per cent of the capacity as determined in 4.3.3.

4.3.4.1. When the storage is underground (see also 4.3.5.1).

4.3.4.2. When the storage is used for nonflammable gas and is suitably isolated from possible envelopment in a fire.

4.3.4.3. When the storage container is used for nonflammable gas and is equipped with suitable water spray or fire extinguishing system.

4.3.5. For insulated containers for liquefied gases, including pressurized liquid argon, nitrogen and oxygen, which meet the requirements of 4.3.6, the minimum required flow capacity of the safety relief device shall be calculated using the formula:

$$Q_a = G_i U A^{0.82}$$

where

Q_a = flow capacity in cubic feet per minute of free air.

G_i = gas factor for insulated containers obtained from Table 4 (in Section B of this chapter), for the kind of gas involved.

U = total thermal conductance of the container insulating material at a mean temperature of 100 F (Btu/(hr) (sq ft) (deg F)). Total thermal conductance = Thermal conductivity in Btu/(hr) (sq ft) (deg F/in.) divided by thickness of the insulation in inches.

A = total outside surface area of the container in square feet (see footnote to 4.3.3).

Note: Graph of A vs. $A^{0.82}$ is shown in Fig. 2 (in Section B of this chapter).

4.3.5.1. The safety relief valve capacity for underground containers may be determined in accordance with 4.3.5 by assigning a value of U for the minimum earth cover (see also 4.3.4.1).

4.3.6. Insulation Requirements.

4.3.6.1. If the insulation remains completely in place in a fire of 30 minutes duration, the U value shall be as defined in 4.3.5. If the insulation deteriorates or remains partly in place, the following procedure shall be followed to determine the flow capacity requirements of the safety relief devices:

(a) Use the formula for uninsulated containers in 4.3.3 or

(b) Measure the total thermal conductance (U) for the insulation system with the outer surface at 1200 F, after exposure to the time-temperature relationship of the assumed fire. This higher value of U shall then be used in the formula in 4.3.5 to determine the minimum required flow capacity of the safety relief device. For the purpose of this standard, a fire produces the time-temperature relationship (1000 F in 5 minutes, 1300 F in 10 minutes, 1550 F in 30 minutes) outlined in NFPA 251, "Fire Tests of Building Construction and Materials" (1961), and produces a heat input of approximately 34,500 Btu/(hr) (sq ft) to a bare tank.

4.3.6.2. Total thermal conductance of insulation shall not exceed .20 Btu/(hr) (sq ft) (deg F) at 100 F, except for liquefied hydrogen.

4.3.6.3. For liquefied hydrogen, the total

* When the surface area is not stamped on the name plate or when the marking is not legible, the area can be calculated by using one of the following formulas:

(1) Cylindrical container with hemispherical heads: Area = over-all length × outside diameter × 3.1416.

(2) Cylindrical container with semiellipsoidal heads: Area = (overall length + .3 outside diameter) × outside diameter × 3.1416.

(3) Spherical container: Area = (outside diameter)^2 × 3.1416.

thermal conductance (U) shall be chosen on the basis that the insulation space is saturated with gaseous hydrogen at atmospheric pressure.

4.3.7. Values are given in Table 4 (in Section B of this chapter) for G_i (for insulated containers) and G_u (for uninsulated containers) for use in formulas $Q_a = G_iUA^{0.82}$ and $Q_a = G_uA^{0.82}$. These values for G_i and G_u shall be used in determining the required flow capacity for a given commodity at 120 per cent of the set pressure shown in Table 4.

4.4 Safety Relief Valves

4.4.1. Safety relief valves shall be of the spring-loaded type. The inlet connection of the valve shall not be less than $\frac{3}{4}$-in. nominal pipe size with physical dimensions for the wall thickness not less than those of Schedule 80 pipe (extra-heavy).

4.4.2. Safety relief valves shall be designed so that the possibility of tampering will be minimized. If the pressure setting or adjustment is external, the safety relief valves shall be provided with suitable means for sealing the adjustment.

4.4.3. If the design of a safety relief valve is such that liquid can collect on the discharge side, the valve shall be equipped with a drain at the lowest point where liquid can collect.

4.4.4. Seats or discs of cast iron shall not be used.

4.5. Frangible Discs

4.5.1. The cross-sectional area of the connection to a container shall be not less than the required relief area of the frangible disc.

4.5.2. Every frangible disc shall have a specified bursting pressure at a specified temperature and shall burst within 5 per cent (plus or minus) of its specified bursting pressure.

Note: It is recommended that the maximum design pressure of the container be sufficiently above the intended operating pressure to provide sufficient margin between operating pressure and frangible disc bursting pressure to prevent premature failure of the frangible disc due to fatigue or creep.

4.5.3. The specified bursting pressure at the coincident operating temperature shall be determined by bursting 2 or more specimens from a lot of the same material and of the same size as those to be used. The tests shall be made in a holder of the same form and pressure area dimensions as that with which the disc is to be used.

4.5.4. A frangible disc may be installed between a spring-loaded safety relief device and the container provided:

4.5.4.1. The maximum pressure of the range for which the disc is designed to burst does not exceed the design pressure of the container.

4.5.4.2. The opening provided through the frangible disc, after breakage, is sufficient to permit a flow equal to the capacity of the attached valve and there is no chance of interference with the proper functioning of the valve; but in no case shall this area be less than the inlet area of the value.

4.5.4.3. The connection between the frangible disc and the valve is so arranged as to form a pocket in which any detached fragment of the disc will be retained, and this space is provided with a pressure gage, telltale device, try cock, or free vent to indicate whether the frangible disc has leaked or burst.

Note: Users are warned that a frangible disc will not burst at its design pressure if back-pressure builds up in the space between the disc and the safety relief valve, which will occur should leakage develop in the frangible disc due to corrosion or other cause.

4.5.5. A frangible disc may be installed on the outlet side* of a spring-loaded safety relief valve which is opened by a direct action of the pressure in the container provided:

4.5.5.1. The valve is so constructed that it will not fail to open at its proper pressure setting regardless of any back-pressure that can

* This use of a frangible disc in series with the safety relief valve is permitted to minimize the loss by leakage through the valve of valuable, or of noxious or otherwise hazardous materials, and where a frangible disc alone or a disc located on the inlet side of the safety valve is impracticable.

accumulate between the valve disc and the frangible disc.

Note: Users are warned that an ordinary spring-loaded safety relief valve will not open at its set pressure if back-pressure builds up in the space between the valve and frangible disc. A specially designed safety relief valve is required, such as a diaphram valve or a valve equipped with a bellows above the disc.

4.5.5.2. The disc is designed to rupture at not more than the design pressure of the container, for the tolerance permitted by 4.5.2.

4.5.5.3. The opening provided through the frangible disc, after breakage, is sufficient to permit a flow equal to the rated flow capacity of the attached safety relief valve; but in no case shall this area be less than the inlet area of the safety relief valve.

4.5.5.4. Any piping beyond the frangible disc cannot be obstructed by the burst frangible disc or fragments.

4.5.5.5. All valve parts and joints subject to stress due to the pressure from the container and all fittings up to the frangible disc are designed for not less than the design pressure of the container.

4.5.5.6. Any small leakage or a larger flow through a break in the operating mechanism that may result in back-pressure accumulation within enclosed spaces of the valve housing other than between the frangible disc and the discharge side of the safety valve, so as to hinder the safety relief valve from opening at its set pressure, will be relieved adequately and safely to atmosphere through telltale vent openings.

4.5.5.7. The contents of the container are clean fluids, free from gumming or clogging matter, so that accumulation in the space between the valve inlet and the frangible disc (or in any other outlet that may be provided) will not clog the outlet.

5. TESTS OF SAFETY RELIEF DEVICES

Tests of safety relief devices shall be in accordance with the Code or the following:

5.1. Pressure Tests of Safety Relief Valves

5.1.1. Each safety relief valve shall be subjected to an air or gas pressure test to determine the following:

5.1.1.1. That the start-to-discharge pressure setting is not less than 100 per cent nor more than 110 per cent of the set pressure marked on the valve.

Note: In setting the valve care must be taken that evidence of start-to-discharge is due to opening of the valve and not due to a defect.

5.1.1.2. That after the start-to-discharge pressure test, the valve closes at a pressure not less than 90 per cent of the start-to-discharge pressure.

5.2. Flow Capacity Test of Safety Relief Valves

5.2.1. The flow capacity of each design and modification thereof of a spring-loaded safety relief valve shall be determined by actual flow test at the flow rating pressure of 120 per cent of the set pressure marked on the valve. Three samples of each size of each valve representative of standard production shall be tested.

5.2.1.1. After pressure testing and without cleaning, removal of parts, reconditioning, or adjusting, each sample valve shall be subjected to an actual flow test at 120 per cent of the set pressure and the rate of flow of air or other gas released by the valve shall be measured. When gases other than air are used for flow test determination, the results shall be converted to equivalent air flow figures. The average rate of flow of air through the valves shall be recorded as the rated flow capacity. The design of the valves shall be considered acceptable, provided the capacities of the 3 valves shall fall within 10 per cent of the highest capacity recorded.

5.2.2. Acceptable Methods of Flow Testing.

5.2.2.1. Safety relief valves may be tested for flow capacities by testing with equipment conforming to the American Gas Association Gas Measurement Committee Report No. 3 "Orifice Metering of Natural Gas," as reprinted with revisions, 1956,[3] or ASME Power Test Code PTC 25-1958.[4]

5.2.2.2. Air or gas shall be supplied to the safety relief valve through a supply pipe provided with a pressure gage and a thermometer for indicating or recording the pressure and temperature of the supply. Observations shall be made and recorded after steady-flow conditions have been established.

5.2.2.3. When the testing method prescribed in 5.2.2.1 is employed, such test may be made by the manufacturer of the valve or a competent testing facility, provided that record of test is maintained by the manufacturer.

5.3. Test of Frangible Discs

Refer to Part 1 of these standards, Paragraph 5.3 through 5.4.1.4.

5.4. Tests of Fusible Alloy

Refer to Part 1 of these standards, Paragraph 5.1 through 5.1.1.4.

5.5. Tests of New or Reconditioned Fusible Plugs

Refer to Part 1 of these standards, Paragraph 5.2 through 5.2.4.

5.6. When a *combination safety relief valve and frangible disc* is used, the valve and the disc shall be tested separately and shall meet all of the requirements prescribed for each type of device.

5.7. *Rejected material* may be reworked, provided the material is subjected to such additional tests, as are required to insure compliance with all requirements of these standards.

6. IDENTIFICATION

6.1. Safety relief valves shall be marked to indicate

(a) manufacturer's name or trademark and catalog number,

(b) the month and year of manufacture,

(c) the set pressure in psig,

(d) the flow capacity in cubic feet per minute of free air.

6.2. For safety relief devices other than spring-loaded safety relief valves, the requirements of Part 1 of these standards shall apply.

7. MAINTENANCE REQUIREMENTS FOR SAFETY RELIEF DEVICES

7.1. Care shall be exercised to avoid damage to safety relief devices. Care shall also be exercised to avoid plugging by paint or other dirt accumulation of safety relief device channels or other parts which would interfere with the functioning of the device.

7.2. Only qualified personnel shall be allowed to service safety relief devices. Any servicing or repairs which require resetting of safety relief valves shall be done only by or with the consent of the valve manufacturer.

7.3. Only replacement parts or assemblies provided by the manufacturer shall be used unless the advisability of interchange is proved by adequate tests.

7.4. Safety relief devices periodically shall be examined externally for corrosion, damage, plugging of external safety relief device channels, and mechanical defects such as leakage or extrusion of fusible metal. Valves equipped with secondary resilient seals shall have the seals inspected periodically. The safety relief device shall be repaired or replaced if there is any doubt regarding the suitability of the device for service.

REFERENCES

1. "Boiler and Pressure Vessel Code of the American Society of Mechanical Engineers," 1949 edition; "Boiler and Pressure Vessel Code of the American Society of Mechanical Engineers," 1950, 1952, 1956, 1959, 1962 and 1965 editions; "The Code for Unfired Pressure Vessels for Petroleum Liquids and Gases of the American Petroleum Institute and the American Society of Mechanical Engineers," 1951 edition (the API-ASME Code, as a joint publication and interpretation service, was discontinued as of Dec. 31, 1956). Available from American Society of Mechanical Engineers, 345 E. 47th St., New York, N. Y., 10017.

2. "Standard for the Storage and Handling of Liquefied Petroleum Gases" (Pamphlet No. 58), National Fire Protection Association, 60 Batterymarch St., Boston, Mass., 02110; "American Standard Safety Requirements for the Storage and Handling of Anhydrous Ammonia" (Pamphlet G-2.1), Compressed Gas Association, Inc.
3. "Gas Measurement Committee Report No. 3, Orifice Metering of Natural Gas" (reprinted with revisions, 1956), American Gas Association, 420 Lexington Ave., New York, N. Y., 10017.
4. "Safety and Relief Valves, Power Test Codes PTC 25-1958," American Society of Mechanical Engineers.

ADDITIONAL REFERENCES

See also Sections A, B, and D of this chapter.

SECTION D

Recommended Practice for the Manufacture of Fusible Plugs

1. INTRODUCTION

These recommendations are intended to guide the manufacturer in the production of fusible plugs to meet the requirements of the "Safety Relief Device Standards" of Compressed Gas Association, Inc. Those standards contain minimum requirement recommendations for operation, application, and testing of safety relief devices for use with compressed gas containers of the types covered by the specifications of the Interstate Commerce Commission and The Board of Transport Commissioners for Canada. These recommendations are suggested minimum requirements to be followed in the manufacture of safety relief devices containing fusible alloys for types CG-2 and CG-3 devices, as defined in the "Safety Relief Device Standards," given in the preceding sections of this chapter.

2. PLUG BODY OR FUSIBLE ALLOY RETAINER

2.1. Material for the fusible plug body or fusible alloy retainer shall be made of brass, steel or other metal suitable for use with the compressed gas concerned, and for the application for which the safety relief device is intended.

2.2. When a taper pipe thread is provided for attaching the fusible plug to a cylinder, valve, or other pressure part, the pipe thread shall conform to standards for series NGT threads as contained in CGA Pamphlet V-1, "American Standard Compressed Gas Cylinder Valve Outlet and Inlet Connections." The "Safety Relief Device Standards," specify that taper pipe threads shall be series NGT when used to attach fusible plugs to cylinders.

2.3. The wall thickness of the fusible plug body provided with an external taper pipe thread shall not be less than the wall thickness of the corresponding size standard iron pipe as measured at any full thread (see Fig. 3a).

2.4. Fusible plugs which are equipped with screw threads and with gasket makeup means to provide the pressure tight joint shall have adequate design strength for the service. Gasket material shall be suitable for use with the compressed gas concerned.

2.5. A suitable wrenching section shall be provided to facilitate installation of the plug. This may be accomplished by means of a hexagon head, a recessed head, or other suitable design which will allow for satisfactory installation. It is recognized that special designs are required for certain applications.

2.6. The plug body shall have provisions to assist in retaining the fusible alloy in place. This may be accomplished by any of the following:

(a) A taper bore increasing in size toward the pressure end (see Fig. 3a).

(b) A step bore resulting from a series of bore diameters increasing progressively in size toward the pressure end (see Fig. 3b).

(c) A straight bore or a taper bore provided with threads. It is recommended that a thread form with rounded roots and crests be used (see Fig. 3c).

2.7. In the design of fusible plugs, due consideration shall be given to the over-all length in relation to the diameter of the bore provided for the fusible alloy. Sufficient testing of the design shall be conducted to show that the fusible alloy is retained securely in place under the service conditions for which the device is intended. These tests shall include subjecting the device to the normal operating pressures and temperatures while in contact with the particular commodity with which the device is to be used. In addition to complying with these recommended minimum standards the completed device shall comply with all the requirements of the "Safety Relief Device Standards."

2.8. All machining shall conform to best practices and procedures.

3. MANUFACTURE OF THE FUSIBLE ALLOY

3.1. Type CG-2 safety device is defined in the "Safety Relief Device Standards" as a fusible plug utilizing a fusible alloy having a yield temperature of not more than 170 F, nor less than 157 F.

3.2. Type CG-3 safety device is defined in the "Safety Relief Device Standards" as a fusible plug utilizing a fusible alloy having a yield temperature of not more than 220 F, nor less than 208 F.

3.3. There are various compositions of fusible alloys which can be used for either of the CG-2 or CG-3 fusible plugs. The correct alloy should be determined by the manufacturer after due consideration is given to the type and length of bore provided in the fusible plug body and the actual service requirements. This selection should be made only after giving consideration to the possible effect of the compressed gas on the fusible alloy itself.

3.4. The composition of the fusible alloy will usually consist of varying percentages of bismuth, lead, tin and cadmium.

3.5. Bismuth should be "technical grade 98-99.5 per cent.

Lead should be "pure chemical lead."

Tin should be "pure bar tin or pure Straits tin."

Cadmium should be "pure 99.5 to 99.9 per cent cadmium."

3.6. The following melting procedure has been found to be useful in minimizing oxidation of materials, which oxidation may affect the satisfactory performance of the device:

(a) After the composition has been determined, measure ingredients accurately and add them to the melt in the order in which they are listed above (see warning note, 3.7).

(b) Stir or agitate the batch frequently after it becomes molten—both before and during casting.

(c) Do not permit the temperature of the batch to exceed 700 F during the mixing of the bismuth and lead. During the addition of the tin and cadmium do not permit the temperature of the batch to exceed 500 F.

3.7. All slag, dirt, metallic oxides, etc., shall be excluded from each cast.

WARNING NOTE: Fumes are hazardous. Follow recognized safety standards for ventilation and handling.

3.8. The fusible alloy may be cast into sticks, ingots, or other shapes. A commonly used stick is 24 in. long and approximately 3/16 in. in diameter. Casting should be done in a continuous operation on the same day, with the molten fusible alloy maintained at a relatively constant temperature just sufficiently above the yield temperature to allow for easy casting.

3.9. Each stick (or other shape) or group thereof of fusible alloy shall be identified adequately by a prominent mark.

3.10. Groups of sticks (or other shapes) of

fusible alloy should be packaged in a convenient sealed package, labeled to identify content.

4. TESTING OF FUSIBLE ALLOY FOR YIELD TEMPERATURE OR MELTING POINT

4.1. To be certain that the fusible alloy is correct, as many melting point tests as are necessary shall be taken from each batch before final production casting. It is recommended that a melting point test be taken from the last lot cast from each batch. If it is impossible to complete the casting on the same day that the batch is compounded, or if it is necessary to remelt it, the melting point tests should be made again before casting is resumed. The fusible alloy may be tested in stick or other form after casting in accordance with the recommendation of Part 5.1 of Section A of these standards.

5. PROCEDURE FOR POURING FUSIBLE ALLOY INTO PLUG BODIES

5.1. The fusible plug body shall be thoroughly cleaned by the use of a suitable solvent immediately before "tinning" and insertion of the fusible alloy. An acid dip to obtain some etching of the bore is recommended. The clean plugs should then be heated to a temperature approximating the yield temperature of the fusible alloy to be used and the bore swabbed with a suitable flux.

5.2. The bore of the body shall now be "tinned" using one of the following methods:

(a) Press a stick of the fusible alloy against the wall of the bore and flow it over the surface while applying just sufficient heat to the body to cause the alloy to flow freely.

(b) Pour the molten fusible alloy into the bore and then empty the plug and inspect.

(c) Dip the entire plug body in a bath of molten fusible alloy. In the case of bodies made of steel, where it is difficult to obtain adequate "tinning," a "tinning" alloy other than the fusible alloy may be used. However, if a "tinning" alloy having a melting point higher than that of the fusible alloy is used, care must be taken in the "tinning" operation to insure a thin even coating. If the metal deposit during the "tinning" operation is too heavy, it could result in variations in the melting point of the completed plugs.

5.3. Place the plug body on a clean surface. The fusible alloy shall now be cast into the plug body using one of the following methods:

(a) Again apply a flux to the "tinned" surface of the bore. Pour the molten fusible alloy into the plug body while maintaining it at a temperature slightly above the melting temperature of the fusible alloy.

(b) Flow the fusible alloy into the plug body by placing the fusible alloy stick inside the bore of the body while simultaneously applying heat to the body.

(c) Cast the fusible alloy into the plug body by means of an induction furnace. After swabbing the bore, a predetermined length or size slug of the fusible alloy is inserted into the bore of the body after which the bodies with the alloy are placed in an induction heating zone and melting of the fusible alloy accomplished by induction heat. This method requires careful control to eliminate overheating.

5.4. When heat is applied to the plug body during the casting operation, it should be done uniformly to insure complete expulsion of flux when completing the cast. To prevent oxidation, excessive heating should be avoided.

5.5. After completing the casting operation the complete assembly shall be treated to remove excess flux.

5.6. The completed fusible plugs should be marked immediately to identify the yield temperature range of the fusible alloy. The marking or coding used shall be consistent with the marking requirements of the "Safety Relief Device Standards."

Note: Various heat treatments are used for various reasons after casting the fusible alloy in the fusible plug. These range from quick chilling at a temperature not exceeding 40 F immediately after the fusible alloy is cast to submerging the completed plugs in a fresh water bath at a temperature of from 85 to 90 per cent of the yield temperature for a period as long as 16 hours. Inasmuch as the exact composition of the fusible alloy used has some bearing on the advantages of the heat treatment no specific recommendation regarding this point is being made.

6. TESTS OF COMPLETED FUSIBLE PLUGS

After the fusible plugs are completed as outlined above, they shall be tested in accordance with the requirements of Part 5.2 of Section A of these standards.

7. OTHER FORMS OF SAFETY DEVICES EMPLOYING FUSIBLE ALLOYS

Section D of these standards is intended basically to cover fusible plugs. It is recognized that there are many variations to a standard fusible plug, including designs which embody the fusible alloy as an integral part of a cylinder valve. It is difficult to anticipate all of the modifications of design which might be used. It is believed, however, that many of the recommended practices included in these standards can be applied in the design and manufacture of the modified forms.

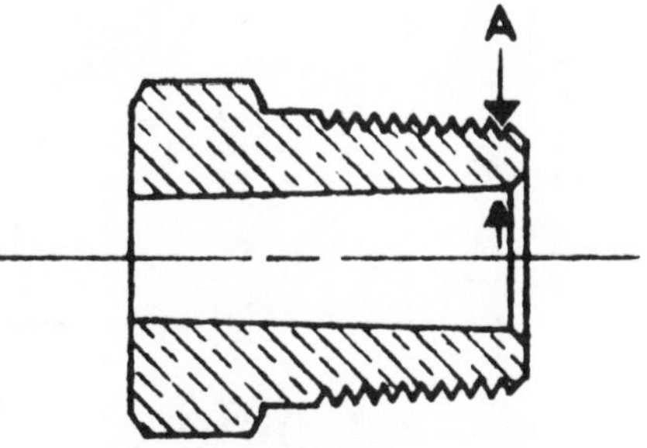

Figure "A"
Plain Taper Bore

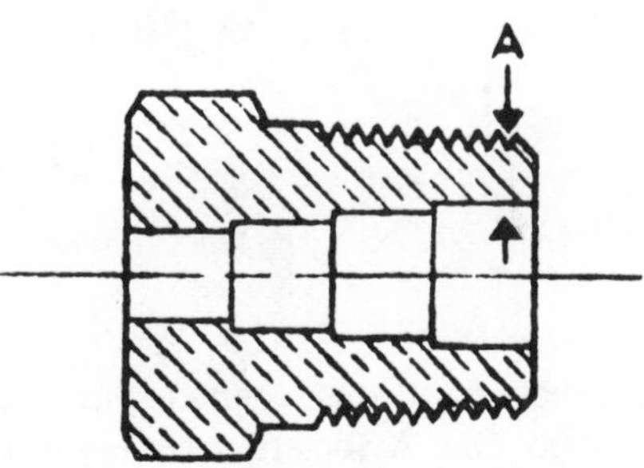

Figure "B"
Step Bore

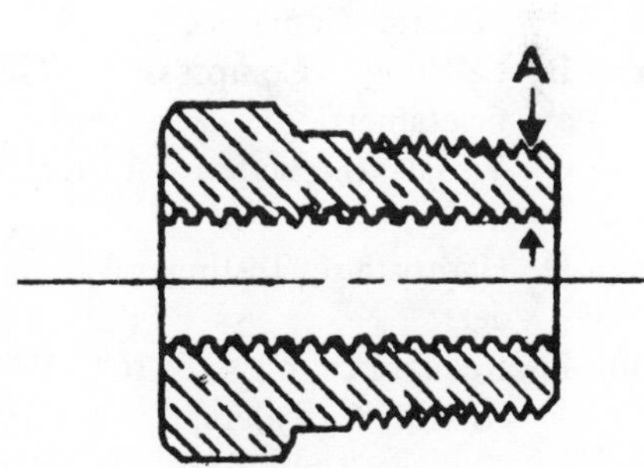

Figure "C"
Taper or Straight Bore Threaded

"A" — Not less than thickness of standard pipe wall at point of first full thread

FIG. 3. Plug Bodies for Types CG-2 and CG-3 Safety Devices:—(*a*) plain taper bore; (*b*) step bore; (*c*) taper or straight bore threaded. Dimension "A"—not less than thickness of standard threaded pipe wall at point of first full thread.

CHAPTER 3

Compressed Gas Cylinders: Marking, Requalifying, Welding and Brazing, Disposition

Each section of this chapter presents a Compressed Gas Association standard that relates primarily to compressed gas cylinders. These are the CGA standards for:

Section A. Marking Portable Compressed Gas Containers to Identify the Material Contained;

Section B. Labeling Compressed Gas Containers;

Section C. Visual Inspection of Cylinders;

Section D. Hydrostatic Testing of Cylinders;

Section E. Requalification of ICC-3HT Aircraft Cylinders;

Section F. Retest Procedures for Seamless, High-Pressure Cylinders ICC-3, ICC-3A, ICC-3AA;

Section G. Welding and Brazing on Thin-Walled Compressed Gas Containers;

Section H. Disposition of Unserviceable Cylinders.

In each section of this chapter where reference is made to cylinder specifications of the Interstate Commerce Commission (ICC), it applies equally to cylinders of the corresponding specifications of the Board of Transport Commissioners for Canada (BTC).

SECTION A

Marking Portable Compressed Gas Containers to Identify the Material Contained

This section presents the "American Standard Method of Marking Portable Compressed Gas Containers to Identify the Material Contained," American Standard Z48.1. Originated by the CGA, the standard was approved May 7, 1954, by the American Standards Association, Inc. It is one of a series of American industrial safety standards.

1. SCOPE

1.1. Requirements for marking portable compressed gas containers, not exceeding 1000 lb water capacity, to identify the material contained.

2. DEFINITIONS

2.1. A portable compressed gas container, for the purposes of these standards, is any container having a water capacity of 1000 lb, or less, that is constructed in accord with the specifications of a recognized authority (such as the specifications of the Interstate Commerce Commission, the "Unfired Pressure Vessel Code" of the ASME, the API-ASME "Unfired Pressure Vessel Code," etc.) and intended to contain a compressed or liquefied gas, as defined in the regulations of the Interstate Commerce Commission. Containers of commodities that are exempt from regulation by federal or state authorities are not included.

3. MARKING OF CONTAINERS

3.1. Compressed gas containers shall be legibly marked with at least the chemical name or a commonly accepted name of the material contained. Marking shall be by means of stenciling, stamping, or labeling, and shall not be readily removable.

3.2. Wherever practical the marking shall be located at the valve end and off the cylindrical part of the body.

3.3. The height of the lettering shall be not less than 1/25 of the diameter of the container with a minimum height of 1/8 in.

4. MARKING OF CONTAINERS FOR USE IN INTERNATIONAL TRADE

4.1. Compressed gas containers for use in international trade shall be legibly marked by means of stenciling, stamping, or labeling and as noted in 3.2, 3.3, 4.2, and 4.3. Such markings shall not be readily removable.

4.2. With the name of the material contained in the language of the country in which the container is charged.

4.3. With the international chemical formula, or other agreed abbreviation, as follows:

Acetylene	C_2H_2
Air	AIR
Ammonia	NH_3
Argon	Ar
Boron trifluoride	BF_3
Butane, propane or their mixtures	LP-Gas or LPG
Carbon dioxide	CO_2
Carbon monoxide	CO
Chlorine	Cl_2
Coal gas	GDV
Cyclopropane	(no formula)
Dichlorodifluoromethane	CCl_2F_2
Difluoroethane	$C_2H_4F_2$
Difluoromonochloroethane	$C_2H_3ClF_2$
Ethane	C_2H_6
Ethyl bromide	C_2H_5Br
Ethyl chloride	C_2H_5Cl
Ethylene	C_2H_4
Ethylene oxide	C_2H_4O
Fluorine	F_2
Helium	He
Hydrogen	H_2
Hydrogen chloride	HCl
Hydrogen cyanide	HCN
Hydrogen fluoride	HF
Hydrogen sulfide	H_2S
Illuminating gas	GDV
Krypton	Kr
Methane	CH_4
Methyl bromide	CH_3Br
Methyl chloride	CH_3Cl
Monochlorodifluoromethane	$CHClF_2$
Natural gas	MET
Neon	Ne
Nitrogen	N_2
Nitrogen peroxide	N_2O_4
Nitrosyl chloride	NOCl
Nitrous oxide	N_2O
Oxygen	O_2
Phosgene	$COCl_2$
Propane, butane or their mixtures	LP-Gas or LPG

Propylene	PRY
Sulfur dioxide	SO_2
Sulfur hexafluoride	SF_6
Town gas	GDV
Vinyl chloride	C_2H_3Cl
Xenon	Xe

5. REGULATIONS

5.1. None of the requirements contained in this standard are intended to conflict with or to supersede federal or state regulations such as the following:

5.1.1. Regulations of the Interstate Commerce Commission for the Transportation of Explosives and Other Dangerous Articles by Land and Water in Rail Freight Service and by Motor Vehicles (Highway and Water);

5.1.2. Regulations of the U. S. Coast Guard for the Transportation, Storage or Stowage of Explosives or Dangerous Articles or Substances and Combustible Liquids on Board Vessels;

5.1.3. The Regulations of the Food and Drug Administration for the Enforcement of the Federal Food, Drug and Cosmetic Act.

SECTION B

Labeling Compressed Gas Containers

1. INTRODUCTION

1.1. The compressed gas industry recognizes the value of precautionary labels on cylinders to warn of known hazards. As a guide for the preparation of adequate cylinder labels, Compressed Gas Association, Inc., has prepared this statement defining general principles and giving sample labels for several types of gases.

1.2. The methods of preparing label precautionary information established by the Manufacturing Chemists' Association have been followed in this guide but modified where necessary to meet the specific labeling needs of the compressed gas industry.

1.3. Precautionary labels are in addition to any labeling required by ICC regulations, which is intended only to alert carrier personnel during transportation.

1.4. The labels suggested in this statement are examples prepared in accordance with the minimum requirements to warn the user of the known hazards involved.

1.5. Individual statutes, regulations or ordinances may require that particular information be included on a label, or that a specific label be affixed to a container. In each case, the requirements of these laws should be observed.

2. GENERAL PRINCIPLES

2.1. In preparing labels for compressed gas containers, the following general principles should serve as a guide.

2.1.1. "Compressed gas containers shall be legibly marked with at least the chemical name or a commonly accepted name of the material contained. Marking shall be by means of stencilling, stamping, or labeling, and shall not be readily removable." (Quoted from American Standard Z48.1*)

2.1.2. For effectiveness all statements on labels should be brief, accurate, and expressed in simple, easily understood terms.

2.1.3. Precautionary information should be used only where necessary. Unnecessary wording on labels, particularly in the case of harmless gases, may develop a disregard for the labels.

2.1.4. It is desirable to employ uniform wording in indicating the same hazards for different gases.

2.1.5. Labels should be designed so that color and shape cannot be confused with or be in conflict with Interstate Commerce Commission labels.

* See preceding section.

2.1.6. Warning statements should be in easily legible type which is in contrast by typography, layout, or color with other printed matter on the label. The label should be in a conspicuous place on the container.

2.2. In addition to the name of the gas (see 2.1.1) the following types of information, where necessary, should be considered for inclusion on the label:

2.2.1. Signal Word.

The signal word is intended to draw attention to the presence and the degree of hazard. Recognized signal words are "DANGER!", "WARNING!", "CAUTION!" The signal word may be omitted where the statement of hazard (see 2.2.2) is itself an effective attention-drawing signal word, such as, "FLAMMABLE" or "POISON."

2.2.2. Statement of Hazard.

This statement should give notice of the hazards present in connection with the customary or reasonably anticipated handling or use of the product.

Examples are: Flammable
Liquid causes burns

2.2.3. Precautionary Measures and Instructions in Case of Contact or Exposure.

These instructions are intended to supplement, if necessary, the statement of hazard by briefly setting forth measures to be taken to avoid injury or damage from stated hazards.

3. ILLUSTRATIVE LABELS

3.1. The following are examples of labels for compressed gas containers prepared in accordance with the general principles given above.

3.1.1. Oxidizing (Nonliquefied).

(*1*) *Name of cylinder content:*	**OXYGEN**
(*2*) *Signal Word:*	**WARNING!**
(*3*) *Statement of Hazard:*	**VIGOROUSLY ACCELERATES COMBUSTION**
(*4*) *Precautionary Measures:*	Keep oil and grease away. Use only with equipment conditioned for oxygen service.

3.1.2. Flammable.

ACETYLENE	**HYDROGEN**	**PROPANE**
FLAMMABLE	**FLAMMABLE**	**FLAMMABLE**
Keep away from heat, flame and sparks.	Keep away from heat, flame and sparks.	Keep away from heat, flame and sparks.

3.1.3. Physiologically Inert (Nonliquefied).

NITROGEN
HELIUM
ARGON

3.1.4. Physiologically Corrosive—Irritant.

AMMONIA
WARNING!
LIQUID CAUSES BURNS
GAS EXTREMELY IRRITATING
Do not breathe gas.
Do not get in eyes, on skin, on clothing.
In case of contact, immediately flush skin with plenty of water for at least 15 minutes. For eye or extensive skin contact call a physician. Affected area should be thoroughly flushed with water.

3.1.5. Liquefied Compressed Gas (High Pressure).

CARBON DIOXIDE

CAUTION!

HIGH PRESSURE

3.1.6. Medical Gas.

NITROUS OXIDE (*Anesthetic Grade*)

CAUTION!

HIGH PRESSURE

Federal law prohibited dispensing without a prescription.

3.1.7. Poison.

PHOSGENE

POISON

Do not breathe gas.
Avoid exposure to skin and eyes to gas or liquid.
Use adequate ventilation.
No exertion should follow exposure. Obtain medical attention promptly.
(Add appropriate statement on antidote where necessary to compiy with local statutes.)

3.1.8. Toxic Gas.

CARBON MONOXIDE

DANGER! HAZARDOUS IF INHALED

FLAMMABLE!

KEEP AWAY FROM HEAT, FLAME AND SPARKS

3.1.9. Radioactive Gas Mixture.

KRYPTON-85

Contents: Krypton-85
*Volume:
*Activity:
*Date:
Caution!

Use only as authorized by Atomic Energy Commission, and in conformance with state and local regulations

*To be filled in as required.

3.1.10. Low Temperature Liquefied Gas.

LIQUID HELIUM

WARNING! Extremely cold liquid. Liquid may cause burns. Use only by experienced personnel.
Keep container upright.
Avoid contact of liquid or cold gas with skin or eyes.

3.1.11. Extreme Pressure.

NITROGEN

CAUTION!

PRESSURE 6000 PSI. Use only with equipment designated for 6000 psig or higher.

4. ALPHABETICAL LISTING OF RECOMMENDED LABELS FOR GASES

Chlorine

DANGER! Hazardous liquid and gas under pressure.
Do not handle or use until safety precautions recommended by supplier have been read and understood.
Do not breathe air containing this gas.
Do not get in eyes or on skin.
Do not heat cylinders.
Have available emergency gas masks approved by U. S. Bureau of Mines for Chlorine service.
In case of exposure, move patient to fresh air, keep warm and quiet, and call a physician.

Cyclopropane

FLAMMABLE!
Keep away from heat, flame and sparks.
Federal Law prohibits dispensing without prescription. (Add other appropriate statements to conform with Federal Food, Drug and Cosmetic Act.)

Ethylene

FLAMMABLE!
Keep away from heat, flame and sparks.
(a) For use as a plant regulator add following statements:
"*Plant Regulator*" *Use.* For use only by

or under supervision of experienced personnel. Exempt from Federal Food and Drug Act residue tolerance requirements when used on fruit and vegetable crops, before or after harvest, in conformity with good agriculture practice. Active Ingredients: Ethylene —% by weight. Inert Ingredients: —% by weight.

(b) For medical use add following statement: Federal law prohibits dispensing without prescription.
(Add other appropriate statements to conform with Federal Food, Drug and Cosmetic Act.)

Fluorine

DANGER!
Extremely Hazardous Gas. Causes severe burns.
Contact with organic or siliceous materials may cause fire.
Contact with water may cause violent reaction.
Do not breathe vapor.
Do not get in eyes, on skin, or clothing.
Store out of sun and away from direct heat.

POISON

First Aid—GET MEDICAL ATTENTION AT ONCE.
Always have on hand a supply of Magnesia Paste (Magnesium Oxide and Glycerine).
In case of contact or suspicion of contact, immediately flush skin with plenty of water (particularly under nails) until whiteness disappears. Apply magnesium paste. Remove and wash clothing before reuse. Eyes —Flush with water for at least 15 minutes. *GET MEDICAL ATTENTION!*

Forming Gas (Containing 5.7% or more hydrogen in nitrogen) (*hydrogen-nitrogen mixture*).
FLAMMABLE!
Keep away from heat, flame, and sparks.
(Pressure precaution optional—see Appendix —Statement No. 11).

Forming Gas (containing less then 5.7% hydrogen) (*Hydrogen-nitrogen mixture*).
(Pressure precaution optional—see Appendix —Statement No. 11).

Hydrogen Sulfide

DANGER! Poisonous Liquid and Gas.
Flammable!
Do not breathe gas.
Keep away from heat, flame, and sparks.
This gas deadens the sense of smell. Do not depend on odor to detect presence of gas.

POISON

No exertion should follow exposure. Obtain medical attention promptly.
(Add appropriate statement on antidote where necessary to comply with local statutes.)

Methane

FLAMMABLE
Keep away from heat, flame, and sparks.

Oxygen, Liquefied

WARNING! Vigorously accelerates combustion. Mixtures of combustible materials and liquid oxygen may explode on ignition or impact.
Keep oil and grease away. Use only with equipment conditioned for Oxygen service.
Extremely cold liquid.
Liquid may cause burns.
Avoid contact of liquid or cold gas with skin or eyes, or on clothing.
Avoid spills. Do not walk on, or roll equipment over spills.

Sulfur Dioxide

WARNING!
Liquid causes burns.
Gas extremely irritating.
Do not breathe gas.
Protect eyes, skin and clothing from contact.

In case of contact, immediately flush eyes or skin with water for at least 15 minutes.

Call physician at once in case of exposure or burns.

APPENDIX

The following statements, where applicable may be considered for inclusion as additional label information. The use of a signal word may be required with the following statements, if not already present on label.

1. Return to supplier for recharging.
2. Keep cylinder away from heat, flame and sparks.
3. Store away from combustible materials and other compressed gases.
4. Do not drop.
5. Keep valve closed when not in use.
6. Use with adequate ventilation.
7. Gas does not support life.
8. Liquid may cause frostbite.
9. Liquid may cause burns.
10. Liquefied gas under pressure.
11. Under (High) Pressure.
12. This gas is for administration only by or under the supervision of a physician, dentist, or other practitioner similarly licensed by law to administer anesthesia gases.
13. Stand away from outlet when opening.

SECTION C

Visual Inspection of Cylinders

1. INTRODUCTION

1.1. Regulations of the Interstate Commerce Commission as well as the Regulations of the Board of Transport Commissioners for Canada require that a cylinder be condemned when it leaks, or when internal or external corrosion, denting, bulging, or evidence of rough usage exists to the extent that the cylinder is likely to be weakened appreciably. Under prescribed conditions of use a formal visual inspection has been authorized in lieu of the periodic hydrostatic retest for certain low pressure cylinders (reference: §73.34(e)(10).

1.2. This statement has been prepared as a guide to cylinder users for establishing their own cylinder inspection procedures and standards. It is of necessity general in nature although some specific limits are recommended. It should be distinctly understood that it will not cover all circumstances for each individual cylinder type and condition of loading. Each cylinder user must expect to modify them to suit his own cylinder design or the conditions of use that may exist in his own service. Rejection, or acceptance for continued use in accordance with these limits, does not imply that these cylinders are or are not dangerous, or subject to impending failure, but represents practice which has been satisfactory to a cross section of the industry.

1.3. Experience in the inspection of cylinders is an important factor in determining the acceptability of a given cylinder for continued service. Users lacking this experience and having doubtful cylinders should return them to a manufacturer of the same type of cylinders for reinspection.

1.4. The information contained in this statement is obtained from sources believed to be reliable and is based on the experience of a number of member companies of Compressed Gas Association, Inc. However, by issuance of the statement, the Association and its members, jointly and severally, make no guarantee of results and assume no liability in connection with the information herein contained or the safety suggestions herein made. These suggestions are not intended to, and it is believed do not, conflict with regulations issued by any authority having jurisdiction.

1.5. The suggestions contained in this statement are not intended to apply to cylinders manufactured under Specification ICC-3HT.

Because of the special provisions of this specification, separate recommendations covering service life and standards for visual inspection of these cylinders have been made (see Section E).

2. GENERAL

2.1. Inspection Equipment

2.1.1. Depth Gages, Scales, etc.

Exterior corrosion, denting, bulging, gouges or digs are normally measured by simple direct measurement with scales or depth gages. In brief, a rigid straight edge of sufficient length is placed over the defect and a scale is used to measure the distance from the bottom of the straight edge to the bottom of the defect. There are also available commercial depth gages which are especially suitable for measuring the depth of small cuts or pits. It is important when measuring such defects to use a scale which spans the entire affected area. When measuring cuts, the upset metal should be removed or compensated for so that only actual depth of metal removed from the cylinder wall is measured.

2.1.2. Supersonic Devices.

There are a variety of commercial supersonic devices available. These can be used to detect subsurface flaws and to measure wall thickness.

2.1.3. Magnetic Particle Inspection.

Magnetic particle inspection can be adapted for cylinder inspection to locate quickly surface faults not readily visible to the naked eye.

2.2. High and Low-Pressure Cylinders

For the purpose of this statement high-pressure cylinders are those with a marked service pressure of 900 psig or greater; low-pressure cylinders are those with a marked service pressure less than 900 psig.

2.3. Minimum Allowable Wall Thickness

For the purposes of this statement the minimum allowable wall thickness is the minimum wall thickness required by the specification under which the cylinder was manufactured.

2.4. Dents

Dents in cylinders are deformations caused by its coming in contact with a blunt object in such a way that the thickness of metal is not materially impaired. A typical dent is shown in Fig. 1.

2.5. Cuts, Gouges or Digs

Cuts, gouges or digs in cylinders are deformations caused by contact with a sharp object in such a way as to cut into or upset the metal of the cylinder, decreasing the wall thickness at that point.

2.6. Corrosion or Pitting

Corrosion or pitting in cylinders involves the loss of wall thickness by corrosive media. There are several kinds of pitting or corrosion to be considered as defined below.

2.6.1. Isolated Pitting.

Isolated pits of small cross section do not effectively weaken the cylinder wall but are indicative of possible complete penetration and leakage. Since the pitting is isolated the original wall is essentially intact.

2.6.2. Line Corrosion.

When pits are not isolated but are connected or nearly connected to others in a narrow band

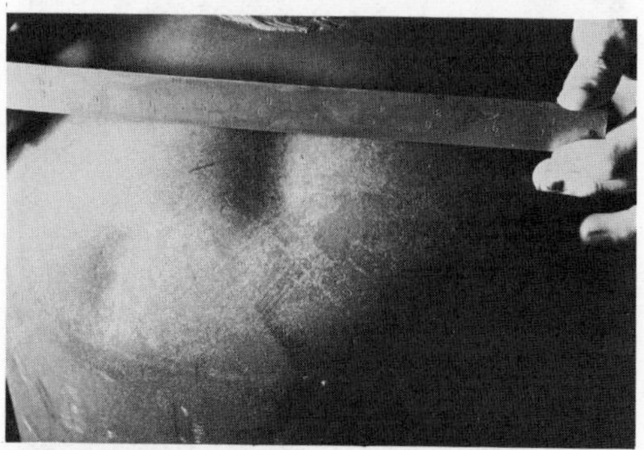

FIG. 1. Measuring the length of a typical dent.

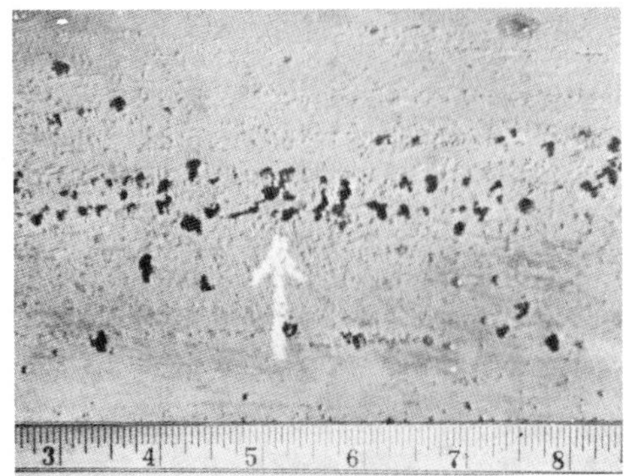

FIG. 2. Line corrosion.

FIG. 3. Crevice corrosion near the cylinder footring.

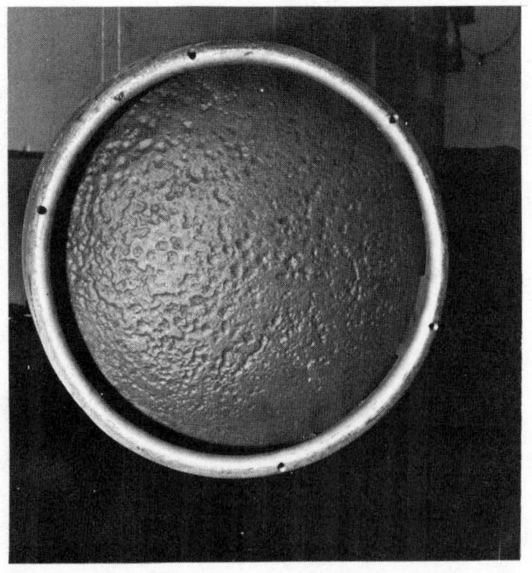

FIG. 4. General corrosion with pitting.

or line, such a pattern is termed "line corrosion." This condition is more serious than isolated pitting. Line corrosion frequently occurs in the area of intersection of the footring and bottom of a cylinder. This is sometimes referred to as "crevice corrosion." An example of line corrosion is shown in Fig. 2, and of crevice corrosion in Fig. 3.

2.6.3. General Corrosion.

General corrosion is that which covers considerable surface areas of the cylinder. It reduces the structural strength. It is often difficult to measure or estimate the depth of general corrosion because direct comparison with the original wall cannot always be made. General corrosion is often accompanied by pitting. This form of corrosion is shown in Figs. 4 and 5.

3. INSPECTION OF LOW-PRESSURE CYLINDERS EXEMPT FROM THE HYDROSTATIC TEST INCLUDING ACETYLENE CYLINDERS

This section covers cylinders exempt from hydrostatic retest requirements of the ICC by virtue of their exclusive use in certain noncorrosive gas service. They are not subject to internal corrosion and do not require internal shell inspection. If, due to unusual circumstances, cylinders of these types suffer internal corrosion they should be inspected in accordance with the procedure in Section 4.

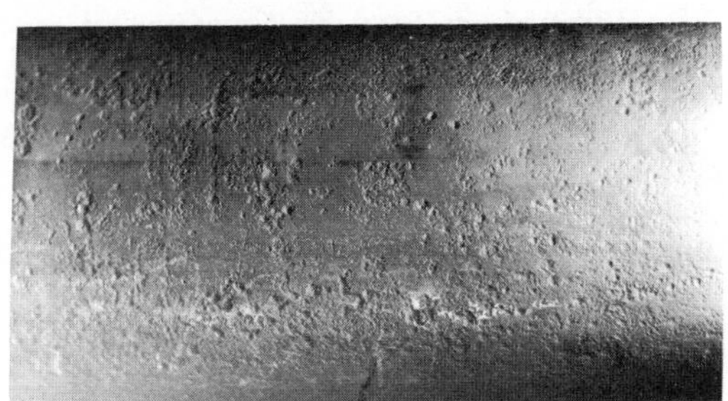

FIG. 5. General corrosion with pitting on cylinder wall.

3.1. Preparation for Inspection

Rust, scale, caked paint, etc., shall be removed from the exterior surface so that the surface can be adequately observed. Facilities shall be provided for inverting the cylinder to facilitate inspection of the bottom. This is important because experience has shown this area to be the most susceptible to corrosion.

3.2. Exterior Inspection

Cylinders shall be checked as outlined below for corrosion, general distortion, or any other defect that might indicate a weakness which would render it unfit for service.

3.2.1. Corrosion Limits.

To fix corrosion limits for all types, designs, and sizes of cylinders, and include them in this statement is not practicable. Such limits are usually established by individual companies for their own cylinders only after much field experience. When specific limits cannot be applied, the removal of cylinders from service is based on judgement. Additionally, in fixing limits of acceptance it is recognized that cylinder design requirements have changed over the years and cylinder wall thicknesses have varied with these design requirements and the use of new materials such as alloy steels. Four general rules are advanced which summarize the field experience of a number of companies. Failure to meet any of these four rules is of itself cause for rejection of a cylinder. The rules follow:

(a) A cylinder shall be rejected when the tare weight is less than 95 per cent of the original tare weight marked on the cylinder. When determining tare weight, be sure that the cylinder is empty.

(b) A cylinder shall be rejected when the remaining wall in an area having *isolated pitting* only is less than 1/3 of the minimum allowable wall thickness.

(c) A cylinder shall be rejected when *line corrosion* on the cylinder is 3 in. in length or over and the remaining wall is less than $\frac{3}{4}$ of the minimum allowable wall thickness or when line corrosion is less than 3 in. in length and the remaining wall thickness is less than $\frac{1}{2}$ the minimum allowable wall thickness.

(d) A cylinder shall be rejected when the remaining wall in an area of *general corrosion* is less than $\frac{1}{2}$ of the minimum allowable wall thickness.

Note 1: Although general corrosion does not always follow a definite pattern, where there is appreciable pitting in areas of general corrosion, the pitted depth (Pd) may usually be considered to be about twice the general corrosion thickness loss (GC) (see Fig. 6).

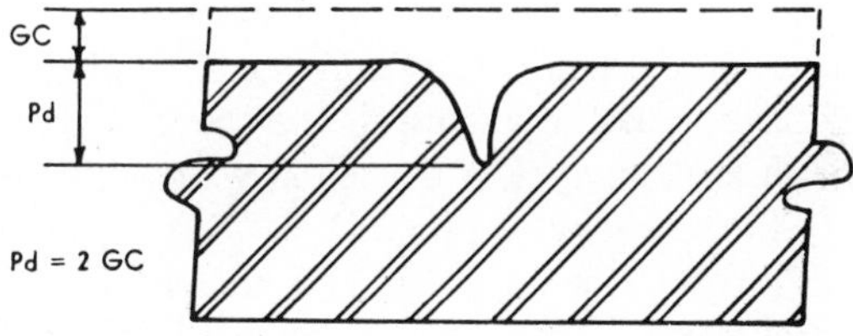

FIG. 6.

Note 2: Pitted depth may be measured by placing a straight edge across the high points in the pitted area and measuring the distance from the bottom of the straight edge to the bottom of the pit. It is recognized that there are certain areas in cylinders of the low-pressure type where this pit depth procedure may be difficult to apply, due to the proximity of the corroded areas to footrings and other appurtenances. In this event, special curved measuring devices can be devised, or, if necessary, putty casts can be made for establishing criteria for acceptance or rejection. Figure 3 shows one method of approximating the depth of such corrosion.

3.2.1.1. To use the above criteria, it is necessary to know the original wall thickness of the cylinder or the minimum allowable wall thickness. Table 1 lists the minimum allowable wall thickness under ICC specifications for a number of common size low pressure cylinders.

3.2.1.2. The following are examples of the calculations necessary to establish actual rejection limits for rules given in 3.2.1. For the examples, the 4B240 cylinder with a 15 in. outside diameter and minimum wall thickness of 0.128 in. is used.

(a) General Corrosion Accompanied by Pitting.

(1) When the wall thickness of the cylinder at manufacture *is not known*, and the actual wall thickness cannot be measured, this cylinder shall be rejected when the inspection reveals that the deepest pit in a general corrosion area exceed 3/64 in. This is arrived at by considering that in no case shall the pitting exceed $\frac{1}{2}$ the minimum allowable wall thickness which is 0.064 in. When a pit measures 0.043 in. (approximately 3/64 in.) in a corrosion area, general corrosion will already have removed 0.021 in. of the original wall and the total pit depth as compared to the initial wall will be 0.064 in. (see Note 1 of 3.2.1).

(2) When the original wall thickness at manufacture *is known*, or the actual wall thickness is measured, this thickness less $1\frac{1}{2}$ times the maximum measured pit depth shall be 0.064 in. or greater. If it is less, the cylinder shall be rejected.

(b) Isolated Pits Not in a General Corrosion Area.

(1) When the original wall thickness *is not known*, and the actual wall thickness cannot be measured, the cylinder shall be rejected if the pit depth exceeds 0.085 in. (0.128 × 2/3 = 0.085 in.). The remaining wall of 0.043 in. is 1/3 of the minimum allowable wall thickness.

(2) When the original wall thickness *is known*, or the actual wall thickness is measured, this thickness less the maximum pit depth shall be 0.043 in. or greater (0.128 in. − (0.128 in. × 2/3) = 0.043 in.). If it is less, the cylinder shall be rejected.

3.2.2. Dents.

Dents are of concern where the metal deformation is sharp and confined, or where they are near a weld. Where metal deformation is not sharp, dents of larger magnitude can be tolerated.

3.2.2.1. Dents at Welds. Where denting occurs so that any part of the deformation includes a weld, the maximum allowable dent depth shall be $\frac{1}{4}$ in.

3.2.2.2. Dents Away from Welds. When denting occurs so that no part of the deformation includes a weld, the cylinder shall be rejected if the depth of the dent is greater than 1/10 of the mean diameter of the dent. A maximum allowable depth is not set here; however, standards of appearance in the industry should provide a reasonable limiting factor.

3.2.3. Cuts, Gouges or Digs.

Cuts, gouges or digs reduce the wall thickness of the cylinder and in addition are considered to be stress raisers. Depth limits are set in the next paragraphs; however, cylinders shall be rejected at $\frac{1}{2}$ of the limit set whenever the length of the defect is 3 in. or more.

3.2.3.1. When the original wall thickness at manufacture *is not known*, and the actual wall thickness cannot be measured, a cylinder shall be rejected if the cut, gouge or dig exceeds $\frac{1}{2}$ of the minimum allowable wall thickness. (*Example:* In a 15 in. O.D. 4B240, 100 lb propane cylinder, the depth of the defect could not exceed $\frac{1}{2}$ of 0.128 in. or 1/16 in. If

TABLE 1.

Cylinder Size (o.d. × length)	ICC Spec. Marking	Nominal Water Capacity (lb)	Minimum Allowable Wall Thickness (in.)
15″ × 46″	4B250*	239	.128
$14\frac{13}{16}$″ × 47″	4E240	239	.140
$14\frac{15}{16}$″ × 46″	4BA240	239	.086
$14\frac{11}{16}$″ × $28\frac{3}{8}$″	4BA240	143	.086
$11\frac{29}{32}$″ × $32\frac{11}{16}$″	4BA240	95	.078
$11\frac{29}{32}$″ × $18\frac{11}{32}$″	4BA240	48	.078

* Without longitudinal seam.

the defect was 3 in. or more in length it could not exceed $\frac{1}{2}$ of 1/16 in. or 1/32 in.)

3.2.3.2. When the original wall thickness at manufacture *is known*, or the actual wall thickness is measured, a cylinder shall be rejected if the original wall thickness minus the depth of the defect is less than $\frac{1}{2}$ of the minimum allowable wall thickness.

3.2.4. Leaks.

Leaks can originate from a number of sources, such as defects in a welded or brazed seam, defects at the threaded opening, or from sharp dents, digs, gouges, or pits.

3.2.4.1. To check for leaks, the cylinder shall be charged and carefully examined. All seams and pressure openings shall be coated with a soap or other suitable solution to detect the escape of gas. Any leakage is cause for rejection.

3.2.4.2. In accordance with §73.34(d) of ICC and BTC regulations, safety relief devices shall be tested for leaks before a charged cylinder is shipped from the cylinder filling plant.

3.2.5. Fire Damage.

Cylinders shall be carefully inspected for evidence of exposure to fire.

3.2.5.1. Inspection for Fire Damage. Common evidences of exposure to fire are (a) charring or burning of the paint or other protective coat, (b) burning or sintering of the metal, (c) distortion of the cylinder, (d) melted out fuse plugs and (e) burning or melting of valve.

3.2.5.2. Evaluation of Fire Damage. ICC and BTC regulations state that, subject to the special provisions of §73.34(f), cylinders which have been in a fire must not again be placed in service until they have been properly reconditioned. The general intent of this requirement is to remove from service cylinders which have been subject to the action of fire which has changed the metallurgical structure or the strength properties of the steel, or in the case of acetylene cylinders caused breakdown of porous filler. This is normally determined by visual examination as covered above with particular emphasis to the condition of the protective coating. If the protective coating has been burnt off or if the cylinder body is burnt, warped, or distorted, it is assumed that the cylinder has been overheated, and §73.34(f) shall be complied with. If, however, the protective coating is only dirtied from smoke or other debris, and is found by examination to be intact underneath, the cylinder shall not be considered affected within the scope of this requirement.

3.2.6. Bulges.

3.2.6.1. Removal from Service. Cylinders are manufactured with a reasonably symmetrical shape. Cylinders which have definite visible bulges shall be removed from service and evaluated as outlined below.

3.2.6.2. Measurement. Bulges in cylinders can be measured in several ways as follows:

(a) Bulges on the cylinder sidewall can be measured by comparing a series of circumferential measurements.

(b) Bulges in the head, and also in some case on the sidewall, can be measured by comparing a series of measurements of the peripheral distance between the valve spud and the center seam (if any) or an equivalent fixed location on the cylinder sidewall.

(c) Variations from normal cylinder contour can be measured directly by (1) measuring the height of a bulge with a scale; (2) comparing templates of bulged areas with similar areas not bulged.

3.2.6.3. Limits. Cylinders shall be rejected when a variation of 1 per cent or more is found in the measured circumferences or in peripheral distances measured from the valve spud to the center seam (or equivalent fixed point). An example for a 15 in. diameter cylinder follows:

Normal cylinder diameter	15″
Cylinder circumference	47.12″
Maximum circumference	
$47.12 + .01\,(47.12) = 47.59''$	
Variation in circumference	0.47″
Equivalent variation in diameter	
$\frac{0.47}{\pi} = 0.15''$	

If the bulge is uniform around the cylinder, the limiting height of the bulge would be 0.15 in./2 = 0.075 in. or about 1/16 in.

3.2.7. Neck Defects.

3.2.7.1. Cylinder necks shall be examined for serious cracks, folds and flaws. Neck cracks are normally detected by testing the neck during charging operations with a soap solution.

3.2.7.2. Cylinder neck threads shall be examined whenever the valve is removed from the cylinder. At manufacture, cylinders have a specified number of full threads of proper form as required in applicable thread standards [National Gas Taper Threads (NGT), formerly National Taper Pipe Threads (NPT)]. Cylinders shall be rejected if the required number of effective threads are materially reduced, or if a gas tight seal cannot be obtained by reasonable valving methods. Gages shall be used to measure the number of effective threads. Common thread defects are worn or corroded crests, broken threads, nicked or cut threads.

3.2.8. General Distortion.

3.2.8.1. Cylinders. Noticeable distortion may be evaluated by reference to the sections in this pamphlet under denting or bulging. If the valve is noticeably tilted, the cylinder shall be rejected.

3.2.8.2. Appurtenances. The footring and headring of cylinders may become so distorted through service abuse that they no longer perform their functions (a) to cause the cylinder to remain stable and upright, (b) to protect the valve. Rings shall be examined for distortion, for looseness, and for failure of welds. Appearances may often warrant rejection of the cylinder. Repair rules for such cylinders are established in ICC and BTC regulations, §73.34.

3.3. Inspection Report Form

ICC and BTC regulations §73.34(e)(10) require that results be recorded and a permanent record kept of all visual inspections made in lieu of the periodic hydrostatic retest. A sample inspection report form is shown on p. 248.

4. LOW-PRESSURE CYLINDERS SUBJECT TO HYDROSTATIC TESTING

Cylinders covered in this section are low-pressure cylinders other than those covered in Section 3. They differ essentially from such cylinders in that they require a periodic hydrostatic retest which includes an internal and external examination. Defect limits for the external examination are prescribed in Section 3 with exceptions for aluminum cylinders shown in 4.3. The additional procedures for internal inspection follow.

4.1. Preparation for Inspection

4.1.1. The provisions outlined in 3.1 shall be followed. Additionally, the interior of the cylinder shall be prepared for inspection by the removal of any internal scale, or any other condition which would interfere with the inspection of the internal surface.

4.1.2. A good inspection light is mandatory for internal inspection. Flammable gas cylinders shall be purged before being examined with a light. Lamps for flammable gas cylinders shall be vapor proof.

4.2. Internal Inspection

Cylinders shall be inspected internally at least every time the cylinder is periodically retested. The examination shall be made with a light of sufficient intensity to clearly illuminate the interior walls.

Note: The Chlorine Institute, New York, N. Y., has published a pamphlet, "Container Procedure," for chlorine cylinders. This procedure includes an internal examination at the time of each filling.

4.2.1. General Corrosion.

Where interior corrosion is general, evaluation is best accomplished by a hydrostatic test combined with careful visual inspection.

Thickness measuring and flaw detection devices of the supersonic type may be used to evaluate specific conditions.

4.2.2. Local Corrosion.

Where corrosion is local, evaluation shall be based on a combination of a hydrostatic test and judgment of the inspector of the seriousness of the local condition. Thickness measuring and flaw detection devices of the supersonic types may be used to evaluate specific conditions.

4.2.3. Miscellaneous Defects.

Interior defects other than corrosion are uncommon. Any such defects can be evaluated to some degree by the following:

4.2.3.1. If the bottom of the defect can be seen it may be possible to evaluate its effect with judgment.

4.2.3.2. Where the bottom of the defect cannot be seen and where its extent cannot be measured by various inspection instruments, the cylinder shall be rejected.

4.3. External Inspection of Aluminum Cylinders.

The inspection requirements of Section 3 shall be met, except as follows:

4.3.1. Aluminum cylinders shall be rejected when impairment to the surface (corrosion or mechanical defect) exceeds a depth where the remaining wall is less than $\frac{3}{4}$ of the minimum allowable wall thickness required by the specification under which the cylinder was manufactured.

4.3.2. Aluminum cylinders subjected to the action of fire shall be removed from service as required by §73.34(f)(4) of ICC regulations.

5. HIGH-PRESSURE CYLINDERS

High-pressure cylinders are those with a marked service pressure of 900 psig or higher. They are seamless; no welding is permitted. The great bulk of such cylinders are of the 3A or 3AA types complying with ICC or BTC specifications.

5.1. Preparation for Inspection

5.1.1. Cylinders shall be cleaned for inspection so that the inside and outside surfaces and all conditions can be observed. This shall include removal of scale and caked paint from the exterior and the thorough removal of internal scale. Cylinders with interior coating shall be examined for defects in the coating. If the coating is defective, it shall be removed.

5.1.2. A good inspection light of sufficient intensity to clearly illuminate the interior wall is mandatory for internal inspection. Flammable gas cylinders shall be purged before being examined with a light. Lamps for flammable gas cylinders shall be vapor-proof.

5.2. Exterior Inspection

5.2.1. Corrosion Limits.

To fix corrosion limits for all types, designs, and sizes of cylinders, and include them in this statement, is not practicable. Such limits are usually established by individual companies for their own cylinders only after much field experience. Considerable judgment is required in evaluating cylinders fit for service. Experience is a major factor, aside from strength considerations, for high pressure cylinders. Three general rules are advanced which summarize the field experience of a number of companies over a period of years, as follows:

(a) The calculated wall stress, based on actual wall thickness measurements, in a corroded section shall not exceed the maximum wall stress limitation published in ICC and BTC regulations, §73.302(c)(3).

(b) The average wall stress, determined from the elastic expansion obtained in the hydrostatic test, shall not exceed the average wall stress limitation published in ICC and BTC regulations, §73.302(c)(3).

(c) The remaining wall in a cylinder which has isolated pitting of small cross section only shall not be less than 2/3 the minimum allowable wall thickness.

5.2.1.1. To use these criteria the cylinder

user must establish minimum wall thicknesses and elastic expansion limits for cylinders in his service (see "Section F, Retest Procedures for Seamless, High-Pressure Cylinders ICC-3, ICC-3A, ICC-3AA").

5.2.1.2. The use of these criteria for a common size cylinder follows:

Size (o.d. × length)	9-1/16″ × 51″
Specification marking	3AA2265
Steel analysis	3AA Alloy
Minimum allowable wall at manufacture	0.221″
Limiting wall thickness in service	0.195″
Limiting elastic expansion in the hydrostatic test	226 cc

(a) Local Pitting or Corrosion or Line Corrosion.

(1) When the original wall thickness of the cylinder *is not known*, and the actual wall thickness cannot be measured, the cylinder is rejected if corrosion exceeds 1/32 in. in depth. This is arrived at by subtracting from the minimum allowable wall at manufacture (0.221 in.), the limiting wall in service (0.195 in.), to give the maximum allowable corrosion limit of 0.026 in., or about 1/32 in.

(2) When the wall thickness *is known*, or the actual wall thickness is measured, the difference between this known wall and the limiting value establishes the maximum corrosion figure. The normal hot forged cylinder of this size will have a measured wall of about 0.250 in. Comparison of this with the limiting wall thickness shows that defects up to about 1/16 in. are allowable, provided, of course, that the actual wall is measured or is known.

(b) General Corrosion.

(1) Cylinders with general corrosion are evaluated by subjecting them to a hydrostatic test. Thus, a cylinder with an elastic expansion of 227 cc or greater would be rejected. If areas of pronounced pitting are included within the general corrosion, the depth of such pitting should also be measured (with the high spots of the actual surface as a reference plane) and the criteria established in the first example apply. Thus, the maximum corrosion limit would be 1/32 in. when the wall was not known.

(c) Isolated Pits.

(1) When the original or actual wall thickness of the cylinder *is not known*, and the actual wall thickness cannot be measured, the cylinder is rejected if the pit depth exceeds 0.074 in, or about 5/64 in., (.221 × 1/3).

(2) When the wall thickness *is known*, or the actual wall thickness is measured, this thickness less the pit depth shall not be less than 0.147 in. (.221 × 2/3).

5.2.2. Cuts, Digs and Gouges.

5.2.2.1. Measurement. Cuts, digs or gouges may be measured with suitable depth gages (any upset metal shall be smoothed off to allow true measurements).

5.2.2.2. Limits. Established by the stress considerations in 5.2.1.(a).

5.2.2.3. General. Any defect of appreciable depth having a sharp bottom is a stress raiser and even though a cylinder may be acceptable from a stress standpoint, it is common practice to remove such defects. After any such repair operation, verification of the cylinder strength and structure shall be made by a hydrostatic test or other suitable means.

5.2.3. Dents.

Dents can be tolerated when the cylinder wall is not deformed excessively or abruptly. Generally speaking, dents are accepted up to a depth of about 1/16 in. when the major diameter of the dent is equal to or greater than 32 times the depth of the dent. Sharper dents than this are considered too abrupt and shall require rejection of the cylinder. On small diameter cylinders these general rules may have to be adjusted. Considerations of appearance play a major factor in the evaluation of dents.

5.2.4. Arc and Torch Burns.

Cylinders with arc or torch burns shall be removed from service. Defects of this nature may be recognized by one of the following conditions:

(a) Removal of metal by scarfing or cratering.

(b) A sintering or burning of the base metal.

(c) A hardened heat affected zone.

5.2.4.1. A simple method for verifying the presence of small arc burns is to file the suspected area. The hardened zone will resist filing as compared to the softer base metal.

5.2.5. Bulges.

Cylinders are normally produced with a symmetrical shape. Cylinders with distinct visual bulges shall be removed from service until the nature of the defect is determined. Some cylinders may have small discontinuities related to the manufacturing process—mushroomed bottoms, offset shoulders, etc. These usually can be identified and are not normally cause for concern.

5.2.6. Fire Damage.

Cylinders shall be carefully inspected for evidence of exposure to fire.

5.2.6.1. Inspection for Fire Damage. Common evidences of exposure to fire are (a) charring or burning of the paint or other protective coat, (b) burning or sintering of the metal, (c) distortion of the cylinder, (d) functioned safety devices and (e) melting of valve parts.

5.2.6.2. Evaluation of Fire Damage. ICC and BTC regulations state that, subject to the special provisions of §73.34(f), cylinders which have been in a fire must not again be placed in service until they have been properly reconditioned. The general intent of this requirement is to remove from service cylinders which have been subject to the action of fire which has changed the metallurgical structure or the strength properties of the steel. This is normally determined by visual examination as covered above with particular emphasis to the condition of the protective coating. If the protective coating has been burnt off or if the cylinder body is burnt, warped, or distorted, it is assumed that the cylinder has been overheated and §73.34(f) shall be complied with. If, however, the protective coating is only mildly blistered or dirtied from smoke or other debris, and is found by examination to be intact underneath, the cylinder shall not be considered affected within the scope of this requirement.

5.2.7. Neck Defects.

5.2.7.1. Cylinder necks shall be examined for serious cracks, folds and flaws. Neck cracks are normally detected by soap testing the neck during charging operations.

5.2.7.2. Cylinder neck threads shall be examined when the valve is removed from the cylinder, at the time of hydrostatic test or any other time the valve is removed. At manufacture, cylinders have a specified number of full threads of proper form as required in applicable thread standards (National Gas Taper Threads (NGT), formerly National Taper Pipe Threads (NPT)). Cylinders shall be rejected if the number of effective threads are materially reduced or if a gas tight seal cannot be obtained by reasonable valving methods. Gages shall be used to measure the number of effective threads. Common thread defects are worn or corroded crests, broken threads, nicked or cut threads.

5.3. Internal Inspection

Cylinders shall be inspected internally at least every time the cylinder is periodically retested. This examination shall be made with a light of sufficient intensity to clearly illuminate the interior walls.

5.3.1. Corrosion and Pitting

5.3.1.1. General Corrosion. Where interior corrosion is general the cylinder shall be evaluated by a hydrostatic test applying expansion limits as given in 5.2.1.

5.3.1.2. Local Corrosion. Where corrosion is local, the acceptance of the cylinder shall be based on the combination of the hydrostatic test and the judgment of the inspector.

APPENDIX A

SAMPLE INSPECTION REPORT FORM

....	....	....
Month Year	COMPANY NAME	Plant
Report No.	**FIVE YEAR CYLINDER INSPECTION REPORT**	Approved Signature

ICC SPEC.	CYLINDER DATA				INSPECTED FOR					PROTECTIVE COATING		INSPECTORS INITIALS	DISPOSITION (See Code)
	TYPE	SERIAL NUMBER	DATE MFD.	REGISTERED SYMBOL	CONDITION OF RINGS	DENT, DIG, BULGE, GOUGE	CORROSION	LEAKAGE	FIRE DAMAGE	TYPE	CONDITION		

Original........................

Duplicate........................

Disposition Code:
1. Returned to Service
2. Returned to Manufacturer for Repair
3. Scrapped

Thickness-measuring and flaw detector devies of the supersonic type are useful in measuring and evaluating any specific defect.

5.3.2. Interior Defects.

Interior defects other than corrosion are uncommon. Any such defects can be evaluated to some degree by the following:

5.3.2.1. If the bottom of the defect can be seen, it may be possible to evaluate the defect with judgment.

5.3.2.2. Where the bottom of the defect cannot be seen and where its extent cannot be measured by various inspection instruments, the cylinder shall be rejected.

5.3.3. Hammer Test.

A hammer test consists of tapping a cylinder a light blow with a suitably sized hammer. A cylinder, emptied of liquid content, with a clean internal surface, standing free, will have a clear ring. Cylinders with internal corrosion will give a duller ring dependent upon the amount of corrosion and accumulation of foreign material. Such cylinders shall be investigated.

5.3.3.1. The hammer test is very sensitive and is an easy, quick and convenient test that can be made without removing the valve before each charging. It is an invaluable indicator of internal corrosion.

SECTION D

Hydrostatic Testing of Cylinders

Hydrostatic testing and retesting of compressed gas cylinders, as specified in ICC and BTC specifications and regulations, require testing by water jacket or other suitable method, operated so as to obtain accurate data. Except when modified by the specifications or regulations, the total expansion, permanent expansion, and per cent permanent expansion must be determined. Apparatus of suitable form must be approved by the Bureau of Explosives, New York, N. Y. The following methods are in use.

(1) **Water Jacket Volumetric Expansion Method.** This method is applicable to all hydrostatic tests when volumetric expansion determinations are required. It consists of enclosing the cylinder in a vessel completely filled with water, measuring in a suitable device attached to the vessel the total and permanent volumetric expansion of the cylinder, by measuring the amount of water displaced by expansion of the cylinder when under pressure and after pressure is released.

(2) **Direct Expansion Method.** This method is applicable to all hydrostatic tests when volumetric expansion determinations are required. It has practical limitations in its use. It consists of forcing a measurable volume of water into a cylinder filled with a known weight of water at a known temperature, and raising the pressure to desired test pressure, measuring the volume of water expelled from the cylinder when pressure is released. Calculating the permanent volumetric expansion of the cylinder by subracting the volume of water expelled from the volume of water forced into the cylinder to raise to desired pressure. Calculating the total volumetric expansion of the cylinder by subtracting the compressibility of the volume of water in the cylinder, when under pressure, from the volume of water forced into the cylinder to raise pressure to desired test pressure.

(3) **Pressure Recession Method.** This method is recognized as being practically obsolete, having been replaced largely by methods 1 and 2. It consists of subjecting the cylinder rapidly to hydrostatic test pressure and immediately cutting off pressure supply, and observing recession of pressure in the cylinder due to permanent expansion.

(4) **Proof Pressure Method.** This method is

permitted where ICC specifications and regulations do not require the determination of total and permanent expansion. It consists of examining the cylinder under test for leaks and defects.

This section presents detailed requirements for operation, equipment, and use of the four test methods.

1. WATER JACKET METHOD

1.1. Scope

1.1.1. This method has been developed from many tests, and years of experience in testing ICC-3A (high-pressure) cylinders. It is the standard method of testing and must be used except when modified by ICC specifications and regulations. It may not be necessary or practicable to apply all details to all types of cylinders. However, the requirements as to installation and checking of testing equipment, care and accuracy in testing, with careful inspection, are applicable to *all* cylinders and should be followed for accurate testing and safety.

1.2. Inspection of Cylinders

1.2.1. Before cylinders are retested under the periodic retest required by ICC regulations:

1.2.1.1. Valves should be removed and careful internal inspection made with suitable light. Scale and sludge deposits (if any) should be removed and if seriously corroded, the cylinders shall be condemned. To avoid a possible flash, cylinders used in flammable gas service should be thoroughly purged before interior inspection light is inserted.

1.2.1.2. Exterior inspection, including the bottom, should be carefully made for corrosion, dents, arc or torch burns, and physical deformation. Seriously corroded cylinders shall be condemned. Cylinders dented, arc or torch burnt, or physically damaged, so as to weaken the cylinder appreciably, shall be condemned.

1.2.2. Note: An empty cylinder before recharging should be examined for physical damage and exterior corrosion, followed by hammer test. Those cylinders which have a dull or peculiar ring when lightly tapped with a suitably sized hammer, should be carefully examined and if necessary, valve removed and interior examined with suitable light and retested before charging.

1.3. Requirements for Accurate Testing

1.3.1. The important element is the ability to determine accurately the total permanent, and per cent permanent expansion of a cylinder as required by ICC specifications and regulations. By simple means the average wall thickness can also be determined. Hence, cylinders with walls that are corroded or otherwise too thin can be eliminated.

1.3.2. For the water jacket cylinder testing method the following conditions are absolutely essential:

1.3.2.1. The apparatus must be so arranged that the water level in the expansion indicator will be the same at zero and when the total expansion and permanent expansion are being read. This water level must be above the highest point of water in the jacket and its connecting piping. Expansion indicator must be accurate to within 1 per cent of the total expansion or 0.1 cc.

1.3.2.2. A dead-weight testing apparatus should be used to verify the accuracy of pressure gages. A master gage checked by a dead-weight tester at frequent intervals, would be an acceptable substitute though less desirable.

1.3.2.3. The pressure gage must agree with a dead-weight testing apparatus within $\frac{1}{2}$ of 1 per cent at the test pressure. Gages must have zero stop pin removed or placed $\frac{1}{2}$ in. below zero and a true zero marked on the dial.

1.3.2.4. A special calibrated cylinder, the expansion of which at known pressure has been determined within 1 per cent by a laboratory test must be used to check the accuracy of the apparatus and the operations of testing. This also affords an additional check on the pressure gages.

1.3.2.5. The cylinder being tested must be suspended from the cover of the jacket or otherwise arranged so that it is free to expand in all directions.

1.3.2.6. Water in cylinder and jacket should be as near the same temperature as possible. (Suitable liquid such as kerosene may be used in place of water with proper precaution.)

1.3.2.7. Apparatus must be constructed and used so as to eliminate the presence of air pockets. Valves, piping, fittings, and all connections must be absolutely tight and free of leaks while cylinder is under test.

1.3.2.8. Any internal pressure applied previous to the test pressure shall not exceed 90 per cent of test pressure. If due to failure of the test apparatus, the test pressure cannot be maintained, the test may be repeated at a pressure increased by 10 per cent, or 100 psig, whichever is the lower value. Test pressure must be maintained for at least 30 seconds and as much longer as may be necessary to secure complete expansion of the cylinder. Before final readings are taken, the burette water level must remain within 1 per cent of the total expansion of the cylinder as determined at the commencement of the 30 second period, or 0.1 cc.

1.4. How to Use Results of Tests

1.4.1. The tests will determine the total and permanent expansion of the cylinders at a given pressure. The total expansion minus the permanent expansion is the elastic expansion, and this elastic expansion at a given pressure is a definite measure of the average wall thickness of the cylinder.

Examples:

	Example 1	*Example 2*
Total expansion	166.0 cc	14.5 cc
Permanent expansion	3.0 cc	0.2 cc
% Permanent expansion	$\frac{3.0}{166.0} = 1.8\%$	$\frac{0.2}{14.5} = 1.4\%$
Elastic expansion	166.0 − 3.0 = 163.0 cc	14.5 − 0.2 = 14.3 cc

1.4.1.1. If the elastic expansion, as determined by the retest of a cylinder, shows a marked increase, when compared with the elastic expansion reported at the time of manufacture, the average wall thickness should be determined to guard against excessive stress at working pressure, due to wall deterioration. An increase of 10 per cent in elastic expansion indicates approximately a 10 per cent reduction in average wall.

1.4.1.2. A cylinder must be condemned if it leaks, or fails to pass the test requirements prescribed by the ICC regulations.

1.4.1.3. A cylinder should be condemned when the elastic expansion exceeds the limit established by the owner or when the maximum wall stress at service pressure exceeds a safe permissible stress in the cylinder.

1.4.1.4. Wall stress calculations for the cylinder tested should be made by the formula specified in the ICC specifications.

1.4.1.5. Minimum wall thickness may be measured; *or* the average wall thickness calculated by the formula:

$$t = \frac{D}{2}\left\{1 - \sqrt{1 - \frac{PKV}{EE}}\right\}$$

1.4.1.6. Elastic expansion may be calculated by the formula:

$$\text{EE} = PKV\left(\frac{D^2}{D^2 - d^2}\right)$$

t = average wall thickness in inches
EE = elastic expansion (total less permanent) in cc
V = internal volume in cc
1 cu in. = 16.387 cc
P = test pressure in pounds per square inch (psig)
D = outside diameter in inches
d = inside diameter in inches = $D - 2t$
K = factor (experimentally determined).

1.4.1.7. Factor K has been determined for all ICC-3A and 3AA high-pressure cylinders. See Table 1 for typical K factors for certain hot drawn, concaved bottom cylinders made of carbon, medium manganese, or alloy steel.

TABLE 1

Type (see note)	K (typical)	Nominal Size—In. (i.d.)	(length)	(o.d.)
220-ft Oxygen	1.30×10^{-7}	$8\frac{1}{2}$	51	$9\frac{1}{16}$
200-ft Oxygen	1.30×10^{-7}	$8\frac{1}{2}$	51	$9\frac{1}{16}$
50-lb CO_2	1.30×10^{-7}	8	51	$8\frac{1}{2}$
110-ft Oxygen	1.27×10^{-7}	$6\frac{5}{8}$	43	$7\frac{1}{16}$
100-ft Oxygen	1.27×10^{-7}	$6\frac{5}{8}$	43	$7\frac{1}{16}$
20-lb CO_2	1.26×10^{-7}	5	51	$5\frac{1}{2}$
150-lb Chlorine	1.31×10^{-7}	10	48	$10\frac{7}{16}$
100-lb Chlorine	1.31×10^{-7}	10	36	$10\frac{7}{16}$

Note: It will be understood that the cylinders described as "oxygen" also include hydrogen, helium, etc.; the "CO_2" cylinders include nitrous oxide, etc., and the "chlorine" cylinders include SO_2, etc.

1.4.1.8. Factor *K* for other sizes and types of cylinders must be determined by the individual users. Additional *K* factors and method of determining *K* factors can be found in the following Section F, "Retest Procedures for Seamless, High-Pressure Cylinders ICC-3, ICC-3A, ICC-3AA."

1.5. Basis of Recommendations

1.5.1. A cylinder in proper condition when it is placed in service remains so, barring accidents, except for the damage done by corrosion. The experiments have shown that there is a direct relation between the elastic expansion of a cylinder and its average wall thickness. In the regular quinquennial tests under the Interstate Commerce Commission Regulations, the total and permanent expansions are determined. The difference of these two figures gives the elastic expansion which can therefore be determined without additional tests, beyond those regularly made. Increase of elastic expansion during the life of a cylinder indicates reduced wall thickness.

1.5.2. The most generally accepted method of determining accurately the elastic expansion at present is the use of the water jacket test. Care must be exercised to make certain that all parts of the equipment are functioning properly and that accurate results are obtained, as otherwise the test is worthless. If the expansion readings obtained are too high, as may be caused by an incorrect pressure gage, cylinders with ample wall thickness may be unnecessarily rejected. If they are too low, which may be caused by the presence of air in the jacket, and if zero, total and permanent expansion are not taken at the same water level, cylinders with dangerously thin wall may be retained in service. Considerable experience with these tests has shown that the number of cylinders which will fail will only be a small fraction of 1 per cent; it is highly desirable that these cylinders be detected, and it is necessary to be certain that every test is correctly made in order to be sure that this is done.

1.5.3. Zero, total and permanent expansion readings shall be taken at the same water level. The rise of level of the water in the burette tube when the pressure is applied to the cylinder under test is liable to cause very considerable error, as it produces a change in pressure within the jacket which tends to expand the jacket and compress any air which may be trapped in it if not leveled. The head of the jacket and all connecting pipes to the burette should be designed with a continuous upward slope with an air vent at the highest point so that no air can remain trapped in the jacket when it is filled with water.

1.5.4. The best way of checking the accuracy of the equipment is by the use of a calibrated cylinder. This cylinder must have been previously tested with an apparatus which is known to be correct so that its elastic expansion at the test pressure has been accurately determined. Each day before retesting this calibrated cylinder should be placed in the jacket and tested in order to verify the accuracy of the test equipment. If its expansion does not agree with the known value, the cause of the error should be determined and corrected before proceeding with the testing.

1.6. Instructions for Use of Calibrated Cylinder

1.6.1. A calibrated cylinder is a cylinder which has been carefully selected and stretched permanently by means of hydraulic pressure higher than the maximum pressure for which it is to be used.

1.6.2. After this is done the cylinder has the property of expanding the same volume for each pressure unit for which it has been calibrated and returning to its original volume when pressure is released. As long as the cylinder does not have a hydraulic pressure applied to it higher than its calibration, or does not change its wall thickness, due to rust or deterioration, it will maintain these properties.

1.6.3. It is to be considered as an instrument for measurement rather than just a cylinder.

1.6.4. A chart shall accompany the cylinder showing the volumetric expansion in cubic centimeters at the various pressures.

1.6.5. In order to check the gage, burette and all connections to a cylinder testing outfit, the cylinder is put into the jacket in the usual way; hydraulic pressure is applied until the volumetric expansion shown in the burrette is equal to the volumetric expansion as shown on the chart, for whatever pressure it is desired to test the outfit. For example, if it is desired to test a water jacket, gage and equipment at 3360 psig and the volumetric expansion for the cylinder at that pressure is shown on the chart as 157, the pressure should be applied until 157 cc expansion is shown on the burette. If the gage, burette and all other parts are accurate the pressure shown on the gage should be 3360 psig at this point. ICC requires gages and burettes to permit reading to accuracy of 1 per cent.

1.6.6. In using a calibrated cylinder with an equipment that has not been tested before, it is well not to make a test at a pressure near the maximum range of the cylinder, for the reason that should a gage read low, a higher pressure than was indicated would be applied to the cylinder and could change its calibration. It is well to check the elastic expansion as compared with the chart furnished with the cylinder, then check the pressure recorded with the expansion, rather than to run the pressure up first and check the expansion last.

1.6.7. When checking the accuracy of hydrostatic testing equipment, the calibrated cylinder should have a volume reasonably comparable to that of the cylinders to be tested.

1.7. Instruction for Care of Calibrated Cylinder

1.7.1. Do not use a calibrated cylinder as a capacitor cylinder to slow down pumping speed when cylinders are being tested or for any purpose other than calibrating or checking the accuracy of the test apparatus.

1.7.2. Leave water and the cylinder test spud in the cylinder between tests.

1.7.3. Do not store a calibrated cylinder outside, or in any location subjected to freezing temperatures. Calibrated cylinders should be kept preferably near test apparatus so that the temperature of the water in the cylinder and the test apparatus will remain approximately the same.

1.8. Common Errors Found in Water Jacket Cylinder Testing Equipment and Operation

1.8.1. Careless operation, in reading gages, burettes, and leveling the water accurately for all readings, disregarding small leaks, and general indifference to doing a precision job of testing.

1.8.2. The absence of a dead-weight tester, master gage, or calibrated cylinder to check gages and equipment.

1.8.3. Gages and burettes set at levels impossible to read correctly. Gages and zero water level of burette should be set 2 to 3 in. below eye level and close together.

1.8.4. Air relief valve not located at the extreme top of jacket to let out all air.

1.8.5. Jacket set too high, making it difficult to operate efficiently. Good operating height approximately 30 in. above floor.

1.8.6. Burette hose so light and installed in such a manner that sharp bends occur when burette is raised and lowered, causing changes in the volume of hose which causes errors in readings.

1.8.7. Jacket made of light material so that the water head pressure of burette will cause deflection in jacket wall.

1.8.8. Pressure valves not designed for hydraulic pressure, often being light and small. Impossible to completely close without excessive fatigue to the operator and breaking of stems, seats, etc.

1.8.9. Valves mounted so that operating strains and vibration are transmitted to pipe joints, causing small and annoying leaks.

1.8.10. Pressure piping, fittings and valves of a material too light to stand pressures, thereby causing leaks and breaks.

1.8.11. Valves installed so that pressure in the cylinder under test is caged under the valve, bonnet and stuffing box, instead of under the seat. Valve seats can be kept free from leak more easily than valve stuffing boxes.

1.8.12. Poor arrangement of gages, burettes, and operating valves, making them difficult to read and operate, causing errors and lack of precision in testing.

1.8.13. Water supply dirty or with a great variation in temperature.

1.8.14. Jackets not kept clean. Dirt compresses and also traps air. Organic matter ferments and causes gas bubbles.

1.9. Schematic Diagrams of Apparatus and Instructions for Operating

Schematic diagrams with instructions of four typical water jacket testing equipments in general use follow. Design and details of equipment to suit individual requirements can be adopted by the individual, but should follow these recommendations for accurate testing and safety.

1.9.1. The Water Jacket Leveling Burette Method for Testing Cylinders. Power Driven Pump. Pressure Connection to Cylinder Outside Water Jacket (see Fig. 1).

1.9.1.1. Scope. The water jacket leveling burette method of testing cylinders consists essentially of enclosing the cylinder in a water jacket and measuring the volume of water forced from the jacket upon application of pressure to the interior of the cylinder, and the volume remaining displaced upon release of the pressure. These volumes represent the total and permanent expansions of the cylinder, respectively. To measure them accurately, a movable burette calibrated in cubic centimeters is positioned to maintain the water level at a uniform height when taking readings.

1.9.1.2. Procedure.

(1) Remove valve from the empty cylinder and fill with water.

(2) Holding cylinder in valving vise, screw connection (C) into neck threads.

(3) Place cover (O) over the shoulder of the cylinder and tighten wing nut (V).

(4) Insert the cylinder in water jacket (B), clamp water jacket head (O) in place, and open petcock (P).

(5) Join cylinder connection (C) to coupling (D) on the pump line. (Up to this point, valves, I, J, K, L, M, Q and U should be kept closed.)

(6) Set zero of burette (T) at eye level adjustable reference point (Z).

(7) Open valve (L) and after water flows from petcock (P), close the petcock.

(8) When the water in burette (T) rises above zero point, close valve (L).

(9) Adjust water level in burette (T) to zero by draining through valve (K).

(10) Open valve (I) and close bypass valve (Q). When the cylinder pressure rises to $\frac{3}{4}$ ICC test pressure, close valve (I) and open bypass valve (Q).

(11) Examine apparatus for leakage (dropping pressure, falling water level in burette, or beads of water at screw threads in cylinder neck indicate leakage).

(12) If no leaks occur, open valve (I), close bypass valve (Q) and raise cylinder pressure to desired test pressure, closing valve (I) and opening bypass valve (Q).

(13) Maintain test pressure for 30 seconds and as much longer as may be necessary to secure complete expansion of the cylinder.

(14) Read the total expansion in burette (T) with water level flush with eye-level zero mark (Z).

(15) Release pressure through valve (J) and read the permanent expansion in burette (T) with water level flush with zero mark (Z).

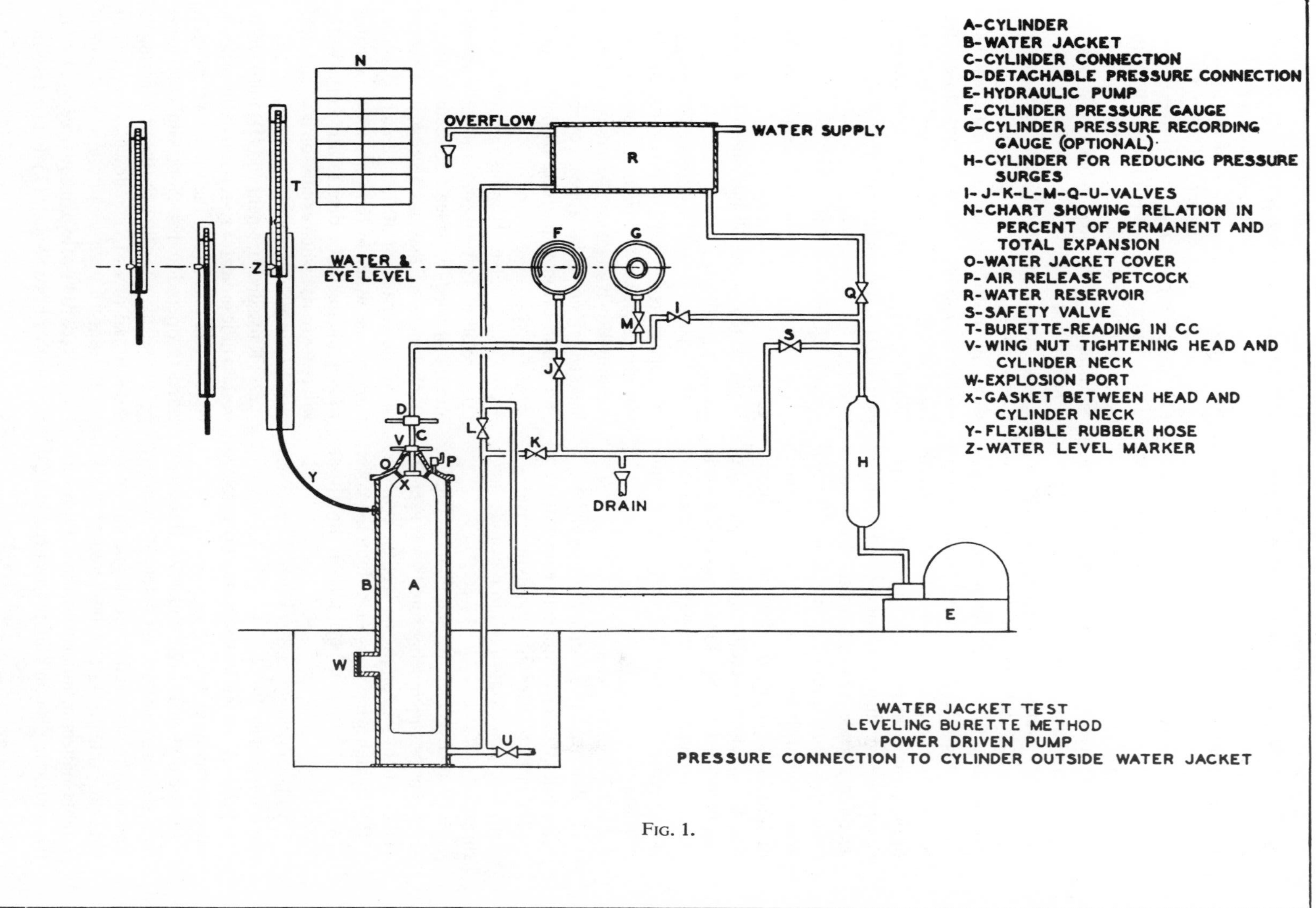
N
OVERFLOW
WATER SUPPLY
WATER & EYE LEVEL
DRAIN
A-CYLINDER
B-WATER JACKET
C-CYLINDER CONNECTION
D-DETACHABLE PRESSURE CONNECTION
E-HYDRAULIC PUMP
F-CYLINDER PRESSURE GAUGE
G-CYLINDER PRESSURE RECORDING GAUGE (OPTIONAL)
H-CYLINDER FOR REDUCING PRESSURE SURGES
I-J-K-L-M-Q-U-VALVES
N-CHART SHOWING RELATION IN PERCENT OF PERMANENT AND TOTAL EXPANSION
O-WATER JACKET COVER
P-AIR RELEASE PETCOCK
R-WATER RESERVOIR
S-SAFETY VALVE
T-BURETTE-READING IN CC
V-WING NUT TIGHTENING HEAD AND CYLINDER NECK
W-EXPLOSION PORT
X-GASKET BETWEEN HEAD AND CYLINDER NECK
Y-FLEXIBLE RUBBER HOSE
Z-WATER LEVEL MARKER
WATER JACKET TEST
LEVELING BURETTE METHOD
POWER DRIVEN PUMP
PRESSURE CONNECTION TO CYLINDER OUTSIDE WATER JACKET

Fig. 1.

1.9.2. The Water Jacket Rod Displacement Method for Testing Cylinders. Hand Pressure Pump. Pressure Connection to Cylinder Outside Water Jacket (see Fig. 2).

1.9.2.1. Scope. The water-jacket rod displacement method of testing cylinders consists essentially of enclosing the cylinder, suspended in a jacket vessel provided with necessary connections and attachments, and measuring the volume of water forced from the jacket upon application of pressure to the interior of the cylinder, and the volume remaining displaced upon release of the pressure. These volumes represent the total and permanent expansions of the cylinder, respectively. To measure them accurately, a displacement rod (H) and a graduated scale (T) are used. The rod is accurately and uniformly machined from a noncorrodible material such as brass, bronze, or stainless steel and the scale is so calibrated that the water displaced by the rod, per unit of length, in the tube (S), is read upon the scale in cubic centimeters. This operation is described in paragraphs 21 and 22 of the procedure following.

1.9.2.2. Procedure.

(1) Remove valve from *empty* cylinder.

(2) Fill cylinder completely with water.

(3) Place cylinder in the valving vise.

(4) Screw connection (C) into cylinder neck threads.

(5) Place cover (O) over shoulder of cylinder, as level as possible, being sure that gasket (X) makes a tight seat on the neck of the cylinder.

(6) Tighten wing nut (V), pulling the cover down on shoulder of cylinder as tightly as possible.

(7) Place cylinder and cover as a unit into the water jacket (B).

(8) Clamp water jacket head (O) in place and open petcock (P).

(9) See that gasket (X) makes an absolutely tight seat against the shoulder of cylinder. This may be observed through openings provided for this purpose in the cover.

(10) Connect cylinder to source of pressure by connecting union (D) to connection (C). (Up to this point valves I, J, K, L, M, R and U should be kept closed.)

(11) Lower displacement rod (H) in the brass tube container (S) until the top of the rod coincides with the zero mark on the scale (T) (see position 1, Fig. 2).

(12) Open valve (L) until water flows from petcock (P) and rises above the water level mark in the glass tube indicator (W).

(13) Close valve (L).

(14) Close petcock (P).

(15) Open valve (K) *slightly*, drawing off water in the tube containers until the water is exactly at water level mark then close valve. (See position 1, Fig. 2.) If water goes below water level mark on glass tube indicator, open petcock (P) and repeat operations 12, 13, 14 and 15.

(16) Open valve (M).

(17) Start clock mechanism of recording gage (G).

(18) Open valve (I) and pump pressure to about $\frac{3}{4}$ final test pressure.

(19) Close valve (I) and immediately examine apparatus for leaks. (Look through opening above gasket for leaks in cover.) Water receding slowly in glass tube indicator denotes leakage. If no leaks appear, open valve (I) and pump cylinder pressure to final requirement.

(20) Close valve (I) immediately to retain this pressure. Test pressure must be applied and *maintained* for at least 30 seconds and as much longer as may be necessary to secure complete expansion of the cylinder. Observations of the water in the glass tube indicator will enable this to be determined as the water will cease to rise when expansion is complete if pressure is maintained.

(21) Raise displacement rod (H) until water in glass tube indicator re-recedes to the original water level mark (i.e., the same level as indicated in operation 15). *Note:* Sufficient time should be taken to allow the water adhering to the rod and tubes to run down and collect at the bottom.

(22) Record total expansion in cc by reading the number of cc on scale (T) that corresponds

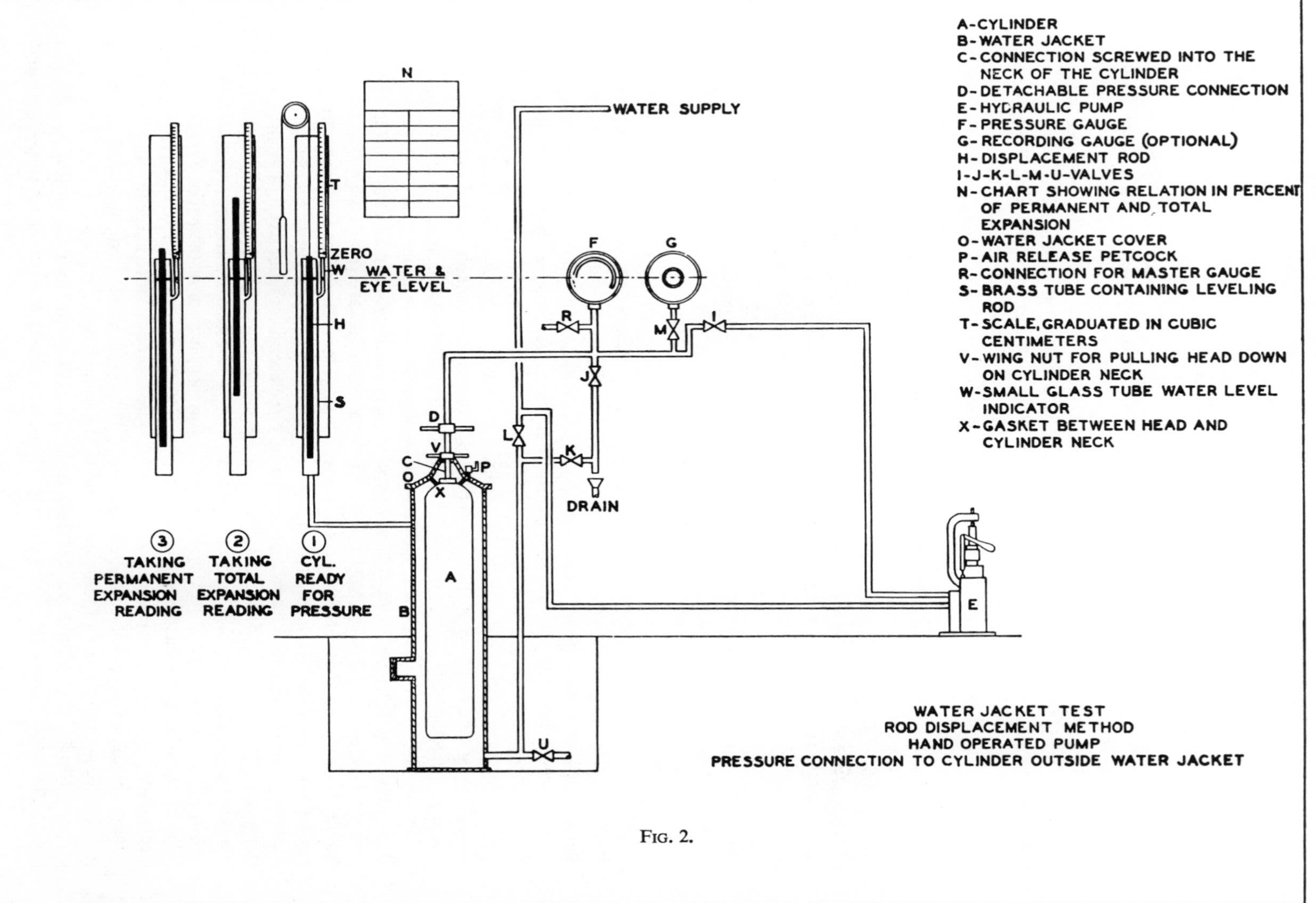
A-CYLINDER
B-WATER JACKET
C-CONNECTION SCREWED INTO THE NECK OF THE CYLINDER
D-DETACHABLE PRESSURE CONNECTION
E-HYDRAULIC PUMP
F-PRESSURE GAUGE
G-RECORDING GAUGE (OPTIONAL)
H-DISPLACEMENT ROD
I-J-K-L-M-U-VALVES
N-CHART SHOWING RELATION IN PERCENT OF PERMANENT AND TOTAL EXPANSION
O-WATER JACKET COVER
P-AIR RELEASE PETCOCK
R-CONNECTION FOR MASTER GAUGE
S-BRASS TUBE CONTAINING LEVELING ROD
T-SCALE, GRADUATED IN CUBIC CENTIMETERS
V-WING NUT FOR PULLING HEAD DOWN ON CYLINDER NECK
W-SMALL GLASS TUBE WATER LEVEL INDICATOR
X-GASKET BETWEEN HEAD AND CYLINDER NECK
WATER JACKET TEST
ROD DISPLACEMENT METHOD
HAND OPERATED PUMP
PRESSURE CONNECTION TO CYLINDER OUTSIDE WATER JACKET
WATER SUPPLY
DRAIN
ZERO
WATER & EYE LEVEL
① CYL. READY FOR PRESSURE
② TAKING TOTAL EXPANSION READING
③ TAKING PERMANENT EXPANSION READING

Fig. 2.

to the level of the top of the rod (H) (see position 2, Fig. 2). This reading should be written on recording gage (G) chart.

(23) Lower displacement rod (H) until the top of the rod coincides with zero on scale (T).

(24) Open valve (J) and release pressure on cylinder to zero as noted on gage (F).

(25) Raise displacement rod (H) until the water in the glass tube indicator recedes to the water level mark (see position 3, Fig. 2) the water level then being the same as it was when readings under steps 15 and 22 were taken.

(26) Record the permanent expansion by reading the number of cc on scale (T) corresponding to the level of the top of rod (H). This reading should be written on recording gage chart. (Observe same note as in step 21.)

(27) The per cent expansion is read from chart (N) or calculated and equals permanent expansion divided by total expansion.

(28) Record per cent expansion on recording gage chart.

(29) Determine elastic expansion by subtracting permanent expansion from total expansion, and record figure on recording gage chart between that for permanent expansion and the per cent expansion.

(30) If the per cent expansion as read from chart (N) exceeds 10 per cent the cylinder has failed to pass the test.

(31) If the *elastic expansion* indicates that the cylinder is thin walled it should be *scrapped.*

Note 1: Use of recording gage (G) is optional. If used, the chart should be kept as a permanent record and it is suggested that total expansion, permanent expansion, per cent expansion and elastic expansion (as indicated in operations 22, 26, 28 and 29) be noted on the chart. If recording gage is not used, a permanent record of these factors should be kept elsewhere.

Note 2: All piping should be sloped upward to avoid air pockets. Use of pipe fittings should be minimized. They should not be used on horizontal lines, if possible.

1.9.3. The Water Jacket Leveling Burette Method for Testing Cylinders. Hand Pressure Pump. Pressure Connection to Cylinder Outside Water Jacket (see Fig. 3).

1.9.3.1. Scope. The water-jacket leveling burette method of testing cylinders consists essentially of enclosing the cylinder, suspended in a jacket vessel provided with necessary connections and attachments, and measuring the volume of water forced from the jacket upon application of pressure to the interior of the cylinder, and the volume remaining displaced upon release of the pressure. These volumes represent the total and permanent expansions of the cylinder, respectively. To measure them accurately, a burette (T) and a flexible rubber hose (Y) are used. The burette will read the expansion directly in cubic centimeters when operated according to the procedures described in the following paragraphs 21 and 22.

1.9.3.2. Procedure.

(1) Remove valve from *empty* cylinder.

(2) Fill cylinder completely with water.

(3) Place cylinder in the valving vise.

(4) Screw connection (C) into cylinder neck threads.

(5) Place cover (O) over shoulder of cylinder, as level as possible, being sure that gasket (X) makes a tight seat on the neck of the cylinder.

(6) Tighten wing nut (V), pulling the cover down on shoulder of cylinder as tightly as possible.

(7) Place cylinder and cover as a unit into the water jacket (B).

(8) Clamp water jacket head (O) in place and open petcock (P).

(9) See that gasket (X) makes an absolutely tight seat against shoulder of cylinder. This may be observed through openings provided for this purpose in the cover.

(10) Connect cylinder to source of pressure by connecting union (D) to connection (C). (Up to this point valves I, J, K, L, M, R, and U should be kept closed.)

(11) Raise burette (T) until the zero of the scale coincides with a fixed water level mark

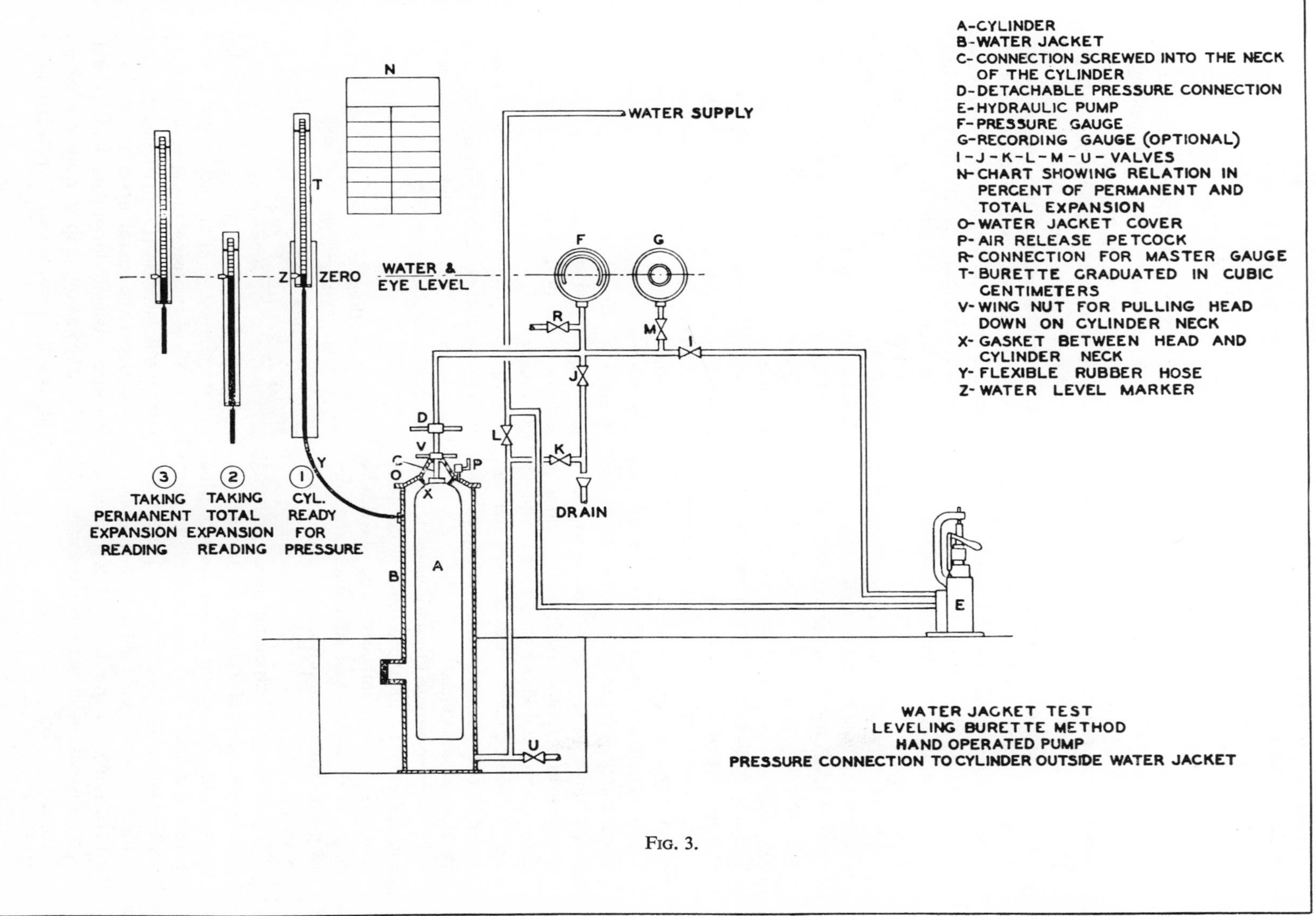
A-CYLINDER
B-WATER JACKET
C-CONNECTION SCREWED INTO THE NECK OF THE CYLINDER
D-DETACHABLE PRESSURE CONNECTION
E-HYDRAULIC PUMP
F-PRESSURE GAUGE
G-RECORDING GAUGE (OPTIONAL)
I-J-K-L-M-U-VALVES
N-CHART SHOWING RELATION IN PERCENT OF PERMANENT AND TOTAL EXPANSION
O-WATER JACKET COVER
P-AIR RELEASE PETCOCK
R-CONNECTION FOR MASTER GAUGE
T-BURETTE GRADUATED IN CUBIC CENTIMETERS
V-WING NUT FOR PULLING HEAD DOWN ON CYLINDER NECK
X-GASKET BETWEEN HEAD AND CYLINDER NECK
Y-FLEXIBLE RUBBER HOSE
Z-WATER LEVEL MARKER
WATER SUPPLY
WATER & EYE LEVEL
ZERO
DRAIN
1 CYL. READY FOR PRESSURE
2 TAKING TOTAL EXPANSION READING
3 TAKING PERMANENT EXPANSION READING
WATER JACKET TEST
LEVELING BURETTE METHOD
HAND OPERATED PUMP
PRESSURE CONNECTION TO CYLINDER OUTSIDE WATER JACKET

FIG. 3.

(Z) adjacent to the scale and at the approximate eye level of the operator.

(12) Open valve (L) until water flows from petcock (P) and rises above the water level marker (Z) adjacent to the burette (T).

(13) Close valve (L).

(14) Close petcock (P).

(15) Open valve (K) *slightly*, drawing off water in the tube containers until the water is exactly at water level mark (Z) then close valve (see position 1, Fig. 3). If water goes below water level mark, open petock (P) and repeat operations 12, 13, 14 and 15.

(16) Open valve (M).

(17) Start clock mechanism of recording gage (G).

(18) Open valve (I) and pump pressure in cylinder to about $\frac{3}{4}$ final test pressure.

(19) Close valve (I) and immediately examine apparatus for leaks. (Look through opening above gasket for leaks in cover.) Water receding slowly in glass tube indicator denotes leakage. If no leaks appear, open valve (I) and pump cylinder pressure to final requirement.

(20) Close valve (I) immediately to retain this pressure. Test pressure must be applied and *maintained* for at least 30 seconds and as much longer as may be necessary to secure complete expansion of the cylinder. Observations of the water in the glass burette will enable this to be determined as the water will cease to rise when expansion is complete if pressure is maintained.

(21) Lower burette (T) until water level in burette is flush with the original water level mark (Z). *Note:* Sufficient time should be taken to allow the water adhering to the sides of the burette to run down and collect at the bottom.

(22) Record total expansion in cc by reading the number of cc of water in the burette (T). This reading should be written on recording gage (G) chart.

(23) Open valve (J) and release pressure on cylinder to zero as noted on gage (F).

(24) Raise burette until water level in burette is flush with the original water level mark (Z).

(25) Record the permanent expansion by reading the number of cc of water in the burette. This reading should be written on recording gage chart. (Observe same note as in operation 21.)

(26) The per cent expansion is read from chart (N) or calculated and equals permanent expansion divided by total expansion.

(27) Record per cent expansion on recording gage chart.

(28) Determine elastic expansion by subtracting permanent expansion from total expansion, and record figure on recording gage chart between that for permanent expansion and the per cent expansion.

(29) If the per cent expansion as read from chart (N) exceeds 10 per cent, the cylinder has failed to pass the test.

(30) If the *elastic expansion* indicates that the cylinder is thin-walled it should be *scrapped.*

Note 1: Use of recording gage (G) is optional. If used, the chart should be kept as a permanent record and it is suggested that total expansion, permanent expansion, per cent expansion and elastic expansion (as indicated in steps 22, 25, 27 and 28) be noted on the chart. If recording gage is not used, a permanent record of these factors should be kept elsewhere.

Note 2: All piping should be sloped upward to avoid air pockets. Use of pipe fittings should be minimized. They should not be used on horizontal lines, if possible.

1.9.4. The Water Jacket Leveling Burette Method for Testing Cylinders. Power Driven Pump. Pressure Connection to Cylinder Inside Water Jacket (see Fig. 4).

1.9.4.1. Scope. The water jacket leveling burette method of testing cylinders consists essentially of enclosing the cylinder in a water jacket and measuring the volume of water forced from the jacket upon application of pressure to the interior of the cylinder, and the volume remaining displaced upon release of the pressure. These volumes represent the total

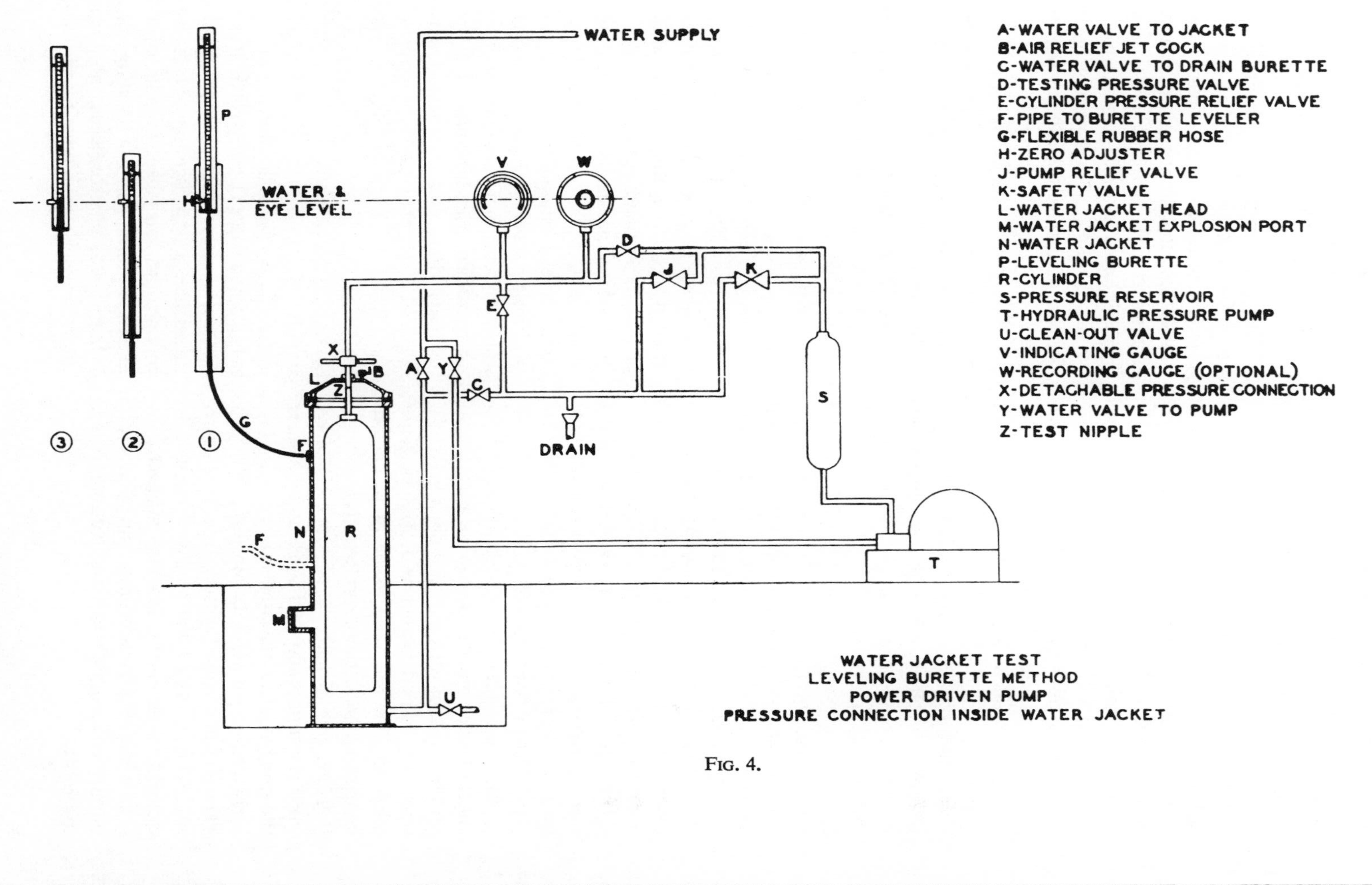
WATER SUPPLY
A-WATER VALVE TO JACKET
B-AIR RELIEF JET COCK
C-WATER VALVE TO DRAIN BURETTE
D-TESTING PRESSURE VALVE
E-CYLINDER PRESSURE RELIEF VALVE
F-PIPE TO BURETTE LEVELER
G-FLEXIBLE RUBBER HOSE
H-ZERO ADJUSTER
J-PUMP RELIEF VALVE
K-SAFETY VALVE
L-WATER JACKET HEAD
M-WATER JACKET EXPLOSION PORT
N-WATER JACKET
P-LEVELING BURETTE
R-CYLINDER
S-PRESSURE RESERVOIR
T-HYDRAULIC PRESSURE PUMP
U-CLEAN-OUT VALVE
V-INDICATING GAUGE
W-RECORDING GAUGE (OPTIONAL)
X-DETACHABLE PRESSURE CONNECTION
Y-WATER VALVE TO PUMP
Z-TEST NIPPLE
WATER &
EYE LEVEL
DRAIN
WATER JACKET TEST
LEVELING BURETTE METHOD
POWER DRIVEN PUMP
PRESSURE CONNECTION INSIDE WATER JACKET

Fig. 4.

and permanent expansions of the cylinder, respectively. To measure them accurately, a movable burette calibrated in cubic centimeters is positioned to maintain the water level at a uniform height when taking readings.

1.9.4.2. Procedure.

(1) Select the proper size burette for the cylinder to be tested and place in burette holder. Adjust rubber connection tight so it will not leak.

(2) With burette in position (1) set zero adjuster (H) with zero of burette.

(3) Select test nipple (Z) to fit valve inlet connection thread of cylinder and screw into the jacket head with pipe wrench; be sure it is *tight*.

(4) Fill the cylinder to be tested with water. Clamp cylinder in vise and screw head and test nipple into the cylinder thread *tight*, using jacket head as wrench.

(5) Place cylinder in jacket by hoist and clamp head tight.

(6) Open valve (A) until water comes out of jet cock (B). Close (B). Fill hose and burette to zero or above. Close (A).

(7) Adjust water level in burette to zero by opening valve (C) with burette in position (1), with valves (E) and (J) open, (D) closed.

(8) Check for leaks. If water falls below zero in burette, jacket or piping is leaking. If water rises in burette, valve (A) is leaking.

(9) Close valves (E) and (J).

(10) Open valve (D) and pump to approximately 75 per cent test pressure (not over 90 per cent). Close valve (D) and open valve (J). Check for leaks indicated by drop in pressure on gage. Pressure leaks inside of jacket are indicated by water rising in burette.

(11) If free of leaks, close valve (J), open valve (D) and pump to desired test pressure. Close valve (D); open valve (J).

(12) Lower burette to position (2) with water in burette level with zero mark on zero adjuster (H). Hold pressure in cylinder 30 seconds or as much longer as necessary to assure complete expansion of the cylinder.

(13) Take total expansion reading in cc, and record.

(14) Release pressure on cylinder by opening valve (E).

(15) Raise burette to position (3) with water in burette level with zero mark on zero adjuster (H).

(16) Take permanent expansion reading in cc and record.

(17) Calculate per cent permanent expansion by dividing permanent expansion by total expansion and record. If over 10 per cent, cylinder fails under ICC regulations.

(18) Example: Total expansion: 166 cc; permanent expansion 3.0 cc; per cent permanent expansion, $\frac{3.0}{166.0} = 1.8$ per cent.

Elastic expansion is equal to total expansion minus permanent expansion (166 cc − 3.0 cc = 163 cc).

(19) *If elastic expansion* indicates that cylinder is thin walled it should be *scrapped.*

Note: All readings of the water in the burette must be taken with the water at the same level when in position (1), (2) and (3) to eliminate errors due to compression of entrapped air in jacket if any is present. Water in the cylinder, jacket, hose and burette should be at approximately the same temperature during test to eliminate errors in burette readings caused by expansion or contraction. When jacket is installed some distance from the burette leveler, and pipe (F) is used, install pipe (F) sloping upward and free from pockets which will entrap air; bends are preferable to pipe fittings. Flexible hose (G) should be heavy-wall hose, and installed so that the bending action when placed in positions (1), (2) and (3) will not change the interior volume. Install so that it will operate on long, easy curve. It is necessary to wash out jacket occasionally. Keep it free of dirt and organic matter. Valve (J) is to relieve pressure on the pump when valve (D) is closed. It is obvious that valves (D) and (J) cannot both be left closed and pump running. By releasing the pressure through valve (J), pump can be left running continuously when testing cylinders—if desired. Relief valve (K) is only intended to relieve pressure on pump if valve (J) is left closed, when valve (D) is closed. It is not intended to operate continuously. Keep long full threads of test nipple, in jacket head, in good condition. Nipple can be cut off and rethreaded when worn.

2. DIRECT EXPANSION METHOD

2.1. Scope

2.1.1. This method may be used when equipment and operation is approved by the Bureau of Explosives. With this method the interior volumetric expansion is determined whereas the water jacket method determines the exterior volumetric expansion.

2.2. Inspection of Cylinders

2.2.1. Before cylinders are retested under the periodic retest required by ICC regulations:

2.2.1.1. Valves should be removed and careful internal inspection made with suitable light. Scale and sludge deposits (if any) should be removed and if seriously corroded, cylinder shall be condemned. To avoid a possible flash, cylinders used in flammable gas service should be thoroughly purged before interior inspection light is inserted.

2.2.1.2. Exterior inspection, including the bottom, should be carefully made for corrosion, dents, arc or torch burns, and physical deformation. Seriously corroded cylinders shall be condemned. Cylinders dented, arc or torch burnt, or physically damaged so as to weaken the cylinder appreciably shall be condemned.

2.2.2. Note: An empty cylinder before recharging should be examined for physical damage and exterior corrosion, followed by hammer test. Those cylinders which have a dull or peculiar ring, when lightly tapped with a suitably sized hammer, should be carefully examined and if necessary, valves removed and interior examined with suitable light and retested before charging.

2.3. Requirements for Accurate Testing

2.3.1. The important element is the ability to accurately determine the total, permanent and per cent permanent expansion of a cylinder as required by ICC specifications and regulations. The average wall thickness can also be determined and used to determine loss from corrosion or wall deterioration.

2.3.2. For the direct expansion method the following conditions are absolutely essential:

2.3.2.1. The apparatus must be so arranged that all air can be eliminated from the apparatus and cylinder under test.

2.3.2.2. When size, location or form of cylinder, such as concave head and bottom make it difficult to vent all air, equipment for removing air and testing for air must be provided. The methods and apparatus applicable to all conditions are too varied and exacting to be included in this statement. Properly trained personnel should be employed.

2.3.2.3. A dead-weight testing apparatus must be used to verify the accuracy of pressure gages. A master gage checked by a deadweight tester at frequent intervals would be an acceptable substitute, though less desirable.

2.3.2.4. The pressure gages must agree with a dead-weight testing apparatus to within $\frac{1}{2}$ per cent at test pressure. Gages must have zero stop pin removed or placed $\frac{1}{2}$ inch below zero, and a true zero marked on the dial.

2.3.2.5. An accurate scale to determine weight of water in cylinder.

2.3.2.6. An accurate thermometer to determine the temperature of water in the cylinder.

2.3.2.7. An accurate and authoritative chart of the compressibility of water at temperature and pressure of water in cylinder under test. (Suitable liquid such as kerosene may be used in place of water with proper precautions and proper compressibility chart for liquid used.)

2.3.2.8. A reservoir graduated in cc and accurate to within a 1 per cent reading, arranged so that water required to pressurize the filled cylinder, and water expelled from cylinder when depressurized, can be accurately determined.

2.3.2.9. Water in reservoir pump and connections must be the same temperature as water in cylinder.

2.3.2.10. Pressure valves, piping and connections to gages and cylinder must be of heavy construction, short as possible and of as small volume as practical, and must be absolutely tight and free of leaks while cylinder is under test. (Elastic expansion and compressibility of water in piping to cylinder under pressure, is an error in the expansion of the cylinder under test, and should be determined for various test conditions.)

2.3.2.11. Any internal pressure applied previous to the test pressure shall not exceed 90 per cent of test pressure. If due to failure of the test apparatus, the test pressure cannot be maintained, the test may be repeated at a pressure increased by 10 per cent, or 100 psig, whichever is the lower value. Test pressure must be maintained for at least 30 seconds and as much longer as may be necessary to secure complete expansion of the cylinder.

2.4. How to Use Results of Tests

2.4.1. The tests will determine the amount of water in cc forced into the filled cylinder to pressurize the cylinder to desired test pressure; the amount of water in cc expelled from the cylinder when cylinder is depressurized. With the weight and temperature of water in cylinder known, the water volume change due to the compressibility of the water can be calculated, and the total and permanent expansion determined. The total expansion minus the permanent expansion is the elastic expansion and can be used for wall thickness determinations.

Compressibility of Water

$C = FWP$

C = volume of water forced into cylinder due to compressibility of water in cc.

F = compressibility factor from Bureau of Explosives curves, Fig. 6, page 267.

W = weight of water in cylinder at test pressure, lb.

P = test pressure, psig.

1 lb water equals 454.5 cc

1 cu in. water equals 16.387 cc

1 lb water equals 27.737 cu in.

Example 1

Test pressure	3360 psig
Weight of water in cylinder zero pressure	251 lb
Temperature of water	60 F
Water forced into cylinder to raise pressure to 3360 psig	1745 cc
Weight of water added to cylinder to pressurize 1745 cc ÷ 454.5 equals	3.8 lb
Total weight of water in cylinder at 3360 psig 251 + 3.8 equals	254.8 lb
Water expelled from cylinder to depressurize	1698 cc
Permanent expansion 1745 cc − 1698 cc equals	47 cc
Compressibility factor for water at 3360 psig 60 F	.001425
Volume of water forced into cylinder due to compressibility of water in cylinder at 3360 psig and 60 F. cc = .001425 × 254.8 × 3360 equals	1219.98 cc
Total expansions 1745 − 1219.98 equals	525 cc
Per cent permanent expansion 47 ÷ 525 equals	8.95%
Elastic expansion 525 − 47 equals	478 cc

Example 2

Test pressure	500 psig
Weight of water in cylinder zero pressure	1657 lb
Temperature of water	73 F
Water forced into cylinder to raise pressure to 500 psig	2736 cc
Weight of water added to cylinder to pressurize 2736 cc ÷ 454.5 equals	6 lb
Total weight of water in cylinder at 500 psig 1657 + 6 equals	1663 lb
Water expelled from cylinder to depressurize	2667.5 cc

Permanent expansion 2736 − 2667.5	68.5 cc
Compressibility factor for water at 500 psig 73 F	.001439
Volume of water forced into cylinder due to compressibility of water in cylinder at 500 psig and 73 F cc = .001439 × 1663 × 500	1196.5 cc
Total expansion 2736 − 1196.5	1539.5 cc
Per cent permanent expansion 68.5 ÷ 1539.5	4.4%
Elastic expansion 1539.5 − 68.5	1471 cc

Note: The expansion of the pipe and fittings and the compressibility of the water in the portion of the pipe between valve (F) and valve (D) and the cylinder (H) (page 266) has been considered to be zero.

2.4.1.1. The elastic expansion as determined by the retest of a cylinder by the direct expansion method can be used for calculating average wall in the same way as calculated under the water jacket method. The direct expansion method determines interior volumetric expansion whereas the water jacket method determines exterior volumetric expansion.

2.4.1.2. Cylinder owners using the direct expansion method of testing should establish their own basis for determining average wall thickness by elastic expansion. The methods and calculations applicable to all conditions are too varied and too exacting to be covered in this statement. Properly trained personnel should be employed.

2.4.1.3. A cylinder must be condemned if it leaks, or fails to pass the test requirements prescribed by the ICC regulations.

2.4.1.4. Cylinders should be condemned when the wall stress at service pressure exceeds a safe permissible stress in the cylinder.

2.5. Basis of Recommendations

2.5.1. A cylinder can be tested by this method for total, permanent and elastic expansion. Care must be exercised to make certain that all parts of the equipment are functioning properly, that cylinder and equipment is free of all air. Temperature measurements and calculations for water compressibility, the measurements of water used to pressurize and depressurize the cylinder must be accurately determined. It is exceedingly important that no leaks in the high or low-pressure parts of the apparatus occur as these will cause errors in the calculations, which must be carefully made.

2.6. Schematic Diagram of Apparatus and Instructions for Operating

2.6.1. Chart for compressibility of water and schematic diagram with instructions for operating a typical direct expansion testing equipment follow. Design and details of equipment to suit individual requirements can be adapted by the individual, but should follow these recommendations for accurate testing and safety.

2.6.2. Procedure (see Fig. 5).

(1) All valves closed, reservoir (N) filled with water.

(2) Weigh cylinder (empty).

(3) Fill cylinder with water, from reservoir (N) or from other source. Water must be at the same temperature as cylinder and room.

(4) Weigh cylinder (filled).

(5) Measure temperature of water in cylinder, record temperature in degrees F.

(6) Determine weight of water in cylinder by subtracting weight of empty cylinder from weight of filled cylinder, call it W_1, in lb and record.

(7) Screw test stem with valve (G) into cylinder (tight).

(8) Open valve (A).

(9) Open valve (C) and fill reservoir (J) half full of water.

(10) Close valve (C).

(11) Open valve (B) until all air is removed, close (B).

(12) Connect cylinder to apparatus with pressure connections (O).

(13) Open valves (G) and (D).

(14) Start pump and fill pump, piping and

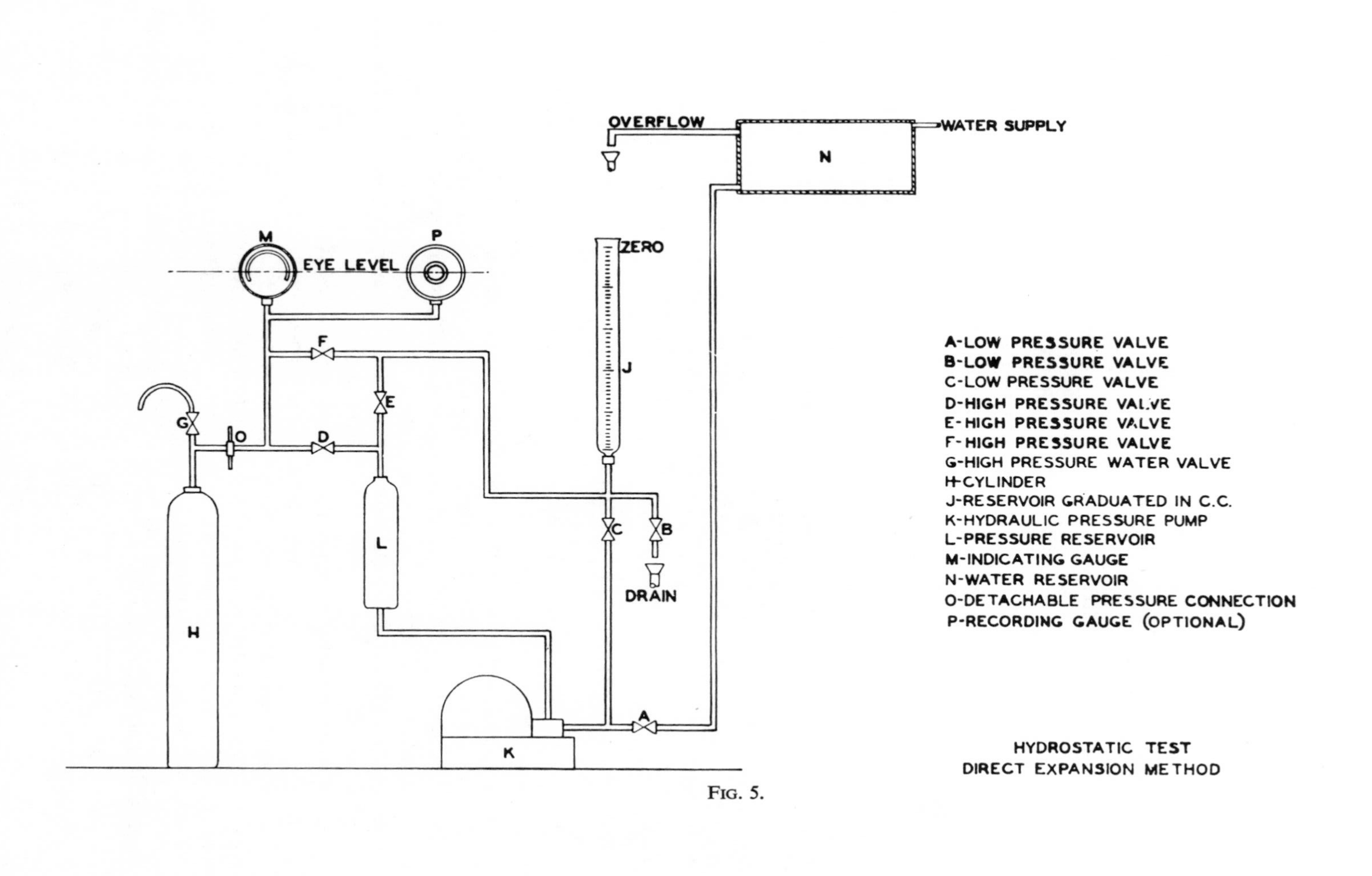

OVERFLOW
N
WATER SUPPLY
M
EYE LEVEL
P
ZERO
J
F
E
G
O
D
L
H
C
B
DRAIN
K
A
A-LOW PRESSURE VALVE
B-LOW PRESSURE VALVE
C-LOW PRESSURE VALVE
D-HIGH PRESSURE VALVE
E-HIGH PRESSURE VALVE
F-HIGH PRESSURE VALVE
G-HIGH PRESSURE WATER VALVE
H-CYLINDER
J-RESERVOIR GRADUATED IN C.C.
K-HYDRAULIC PRESSURE PUMP
L-PRESSURE RESERVOIR
M-INDICATING GAUGE
N-WATER RESERVOIR
O-DETACHABLE PRESSURE CONNECTION
P-RECORDING GAUGE (OPTIONAL)
HYDROSTATIC TEST
DIRECT EXPANSION METHOD

FIG. 5.

high pressure reservoir (L) from reservoir (N). When water free of air comes out of overflow pipe on top of cylinder, close (G).

(15) Open valve (F). When water free of air rises in reservoir (J), close (F).

(16) Open valve (E). When water free of air rises in reservoir (J), close (E). Allow pressure to build up not over 90 per cent of test pressure.

(17) Open valve (G) to release pressure and purge all air out of top of cylinder and test stem.

(18) Stop pump and close valve (G).

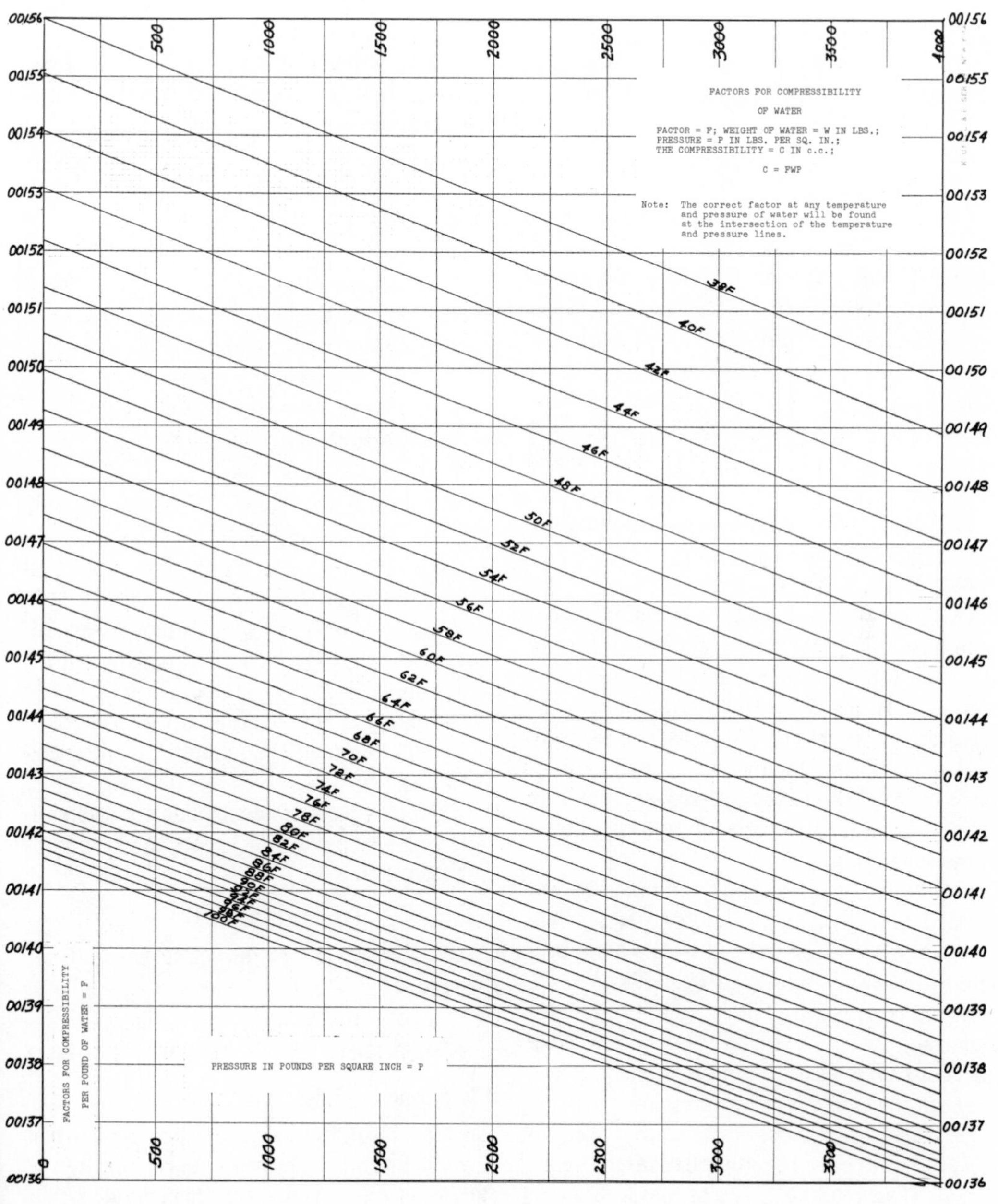

FIG. 6.

(19) Repeat if necessary to expel all air from top of cylinder and test stem.

(20) When operating new equipment for the first time break gage connection and purge all air from pipe connecting gage to equipment.

(21) Close valve (D).

(22) Close valve (A).

(23) Open valve (C).

(24) Open valve (E).

(25) Start pump and circulate water from reservoir (J) through pump, reservoir (L), and valve (E), back to (J).

(26) Stop pump.

(27) Close valve (E).

(28) Fill reservoir (J), and adjust water level to zero with valves (A) and (B).

(29) Close valves (A) and (B) with water at zero mark. Cylinder and system are now full of water free of air and ready for test, valve (C) open and all other valves closed.

(30) Open valve (D), start pump and raise pressure to approximately 75 per cent of test pressure, not over 90 per cent. Stop pump and check for leaks.

(31) If free of leaks, start pump and pressurize cylinder to test pressure.

(32) Close valve (D).

(33) Open valve (E) and stop pump; water compressed in pump and system will return to reservoir (J). Test pressure will be retained in cylinder. Hold pressure in cylinder for 30 seconds, or as much longer as necessary to assure complete expansion of cylinder. If pressure gage shows loss of pressure, pressure recessions must be made up until cylinder stops expanding.

(34) When cylinder has completely expanded take reading in cc of water level in reservoir (J), record and call it X. This represents total water added to cylinder to pressurize to test pressure and is equal to the total expansion of the cylinder plus the volume added to take care of compressibility of the water in the cylinder at test pressure.

(35) Depressurize the cylinder by opening valve (F); water will be returned to reservoir (J).

(36) Take reading in cc of water level in reservoir (J), and record. Call it P. E. This represents the permanent expansion of the cylinder.

(37) Determine the weight of water added to pressurize the cylinder by dividing cc recorded X in step 34 by 454.5, the number of cc in 1 lb of water. Call it W_2.

(38) Determine total weight of water in cylinder at test pressure by adding W_1, step 6, to W_2, step 37. Call it W and record.

(39) Determine compressibility factor from compressibility curves for test pressure and temperature recorded in step 5.

(40) Calculate the volume of water added to take care of compressibility of water in cylinder at test pressure by formula:

$$C = FWP$$

C = volume of water forced into cylinder due to compressibility of water in cc

F = compressibility factor from curve for test temperature and pressure

W = weight of water in cylinder at test pressure, determined, step 38

P = test pressure psig.

(41) Determine total expansion of the cylinder by subtracting cc as determined in step 40 from X as determined in step 34. Call it T.E. and record.

(42) Calculate per cent permanent expansion by dividing permanent expansion (P.E.) found in step 36 by total expansion (T.E.) found in step 41, and record. If over 10 per cent, cylinder fails under ICC regulations and specifications.

(43) Elastic expansion equals total expansion (T.E.) step 41, minus permanent expansion (P.E.) step 36.

$$\text{T.E.} - \text{P.E.} = \text{E.E.}$$

(44) Record all readings and calculations for record.

3. PRESSURE RECESSION METHOD

3.1. Scope

3.1.1. This method may be used when equipment and operation is approved by the Bureau of Explosives for testing cylinders that require a test pressure of 2000 psig or more.

The method consists essentially in subjecting the cylinder to the required hydrostatic test pressure, then to immediately cut off further pressure supply, and to observe for a reasonable period whether or not there occurs a recession of the pressure in the cylinder. If the pressure does not recede this is positive proof that at the test pressure the cylinder does not show any permanent expansion. Should the pressure recede (not due to leakage), the amount of recession can be utilized in certain calculations to determine whether a permanent expansion occurs in excess of the amount allowed by ICC regulations.

3.2. Inspection of Cylinders

3.2.1. Before cylinders are retested under the periodic retest required by ICC regulations:

3.2.1.1. Valves should be removed and careful internal inspection made with suitable light. Scale and sludge deposits (if any) should be removed and if seriously corroded, the cylinder shall be condemned. To avoid a possible flash, cylinders used in flammable gas service should be thoroughly purged before the interior inspection light is inserted.

3.2.1.2. External inspection, including the bottom, should be carefully made for corrosion, dents, arc or torch burns, and physical deformation. Seriously corroded cylinders shall be condemned. Cylinders dented, arc or torch burnt, or physically damaged so as to weaken the cylinder appreciably shall be condemned.

3.2.2. Note: An empty cylinder before recharging should be examined for physical damage and exterior corrosion, followed by hammer test. Those cylinders which have a dull or peculiar ring, when lightly tapped with a suitably sized hammer, should be carefully examined and if necessary, valves removed and interior examined with suitable light and retested before charging.

3.3 Requirements for Accurate Testing

3.3.1. The important element is the ability to accurately determine the permanent expansion of a cylinder as required by the ICC specifications and regulations. Hence, cylinders which are corroded or otherwise too thin can be determined.

3.3.2. For the pressure recession method the following conditions are absolutely essential:

3.3.2.1. The apparatus must be so arranged that all air can be eliminated from the apparatus and cylinder under test.

3.3.2.2. The apparatus must be so arranged that careful inspection, including the bottom, can be made while cylinder is under test pressure.

3.3.2.3. A dead-weight testing apparatus must be used to verify the accuracy of pressure gages. A master gage checked by a deadweight tester at frequent intervals would be acceptable substitute, though less desirable.

3.3.2.4. The pressure gages must agree with a dead-weight testing apparatus to within $\frac{1}{2}$ per cent at test pressure. Gages must have Zero Stop Pin removed or placed $\frac{1}{2}$ in. below zero and a true zero marked on the dial.

3.3.2.5. The pressure gages should be graduated to at least 4000 psig with subdivisions of not more than 25 psig.

3.3.2.6. The recording gage should be equipped with clock making one revolution in one hour and stop and start device. Chart should be not less than 10 in. in diameter graduated over a range of 2000 to 4000 psig with 20 psig subdivisions.

3.3.2.7. An accurate scale to determine weight of water in cylinder.

3.3.2.8. An accurate thermometer to determine the temperature of water in the cylinder.

3.3.2.9. An accurate and authoritative chart of the compressibility of water at temperature and pressure of water in cylinder under test.

3.3.2.10. A measure graduated in cc and readable to an accuracy of 1 per cent, arranged so that water expelled from cylinder, when depressurized, can be accurately determined.

3.3.2.11. Water in the reservoir pump and connections must be the same temperature as the water in cylinder. The temperature of the water in the cylinder and the temperature of the cylinder shall be the same.

3.3.2.12. Pressure valves, piping and connections to gages and cylinder must be heavy construction, short as possible and of as small volume as practical, of such strength as to stand 4000 psig internal pressure without leakage, and must be absolutely tight and free of leaks while the cylinder is under test.

3.3.2.13. Sufficient capacity of the pump or capacitor to increase the pressure from 90 per cent of test pressure to test pressure in less than 5 seconds. The pressure medium must be a liquid because compressed air or other gases will not give correct results.

3.4. How to Use Results of Test

3.4.1. First Test.

The first test will determine whether or not any permanent expansion has occurred within the cylinder. Permanent expansion, provided there are no leaks in the test system, is indicated by recession of the pressure gage pointers.

3.4.1.1. If no pressure recession occurs, then there has been no permanent expansion and the cylinder has passed the test.

3.4.1.2. If the pressure recession has not exceeded 25 psig (15 psig for cylinders less than 5 in. outside diameter) then the cylinder has passed the test.

3.4.1.3. If the pressure recession has exceeded 25 psig (15 psig for cylinders less than 5 in. outside diameter), then the cylinder has not passed the test. The pressure must be released, and the cylinder set aside for at least 24 hours and then subjected to a second test as described later.

3.4.1.4. If on account of leakage in apparatus or other reasons it is believed that the results of the first test are not reliable, then further tests may be made. These must be made in the same manner as the first test, but each successive test must be made at a test pressure of at least 100 psig greater than that used in the last preceding test.

3.4.2. Second Test.

The second test of the cylinder must be made at a test pressure at least 100 psig higher than the previous test pressure applied and must be made in a manner to accurately measure the total expansion under test pressure and the permanent expansion resulting therefrom. The permanent expansion must not exceed 10 per cent of the total expansion. The water jacket volumetric method can be used to calculate the elastic and permanent expansion at required test pressure, or the following method can be used to determine the pressure recession as an indication of the permanent expansion of the cylinder at test pressure.

3.4.2.1. The allowable pressure recession for varying values of e' is shown in the following table:

e'*	Allowable Pressure Recession (psig)
.01	11
.02	21
.03	31
.04	40
.05	48
.06	56
.07	64
.08	71
.09	78
.10	84
.11	90
.12	96
.13	102
.14	107
.15	112
.16	117
.17	122

*The designation e' is the elastic expansion per psig of pressure in terms of hundred thousandths of the volume capacity of the cylinder.

Calculate the value of e' from the following formula:

$$e' = \frac{100{,}000\, e}{V}$$

where

e = elastic expansion of cylinder per psig of pressure

V = volume of cylinder (in cubic centimeters)

Note: 1 lb water at 60 F equals 454.5 cc.
1 cu in. equals 16.387 cc.
1 lb water at 60 F equals 27.737 cu in.

The elastic expansion of the cylinder per psig of pressure (e), which is in cubic centimeters is determined by using the following formula:

$$e = \frac{R - Vpm}{p + p^2m}$$

where

E = total elastic expansion of the cylinder at pressure p.

V = volume of cylinder (in cubic centimeters).

p = pressure applied (1500 psig or 2500 psig as the case may be).

m = factor of compressibility of water as determined from the following table:

Temperature of Water (°C)	m (1500 psig)	m (2500 psig)
0	.000003469	.000003423
2	.000003456	.000003412
4	.000003437	.000003393
6	.000003414	.000003370
8	.000003386	.000003341
10	.000003353	.000003307
12	.000003313	.000003266
14	.000003270	.000003224
16	.000003231	.000003185
18	.000003193	.000003151
20	.000003160	.000003118
22	.000003128	.000003088
24	.000003097	.000003060
26	.000003068	.000003033
28	.000003041	.000003008
30	.000003016	.000002985

R = amount of water forced into cylinder by the pressure applied (in cubic centimeters). This may be found as follows: Raise pressure in cylinder to p psig, close valve (F), release pressure by opening valve (G), and measure accurately in cubic centimeters the amount of water expelled through valve (G). This will be R. Then as this represents the elastic expansion of the cylinder plus the compressibility of the water we will have the relation $R + V = (V + E)\ (1 + pm)$ from which the above formula is deduced. E is equal to $e \times p$.

3.4.2.2. The only condition under which further tests may be made after the conclusion of the second test is when leakage in the apparatus has occurred during test. Each subsequent test must be made at a pressure at least 100 psig higher than the last preceding test and in the same manner as the second test except that only $\frac{1}{2}$ of the allowable pressure recession as determined from the cylinder's value of e is allowed to be utilized in determining whether the cylinder passes the test.

3.5. Basis of Recommendations

3.5.1. A cylinder can be tested by this method of pressure recession. Care must be exercised to make certain that there are no leaks in the test system and that measurements of cylinder volume and temperature of water in the test system are measured accurately. Care must also be exercised in measuring the amount of water R expelled from the cylinder when it is depressurized, and in determining the compressibility of water, or errors will occur in the calculations of the allowable pressure recession.

3.6. Schematic Diagram of Apparatus and Instructions for Operating

3.6.1. A schematic diagram with instructions of two methods of test follow. Refer to Fig. 6, p. 267. Design and details of equipment to suit individual requirements can be adopted by the individual but should follow these recommendations for accurate testing and safety.

3.6.1.1. No pressure in excess of 90 per cent test pressure shall be applied to the cylinder previous to the tests.

3.6.1.2. Vessel (B) serves to retain air or gases coming through pressure line. It need not have an inside diameter of more than 2 in. but should have at least 1 qt capacity. It must be of such strength as to stand 4000 psig internal pressure without permanent expansion.

3.6.1.3. Valves (F) and (G) must be placed as shown close to the connection to the cylinder.

3.6.1.4. Valve (D) is for the purpose of releasing the air which will accumulate in vessel (B) and this should frequently be done.

3.6.1.5. The pressure may be applied either directly from a force pump or from an accumulator. The latter is preferable. If a pump is used directly it must be capable of raising the pressure from 2700 lb to 3000 lb within 5 seconds. The pressure medium must be a liquid as compressed air or other gases will not give correct results.

3.6.2. Procedure (see Fig. 7).

3.6.2.1. Preliminary Preparations.

(1) Cylinder, apparatus, and all connections must be filled with water free of air.

(2) Valves (D) and (K) must be closed.

(3) Valves (C), (E), (H), (J) and (F) must be open.

(4) Close valve (F), raise pressure to about 500 psig and open valve (D). Continue this until water free of air flows from valve (D). Then close valve (D) and open valve (F).

3.6.2.2. First Test.

(1) Attach cylinder with valve (G) closed.

(2) Raise pressure to 500 psig. Close valve (E) and open valve (G). Repeat as necessary until water free from air bubbles flows from valve (G).

(3) Open valve (E) and close valve (G). Start the clockwork in recording gage (M). Raise pressure to about 50 psig less than 90 per cent of the required test pressure and close valve (E).

(4) Examine for leaks. The pressure as indicated by recording gage (M) should remain steady without dropping at all. If leaks occur or other contingencies arise, release the pres-

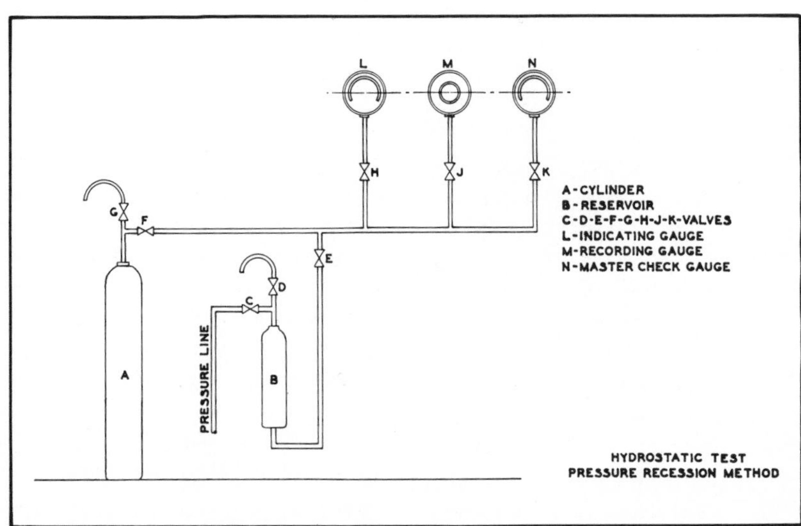

Fig. 7.

sure and repair as may be necessary and again apply pressure until it will remain steady. (Under no circumstances apply more than 90 per cent of the required test pressure up to this point of the test.)

(5) Open valve (E), allowing pressure to rise quickly (in less than 5 seconds) to the required test pressure and close valve (E) immediately.

(6) Leave apparatus undisturbed for 2 minutes or until needle of recording gage (M) ceases to indicate a drop in pressure.

(a) If no drop is indicated on the recording gage (M) or if the drop does not exceed 25 psig (15 psig if the cylinder is less than 5 in. outside diameter) the cylinder may be considered to have passed the test.

(b) If the drop is more than 25 psig (15 psig if the cylinder is less than 5 in. outside diameter) the cylinder has not passed the test and must be set aside for at least 24 hours and then subjected to a second test as described later.

(7) Open valve (G), thus releasing the pressure. Stop the clockwork in recording gage (M) and remove cylinder.

(8) Records. Mark opposite the tracing on the chart of the recording gage the serial number of the cylinder and the date of the test. When the chart is full it must be filed for future reference.

3.6.2.3. Second Test. The second test of the cylinder must be made at a test pressure at least 100 psig higher than the previous test pressure applied and must be made in a manner to accurately measure the total expansion under test pressure and the permanent expansion resulting therefrom. The permanent expansion must not exceed 10 per cent total expansion. The regular water jacket volumetric expansion method may be used and also the following:

(1) Determine (in cubic centimeters) the water capacity of the cylinder.

(2) Accurately measure the temperature of the water in the test system and cylinder.

(3) Determine (in cubic centimeters) the elastic expansion of the cylinder per psig of pressure applied, and then determine the allowable pressure recession of the cylinder at test pressure. (If the test pressure is to be 2000 to 2800 psig the elastic expansion should be found at 1500 psig. If the test pressure is to be 2800 psig or over the elastic expansion should be found at 2500 psig.)

(4) Apply the required test pressure (at least 100 psig greater than the first test) to the cylinder (the last 300 psig increase in pressure must be applied in less than 5 seconds) and then immediately cut off further pressure supply.

(5) Observe the pressure gages until the pressure recession ceases (this may take 10 to 30 minutes). If the pressure recession shown by the gages does not exceed the allowable pressure recession as determined before, then the permanent expansion of the cylinder does not exceed 10 per cent, and the cylinder has passed the test. If the pressure recession exceeds the allowable pressure recession, the cylinder has failed and must be condemned.

(6) The tracing on the chart of the recording gage must be marked "2nd Test" including the serial number of the cylinder tested and the date of the test. When the chart is full it must be filed for future reference.

(7) Records. Chart of the recording gage with tracing marked with the serial number of the cylinder tested and date of test and the words "2nd Test." Also the following data:

V = volume of cylinder (in cubic centimeters)
R = amount of water forced into the cylinder in determining the elastic expansion in third operation (in cubic centimeters)
T = temperature of water during test
e' = as determined by calculation.

When the chart is full, it must be filed for future reference.

3.6.2.4. Further Tests. The only condition under which further tests may be made is when leakage in the apparatus has occurred during previous tests. Each subsequent test must be made at a pressure at least 100 psig higher than the last preceding test and in the same manner as the second test before described, except that only $\frac{1}{2}$ of the allowable pressure recession as determined in the third operation of the second test is allowed to be utilized in determining whether the cylinder passes the test.

3.7. Simplified Method Detailed and Abbreviated for Use by Nontechnical Persons

3.7.1. The pressure recession method of testing described earlier involves obtaining the temperature of the water used in the test and also rather complicated calculations. For those who are not able or do not desire to make the tests in that manner, the following method is detailed. It is an approximation only, and therefore the latitude allowed is decreased in order to eliminate the possibility of inefficient or dangerous cylinders passing the test. No pressure in excess of 90 per cent of the test pressure shall be applied to the cylinders previous to the tests.

3.7.2. Apparatus.

The necessary apparatus is the same as before described (see Fig. 7).

3.7.3. First Test.

The preliminary preparations, operation of test, and the records to be recorded, are the same as described in *3.6.2.*

3.7.4. Second Test.

(1) Weigh cylinder empty. Weigh cylinder full of water. Find water capacity of cylinder in pounds of water.

(2) Attach cylinder with valve (G) closed.

(3) Raise pressure to 500 psig. Close valve (E) and open valve (G). Repeat if necessary until water is free from air bubbles flows from valve (G).

(4) Open valve (E) and close valve (G).

(5) Raise pressure to 2500 psig exactly and close valve (F). (If required test pressure is to be less than 2800 psig, then raise pressure only to 1500 psig in this operation.)

(6) Draw off water from valve (G) and measure accurately in cubic centimeters.

(7) Find value of e'' from the formula:

$$e'' = \frac{1000R}{115W}$$

(If 1500 psig pressure was used in Operation 5 described above, then find e'' from the formula:

$$e'' = \frac{10{,}000R}{685W}$$

where

W is the water capacity of the cylinder in pounds as found in Operation 1.

R is the water drawn off in Operation 6, and is in cubic centimeters.)

(8) From the following table find "allowable pressure recession" corresponding to the value of e'' just obtained:

e''*	Allowable Pressure Recession (psig)
32	7
33	14
34	21
35	27
36	32
37	37
38	42
39	47
40	52
41	56
42	60
43	64
44	68
45	71
46	74
47	78
48	81

*The designation e'' is the elastic expansion per psig of pressure in terms of hundred thousandths of the volume capacity of the cylinder.

(9) Open valve (F) and close valve (G), and start clockwork in recording gage (M).

(10) Raise pressure to the required test pressure which must be at least 100 psig higher than that used in the first test, and close valve (E) immediately. (The last 300 psig raise of pressure must be made in less than 5 seconds.)

(11) Leave apparatus undisturbed for 2 minutes or until needle of recording gage (M) ceases to indicate a drop of pressure.

(12) If the drop in pressure indicated by the recording gage (M) does not exceed the "allowable pressure recession" found from the table in step 8, then the cylinder may be considered to have passed the test. Otherwise, the cylinder has failed and must be condemned.

(13) Open valve (G), thus releasing the pressure. Stop the clockwork in recording gage and remove the cylinder.

(14) Mark opposite the tracing on the chart the serial number of the cylinder, the date of the test and also the words, "2nd Test."

3.7.5. Records

3.7.5.1. First Test. Chart of recording gage with tracing marked with serial number of the cylinder tested and date of test.

3.7.5.2. Second Test. Chart of recording gage with tracing marked with the serial number of the cylinder tested and date of test and the words, "2nd Test." Also the following data:

W = water capacity of cylinder (lb)
R = amount of water drawn off from cylinder (cu cm)
e'' = as found by calculation.

4. PROOF PRESSURE METHOD

4.1. Scope

4.1.1. This method may be used when ICC specifications and regulations do not require the determination of total and permanent volumetric expansion of a cylinder.

4.2. Inspection of Cylinders

4.2.1. Before cylinders are retested under the periodic retest required by ICC regulations:

4.2.1.1 Valves should be removed and careful internal inspection made with suitable light. Scale and sludge deposits (if any) should be removed and if seriously corroded, the cylinder shall be condemned. To avoid a possible flash, cylinders used in flammable gas service should be thoroughly purged before the interior inspection light is inserted.

4.2.1.2. Exterior inspection, including the bottom, should be carefully made for corrosion, dents, arc or torch burns, and physical deformation. Seriously corroded cylinders shall be condemned. Cylinders dented, arc or torch burnt, or physically damaged so as to weaken the cylinder appreciably shall be condemned.

4.2.2. Note. An empty cylinder before recharging should be examined for physical damage and exterior corrosion, followed by the hammer test. Those cylinders which have a dull or peculiar ring, when lightly tapped with a suitably sized hammer, should be carefully examined and if necessary, the valves removed and the interior examined with suitable light and tested before charging.

4.2.3. ICC regulations require that ICC-4B, ICC-4BA and ICC-26 cylinders used exclusively in noncorrosive gas service, tested by the proof pressure method, and marked with the letter S after the date of test, to be carefully examined at *each filling*. They must be rejected if evidence is found of bad dents, corroded areas, a leak or other conditions that indicate possible weakness which would render the cylinder unfit for service.

4.3. Requirements for Accurate Testing

4.3.1. The important element is a careful inspection for leaks and defects while the cylinder is under test pressure.

4.3.2. For the proof pressure cylinder testing method, the following conditions are absolutely essential:

4.3.2.1. The apparatus must be so arranged that careful inspection, including the bottom, can be made while cylinder is under test pressure.

4.3.2.2. A dead-weight testing apparatus must be used to verify the accuracy of the pressure gage. A master gage checked by a dead-weight tester at frequent intervals, would be an acceptable substitute though less desirable.

4.3.2.3. The pressure gage must agree with a dead-weight testing apparatus within $\frac{1}{2}$ of 1 per cent at the test pressure. Gages must have zero stop pin removed or placed $\frac{1}{2}$ in. below zero, and a true zero marked on the dial.

4.3.2.4. Test pressure must be maintained for at least 30 seconds and as much longer as may be necessary.

4.3.2.5. Cylinders may be manifolded and tested in groups.

4.3.2.6. The testing medium may be water or any suitable liquid such as kerosene.

4.3.2.7. Any suitable pressure source may be used. If air or gas is used, proper precautions must be made to protect personnel.

4.4. Results of Tests

4.4.1. A cylinder must be condemned if it leaks, ruptures, or shows evidence of bad dents, corroded areas or other conditions that indicate possible weakness which would render the cylinder unfit for service.

4.5. Schematic Diagrams and Instructions for Operating

4.5.1. Design and details of equipment to suit individual requirements can be adopted by the individual, but should follow these recommendations for accurate testing and safety.

4.5.2. For schematic diagram for applying the proof pressure testing method, to meet ICC modified retesting requirements for ICC-4B, ICC-4BA and ICC-26 cylinders, used exclusively in noncorrosive service, refer to Fig. 8. Operating instructions follow:

4.5.2.1. The proof pressure method may be used instead of the volumetric expansion method for retesting cylinders if applied within: (a) 12 years after cylinder was tested by the volumetric expansion method, or (b) 7 years after cylinder was tested by the proof pressure method. The test consists essentially of applyan internal hydrostatic pressure at least 2 times the cylinder service pressure and holding the pressure for a period of at least 30 seconds. If the cylinder does not leak or does not rupture it passes the test. However, the cylinder cannot be returned to service until it has been carefully examined externally and no evidence has been found of bad dents, corroded areas, or other conditions which would indicate weakness and render the cylinder unfit for service.

4.5.2.2. Two types of pressure source may be used: (a) a hydraulic pump discharging directly into the cylinder to be retested, and (b) a source of high pressure gas (air, nitrogen or carbon dioxide) applied to the top of a liquid pressure reservoir containing the liquid pressurizing medium which is connected to the cylinder to be retested.

4.5.3. Procedure (see Fig. 8, page 277).

(1) Fill cylinder (A) full of liquid which may be water or some other fluid such as kerosene.

(2) Close valve (G) to drain.

(3) Open valve (H).

(4) Free system of gas or air by purging through connector (B).

(5) Make connector (B) tight to liquid full cylinder (A).

(6) Operate pump or apply pressure from suitable source until required test pressure is indicated on gage (C). Close valve (H).

(7) Test pressure must be maintained for at least 30 seconds.

(8) The cylinder exterior shall be carefully inspected while the test pressure is maintained on the cylinder. Special attention should be given to inspection of cylinder bottoms. If valve (G), is closed and the pressure on the gage falls, it is an indication of a leak, deformation of cylinder or change in temperature of testing liquid.

(9) Release the pressure by opening valve (G) to the drain or by returning the liquid to reservoir by opening valve (H).

(10) Remove connector (B) from cylinder (A).

(11) Drain cylinder (A).

(12) Exterior inspection, including the bottom, should be carefully made for corrosion, dents, arc or torch burns, and physical deformation. Seriously corroded cylinders should be condemned. Cylinders dented, arc or torch burnt, or physically damaged, to

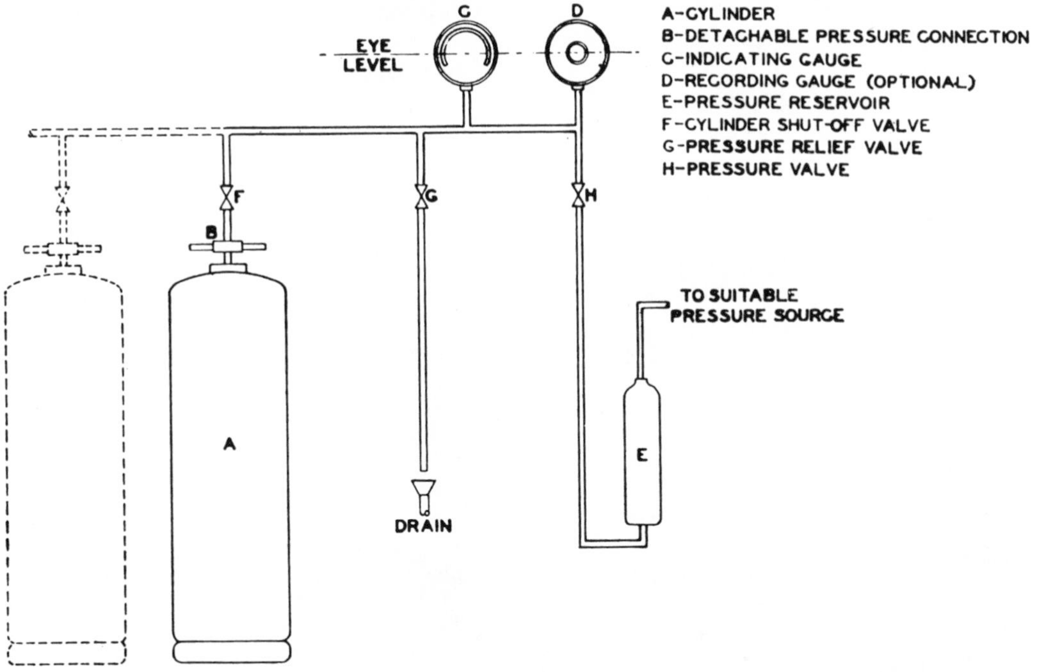

FIG. 8.

cause serious concentrated stresses, should be condemned.

(13) If cylinder has successfully passed the hydrostatic test and external inspection, it shall be stamped with date of test. After this date the letter "S" must be added to indicate that the cylinder was retested by a "special modified" hydrostatic test.

(14) A permanent record must be made of all cylinders tested indicating the type, registered symbol, and serial number of the cylinder together with date of test, operator, test pressure and test inspection data.

Note: As far as practicable, all the air or gas should be removed from the cylinder and system before applying the test pressure. While the presence of air or gas may not affect the accuracy or effectiveness of the test nevertheless, the presence of air or gas may constitute a definite hazard in case of the rupture of a cylinder or any other part of the system.

SECTION E

Requalification of ICC-3HT Aircraft Cylinders

1. INTRODUCTION

1.1. Regulations of the Interstate Commerce Commission as well as the regulations of the Board of Transport Commissioners for Canada require that a cylinder be condemned when it leaks, or when internal or external corrosion, denting, bulging, or evidence of rough usage exists to the extent that the cylinder is likely to be weakened appreciably.

1.2. Cylinders made in accordance with

ICC-3HT specifications are similar to those made under specification ICC-3AA in function and methods of fabrication. The major difference is that ICC-3HT cylinders are for aircraft use and are made to a higher allowable design stress and, therefore, have a thinner wall thickness and a lower weight. Several additional tests and added quality features are imposed to insure that the general quality of 3HT cylinders is maintained at a higher level than for 3A or 3AA cylinders. Most of these requirements are the responsibility of the cylinder manufacturer; however, some of the burden for maintaining a high quality cylinder rests with the cylinder user, shipper, charging facility and retest agency. Some of these added requirements are in ICC specifications and regulations, while others are implied.

1.3. This standard has been prepared as a guide to cylinder users for establishing their own cylinder inspection procedures and standards. It is of necessity general in nature although some specific limits are recommended. Each cylinder user must expect to modify them to suit his own cylinder design or the conditions of use that may exist in his own service. Rejection, or acceptance for continued use in accordance with these limits, does not imply that these cylinders are or are not dangerous, or subject to impending failure, but represents practice which has been satisfactory to a cross section of the industry.

1.4. Experience in the inspection of cylinders is an important factor in determining the acceptability of a given cylinder for continued service. Users lacking this experience and having doubtful cylinders should return them to a manufacturer of the same type of cylinders for reinspection.

1.5. The information contained in this statement is obtained from sources believed to be reliable and is based on the experience of a number of member companies of Compressed Gas Association, Inc. However, by issuance of the statement, the Association and its members jointly and severally, make no guarantee of results and assume no liability in connection with the information herein contained or the safety suggestions herein made. These suggestions are not intended to conflict with regulations issued by any authority having jurisdiction.

2. REGULATIONS APPLICABLE TO 3HT CYLINDERS

2.1. Outside Shipping Containers

It is mandatory that 3HT cylinders be shipped in suitable outside containers when in the pressurized condition. Corrugated cartons or wooden boxes are considered sufficient to protect the cylinders from adverse effects of rough handling during shipment. It is strongly recommended that 3HT cylinders be stored and transported in this same type outside container, whether full or empty, to prevent damage to the cylinder.

2.2 Tri-Annual Retest

It is mandatory that 3HT cylinders be subjected, at least once in three years, to a test by hydrostatic pressure in a water jacket for the determination of the expansion of the cylinder. A cylinder must be condemned if the permanent volumetric expansion exceeds 10 per cent of the total expansion, or if the elastic expansion on retest exceeds the original elastic expansion by more than 5 per cent. The original elastic expansion, in cubic centimeters, is stamped on the cylinder near the date of test.

2.2.1. Retest dates shall be applied by low stress type steel stamping to a depth no greater than that of the original marking at time of manufacture. Stamping on the sidewall not authorized.

2.3. Service Life Limitation

It is mandatory that 3HT cylinders be removed from service and condemned at the termination of a 12-year period following the date of the original test marked on the cylinder. The cylinder must also be condemned after 4380 pressurizations (12 yr $\times$ 365 $=$ 4380 days). If a cylinder is recharged more than once a day, an accurate record of the number of such rechargings must be maintained by the user.

3. GENERAL

3.1. Inspection Equipment

3.1.1. Depth Gages. Scales, Etc. Exterior corrosion, denting, bulging, gouges or digs are normally measured by simple direct measurement with scales or depth gages. In brief, a rigid straight edge of sufficient length is placed over the defect and a scale or feeler gage is used to measure the distance from the bottom of the straight edge to the bottom of the defect. There are also available commercial depth gages which are especially suitable for measuring the depth of small cuts or pits. It is important when measuring cuts that the upset metal be removed or compensated for so that only actual depth of metal removed from the cylinder wall is measured.

3.1.2. Supersonic Devices.

There are a variety of commercial supersonic devices available. These can be used to detect subsurface flaws and to measure wall thickness.

3.1.3. Magnetic Particle Inspection.

Magnetic particle inspection can be adapted for cylinder inspection to locate quickly faults not readily visible to the naked eye.

3.1.4. Inspection Lamps.

A good inspection light of sufficient intensity to clearly illuminate the interior wall is mandatory for internal inspection. Flammable gas cylinders shall be purged before being examined with a light. Lamps for flammable gas cylinders shall be vapor-proof.

3.2. Minimum Allowable Wall Thickness

For the purposes of this standard the minimum allowable wall thickness is the minimum wall thickness required by the specification.

3.3. Dents

Dents in cylinders are deformations caused by the cylinder coming in contact with a blunt object in such a way that the thickness of metal is not materially impaired.

3.4. Cuts, Gouges or Digs

Cuts, gouges or digs in cylinders are deformations caused by contact with a sharp object in such a way as to cut into or upset the metal of the cylinder, decreasing the wall thickness at that point.

3.5. Corrosion or Pitting

Corrosion or pitting in cylinders involves the loss of wall thickness by corrosive media. Corrosion in cylinders can usually be classed as (1) isolated pitting, (2) line corrosion, or (3) general corrosion.

3.5.1. Isolated Pitting.

Isolated pits have the least weakening effect on the cylinder walls of the three types listed. Pits have a stress raising effect which is more significant in 3HT cylinders with a high design stress than it is in cylinders with lower design stress.

3.5.2. Line Corrosion.

Line corrosion consists of a series of connected or nearly connected pits and is more serious than isolated pitting.

3.5.3. General Corrosion.

General corrosion is that which covers considerable surface area of the cylinder. It reduces the structural strength. It is often difficult to measure or estimate the depth of general corrosion because direct comparison with the original wall cannot always be made. General corrosion is often accompanied by pitting.

4. VISUAL INSPECTION

4.1. Preparation for Inspection

Cylinders shall be cleaned for inspection so that the inside and outside surfaces and all conditions can be observed. This shall include removal of scale and caked paint from the exterior and the thorough removal of internal scale. Cylinders with interior coating shall be examined for defects in the coating. If the coating is defective, it shall be removed.

4.2. General

3HT cylinders are more highly stressed than 3A or 3AA cylinders. For this reason the quality level and inspection standards for 3HT cylinders must be higher. The surface condition of newly manufactured cylinders is carefully inspected both visually and by the magnetic particle method, at the time of manufacture, to insure a high quality surface. It follows that the users, shippers and testers of these cylinders should exercise a high degree of care to insure the maintenance of surface quality, because of the local stress raising effect of nicks, scratches, and gouges which may have a detrimental effect on the cycle life of the cylinder.

4.3 Interior Inspection

The interior is carefully inspected both visually and by the magnetic particle method, for mechanical defects at the time of manufacture. Under normal usage there should not be any additional mechanical defects on the inside surface. Close attention should be paid to the general surface condition and in particular to the amount and type of corrosion present. Superficial corrosion which is deposited on the surface by rusty water or localized shallow pit type corrosion is not cause for rejection but steps should be taken to determine the cause and to prevent progressive corrosion. In most cases, it will not be practical to measure the loss in wall thickness due to corrosion, and the decision concerning the acceptability of the cylinder must be left to the experience and judgment of the competent inspector. It should be noted in this respect that the hydrostatic test by its nature can only determine the effective average corrosion and will, therefore, not serve to assess the extent of localized corrosion.

4.4. Exterior Surface

The outside of the cylinder is likely to take the brunt of any rough handling damage. The condition of the external finish (paint or plating) is a good indication of the usage that the cylinder has had. It should be recognized that it is a poor practice to mask the defects. Inspection should be performed before painting. Removal of the paint is desirable for a thorough inspection; however, it is recognized that this is not usually practical. Exterior inspections should be made for the following defects:

4.4.1. Dents.

Dents are cause for rejection. Care should be exercised to distinguish between a dent and the normal minor irregularities which are inherent in cylinder manufacturing processes. The good judgment of the inspector must be relied upon to determine what constitutes a dent and what constitutes safe wear as a result of normal usage.

4.4.2. Mechanical Defects.

Scratches, cuts, digs or gouges, are causes for rejection provided they are not superficial in nature. Defects which are sharp are considered more serious than those which are smooth. The cumulative effect of many defects must be evaluated by the inspector. A cylinder shall be rejected if the wall thickness has been reduced more than 5 per cent of the actual wall thickness by mechanical defects. If the actual wall is not known the cylinder shall be rejected if the defect reduces the wall thickness more than .005 in.

4.4.3. Bulges.

Bulged cylinders shall be removed from service. Bulging is usually caused by pressurization above the yield point of the cylinder material. This usually indicates either that the cylinder has been subjected to abnormally high pressures, normally during test, or that the cylinder material has been weakened to the point where it is not sufficiently strong to withstand normal pressures. The latter can generally be ruled out unless the cylinder has been exposed to a fire or has been severely corroded.

4.4.4. Fire Damage.

Fire damage to cylinders is usually indicated by (a) charring or burning of the paint or other protective coat, (b) burning or sintering of the metal, (c) distortion of the cylinder, (d) functioned relief devices, (e) melting of valve parts. Cylinders showing evidence of fire damage shall be rejected as required by §73.34(f) of ICC and BTC Regulations.

4.4.5. Corrosion.

External corrosion should be inspected with the same standards described above for cylinder interiors.

SECTION F

Retest Procedures for Seamless, High-Pressure Cylinders ICC-3, ICC-3A, ICC-3AA

1. INTRODUCTION

1.1. Compressed gas cylinders constructed in accord with the specifications of the Interstate Commerce Commission (ICC) start out in life as economically safe as 20th century industry can make them. They cannot be expected to stay "new" forever, but they should be expected to remain sufficiently safe so that regulatory bodies will continue to approve the use of charged compressed gas cylinders. After several decades of developing new cylinders and new cylinder materials, it is not surprising that the original service life controls are no longer completely applicable under all conditions.

1.2. This statement contains detailed methods of estimating wall thickness that can be applied with accuracy and simplicity to the retesting of compressed gas cylinders in order to determine their suitability for continued service. It applies to seamless, high-pressure cylinder specifications ICC-3, ICC-3A and ICC-3AA. All of the retest procedures outlined herein have been used successfully although there is no guarantee that the choice of any one method will assure its successful application. The responsibility for the ultimate success of whatever method of control is chosen will always rest upon the shoulders of the user.

1.3. The intent of this statement is to make available to all compressed gas cylinder owners those unofficial retest procedures that are believed to share with the official retest requirements the responsibility for the present excellent cylinder service life record. The major portion of these retest procedures has been written into §73.302(c) of ICC regulations covering the 110 per cent service pressure filling of certain ICC cylinders, and this statement should be useful for those whose cylinders are to be filled in accordance with the requirements of that section.

2. PRACTICES AND RECOMMENDATIONS IN CONTROLLING CYLINDER SERVICE LIFE

2.1. Compressed gas cylinders have been specified by ICC regulations to conform to acceptable standards recommended by Compressed Gas Association, Inc. The cylinders were characterized by the assumption that they should be acceptable for any of the gas services for which the cylinder is specified. In addition, the cylinder was assumed to be unacceptable for further service when the hydrostatic retest indicated the wall thickness has thinned out through corrosion or other means to the point where stresses in excess of the yield strength were developed under the hydrostatic test pressure.

2.2. Originally, the cylinders were made with steel of low yield-tensile ratios (below 70 per cent), and it was reasonable to expect that excessive corrosion would eventually be caught in the hydrostatic test when the permanent expansion exceeded 10 per cent of the total expansion and the cylinder would then stand condemned for further service.

2.3. In later years, about 1937, with the introduction of the alloy steels (e.g., Chrome-moly), the yield ratios went above 70 per cent, and the specified regulating control on service life established by the 10 per cent expansion limit was seriously weakened; the higher yield pressures of the alloy cylinders permitting a considerable increase in the amount of corrosion that must occur for any given wall thickness before the 10 per cent permanent expansion test rejected the cylinder. This helped to focus attention on methods of wall measurement.

2.4. During World War II, a 10 per cent increase in filling pressures was permitted with no change in the test pressure or safety device setting, and a successful record of performance was turned in by industry.

2.5. When the time came to consider the permanent adoption of the increased filling pressure, a study was made of the retest limitations that industry had been applying to high pressure cylinders during this period, as compared with those required by ICC regulations.

2.6. During this period it was found that the majority of cylinders utilizing this emergency 110 per cent filling authorization had been following in one way or another the wall stress limitations first publicized by Compressed Gas Association, Inc., in 1929 for use with carbon and medium manganese steel cylinders of standard size and based upon the use of the hydrostatic test. For the alloy cylinders, higher wall stress limitations were being followed. In addition, a reliable unbacked frangible disc safety device was in use that gave better protection to the cylinder in a fire. These additional retest limitations led to the present provisions of §73.302(c) of ICC regulations authorizing the continued use of the 110 per cent filling providing that wall stress limitations based upon water jacket hydrostatic tests are followed and provided that the unbacked frangible disc safety device is used (without change in setting).

2.7. In testing cylinders to be charged to 110 per cent of the service pressure, the cylinder retest station determines the elastic expansion in a water jacket and, ordinarily, if the elastic expansion reading at minimum test pressure does not exceed a predetermined elastic expansion limit, the cylinder may then be examined internally and externally for objectionable defects and stamped with a plus (+) sign following the test date to signify conformance of the cylinder to the wall stress retest requirements of §73.302(c) of ICC regulations. Standards for visual inspection of compressed gas cylinders appear in Section C.

3. ALLOWABLE ELASTIC EXPANSIONS BASIC CONSIDERATIONS

3.1. Service life control by interior hydrostatic pressure test in a water jacket or other approved apparatus is a requirement of ICC regulations, but the regulation which requires that limiting wall stresses shall not be exceeded in service has been applied only to cylinders being charged to 110 per cent of service pressure markings. It is believed that the control of service life by limiting wall stresses should be applied to all ICC-3, 3A and 3AA cylinders.

3.2. It is intended at this time to provide as much information as is available regarding elastic expansion limits of cylinders in present use to facilitate the adoption of elastic expansion limits in retesting. It should be remembered that the reliability of the elastic expansion limit, insofar as it applies to the size of cylinder and type of cylinder end closure, is the responsibility of the cylinder owner. The Compressed Gas Association, Inc., in its bulletin published in 1929, gave recommendations to the industry on elastic expansion limits for certain cylinders. Those recommendations have since been followed by many in the industry. While the basic responsibility is that of the cylinder owner, Compressed Gas Association, Inc., desires to submit to the industry such later information as will aid in the adoption of elastic expansion limits covering present principal types of cylinders in use. A discussion of the effects of variables on elastic expansion limits (and K factors) will be given later in this statement.

3.3 The provisions of §73.302(c) of ICC regulations appear in the appendix. These provisions infer that a cylinder is no longer fit for an additional period of service when the wall stresses due to the required test pressure exceed a predetermined limiting figure. The limiting wall stress has been established at minimum test pressure, and the allowable minimum wall thickness which will develop this wall stress can be readily computed from the Bach formula which is given below:

$$S = P\left(\frac{1.3D^2 + 0.4d^2}{D^2 - d^2}\right)$$

where

S = wall stress at test pressure, psig
P = test pressure, psig
D = outside diameter, in.
d = inside diameter, in.

This formula may also be written:

$$\frac{S}{P} = \frac{1.3D^2 + 0.4d^2}{D^2 - d^2} = \frac{1.3\,(D/d^2 + 0.4}{(D/d)^2 - 1}$$

The computations for this formula have already been worked out, and the results are tabulated in Table 1 for the major part of the range covered by ICC cylinders.

3.4. The most direct method of applying this limitation would be to measure the wall thickness with a mechanical caliper with one leg inserted through the neck of the cylinder. Although this method is the most direct, it is not the simplest. It can be done with reasonable accuracy even on 55 in. cylinders and is being done with the calipers hung vertically with magnetic contact points, but it is not yet recommended for general use.

3.5. There are several electronic, ultrasonic and radiographic devices commercially available which can measure metal thicknesses. Of these, the resonance-type ultrasonic devices and the reflection-type ultrasonic devices have been used to measure the sidewall of closed cylinders with excellent results. These devices must be properly calibrated and must be operated by competent and experienced personnel, in order to provide reliable data of necessary accuracy. It should be remembered also that these devices do not show deep localized pits, which must be detected visually.

3.6. The most practical method of determining limiting wall stresses at the present time, and the method recommended for use by the average cylinder owner is the indirect method based upon computing the limiting wall stress from the elastic expansion at test pressure by means of the water jacket hydrostatic test.

3.6.1. It is not necessary to detail the water jacket test in these pages as it has been adequately covered in section D, "Hydrostatic Testing of Cylinders." It should be sufficient to point out that the amount a cylinder with stiff head or base will expand is inversely proportional to some function of the side wall thickness. The heavier the side walls, the stiffer the cylinder, and the lower will be the elastic expansion under pressure. As the wall thickness drops from corrosion or wear, the elastic expansion of the cylinder increases.

3.6.2. A simplified application of elastic expansion to indicate the serviceability of cylinders is described in Section 6. This application requires that the elastic expansion of the cylinder as originally measured at the time of manufacture be known. It is based upon the same equations covered in the following paragraphs, so an understanding of those sections will be desired even though the method of paragraph 6 below is to be employed.

3.6.3. The water jacket test is performed by pressurizing the cylinder and measuring the volumetric expansion of the cylinder by en-

TABLE 1. SOLUTIONS OF BACH STRESS FORMULA

$$\frac{S}{P} = \frac{1.3D^2 + 0.4d^2}{D^2 - d^2} = \frac{1.3(D/d)^2 + 0.4}{(D/d)^2 - 1}$$

D/d	S/P at intervals of .0001 D/d									
	.0000	.0001	.0002	.0003	.0004	.0005	.0006	.0007	.0008	.0009
1.010	85.877	85.035	84.210	83.401	82.607	81.829	81.065	80.316	79.580	78.858
1.011	78.150	77.453	76.770	76.098	75.438	74.790	74.153	73.527	72.911	72.306
1.012	71.710	71.125	70.549	69.983	69.426	68.877	68.337	67.806	67.283	66.769
1.013	66.262	65.763	65.271	64.787	64.310	63.840	63.377	62.921	62.472	62.029
1.014	61.592	61.161	60.737	60.318	59.905	59.498	59.097	58.701	58.310	57.925
1.015	57.544	57.169	56.799	56.433	56.073	55.716	55.365	55.018	54.675	54.337
1.016	54.002	53.673	53.347	53.025	52.707	52.393	52.083	51.776	51.473	51.174
1.017	50.878	50.576	50.297	50.011	49.729	49.450	49.174	48.901	48.631	48.364
1.018	48.101	47.840	47.582	47.326	47.074	46.824	46.577	46.333	46.091	45.852
1.019	45.615	45.318	45.149	44.920	44.693	44.468	44.246	44.026	43.808	43.592
1.020	43.379	43.167	42.958	42.751	42.545	42.342	42.141	41.942	41.744	41.549
1.021	41.355	41.163	40.973	40.785	40.599	40.414	40.231	40.050	39.870	39.692
1.022	39.515	39.341	39.167	38.996	38.826	38.657	38.490	38.324	38.160	37.997
1.023	37.836	37.676	37.517	37.360	37.204	37.050	36.896	36.744	36.594	36.444
1.024	36.296	36.149	36.004	35.859	35.716	35.574	35.433	35.293	35.154	35.016
1.025	34.880	34.774	34.610	34.477	34.344	34.213	34.083	33.954	33.826	33.698
1.026	33.572	33.447	33.323	33.199	33.077	32.956	32.835	32.715	32.597	32.479
1.027	32.362	32.245	32.130	32.016	31.902	31.789	31.677	31.566	31.456	31.346
1.028	31.238	31.129	31.022	30.916	30.810	30.705	30.601	30.497	30.394	30.292
1.029	30.191	30.090	29.990	29.891	29.792	29.694	29.597	29.500	29.404	29.309
1.030	29.214	29.120	29.027	28.934	28.841	28.750	28.659	28.568	28.478	28.389
1.031	28.300	28.212	28.125	28.038	27.951	27.865	27.780	27.695	27.611	27.527
1.032	27.444	27.361	27.279	27.197	27.116	27.035	26.955	26.875	26.796	26.717
1.033	26.639	26.561	26.484	26.407	26.331	26.255	26.179	26.104	26.029	25.956
1.034	25.882	25.808	25.735	25.663	25.591	25.519	25.448	25.377	25.307	25.237
1.035	25.168	25.098	25.030	24.961	24.893	24.826	24.758	24.692	24.625	24.559
1.036	24.493	24.428	24.363	24.298	24.234	24.170	24.106	24.043	23.980	23.917
1.037	23.855	23.793	23.732	23.670	23.610	23.549	23.489	23.429	23.369	23.309
1.038	23.251	23.192	23.134	23.076	23.018	22.960	22.903	22.846	22.790	22.734
1.039	22.678	22.622	22.566	22.510	22.456	22.402	22.347	22.293	22.240	22.186
1.040	22.133	22.080	22.027	21.975	21.923	21.871	21.819	21.767	21.716	21.665
1.041	21.615	21.564	21.514	21.464	21.415	21.365	21.316	21.267	21.218	21.170
1.042	21.121	21.073	21.025	20.978	20.930	20.883	20.836	20.790	20.743	20.697
1.043	20.651	20.605	20.559	20.514	20.469	20.424	20.379	20.334	20.290	20.246
1.044	20.202	20.158	20.115	20.071	20.028	19.985	19.942	19.899	19.857	19.815
1.045	19.773	19.731	19.689	19.648	19.606	19.565	19.524	19.484	19.443	19.403
1.046	19.362	19.322	19.282	19.243	19.203	19.164	19.125	19.085	19.047	19.008
1.047	18.969	18.931	18.893	18.855	18.817	18.779	18.742	18.704	18.666	18.630
1.048	18.593	18.556	18.514	18.483	18.447	18.410	18.374	18.338	18.303	18.267
1.049	18.232	18.196	18.161	18.126	18.091	18.057	18.022	17.987	17.953	17.919
1.050	17.885	17.852	17.818	17.785	17.752	17.719	17.685	17.652	17.619	17.585
1.051	17.552	17.520	17.488	17.456	17.424	17.392	17.359	17.327	17.295	17.263
1.052	17.231	17.200	17.169	17.139	17.108	17.077	17.046	17.015	16.985	16.954
1.053	16.923	16.893	16.864	16.834	16.804	16.775	16.745	16.715	16.685	16.656
1.054	16.626	16.597	16.569	16.540	16.512	16.483	16.454	16.426	16.397	16.369
1.055	16.340	16.313	16.285	16.258	16.230	16.203	16.175	16.148	16.120	16.093
1.056	16.065	16.038	16.012	15.985	15.959	15.932	15.905	15.879	15.852	15.826
1.057	15.799	15.773	15.748	15.722	15.696	15.671	15.645	15.619	15.593	15.568
1.058	15.542	15.517	15.492	15.467	15.442	15.418	15.393	15.368	15.343	15.318
1.059	15.293	15.269	15.245	15.221	15.197	15.173	15.149	15.125	15.101	15.077
1.060	15.053	15.030	15.007	14.983	14.960	14.937	14.913	14.890	14.867	14.844
1.061	14.822	14.800	14.777	14.755	14.732	14.710	14.687	14.665	14.642	14.620
1.062	14.597	14.575	14.554	14.532	14.510	14.489	14.467	14.445	14.423	14.402
1.063	14.380	14.359	14.338	14.317	14.296	14.275	14.253	14.232	14.211	14.190
1.064	14.169	14.149	14.128	14.108	14.087	14.067	14.047	14.026	14.006	13.985
1.065	13.965	13.945	13.925	13.906	13.886	13.866	13.846	13.826	13.807	13.787
1.066	13.767	13.748	13.729	13.709	13.690	13.671	13.652	13.633	13.614	13.594
1.067	13.575	13.556	13.538	13.519	13.500	13.482	13.463	13.444	13.425	13.407
1.068	13.388	13.370	13.352	13.334	13.316	13.298	13.280	13.262	13.244	13.226
1.069	13.208	13.190	13.173	13.155	13.138	13.120	13.102	13.085	13.067	13.050
1.070	13.032	13.015	12.998	12.981	12.964	12.947	12.929	12.912	12.895	12.878
1.071	12.861	12.844	12.828	12.811	12.795	12.778	12.761	12.745	12.728	12.712
1.072	12.695	12.679	12.663	12.646	12.630	12.614	12.598	12.582	12.565	12.549
1.073	12.533	12.517	12.502	12.486	12.470	12.455	12.439	12.423	12.407	12.392
1.074	12.376	12.361	12.345	12.330	12.315	12.300	12.284	12.269	12.254	12.238
1.075	12.223	12.208	12.193	12.178	12.163	12.149	12.134	12.119	12.104	12.089
1.076	12.074	12.060	12.045	12.031	12.016	12.002	11.987	11.973	11.958	11.944
1.077	11.929	11.915	11.901	11.887	11.873	11.859	11.844	11.830	11.816	11.802
1.078	11.788	11.774	11.760	11.747	11.733	11.719	11.705	11.691	11.678	11.664
1.079	11.650	11.637	11.623	11.610	11.596	11.583	11.570	11.556	11.543	11.529
1.080	11.516	11.503	11.490	11.477	11.464	11.451	11.437	11.424	11.411	11.398
1.081	11.385	11.372	11.359	11.347	11.334	11.321	11.308	11.295	11.283	11.270
1.082	11.257	11.245	11.232	11.220	11.207	11.195	11.182	11.170	11.157	11.145
1.083	11.132	11.120	11.108	11.096	11.084	11.072	11.059	11.047	11.035	11.023
1.084	11.011	10.999	10.987	10.975	10.963	10.952	10.940	10.928	10.916	10.904

closing it in a water jacket connected to an accurate measuring tube such as a glass burette open at one end to the atmosphere with the water level adjusted so that cylinder expansions show up as increases in the water level in the burette. Since the water jacket and the burette are exposed only to the water head in the jacket which is kept constant by taking all expansion readings at a fixed level, the sensitivity of measurement with a constant leveling burette is controlled by the diameter of the burette bore. The total expansions are composed of reversible (elastic) expansions and the irreversible (permanent) expansions, and when the latter are subtracted from the total expansions, the net or elastic readings are obtained, and these can be used to calculate the average wall thickness of the cylinder being pressurized.

3.6.4. The Clavarino formula expresses the relationship between the elastic expansion and the cylinder outside diameter and inside diameter and from this, the wall thickness can be computed.

$$\text{EE} = PKV\,(16.387)\left(\frac{D^2}{D^2 - d^2}\right)$$

where

EE = elastic expansion, total less permanent, cc
P = test pressure, psig
K = factor (experimentally determined)
V = internal volume, cu in.
16.387 = number of cc in 1 cu in.
D = outside diameter, in.
d = inside diameter, in.
t = wall thickness = $\frac{1}{2}(\text{D} - \text{d})$.

3.6.5. The Clavarino formula relating elastic expansions to cylinder dimensions will establish an elastic expansion reading that is indicative of the average wall (or effective wall, as it has been called), providing the proportionality constant or K factor has been properly determined. Since the K factor is the proportionality constant indicating the rate of flexibility of both side walls and the cylinder ends, it becomes doubly important in using the K factor to segregate cylinders of different base design (flexibility) to assure that the experimentally determined K factor is applicable to the cylinders involved. In general, plate drawn cylinders have K factors up to 2 per cent higher than billet pierced cylinders because of different bottom thickness. The original theory of elastic expansion contemplated relatively stiff cylinder ends in which manufacturing variations in end thickness would not affect the K factor, so that the proportionality constant related elastic expansions to side wall thickness only. Where the ends are more flexible thereby introducing larger elastic expansion readings, manufacturing variations in end thickness may affect this K factor and thereby reduce the accuracy with which the side wall thickness or stress can be estimated from the elastic expansion. The proof of the accuracy of the K factor and subsequent stress determination is the responsibility of the cylinder owner or user.

3.7. In applying the average wall stress limitation, the rejection average wall thickness is computed from the maximum allowable average wall stress using the Bach formula; this rejection average wall thickness is used as the service life control by measuring the elastic expansion of each cylinder and computing the equivalent average wall thickness for comparison with this limiting average wall thickness. *Practically, this is done by setting up a maximum allowable elastic expansion according to the Bach and Clavarino formulas noted herein and rejecting any cylinders that show elastic expansions at test pressure that exceeded the computed rejection limit.*

3.8. Example

As an example, let us consider a cylinder as follows:

Medium manganese steel
ICC-3A-2015
$8\frac{1}{2}$ in. ID
51 in. long
Volume, 2640-2700 cu in.

ICC limitation (for 110 per cent filling of ICC-3A cylinders): 53,000 psig average wall stress

K (from tables on previous experiments) = 1.31×10^{-7}

Test pressure, 3360 psig ($5/3 \times 2015$)

Use these values as shown below in the Bach formula:

$$S = P\left(\frac{1.3D^2 + 0.4d^2}{D^2 - d^2}\right)$$

$$53{,}000 = 3360\left(\frac{1.3D^2 + 0.4\,(8.5)^2}{D^2 - (8.5)^2}\right)$$

where

D = 8.9854 in.
t = 0.243 in.
t = wall thickness = $(D - d)/2$

Instead of the laborious computation involved, a simpler method is to use the figures of Table 1 as follows:

$$\frac{S}{P} = \frac{53{,}000}{3360} = 15.7738$$

from Table 1

$$\frac{D}{d} = 1.0571$$

D = 8.9853
t = 0.243 in.

To determine the rejection elastic expansion limit,

$$\text{EE} = PKV\,(16.387)\left(\frac{D^2}{D^2 - d^2}\right)$$

$$= (3360)(1.31 \times 10^{-7})(2640)(16.387) \times \left(\frac{(8.9853)^2}{(8.9853)^2 - (8.5)^2}\right)$$

$$= 180 \text{ cc (184 cc for volume of 2700 cu in.)}$$

Cylinders that exceed the 180 to 184 cc limitation on quinquennial retest are therefore considered as thin wall rejects and are rejected for further service at the marked service pressures.

3.8.1. A simplification of the Bach and Clavarino formulas is as follows:

$${}^*\text{EE} = KV\,(16.387)\left(\frac{S + .4\,P}{1.7}\right)$$

$$\text{EE} = (1.31 \times 10^{-7})\,(2640)\,(16.387) \times \left(\frac{53{,}000 + .4\,(3360)}{1.7}\right)$$

$$= 180 \text{ cc (184 cc for volume of 2700 cu in.)}$$

It will be noted from this formula that the elastic expansion rejection limit seems to be independent of the cylinder diameter. Actually, the elastic expansion limit is not independent of diameter or size since the K factor is dependent upon the cylinder diameter.

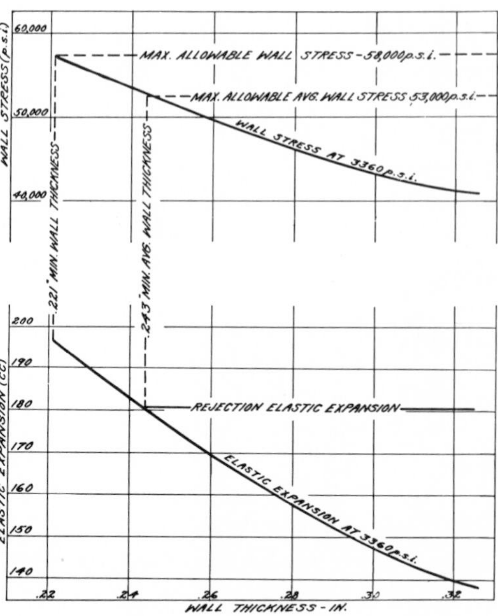

FIG. 1. Elastic expansion rejection limit.

This limit is the same for all cylinders of the following type and is based upon the minimum allowable average wall thickness at minimum test pressure.

*This equation is a rearrangement of the equation appearing in Note 1 of the Appendix.

ICC-3A Cylinder:

Capacity	220 cu ft oxygen
Size	8.5 in. I.D. × 51 in. long
Volume	2640 cu in. minimum
Material	Medium Manganese Steel (plate drawn)
Service Pressure	2015 psig
Minimum Test Pressure	3360 psig
K Value	1.31×10^{-7} (estimated)

3.9. Chart

It will be of interest to show on a chart (see Fig. 1) the values of the example above and certain further values showing the relationship between wall thickness, wall stress, and elastic expansion. (Slight numerical differences from the values of paragraph 3.8 result from the rounding of figures.) The following steps are taken:

3.9.1. Stress at test pressure is plotted in the upper part against average wall thickness determined by the following calculations using Table 1 based upon the Bach formula. These points are shown on the upper curve of wall stress at 3360 psig test pressure. The horizontal line at 53,000 psig indicates the limiting average wall stress permitted by §73.302(c). The horizontal line at 58,000 psig represents the limiting maximum wall stress permitted by §73.302(c) for any minimum wall thickness.

$$P = 3360 \text{ psig} \qquad d = 8.5 \text{ in.}$$

S	S/P	D/d	D	$D - d = 2t$	t
58,000	17.262	1.0519	8.941	.441	.221
55,000	16.369	1.0549	8.967	.466	.233
53,000	15.774	1.0571	8.986	.486	.243
50,000	14.881	1.0607	9.016	.516	.258
47,500	14.137	1.0642	9.046	.546	.273
45,000	13.393	1.0680	9.078	.578	.289
42,500	12.649	1.0723	9.115	.615	.307
40,000	11.905	1.0772	9.157	.657	.328

3.9.2. Calculate elastic expansion for wall thicknesses from formula

$$\text{EE} = PKV\,(16.387)\left(\frac{D^2}{D^2 - d^2}\right)$$

as shown in the table below when

$$P = 3360 \text{ psig test pressure}$$
$$V = 2640 \text{ cu in. volume}$$
$$16.387 = \text{Number of cc in 1 cu in.}$$
$$K = 1.31 \times 10^{-7} = 0.000000131$$
$$PKV\,(16.387) = 19.04206$$

t	d	D	$\left(\frac{D^2}{D^2 - d^2}\right)$	EE
.221	8.5	8.941	10.3935	197.9
.233	8.5	8.966	9.8768	188.1
.243	8.5	8.986	9.5018	180.9
.258	8.5	9.016	8.9938	171.3
.273	8.5	9.046	8.5417	162.7
.289	8.5	9.078	8.1112	154.5
.307	8.5	9.115	7.6693	146.0
.328	8.5	9.157	7.2281	137.6

3.9.3. Plot average wall against elastic expansion to form elastic expansion curve at 3360 psig test pressure.

3.9.4. The minimum manufactured wall thickness (.221 in.) established by the maximum allowable wall stress (58,000 psig) determines the maximum permissible elastic expansion (197.9 cc) on the elastic expansion curve. It represents the maximum elastic expansion reading permissible if the cylinder had perfectly uniform (such as concentric machined) walls of the minimum thickness.

3.9.5. The minimum average wall thickness (.243 in.) established by the maximum average wall stress (53,000 psig) determines the maximum elastic expansion of the cylinders as manufactured (180.9 cc), and represents the maximum permissible elastic expansion based upon average wall thickness developing 53,000 psig stress at test pressure. This agrees substantially with the rejection limit derived in paragraph 3.8.

3.10. Additional elastic expansion limits for other combinations of cylinder design, service pressure, and allowable wall stresses can be developed in a similar manner. In Table 2, the data is tabulated which led to the development of the original elastic expansion limits proposed by Compressed Gas Association's cylinder testing committee in December, 1929. It was recommended "that any cylinder

TABLE 2. ELASTIC EXPANSION REJECTION LIMITS*
(Reference: Report of Cylinder Testing Committee, December 1929)

Type of Cylinders	Carbon Dioxide		Oxygen		Chlorine***	
Gas Capacity	20 lb	50 lb	110 CF	220 CF	100 lb	150 lb
Nominal I.D., in.	5.00	8.00	6.62	8.50	10.00	10.00
Nominal length, in.	51	51	43	51	36	48
Minimum volume, cu in.	941	2327	1343	2640	2465	3380
Test pressure, psig	3000	3000	3360	3360	1000	1000
**Rejection effective wall stress, psig	36,400	53,200	49,200	53,300	24,300	24,400
**Rejection effective wall thickness, in.	.195	.206	.210	.247	.185	.185
"K" factor $\times 10^7$	1.26	—	1.27	1.30	1.31	1.31
Elastic expansion limit, cc	42	157	82	180	77	105

*Cylinders tested were manufactured prior to 1926 of carbon steel, hot drawn with "bumped in" bottoms.
**Effective thickness and stresses are now considered to represent average side wall thicknesses and the average transverse or hoop stress.
***This chlorine cylinder was of low pressure design, but was included because of interest in K values.

exceeding in the water jacket test the elastic expansion limit given—for that type, shall be withdrawn from service because of too thin wall."

4. *K* FACTORS

4.1. For those who experimentally determine the K factor in the Clavarino formula or check the factor they have been using, the following procedure has been found useful:

$$EE = PKV(16.387)\left(\frac{D^2}{D^2 - d^2}\right)$$

where

EE = elastic expansion (total less permanent), cc
P = test pressure, psig
K = factor (experimentally determined)
V = internal volume, cu in.
16.387 = Number of cc in 1 cu in.
D = Outside diameter, in.
d = Inside diameter, in.

4.2. To determine accurately the proportionality constant K relating elastic expansions to cylinder dimensions, it is necessary to measure the remaining variables accurately.

4.2.1. The *Elastic Expansion* (EE) is measured by use of water jacket and is capable of unusual accuracy if properly performed (see Section D). A useful check on the reliability of the test procedure by proving the freedom from gross errors due to temperature difference, leakage, and changes in shape may be made as follows:

(1) Pressurize to 90 per cent of the minimum test pressure.

(2) Release the pressure.

(3) Set the scale reading to zero.

(4) Repressurize once more to the previous pressure.

(5) Release the pressure.

(6) If the water level with a large cylinder returns to within $\frac{1}{2}$ cc of the original zero reading (approximately 1 part in 400), proceed with the test.

4.2.2. The *pressure P* must be applied and maintained at test pressure until all permanent expansion is removed as indicated by the water level becoming steady. Further, the pressure gage should be accurate to within $\frac{1}{2}$ of 1 per cent and preferably calibrated directly against a dead-weight tester reading in pounds per square inch.

4.2.3. The *internal volume V* of the cylinder in cubic inches can be readily determined by weighing the cylinder empty and then full of water and computing the internal volume from the water capacity in pounds by multiplying by the specific volume of water. (27.71 cu in. per lb at 60 F.)

4.2.4. The *outside diameter D* can best be determined by measuring the circumference carefully in several places with a flexible metal rule reading in inches and computing the diameter by dividing by 3.1416.

4.2.5. The *inside diameter d* is usually computed by subtracting $2 \times t$ (wall thickness in inches) from the outside diameter. When used to compute an elastic expansion, the average inside diameter is employed based upon the average wall thickness.

4.2.6. The most sensitive variable in the determination of the *K* factor is the *wall thickness* measurement. With new cold drawn cylinders, pointed or ball and flat micrometers are satisfactory. With spun, plate drawn, or billet pierced cylinders, all hot working processes, the only reliable method is believed to be by computation from a weight measurement. Since the density of the steel can be assumed fairly accurately as 0.2833 pounds per cubic inch, the average thickness can be readily determined by cutting the heads and bases off the cylinder at a distance from extreme top and bottom approximately equal to the diameter (to avoid any thickening at the ends) and then machining the ends to leave a cylindrical shell, weighing the shell and measuring its outside diameter and length and computing the average thickness as follows:

$$d^2 = D^2 - \left(\frac{W}{.2225L}\right)$$

where

$$.2225 = \frac{3.1416}{4} \times .2833$$

$$t = \frac{D - d}{2}$$

TABLE 3. TYPICAL *K* VALUES

PART I

Rated Oxygen Capacity (cu ft.)	Diameter (in.) ID except OD as shown	Length (in.)	Typical Minimum Wall (new cylinder) (in.)	Nominal Volume (cu in.)	ICC Marking	*Type of Bottom	**Typical *K*	***Range of $K \times 10^7$
1500	$9\frac{5}{8}$ O.D.	20′ $6\frac{1}{4}$	.300	15,000	3A2400	—	1.250×10^{-7}	1.228-1.298
400	10	56	.282	4000	3AA2400	BP or PD	1.300×10^{-7}	—
300	$8\frac{3}{4}$	55	.242	3025	3AA2400	BP or PD	1.300×10^{-7}	—
250	$8\frac{1}{2}$	51	.222	2650	3AA2265	BP or PD	1.300×10^{-7}	1.30-1.32
220	$8\frac{1}{2}$	51	.260	2650	3A2015	BP or PD	1.300×10^{-7}	1.29-1.31
220	$8\frac{1}{2}$	51	.200	2650	3AA2015	BP or PD	1.300×10^{-7}	1.29-1.31
195	8	51	.190	2370	3AA2400	BP or PD	1.290×10^{-7}	1.29-1.31
110	$6\frac{5}{8}$	43	.205	1325	3A2015	BP or PD	1.265×10^{-7}	1.265-1.30
110	$6\frac{5}{8}$	43	.158	1325	3AA2015	BP or PD	1.265×10^{-7}	1.265-1.28
27	$4\frac{1}{8}$	$26\frac{1}{2}$	.150	326	3A2400	BP or PD	1.250×10^{-7}	—
14.5	$3\frac{15}{16}$	$16\frac{3}{4}$	.150	174	3A2400	BP or PD	1.250×10^{-7}	—
PART II								
22.2	$3\frac{15}{16}$	$25\frac{3}{4}$	.110	293	3AA2015	SP	1.330×10^{-7}	1.27-1.33
22.2	$3\frac{15}{16}$	$25\frac{3}{4}$	.110	293	3AA2015	CD	1.305×10^{-7}	—
12.7	$3\frac{15}{16}$	$16\frac{3}{4}$	.110	175	3AA2015	SP	1.275×10^{-7}	1.27-1.33
12.7	$3\frac{15}{16}$	$16\frac{3}{4}$	.110	175	3AA2015	CD	1.330×10^{-7}	—
5.35	$3\frac{1}{8}$	$12\frac{1}{2}$	.085	86	3AA2015	CD	1.330×10^{-7}	—
20 lb CO_2	7.75 O.D.	$23\frac{1}{2}$	.155	855	3A1800	CD and SP	1.43×10^{-7}	—

Notes: Part I: *K* values determined by measuring *t* with micrometers and by computing from weight.
Part II: *K* values determined by computing *t* from weight.

*BP—Billet pierced; PD—Plate drawn; CD—Cold drawn; SP—Spun.

***K* values for cylinders of different end design may differ by as much as 25 per cent.

***The figures listed represent reported values from only a few sources. The actual ranges probably are considerably wider and possibly are higher or lower than the typical *K* values shown.

It should be remembered in applying *K* factors and elastic expansion limits, that Compressed Gas Association, Inc., can accept no responsibility for the reliability of published values, but it is publicizing them for information and guidance with the recommendation that they be rechecked by each user until agreement on individual values by the majority of users establishes the factors beyond question.

4.3. Using procedures somewhat similar to those followed by the early cylinder testing committee, the various cylinders listed in Table 3 have been tested by cylinder manufacturers and gas charging plants and the *K* factors determined for use in establishing elastic expansion limits and tabulated herein.

5. ELASTIC EXPANSIONS—ALLOWANCE FOR NONUNIFORM WALLS

5.1. Although corrosion is usually general, it is recognized that it can never be absolutely uniform and that corrosion can be more severe in some areas than in others. Experience has shown that for the purposes of this statement, it has been safe to assume that the maximum loss of wall thickness is ordinarily twice the average loss.

5.2. Unless a correction is made for the nonuniformity of corrosion, a considerable error can be made in applying elastic expansion limits to overdesigned cylinders. As an example, a cylinder which requires an average wall of 0.200 in. to pass the hydrostatic test at 53,000 psig at test pressure as required by ICC regulations §73.302(c) for medium manganese steel, might have been made to 0.300 in. average wall when manufactured. On the basis that the maximum corrosion may be twice the average corrosion, the minimum wall of this cylinder might be 0.100 in. when the average wall was 0.200 in. Under these circumstances, several things might happen to indicate that this cylinder had developed an excessively thin (minimum) wall.

5.2.1. The stress at 0.100 in. minimum wall might be as high as 129,500 psig since we assumed the cylinder showed a stress at 0.200 in. average wall of 53,000 psig. This local stress might exceed the tensile strength in which case the cylinder would develop a leak or rupture under test pressure.

5.2.2. The elastic limit of the steel would probably be exceeded in the highly stressed deeply corroded area. This would produce an amount of permanent expansion dependent upon the area corroded to 0.100 in. in thickness. If the area corroded were large enough, the permanent expansion would exceed 10 per cent of the total expansion, so that the cylinder would be rejected for this reason.

5.2.3. If the area corroded to 0.100 in. wall were small enough so that the permanent expansion did not exceed 10 per cent of the total expansion and if the tensile strength of the steel were not exceeded, the cylinder might pass the retest (if internal inspection did not question the localized corrosion) and may be used at a pressure 10 per cent in excess of its marked service pressure. However, the elastic expansion for such a cylinder would show a marked increase and the weight of the cylinder would probably show a marked decrease.

5.3. Where cylinders have corroded unevenly, it becomes desirable to pay more attention to the minimum wall thickness and to modify the elastic expansion allowed in such a way as to assure adequate wall thickness even in corroded areas. This method is illustrated in the following paragraphs.

5.4. Fig. 2 has been developed by superimposing on Fig. 1 a third curve, OABC, which was developed in the belief that it is more reliable to estimate the limiting minimum wall where a cylinder has corroded nonuniformly.

5.5. Based upon the assumption that the maximum loss of wall thickness from corrosion is twice the average wall loss, a cylinder with average wall thickness in excess of the minimum allowable wall thickness is allowed to corrode until half this excess is lost.

5.6. In Fig. 2, line OABC is developed as follows:

5.6.1. From elastic expansion of cylinder as manufactured, determine average new wall thickness. (Point C_1 indicates 0.321 in. average wall at 140.7 cc elastic expansion.)

5.6.2. Estimate the maximum allowable reduction in wall thickness if cylinder is uniform by subtracting minimum service wall from average new wall. (0.321 − 0.221 in. = 0.100 in.) Note that the minimum wall comes from a limiting stress of 58,000 psig in table of 3.9.1.

5.6.3. Estimate the allowable change in

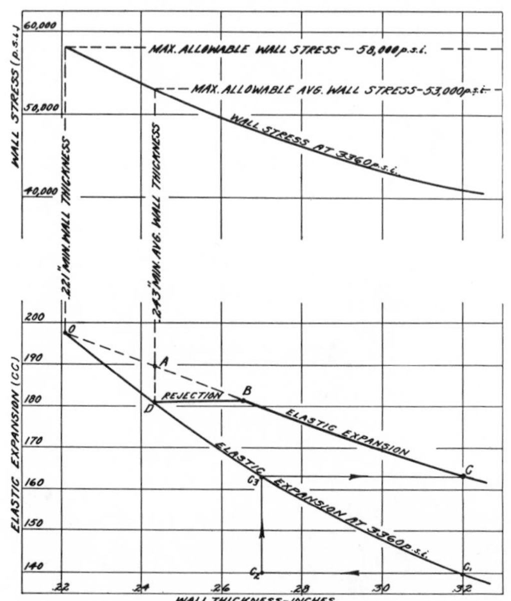

FIG. 2. Elastic expansion rejection limits. These limits for cylinders of the same type described in Fig. 1 differ from each other according to the values of the original elastic expansion readings at minimum test pressure.

average wall thickness by allowing only $\frac{1}{2}$ the difference between the minimum service wall and the average new wall. ($\frac{1}{2} \times 0.100 = 0.050$ in.)

5.6.4. Determine the rejection average wall by subtracting half of maximum possible wall loss from average new wall. (0.321 – 0.050 in. = 0.271 in. locating point C_2.)

5.6.5. Determine rejection elastic expansion corresponding to rejection average wall. (Point C_3 indicates 163.7 cc rejection elastic expansion at 0.271 in. average wall.)

5.6.6. Plot the rejection elastic expansion directly over the average new wall to simplify reading the graph. (Point C indicates that the cylinder should be rejected at an elastic expansion reading of 163.7 cc which would be reached when the original average wall of 0.321 in. corroded to a rejection average wall of 0.271 in. theoretically corresponding to a rejection minimum wall of 0.221 in.)

5.7. Since the curve OABC in the region AB actually provides for elastic expansion rejection limits in excess of the maximum elastic expansion based upon average wall thickness, the rejection curve is modified in this region by adhering to the maximum elastic expansion of point D. The rejection limits thus become the curve DBC.

5.8. The reliability of this rejection limit is based upon several factors:

5.8.1. The assumption that the maximum local corrosion loss in a cylinder is normally not more than twice as great as the average corrosion loss.

5.8.2. The assumption that either the original cylinder was of uniform wall thickness so that the maximum wall stress based upon minimum wall thickness can be used as the limiting factor, or the assumption that the maximum corrosion did not occur in the original thinner wall section.

5.8.3. Although the assumptions are not exactly true, experience has shown that they do represent a reasonable guide for the conservative reduction of allowable increases in elastic expansion to prevent excessive corrosion in the heavy wall cylinders. Paragraph 5.7 presents a summary of the method as well as a further adjustment to maintain conformity to limits covering both maximum and average wall stresses.

6. CYLINDER SERVICE LIFE BASED UPON ORIGINAL ELASTIC EXPANSION

6.1. Although industrial cylinder recharging plants have been successfully using elastic expansion limits and *K* factors in accordance with the aforementioned procedure, there is a modification of this procedure that eliminates some requirements for accurately determining the *K* factor and that simplifies the problem of controlling an inventory of cylinders made by the various manufacturing procedures in past and present use.

6.2. In this modification, each new cylinder is stamped with its original elastic expansion and the cylinder is rejected when its elastic expansion increases some predetermined amount.

6.3. From the Clavarino formula,

$$EE = PKV\,(16.387)\left(\frac{D^2}{D^2 - d^2}\right)$$

After retest, the new

$$EE = P_rK_rV_r\,(16.387)\left(\frac{D_r^2}{D_r^2 - d_r^2}\right)$$

where

$K = K_r$ (same elastic constant)
$D = D_r$ (same outside diameter)
$P = P_r$ (same test pressure)
$V = V_r$ (since increase is sufficiently small to be neglected)

Therefore, change in elastic expansion or

$$EE_r - EE = PKV(16.387)D^2 \times \left(\frac{1}{(D^2 - d_r^2)} - \frac{1}{D^2 - d^2}\right)$$

and since P, K, V, and D are constants, it can be seen that the change in elastic expansion is inversely proportional only to a function of the inside diameter or the extent of corrosion.

6.4. Various systems can be developed to use the amount of increase in a cylinder's elastic expansion as a measure of when that cylinder is no longer fit for further service. These systems require the construction of curves from which tables are developed and rejection is on the basis of the tables.

6.4.1. Using this system, a single table or graph can be made up for each cylinder which will list the rejection elastic expansion reading for each original elastic reading. Such a table will be valid and applicable to any test pressure used or value of K estimated. Such a system has only one remaining variable to consider and that is the "uniformity of corrosion."

6.4.2. A rejection table may be prepared from the curve DBC of Fig. 2. This is shown below as Table 4.

Table 4. Example of Elastic Expansion Service Rejection Limits

(Based upon Original Elastic Expansion Readings at Minimum Test Pressure)

ICC-3A Cylinder:
Capacity ..... 220 cu ft
Size ..... 8.5 in. I.D. × 51 in. long
Volume ..... 2640 cu in. minimum
Material ..... Medium manganese steel (plate drawn)
Service pressure ..... 2015 psig
Minimum test pressure ..... 3360 psig

Original Elastic Expansion (cc)	Rejection Elastic Expansion (cc)
	197.9*
181.3	181.3
180	181.3
175	181.3
170	181.3
165	180
160	177
155	174
150	170
145	167

*Maximum service limit.

6.5. Such systems as above have the advantage of removing some variables encountered because the results obtained for a cylinder are compared only with other results obtained on the same cylinder. Thus there need be considered no variations from one cylinder to another for pressure, K factor, volume or outside diameter since these quantities are either constant or, in the case of volume, change a negligible amount from corrosion.

6.6. Where corrosion is highly localized as from a stagnant water level, the elastic expansion formula is no longer useful in estimating wall thickness of this zone. For this reason, cylinders must be inspected internally as well as hydrostatically and where localized corrosion is severe, the containers must be rejected on the basis of visual inspection alone.

7. ESTIMATING MINIMUM WALL THICKNESS

7.1. Although the elastic expansion test when properly performed will determine the average wall thickness quite accurately for the great majority of cylinders, there are exceptional cases when limiting the average wall thickness may not provide the best guarantee for further service. Cylinders may be excessively eccentric, or locally corroded, may have scale pockets on the inner surface or may have been machined or ground locally so that minimum wall thicknesses are not sufficient for the service pressure intended.

7.2. Theoretically, it would even be preferable to establish a single minimum wall thickness limitation for any size cylinder and type material, but the practicability of testing cylinders for minimum wall thicknesses prevents a single control of this type at this time. Accordingly, while the cylinder walls are basically limited to a stress based upon average wall thickness, §73.302(c) of ICC regulations also recognizes that limitations can be properly based upon a higher maximum allowable wall stress computed for a minimum wall thickness (see Appendix to this section).

7.3. Consequently, the user has the choice of either the average wall stress limitation or the maximum wall stress limitation and can either (1) determine the average wall in the elastic expansion test, or (2) estimate the minimum wall by correcting the elastic expansion average wall thickness for nonuniform corrosion, or for eccentricity, or (3) measure the minimum wall by taking local measurements in ground or corroded spots with mechanical calipers or electrical measuring devices. Whatever the means, it should be remembered that it is the cylinder owner's responsibility to apply the proper inspection test and rejection limit to his cylinder inventory and that the limits he establishes should be checked experimentally whenever possible.

APPENDIX

Extract from ICC and BTC regulations

§73.302 Filling limits.

(c) Specs. 3A and 3AA cylinders may be charged with compressed gases, other than liquefied, dissolved, poisonous, or flammable gases to a pressure 10 per cent in excess of their marked service pressure, provided:

(1) That such cylinders are equipped with frangible disc safety devices (without fusible metal backing) having a bursting pressure not exceeding the minimum prescribed test pressure.

(2) That the elastic expansion shall have been determined at the time of the last test or retest by the water jacket method.

(3) That either the average wall stress or the maximum wall stress shall not exceed the wall stress limitation shown in the following table: (See Notes 1 and 2.)

Type of Steel	Average Wall Stress Limitation	Maximum Wall Stress Limitation
Plain carbon steels over 0.35 carbon and medium manganese steels	53,000	58,000
Steels of analysis and heat-treatment specified in spec. 3AA	67,000	73,000
Plain carbon steels less than 0.35 carbon made prior to 1920	45,000	48,000

Note 1: The average wall stress shall be computed from the elastic expansion data using the following formula:

$$S = \frac{1.7EE}{KV} - 0.4P$$

where

S = wall stress, pounds per square inch;
EE = elastic expansion (total less permanent) in cubic centimeters;
K = factor·x 10^{-7} experimentally determined for the particular type of cylinder being tested;
V = internal volume in cubic centimeter (1 cubic inch = 16.387 cubic centimeters);
P = test pressure, pounds per square inch.

Formula derived from formula of Note 2 and the following:

$$EE = PKV \times \frac{D^2}{D^2 - d^2}$$

Note 2: The maximum wall stress shall be computed from the formula:

$$S = P\frac{(1.3D^2 + 0.4d^2)}{D^2 - d^2}$$

where

S = wall stress, pounds per square inch;
P = test pressure, pounds per square inch;
D = outside diameter, inches;
$d = D - 2t$, where t = minimum wall thickness determined by a suitable method.

(4) That an external and internal visual examination made at the time of test or retest shows the cylinder to be free from excessive corrosion, pitting, or dangerous defects.

(5) That a plus sign (+) be added following the test date marking on the cylinder to indicate compliance with subparagraphs (c) (2), (3), and (4) of this section.

SECTION G

Welding and Brazing on Thin-Walled Compressed Gas Containers

1. INTRODUCTION

1.1. Purpose of Standards

1.1.1. Due to the lack of other applicable standards and in order to promote uniformity and serviceability of compressed gas containers produced under the specifications of the Interstate Commerce Commission, technical committees of Compressed Gas Association, Inc., have established these standards for the manufacturers of welded or brazed compressed gas containers and for the inspectors who may be called upon to qualify welding or brazing on such containers. Establishment of standards for procedure and operator qualification and for radiographic inspection was necessary to cover the gage of materials used in compressed gas containers of 900 psig maximum service pressure.

1.2. Scope of Standards

1.2.1. These standards are recommended as being applicable to ICC specification containers of 900 psig maximum service pressure, 1000 lb maximum water capacity, and minimum wall thickness under $\frac{3}{8}$ inch.

1.2.2. These standards include the following subjects:

(a) Procedure Qualification. Standards relating to welding procedure require that each procedure be qualified by a minimum of tests to demonstrate that adequate soundness, strength, and ductility of joints are obtainable under the welding conditions specified.

(b) Operator Qualification. Standards relating to welding operator require that each operator be qualified by a minimum of tests, adhering to an approved welding procedure, to demonstrate that he possesses and has not lost his welding ability.

(c) Radiographic Inspection. Standards relating to radiographic inspection apply to details of radiographic inspection technique which conform to sound, current industrial practice and which are not found in container specifications.

(d) Container Repair. Standards relating to container repair apply to details of repair technique which conform to sound, current industrial practice and which are not found in container specifications.

1.2.3. Standards contained herein for procedure and operator qualification were adapted from Section IX of the ASME Boiler and Pressure Vessel Code, "Qualification Standard for Welding and Brazing Procedures, Welders, Brazers, and Welding and Brazing Operators," which in turn are patterned after American Welding Society Standards. Standards for radiographic inspection are almost identical to provisions of the ASME Boiler and Pressure Vessel Code for "Unfired Pressure Vessels."

2. PROCEDURE QUALIFICATION

2.1. Applicability

2.1.1. Procedure for all welding of pressure parts of compressed gas containers referred to in 1.2.1. shall be qualified in accordance with these standards.

2.1.2. If certain changes are made in a procedure specification, requalification of procedure shall be required as set forth in 2.9.

2.2. Procedure Specification

2.2.1. The procedure of welding to be followed in construction shall be established and recorded by the manufacturer as a procedure specification, and in the investigation to qualify this procedure, the procedure specification shall be followed. Recommended forms for the procedure specification are given in 7.2, 7.3, 7.4 and 7.5. It is not necessary that these exact forms be used, but the information contained therein should be set forth in any alternate form which is adopted.

2.2.2. If any changes are made in a procedure, the procedure specification shall be revised or amended to show these changes.

2.3. Welding Procedure

2.3.1. Welding procedure on specimens prepared for qualification tests shall comply in all respects with the procedure specification. For qualification of a manual welding process the container shall be manually welded. For qualification of an automatic welding process the container shall be welded by the production procedure using equipment duplicating that to be used in production with respect to position of work, protection of hot metal, automatic control of preheat, feed, speed, current, oscillation, interruption, rate of cooling, etc. Qualification of automatic welding on one type of equipment shall not qualify the process on another type of equipment which lacks any single element of control provided on the first.

2.4. Base Material and Its Preparation

2.4.1. Base material and its preparation for welding of qualification test specimens shall comply with the Procedure Specification.*

2.4.2. The material to be used for qualification shall be formed to within ±15 per cent of the container diameter.

2.4.3. The thickness of the material shall be the same as or within −15 and +30 per cent of the nominal thickness to be used in production, unless companion bar test specimens are permitted as alternates. In multiple qualification tests, using identical composition rod and base material, where the nominal gage thickness variation of the plates tested has not exceeded $\frac{1}{8}$ in., two such satisfactorily approved qualification tests shall simultaneously qualify both operator and procedure for all intermediate thickness gages existing between the two tests (see Example). Necessary adjustment of

Example:

<table>
<tr><th>Nominal thickness of test plate, in.</th><th>1/16 (.0625)</th><th>3/16 (.1875)</th><th>5/16 (.3125)</th></tr>
<tr><td>Qualification thickness range individual test, in.</td><td>.053-.081</td><td>.160-.244</td><td>.266-.406</td></tr>
<tr><td>Qualification thickness range multiple test</td><td colspan="2">.053-.244</td><td></td></tr>
<tr><td>Qualification thickness range multiple test, in.</td><td></td><td colspan="2">.160-.406</td></tr>
</table>

*To avoid misleading results it is desirable that the base material used in the qualification of a procedure contain amounts of carbon, manganese, chromium, molybdenum, and other alloys approaching the maximum quantities of these elements which may be present in the materials that will be welded in actual construction.

current, voltage and welding speeds are permissible between the limits of the two qualifying test results to produce the best practice, and depth of penetration must be determined on all intermediate weld joints by radiographic inspection or at least two tests of the production item by macro etching. In order to pass the test, visual examination of the cross-section of the weld or of the radiograph shall show freedom from cracks and show complete fusion at the root.

2.4.4. The welded parts shall be heat treated in a manner similar to the finished container.

2.5. Positions of Welding

2.5.1. Positions of welding shall be classified as being flat, horizontal or vertical in accordance with the definitions of welding positions given in 6.1 and 6.2.

2.5.2. Positions of welding qualification test specimens and production containers shall be restricted as follows:

Type of Joint	Permitted Positions for Welding
Butt	Flat
Edge	Flat and vertical
Lap	Flat and horizontal

2.6. Type, Number, and Method of Preparation of Test Specimens

2.6.1. Each type of welded and brazed joint in a container shall be tested for each procedure to be qualified.

2.6.2. Types of test specimens shall be taken for each type of joint as specified in Table 1

TABLE 1. PROCEDURE QUALIFICATION TEST SPECIMENS

Type of Joint	Type of Test Required	Purpose of Test	Number of Specimens	Method of Preparation, Testing and Test Results Required
Longitudinal groove welded butt	Tension (transverse to weld)	Strength	2	See 6.3
	Tension (longitudinal to weld)	Ductility	2	See 6.4
	Standard guided-bend root-bend	Soundness	2	See 6.9
	Standard guided-bend face-bend	Soundness	2	See 6.9
Butt weld of seam in sphere	Companion bar reduced section tension (transverse to weld)	Strength	2	See 6.3
	Flattening	Ductility	1	See 6.7
	Standard guided-bend root-bend**	Soundness	2	See 6.9
	Standard guided-bend face-bend**	Soundness	2	See 6.9
Longitudinal brazed lap	Lap joint tension	Strength	2	See 6.5
	Root-break	Soundness	2	See 6.8
Circumferential welded butt	Flattening	Ductility	1*	See 6.6
	Standard guided-bend root-bend**	Soundness	2	See 6.9
	Standard guided-bend face-bend**	Soundness	2	See 6.9
Circumferential welded lap or joggle butt	Flattening	Ductility	1*	See 6.6
	Alternate guided-bend root-bend**	Soundness	2	See 6.11
Circumferential welded edge or brazed lap	Flattening	Ductility	1*	See 6.6
	Root-break	Soundness	2	See 6.8
Welded attachments on pressure parts	Root-break	Soundness	2	See 6.8

*When weld segments are bent and flattened, 2 specimens shall be chosen at least 180° apart and bent and flattened.

**For joints where guided-bend tests cannot be made, a root-break test or radiographic inspection may be substituted.

Note: Order of Removal of Test Specimens. On circumferential seams each pair of test specimens shall be taken 180° apart. On longitudinal seams each pair of test specimens shall be taken not less than 50 per cent of the seam length apart.

to determine tensile strength, ductility, degree of soundness of welded and brazed joints made under a given Procedure Specification.

2.6.3. Number of test specimens and method of preparation of specimens shall be in accordance with applicable references in Table 1.

2.7. Methods of Testing Specimens and Acceptable Test Results

2.7.1. Method of testing specimens and acceptable test results shall be in accordance with applicable references in Table 1.

2.8. Records

2.8.1. Records of test results shall be kept by the manufacturer and shall be available to those authorized to examine them. A recommended form for recording the results of both procedure and operator qualification tests is given in 7.1.

2.9. Changes in Procedure Specification Necessitating Requalification

2.9.1. The welding procedure must be set up as a new procedure specification and must be completely requalified when any of the changes listed below are made in the procedure. Changes other than those given below may be made in a procedure without the necessity for requalification, provided the procedure specification is revised to show these changes.

V-1. A change in the specification of either or both of the base metals to be welded from one P-Number in 6.13.1 to another P-number. Joints involving two base metals of different P-Numbers shall be qualified even where procedure qualification tests on each of the two base metals to itself have previously been made.

V-2. A change in filler metal analysis or type shall require requalification under the following conditions.

a-1. For metal arc-welding with covered electrodes a change from one F-Number in 6.14 to any other F-Number (see Notes 3 and 4 in 6.14).

a-2. For metal arc-welding with covered electrodes a change in the chemical composition of the weld deposit from the one A-Number to any other A-Number in 6.15 or to a deposit analysis not listed in the table, except as permitted in Note 5 of 6.14 and Notes 2 and 3 of 6.15.

b-1. For gas welding a change from a GAXX to a GBXX type of filler metal and vice-versa.

b-2. For gas welding a change from a silicon-killed to an aluminum-killed type of filler metal and vice versa.

b-3. For gas welding a change in weld metal composition from one A-Number in 6.15 to any other A-Number. Deposit analyses not listed in 6.15 shall require separate qualification. Note 5 of 6.14 and Notes 2 and 3 of 6.15 do not apply.

c-1. For inert-gas metal arc welding a change in electrode from one type to another (such as carbon electrode to tungsten electrode) or from a nonconsumable electrode to a consumable electrode and vice-versa.

c-2. For inert-gas metal arc welding a change in weld metal composition from one A-Number in 6.15 to any other A-Number or to a deposit analysis not listed in 6.15. An increase in silicon content up to $1\frac{1}{2}$ per cent silicon shall not require separate qualification. Note 5 of 6.14 and Notes 2 and 3 of 6.15 do not apply.

d-1. For submerged arc welding a change from a filler metal containing 1.75 to 2.25 per cent manganese to a filler metal containing less than 1.00 per cent manganese or vice versa shall require requalification. The presence or absence of $\frac{1}{2}$ per cent molybdenum in the filler metal analysis shall not require requalification.

d-2. For submerged arc welding a change in filler metal analysis from one A-Number in 6.15 to any other A-Number or to any analyses of weld deposit not listed in the table, except as permitted in Note 5 of 6.14.

e-1. For any other welding process (such as "Thermit" welding or others) a change in the composition of the deposited weld metal from one A-Number in 6.15 to any other A-Number or to an analysis not listed in the table.

V-3. In submerged arc welding, a change in the nominal composition or type of flux used (requalification is not required for a change in flux particle size).

V-4. The addition of other welding positions than those already qualified (see 2.5).

V-5. A decrease of 50 F or more in the minimum specified preheating temperature.

V-6. A change in the heat-treating temperature and time cycle range.

V-7. In metal arc welding, the omission of the backing strip in welding single-welded butt joints; and in gas welding, the addition of the backing strip in welding single-welded butt joints.

V-8. In machine welding, a change from multiple pass per side to single pass per side.

V-9. In machine welding, a change from single arc to multiple arc, or vice versa.

V-10. In inert gas-metal arc welding a change from one type of inert gas to another.

V-11. A change from one welding process to any other welding process.

3. OPERATOR QUALIFICATION

3.1. Applicability

3.1.1. Each operator welding on pressure parts of compressed gas containers referred to in 1.2.1 shall be qualified in accordance with these standards.

3.1.2. If certain changes are made in a Procedure Specification, requalification of operator shall be required as set forth in 3.10.

3.1.3. A welding machine operator shall be considered to be any individual who may make a mechanical or electrical adjustment on the welding machine which may materially affect the quality of the resultant weldment.

3.2. Welding Procedure

3.2.1. The operator in welding on specimens prepared for qualification tests shall comply in all respects with the Procedure Specification.

3.3. Base Material and Its Preparation

3.3.1. Base material and its preparation for welding of qualification test specimens shall comply with the Procedure Specification.

3.3.2. The material used for qualification shall be formed to container dimensions.

3.3.3. The thickness of the material shall be the same as or within -15 and $+30$ per cent of the nominal thickness to be welded by the operator in production. In multiple qualification tests, using identical composition rod and base material, where the nominal gage thickness variation of the plates tested has not exceeded $\frac{1}{8}$ in., two such satisfactorily approved qualification tests shall simultaneously qualify both operator and procedure for all intermediate thickness gages existing between the two tests.* Necessary adjustment of current, voltage and welding speeds are permissible between the limits of the two qualifying test results to produce the best practice, and depth of penetration must be determined on all intermediate weld joints by radiographic inspection or at least two tests of the production item by macro etching. In order to pass the test, visual examination of the cross-section of the weld or of the radiograph shall show freedom from cracks and show complete fusion at the root.

3.4. Positions of Welding

3.4.1. Positions of welding for operator qualification shall be the same as those to be used by the operator in production.

3.5. Type, Number, and Method of Preparation of Test Specimens

3.5.1. The operator shall be tested on each type of welded and brazed joint which he is to weld in production.

3.5.2. Types of test specimens shall be taken for each type of manually welded joint for which the operator is to be qualified as specified in Table 2 to determine the degree of soundness of joints welded by the operator under a given Procedure Specification.

*See Example in 2.4.

TABLE 2. OPERATOR QUALIFICATION TEST SPECIMENS

Type of Joint	Type of Test Required	Number of Specimens	Method of Preparation, Testing, and Test Results Required
Butt	Standard guided-bend root-bend*	2	See 6.9
	Standard guided-bend face-bend*	2	See 6.9
Welded lap and joggle butt	Alternate guided-bend root-bend	2	See 6.11
Brazed lap and welded edge	Root break	2	See 6.8
Welded attachments on pressure parts	Root break	2	See 6.8

*For joints where guided-bend tests cannot be made, a root-break test or radiographic inspection may be substituted.

Note: Order of Removal of Test Specimens. On circumferential seams each pair of test specimens shall be taken 180° apart. On longitudinal seams each pair of test specimens shall be taken not less than 50 per cent of the seam length apart.

3.5.3. Machine welded joints for which the operator is to be qualified shall be radiographically inspected and shall satisfy the acceptability requirements of Section 4—Radiographic Inspection.

(a) The joint so inspected shall be at least 3 ft long. For girth welds on small diameter containers when weld length is less than 3 ft two containers shall be welded and radiographically inspected to satisfy requirements.

3.5.4. Number of test specimens and method of preparation of specimens shall be in accordance with applicable reference in Table 2.

3.6. Methods of Testing Specimens and Acceptable Test Results

3.6.1. Method of testing specimens and acceptable test results shall be in accordance with applicable references in Table 2.

3.7. Retests

3.7.1. In case an operator fails to meet the requirements of one test weld a retest may be allowed under the following conditions:

(a) An immediate retest may be made which shall consist of two test welds of each type on which he failed, all of which shall meet all the requirements specified for such welds.

(b) An additional retest may be made provided there is evidence that the operator has had further training or practice. In this case a complete retest shall be made.

3.8. Period of Effectiveness

3.8.1. The operator qualifications test herein specified shall be considered as remaining in effect indefinitely unless the welding operator is not engaged in a given process of fusion or resistance welding for a period of three months or more,* or there is some specific reason to question an operator's ability.

3.9. Records

3.9.1. Copies of the record for each qualified welding operator shall be kept by the manufacturer and shall be available to those authorized to examine them. A recommended

* The intent of this statement is as follows: An operator who has been qualified for metal arc welding under any given Procedure Specification is not required to requalify for that Procedure unless he has done no metal arc welding for a period of three months or more. In a similar manner the qualification of an operator for gas welding is not considered as having expired unless he has done no gas welding for a period of three months or more.

form for recording the results of both procedure and operator qualification tests is given in 7.1.

3.10. Essential Variables

3.10.1. A welder must be requalified whenever one or more of the changes listed below are made in the performance specification. Other changes other than those listed do not require requalification. A welder that prepares welding procedure qualification test plates meeting the requirements of 3.5 is thereby qualified.

W-1. A change from the filler metal used in the performance qualification to a filler metal having a different F-Number in 6.14 or to a filler metal not included in 6.14, provided that qualification under any F-Number up to and including F4 shall qualify a welder for all lower F-Numbers.*

W-2. The addition of other welding positions than those already qualified (see 3.4).

W-3. A change from upward to downward or from downward to upward in the progression specified for any pass of a vertical weld, other than a wash pass.

W-4. The omission of the backing strip in arc welding single-welded butt joints.

W-5. The addition of the backing strip in gas welding.

W-6. A change from one welding process to any other welding process.

4. RADIOGRAPHIC INSPECTION

4.1. Preparation

4.1.1. Technique of radiographic inspection, where required by ICC container specifications, shall conform to the standards set forth in this section.

4.1.2. All welded joints to be radiographed shall be prepared as follows. The weld ripples or weld surface irregularities, on both the inside and outside, shall be removed by any suitable mechanical process where necessary, so that the resulting radiographic contrast due to any remaining irregularities cannot mask or be confused with that of any objectionable defect. Also the weld surface shall merge smoothly into the plate surface. The finished surface of the reinforcement may have a crown of approximately uniform amount not to exceed 1/16 in.

4.1.3. Single-welded butt joints made the equivalent of double-welded butt joints, may be radiographed without removal of backing strip, provided the backing strip image will not interfere with the interpretation of resultant radiographs.

4.2. Terminology

The following definitions are applicable:

4.2.1. Radiographic Inspection.

The use of x-rays, gamma rays, or both, to detect discontinuities in material by presenting their images on a recording medium suitable for interpretation.

4.2.2. Recording Medium.

Film or a detector which converts radiation into a visible image. The films obtained by the use of x-rays shall be known as "exographs," and those obtained by the use of gamma rays as "gammagraphs." Both types of film shall be generally termed radiographs.

4.2.3. Radiograph.

A visual image on film produced by the penetration of radiation through the material being tested. When two superimposed films are exposed simultaneously in the same film holder, to be viewed later as a superimposed pair, the superimposed pair of exposed film constitutes the radiograph.

4.2.4. Intensifying Screens.

Sheets of lead or layers of fluorescent crystals between which the film is placed to decrease the exposure time and to improve image quality.

*For example, a welder who qualified with Number F4 electrodes is thereby qualified to weld with electrodes listed under Numbers F1, F2, and F3. Independent qualifications are required for Numbers F5 and F6.

4.2.5. Film Holders or Cassettes.

Lightproof containers for holding radiographic film with or without intensifying screens. These film holders or cassettes may be rigid or flexible.

4.2.6. Filters.

Sheets of lead or other materials placed in the radiation beam, either at the x-ray tube, between the specimen and the film or behind the film to improve image quality by selectively removing low energy components from the radiation beam and absorbing scattered radiation.

4.2.7. Penetrameter.

A device whose image in a radiograph is used to determine radiographic quality level. It is not intended for use in judging the size nor for establishing accepting limits of discontinuities.

4.2.8. Penetrameter Sensitivity.

An indication of the ability of the radiographic procedure to demonstrate a certain difference in specimen thickness (usually 2 per cent). It is the ratio (expressed as a percentage) of the thickness of a penetrameter whose outline is descernible in a radiograph to the material thickness (T) of the specimen radiographed.

4.2.9. Radiographic Film Density.

A quantitative measure of film blackening, defined by the equation:

$$d = \log \left(\frac{I_o}{I_t}\right)$$

where

d = density
I_o = the light intensity incident on the film
I_t = the light intensity transmitted through the film.

4.2.10. Radiographically Similar Materials.

Materials which have similar x-ray or gamma-ray absorption characteristics, regardless of chemical composition.

4.2.11. Source.

A machine or radioactive material which emits penetrating radiation.

4.2.12. Source-to-film Distance.

The distance between the radiation producing area of the source and the film.

4.2.13. Material Thickness.

The thickness of material (t) upon which the penetrameter is based. For welds, including repair welds, the material thickness shall be the nominal thickness or actual thickness, if measured, of the strength member, and shall not include reinforcements, backing rings or strips. The strength member is defined as the thinner of the sections being joined.

4.2.14. Energy.

A property of radiation which determines its penetrating ability. In x-ray radiography, energy is usually determined by the accelerating voltage applied to the anode and is expressed as kilovolts (kv) or million electron volts (Mev). In gamma ray radiography, energy is a characteristic of the source and is measured in either kv or Mev.

4.2.15. Maximum Effective Radiation Source Dimension.

The maximum source of focal spot dimension projected on the center of the radiographic film. For example, a cylindrical isotope source whose length is greater than its diameter will have a greater effective radiation source dimension when oriented coaxially in the center of a pipe for a panoramic exposure than when the axis of the source is positioned at right angles to the pipe.

4.3. Penetrameters

4.3.1. The weld shall be radiographed with a technique which will determine quantitatively the size of defects with thicknesses equal to 2 per cent of the base metal ±.001 in., but in no case less than .005 in. To determine whether the radiographic technique employed is detecting defects of a thickness equal to and greater than the percentage specified thickness, gages or penetrameters of the following type shall be placed on the side of the plate nearest the source of radiation except where in cases of necessity the penetrameters shall be located on the side of the weld remote from the source of radiation and used as directed:

(a) The material of the penetrameter shall be radiographically similar to that of the filler metal under examination. For example, any steel, preferably stainless, may be used for steel; and aluminum for aluminum, etc. If the filler metal is not radiographically similar to the base material, then the penetrameter may be placed over the filler metal.

(b) The thickness of the penetrameters shall not be more than 2 per cent of the thickness of the plate ±.001 in. but in no case less than .005 in.

(c) In each penetrameter there shall be three holes of diameters equal, respectively, to two, three, and four times the penetrameter thickness but in no case less than 1/16 in., except when gamma rays are used as a source of radiation the minimum hole need not be less than 3/32 in. The smallest hole must be distinguishable on the radiograph.

(d) Each penetrameter shall carry an identifying number representing, to at least two significant figures, the minimum thickness of the plate for which it may be used. See Figure 1:

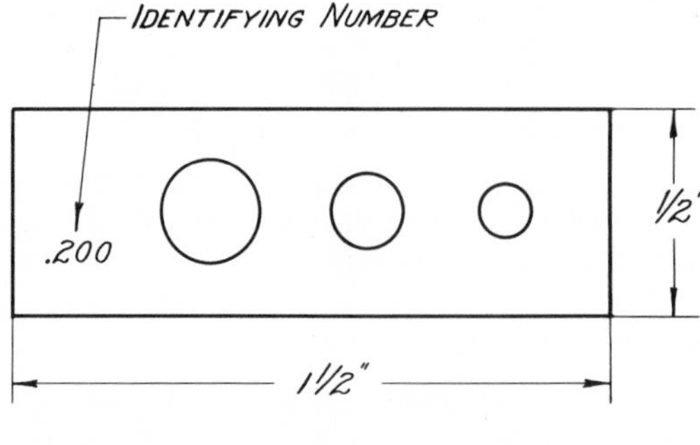

FIG. 1.

(e) The images of these identifying numbers shall appear clearly on the radiograph.

(f) At least one penetrameter shall be used. It shall be placed at one end of the exposed length, parallel and adjacent to the weld seam with the small holes at the other end. If there is any difference between the angularity of the radiation at the two ends, the penetrameter shall be placed at the end of maximum angularity. In special cases the inspector may require the employment of two penetrameters, one at each end of the exposed region.

(g) With the weld reinforcement or the backing strip or both are not removed, a shim shall be placed under the penetrameter such that the total thickness through the weld including the backing strip shall equal the total thickness of the penetrameter, shim and shell within ±5 per cent.

4.4. Film Identification

4.4.1. There should be a plain indication on each film with suitable designation of job number, the container, and seam, as well as the manufacturers' identification symbol or name, radiographer's symbol or name and the date.

4.5. Weld Identification

4.5.1. Identification markers, the images of which will appear on the film shall be placed adjacent to the weld and their locations accurately marked near the weld on the outside surface of the container, so that a defect appearing on the radiograph may be accurately located in the actual weld.

4.6. Film Processing

4.6.1. Film.

Radiographs shall be made of fine grain or extra fine grain film.

4.6.2. Film Quality.

Radiographs presented for interpretation shall be free from blemished or film defects which might mask or be confused with defects in the material being examined. If doubt exists concerning the true nature of an indication on the film, the radiograph shall be rejected. Typical blemishes are as follows:

(a) Fogging caused by light leaks in the processing room or cassettes, defective safelights, exposure marks caused by improper processing, or old film.

(b) Processing defects such as streaking, air bells, water marks, or chemical stains.

(c) Blemishes caused by dirt in cassettes, particularly between intensifying screens and the film.

(d) Pressure or lead marks, scratches, gouges, finger marks, crimp marks, or static electricity marks.

(e) Loss of detail caused by poor film to screen contact in localized areas.

4.6.3. Film Density.

Where single film viewing is used, the density of individual films shall be between 1.3 and 3.0 in the area being examined. Where the superimposed film viewing is used, the density of the superimposed films shall be between 1.8 and 3.0 in the area being examined. When the thickness of the part varies considerably in the area under examination, two films either of equal or different speeds as employed during procedure qualification may be exposed simultaneously in the same film holder and the resultant radiograph interpreted either as single or superimposed film, whichever is better suited for the interpretation of any small portion of the area covered by the exposure. For the small portion of the area under immediate examination, the density of either the single or the superimposed film shall be in accordance with the above requirements.

4.6.4. Darkroom Facilities.

Darkroom facilities, including equipment and materials, shall be capable of producing uniform, blemish-free radiographic negatives.

4.6.5. Film Viewing Facilities.

Viewing facilities shall be so constructed as to afford the exclusion of objectionable background lighting of an intensity that may cause reflection of the radiographic film.

4.6.5.1. Equipment used for radiographic interpretation shall provide the following minimum features:

(a) A light source of sufficient intensity and suitably controlled to allow the selection of optimum intensities for viewing film densities specified in 4.6.3. The required intensity range may be provided by the use of a separate high intensity viewing port. The light enclosure shall be so designed as to provide a uniform level of illumination over the entire viewing surface.

(b) A suitable fan, blower or other cooling device to provide stable temperature at the viewing port such that film emulsions shall not be damaged during 1 minute of continuous contact with the viewing surface.

(c) An opal glass front in each viewing port, except for high intensity viewers used for high density film.

(d) A set of opaque masks to suit the sizes of radiographs to be viewed.

(e) Densitometers shall be provided for assuring compliance with film density requirements.

4.7. Inspection Data

4.7.1. The radiographs shall be submitted to the inspector with such information regarding the radiographic technique as he may request.

4.7.2. Suggested information on a suitable form would be:

(a) Number of films, and type.

(b) Location of each film on the welded joint.

(c) Placement of location markers.

(d) Location of radiation source including angle and source-to-film distances.

(e) Voltage or isotope type and intensity.

(f) Focal spot size or physical dimensions of source.

(g) Type and thickness of intensifying screens and filters.

(h) Material and thickness of area to be radiographed.

(i) Material and thickness of penetrameters and shims.

4.7.3. The following is suggested as a suitable form:

PART NO. .... PART NAME .... CLASS ....

MATERIAL .... X-RAY MACHINE .... SET-UP ....

MARKING ....

FILM TYPE ....

VIEW NO. ....

KV ....

MA ....

EXPOSURE
TIME ....

PENETRAMETER ....

APPROX. GAGE ....

SCREEN TYPE
AND FRONT
GAGE ...... BACK ....

FILTERS ....

FFD ....

ANGLE OF VIEWS ....

LOCATION OF
FILM MARKERS ....

SIZE OF FOCAL SPOT ....

DEVELOPMENT
TIME OF FILM ....

FILM DISTANCE FROM
SURFACE BEING RADIOGRAPHED ....

This form to be completely filled in and submitted in triplicate

LABORATORY .... CERTIFICATION NO. ....

APPROVAL .... DATE ....

4.7.4. The inspector shall be sufficiently familiar with pressure vessel welding to accurately interpret the imperfections and to appraise the seriousness of any defects on the over-all strength and functional use of the container.

4.8. Radioactive Source

4.8.1. When radiographing a circumferential joint by placing a capsule inside the container, the penetrameters may be placed on the film side of the circumferential joint, provided the manufacturer satisfies the inspector that the technique followed in doing the work is known to be adequate.*

4.8.2. When the capsule is placed on the axis of the joint and the complete circumference radiographed with a single exposure, four penetrameters uniformly spaced shall be employed.

4.9. Limits of Acceptability

4.9.1. Welds that are shown by radiography to have any of the following types of imperfections shall be judged unacceptable.

(a) Any zones of incomplete fusion or penetration.

(b) Any type of crack.

(c) Porosity (cavities or slag inclusions) in the weld, exclusive of reinforcements, if the length of any such imperfection L is greater than $1/3\ T$ where T is the thickness of the weld. Weld thickness includes weld reinforcement which is limited by the lesser value of 1/16 in. or $\frac{1}{2}$ the thickness of the plate being welded.

(d) Any group of slag inclusions or cavities in line that have an aggregate length greater than T in a length of $12\ T$ except when the distance between the successive imperfections exceed $6L$ where L is the length of the longest imperfection in the group. These imperfections may be judged acceptable or unacceptable by comparison with porosity standards shown in 6.14 and 6.15 of the current CGA Pamphlet C-3 ("Standards for Welding and Brazing on Thin-Walled Containers," Pamphlet C-3, Compressed Gas Association, Inc.), which indicate maximum permissible defects for the range of plate thickness shown.

(e) Any group of slag inclusions or cavities existing entirely in the reinforcement of the weld but which break out into the surface.

4.10. Retests

4.10.1. Should a radiographed container fail to meet requirements for porosity, penetration and freedom from slag inclusions and cavities, additional containers may be selected for test as provided in applicable container specifications.

4.11. Retention of Radiographs

4.11.1. A complete set of radiographs for each job shall be retained by the manufacturer and kept on file for a period of at least five years.

5. CONTAINER REPAIR

5.1. Authorized repair of welded or brazed joints in compressed gas containers shall be made by welding or brazing. Such repairs must be made by a manufacturer of the same type of ICC container and by a process similar to that used in its manufacture and in accordance with the standards of this section.

5.2. Defects in welded joints in or on pressure parts must be completely removed prior to rewelding.

5.3. Defects in brazed joints must be re-

*A suggested method of proving the adequacy of the radioactive source method of radiography is as follows: A preliminary radiograph should be made with a piece of pipe with penetrameters on both the inside and outside. The diameter of the pipe employed in making this proof radiograph should be substantially the same as that of the job in hand, and its wall thickness the practical equivalent of the over-all thickness of the joint to be radiographed, including both backing ring and reinforcement if these are present in the joint to be examined. The capsule employed in making this proof radiograph, together with all other items of technique such as the location of the capsule and the time of the exposure, should be the same as employed on the actual job. Each penetrameter should be provided with a marker which will show clearly on the film and which will indicate the side of the joint on which it is located: F for the film side, and R for the radiation side.

paired by rebrazing. Brazing is required for replacement of authorized parts that were originally brazed, if the joint is at its prior location.

5.4. Containers during rewelding must be free of materials in contact with the welded joint that may impair the serviceability of the metal in or adjacent to the weld. (Precautions must be taken to prevent acetylene cylinder steels from picking up carbon during repair.)

5.5. Walls, heads, or bottoms of containers with injurious defects or leaks, in base metal shall not be repaired, but may be replaced as provided for in ICC regulations.

5.6. Neckrings, footrings, or other nonpressure attachments authorized by the specification may be replaced or repaired.

5.7. After removal of any parts or attachments, containers must be inspected and defective ones rejected, repaired, or rebuilt.

5.8. Repair of containers must be followed by reheat treatment, testing, inspection, and reporting when and as prescribed by the specification covering their original manufacture in the following circumstances (except as in 5.9.):

(a) When welding or brazing seams in a pressure part of a container.

(b) When welding or brazing on pressure parts of containers of plain carbon steels with carbon over 0.25 per cent or manganese over 1.00 per cent or of alloy steels.

5.9. Each repaired cylinder must be uniformly and properly heat treated prior to test by the applicable method. Heat treatment must be accomplished after all forming and welding operations, except that when brazed joints are used heat treatment must follow any forming and welding operations, but may be done before, during or after the brazing operations.

5.10. Repair of containers must be followed by a proof pressure leakage test at prescribed test pressure and visual examination for weld quality.

5.11. Repair of nonpressure attachments by welding or brazing without affecting a pressure part of the container must be followed by visual examination for weld quality.

6. FIGURES AND TABLES

6.1. Positions of Groove Welds

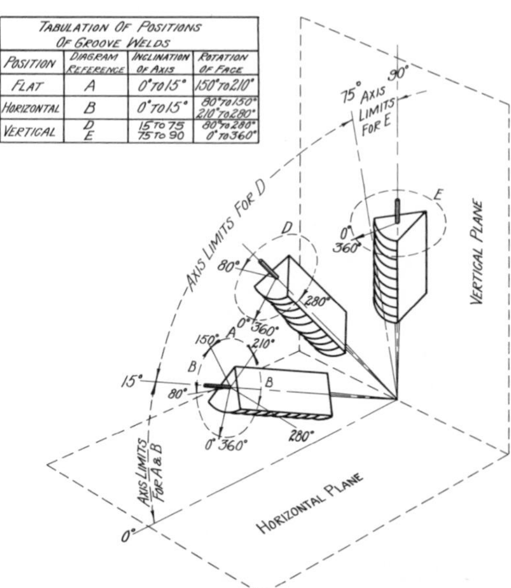

TABULATION OF POSITIONS OF GROOVE WELDS			
POSITION	DIAGRAM REFERENCE	INCLINATION OF AXIS	ROTATION OF FACE
FLAT	A	0° TO 15°	150° TO 210°
HORIZONTAL	B	0° TO 15°	80° TO 150° 210° TO 280°
VERTICAL	D E	15 TO 75 75 TO 90	80° TO 280° 0° TO 360°

FIG. 2.

6.2. Positions of Fillet Welds

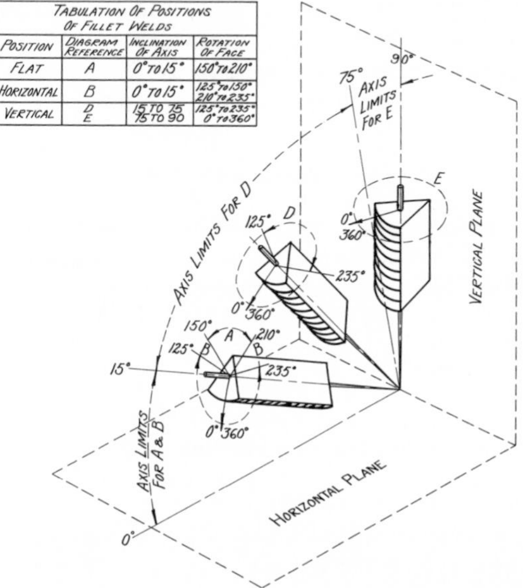

TABULATION OF POSITIONS OF FILLET WELDS			
POSITION	DIAGRAM REFERENCE	INCLINATION OF AXIS	ROTATION OF FACE
FLAT	A	0° TO 15°	150° TO 210°
HORIZONTAL	B	0° TO 15°	125° TO 150° 210° TO 235°
VERTICAL	D E	15 TO 75 75 TO 90	125° TO 235° 0° TO 360°

FIG. 3.

6.3. Tension Test (Transverse to Weld)

6.3.1. Method of Preparation.

Specimen shall be prepared in conformance with the following drawing. Specimen shall be straightened by gradual application of force, not by blows.

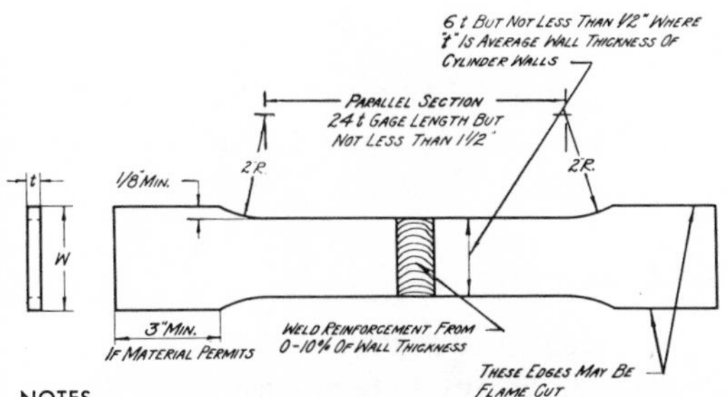

NOTES:

For small cylinders, where insufficient metal for grips is available, it is permissible to increase "W" and weld grip ends to test specimens.

As alternates to the dimensions given above, the length of the parallel section and the width of the specimen may be 8" x 1½" respectively or 2" x 1½" instead of 24t x 6t.

FIG. 4.

6.3.2. Method of Testing.

Before testing, the base metal thickness and least width of the section shall be measured in inches. Specimen shall be ruptured under tensile load and the maximum load in pounds shall be determined.

6.3.3. Acceptable Test Results.

The specimen shall fail at a stress calculated on the parent metal area of not less than two times the maximum stress developed at the minimum prescribed test pressure by the minimum calculated wall thickness.

6.4. Tension Test (Longitudinal to Weld)

6.4.1. Method of Preparation.

The specimen shall be prepared in conformance with the following drawing. Specimen shall be straightened by gradual application of force, not by blows.

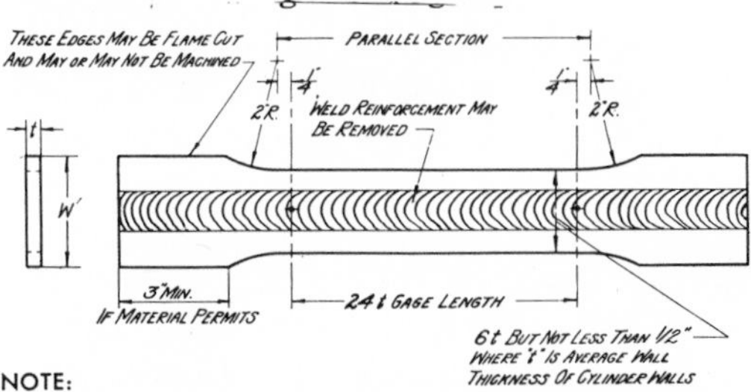

NOTE:

For small cylinders where sufficient metal for grips is not available, it is permissible to increase "W" and weld grip ends to test specimens.

FIG. 5.

6.4.2. Method of Testing.

Gage marks shall be scribed along the weld axis as indicated in the above drawing. The specimen shall be stretched until the minimum required elongation is obtained, or until specimen is ruptured. Measurements for determination of elongation shall be made to nearest 0.01 in.

6.4.3. Acceptable Test Results.

For steels authorized for wall stress not exceeding 25,000 psi at minimum test pressure, elongation must be at least 40 per cent for 2 in. gage length or at least 20 per cent in other cases. For steels authorized for wall stress exceeding 25,000 psi at minimum test pressure, the elongation percentages may be reduced numerically by 2 for 2 in. gage lengths, and by 1 in other cases for each 7,500 psi increment of tensile strength above 50,000 psi to a maximum of 4 such increments.

6.5. Brazed Lap Joint Tension Test

6.5.1. Method of Preparation.

The specimen shall be prepared in conformance with the drawing on the following page. The specimen shall be straightened by gradual application of force, not by blows.

6.5.2. Method of Testing.

The specimen shall be ruptured in tension.

6.5.3. Acceptable Test Results.

The specimen shall fail outside the brazed joint.

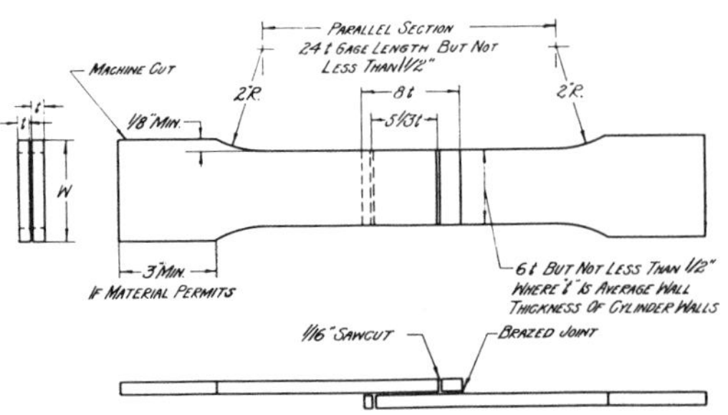

NOTES:

For small cylinders, where insufficient metal for grips is available, it is permissible to increase "W" and weld grip ends to test specimens.

As an alternate to the dimensions given above, the length of the parallel section and the width of specimen may be 4" x 1½" respectively instead of 24t x 6t.

FIG. 6.

6.6. Flattening Test (On Cylindrical Section)

6.6.1. Method of Preparation.

Specimens shall be prepared so as to contain the weld and shall be at least $1\frac{1}{2}$ in. wide.

6.6.2. Method of Testing.

The specimen shall be bent or flattened so that the axis of the weld is bent to a radius with the weld in tension in such a manner that the distance between the outside faces of the weld (without reinforcement) shall be not more than 6 times the total thickness of the overlapping plates.

6.6.3. Acceptable Results.

The specimen shall not crack when flattened as required.

6.7. Flattening Test (On Sphere)

6.7.1. Method of Preparation.

The specimen consists of a sphere. No preparation is required, except that projecting appurtenances may be cut off by mechanical means prior to flattening.

6.7.2. Method of Testing.

By flattening between parallel steel plates on a press with welded seam at right angles to the plates.

6.7.3. Acceptable Test Results.

The sphere shall be capable of being flattened to 50 per cent of the original outside diameter without cracking.

6.8. Root Break Test

6.8.1. Method of Preparation.

Specimen shall be prepared so as to contain at least 1 in. of weld and shall include 3 in. of parent metal to either side of the weld.

6.8.2. Method of Testing.

Specimen shall be broken through the root of the weld by any suitable means.

6.8.3. Acceptable Test Results.

6.8.3.1. Examination shall show brazed joints to be free of objectionable defects.

6.8.3.2. Examination of the fracture of welds shall show neither cracks nor lack of fusion. Gas pockets and slag inclusions shall be permissible only:

(a) When the width of any single slag inclusion substantially parallel with the plate surface is not greater than $\frac{1}{2}$ the width of the weld metal where the slag inclusion is located;

(b) When the total thickness of all of the slab inclusions in any plane at approximately right angles to the plate surface is not greater than 10 per cent of the thickness of the plate;

(c) When there are gas pockets that do not exceed 10 per cent of the wall thickness in greatest dimension and when there are no more than (6) gas pockets of this maximum size per square inch of the weld metal.

6.9. Standard Guided-Bend Test (Root-Bend and Face-Bend)

6.9.1. Method of Preparation.

The specimen shall be prepared in conformance with the following drawing. The specimen may be straightened.

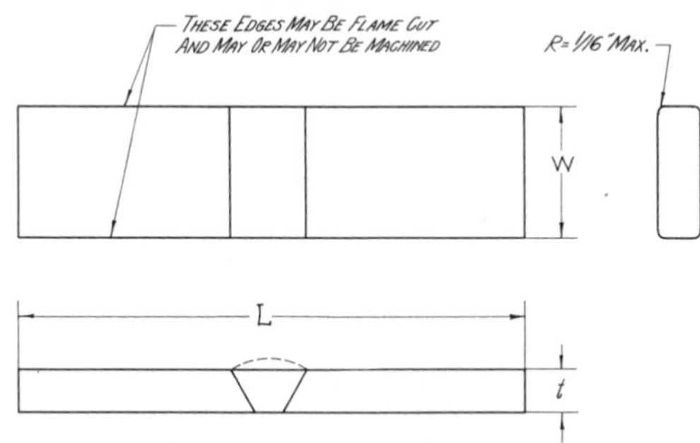

FIG. 7.

Thickness (t) of Specimen, in.	L in.	W in.
Up to .060	100 t	25 t
.060 and over	6 min	$1\frac{1}{2}$

Note: Weld reinforcement and backing strip, if any, shall be removed flush with the surface of the specimen.

6.9.2. Method of Testing.

The specimen shall be tested on the standard guided-bend test jig illustrated in 6.10. The specimen shall be placed on the die with the weld at midspan. Face-bend specimens shall be placed with the face of the weld directed toward the gap; root-bend specimens shall be placed with the root of the weld directed toward the gap. The specimen shall be forced into the die by applying load on the plunger until the curvature of the specimen is such that a 1/32 in. diameter wire cannot be inserted between the die and the specimen.

6.9.3. Acceptable Test Results.

The convex surface of the specimen shall be examined for the appearance of cracks or other open defects. Any specimen in which a crack or othe. defect is present after the bending, exceeding $\frac{1}{8}$ in. measured in any direction, shall be considered as having failed. Cracks occurring on the corners of the specimen during testing shall not be considered unless there is definite evidence that they result from slag inclusions or other internal defects.

6.10. Standard Guided-Bend Test Jig

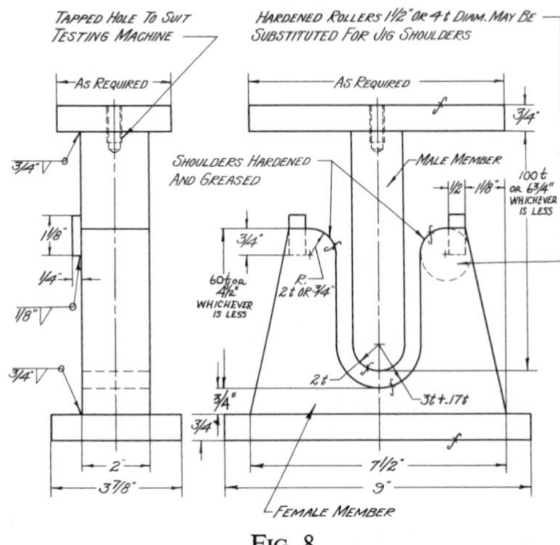

FIG. 8.

Note: For aluminum, use test jig details for testing aluminum welds, as published by The American Welding Society, New York, N. Y.

6.11. Alternate Guided-Bend Root-Bend Test

6.11.1. Method of Preparation.

The transverse weld specimen shall be prepared with weld reinforcement removed. The specimen may be straightened.

6.11.2. Method of Testing.

The specimen shall be tested on the alternate guided-bend test jig illustrated in 6.12. The specimen shall be bent across the weld as shown in Fig. 9, A or B for lap joints and in Fig. 9, C or D for joggle butt joints.

6.11.3. Acceptable Test Results.

The weld shall be examined for the appearance of cracks or other open defects. Any specimen in which a crack or other defect is present after bending, exceeding $\frac{1}{8}$ in. measured in any direction, shall be considered as having failed. Cracks occurring on the corners of the specimen during testing shall not be considered.

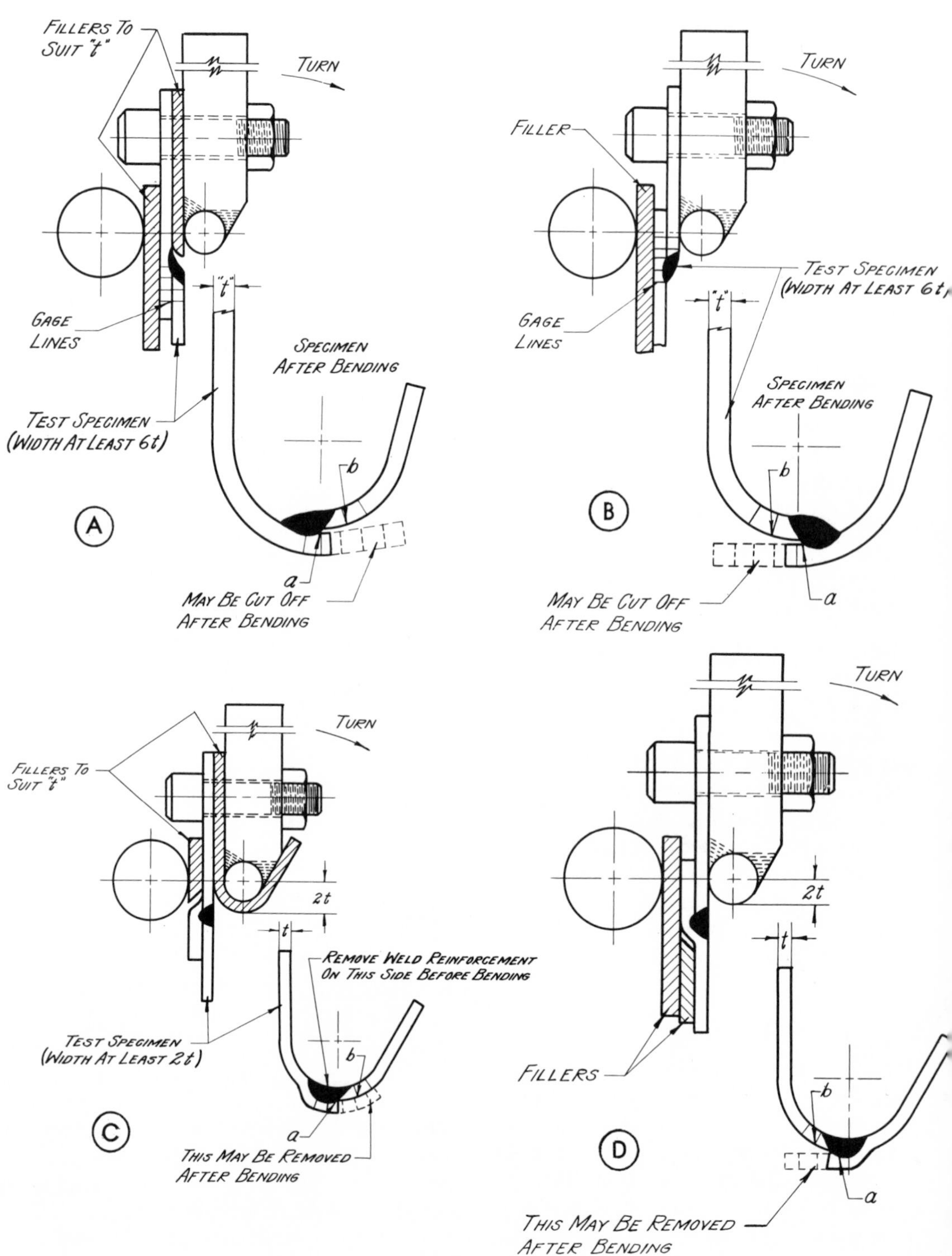

FIG. 9. A, B, C, and D illustrating Section 6.11.2.

6.12. Alternate Guided-Bend Test Jig

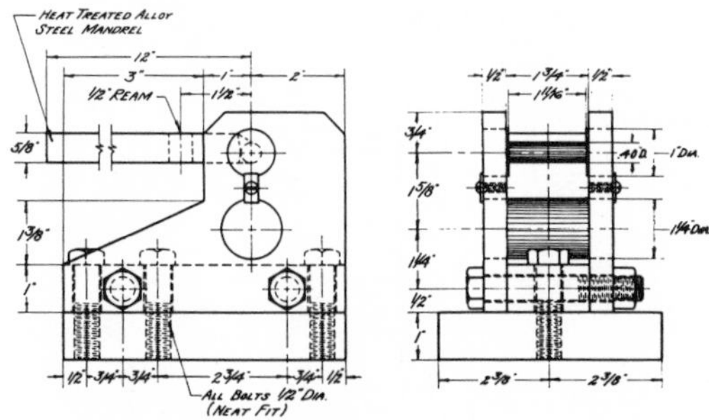

FIG. 10.

6.13. Grouping of CGA Materials For Procedure Qualification

6.13.1. The grouping of materials and electrodes in this section of the welding standards as to P-Number and F-Number classification is made on the basis of hardenability characteristics, where this can logically be done, to reduce the number of welding procedure qualifications required. The grouping does not imply that the base materials or filler metals of different analyses within a group may be indiscriminately substituted for a material which was used in the qualification test, without the consideration of the compatibility of the base materials and filler metals from the standpoint of metallurgical properties, post heat-treatment, design and service requirements, and mechanical properties.

P-Number 1 CGA Material Spec.	P-Number 3 CGA Material Spec.
C-10	LA-1
C-15	LA-2
C-20	LA-3
C-25	LA-4
M-15	LA-5
M-20	LA-6
M-21	LA-7
M-25	LA-8

P-Number 4 CGA Material Spec.	P-Number 8 CGA Material Spec.
A-10	S-10
	S-15
	S-20

Precautionary Note: It is recommended that no welding of any kind should be done when the temperature of the base metal is lower than 0 F. At temperatures between 32 F and 0 F, the surface of all areas within 3 in. of the point where a weld is to be started should be heated to a temperature at least warm to the hand before welding is started.

TABLE 3. CHEMICAL ANALYSIS LIMITS OF CGA MATERIALS LISTED IN 6.13.1

Type Designation	Carbon	Manganese	Phosphorus	Sulfur max.	Silicon	Chromium	Molybdenum	Nickel	Copper	Other Elements
GROUP P-NUMBER 1										
C-10	.05/.15	.30/.60	.045 max.	.050	.15/.35					Note 3
C-15	.10/.20	.30/.60	.045 max.	.050	.15/.35					Note 3
C-20	.15/.25	.30/.60	.045 max.	.050	.15/.35					Note 3
C-25	.20/.30	.30/.60	.045 max.	.050	.15/.35					
M-15	.10/.20	1.10/1.65	.045 max.	.050	.15/.35				.40 max.	
M-20	.15/.25	1.30/1.65	.045 max.	.050	.15/.35					
M-21	.15/.25	1.00/1.30	.045 max.	.050	.15/.35					
M-25	.20/.30	1.30/1.65	.045 max.	.050	.15/.35					
GROUP P-NUMBER 3										
LA-1	.12 max.	.50/.90	.05/.12	.050	.15 max.		.08/.18	.45/.75	.95/1.30	Al .12/.27
LA-2	.12 max.	.50/1.00	.12 max.	.050	.10/.50	.40/1.00		.50/1.00	.20/.50	
LA-3	.20 max.	.45/.75	.045 max.	.050	.50/.90	.45/.70				Zr .05/.25
LA-4	.12 max.	.20/.50	.07/.15	.050	.25/.75	.50/1.25		.65 max.	.25/.55	
LA-5	.15 max.	.90/1.40	.090/.135	.040	.10 max.				.30/.70	Note 4
LA-6	.12 max.	.50/1.00	.040 max.	.050			.10/.30	.50/1.20	.50/1.00	Notes 4 & 5
LA-7	.20 max.	.60/1.00	.045 max.	.045	.15/.30	.15/.50	.15/.35		.20/.50	
LA-8	.15 max.	.30/.60	.040 max.	.050				1.50/2.00	.75/1.25	Notes 4 & 5
GROUP P-NUMBER 4										
A-10	.25/.35	.40/.90	.040 max.	.050	.20/.35	.80/1.10	.15/.25			
GROUP P-NUMBER 8										
S-10	.08 max.	2.00 max.	.030 max.	.030	.75 max.	18.0/20.0		8.00/11.00		
S-15	.08 max.	2.00 max.	.030 max.	.030	.75 max.	17.0/20.0		9.00/13.00		Ti 5 x C min. .60 max.
S-20	.08 max.	2.00 max.	.030 max.	.030	.75 max.	17.0/20.0		9.00/13.00		Cb 10 x C min. 1.00 max.

NOTES 1: Chemical analysis limits are given in per cent.
2: Addition of unspecified elements to obtain alloying effect is not permitted.
3: Specifications for silicon content apply only when silicon killed steel is specified on the order.
4: Grain size 6 or finer according to ASTM Specification E19-46.
5: Only fully killed steel authorized.

6.14. Electrodes and Welding Rods for Qualification

6.14.1. F-Number Grouping of Electrodes and Welding Rods for Procedure Qualification

Weld Metal Type From 6.15			Electrode Classification Number				Welding Rod Classification Number				
								Types GA and GB			
Applicable SA-Spec.	(A-No. Ref.)	(P-No. Ref.)	E-XX20 E-XX24 E-XX27 E-XX28 E-XX30	E-XX12 E-XX13 E-XX14	E-XX10 E-XX11	E-XX15 E-XX16 E-XX18	Type ER	GX-45	GX-50	GX-60	GX-65
SA-233	(A-1)	(P-1)	F1	F2	F3	F4					
SA-316	(A-2)	(P-3)	F1	F2	F3	F4					
SA-316	(A-3)	(P-4)	F1	F2	F3	F4					
SA-316	(A-4)	(P-5)	F1	F2	F3	F4					
SA-298	(A-4)	(P-5)				F4					
SA-298	(A-5)	(P-6)	Stainless Chrome Electrodes			F4					
SA-298	(A-6)	(P-7)				F4					
						F4					
SA-298	(A-7)	(P-8)				F5					
SA-298	(A-8)	(P-8)	Stainless Chrome-Nickel Electrodes			F5					
SA-251	(A-1)	(P-1)						F6	F6	F6	F6
SA-251	(A-2)	(P-3)						F6	F6	F6	F6
SA-371	(A-4)	(P-5)					F7				
SA-371	(A-5)	(P-6)					F7				
SA-371	(A-6)	(P-7)					F7				
SA-371	(A-7)	(P-8)					F7				
SA-371	(A-8)	(P-8)					F7				

Note 1: E-45 series electrodes are not permitted for welding under these standards.

Note 2: For procedure qualification a change from one F-Number to any other F-Number shall require requalification.

Note 3: Covering and filler metal types not listed above may be used but shall require separate qualification for procedure.

Note 4: Qualification of a welding procedure with an electrode up to and including A-5 (SA-298) shall also qualify the procedure for welding with any lower A-Number weld metal of Spec. SA-316 or SA-233. A change in weld metal composition to a higher A-Number within the Group A-1 to A-5 shall require requalification of the procedure. Weld metal Types A-6, A-7, and A-8 shall require separate procedure qualification (see 2.9.1, V-2).

6.14.2. F-Number Grouping of Electrodes and Welding Rods for Performance Qualification

Applicable SA-Spec.	(Equivalent P-No. Ref.)	Electrode Classification Number				Welding Rod Classification Number				
		E-XX20 E-XX24 E-XX27 E-XX28 E-XX30	E-XX12 E-XX13 E-XX14	E-XX10 E-XX11	E-XX15 E-XX16 E-XX18	Type ER	Types GA and GB			
							GX-45	GX-50	GX-60	GX-65
SA–233	(P–1)	F1	F2	F3	F4					
SA–316	(P–3)	F1	F2	F3	F4					
SA–316	(P–4)	F1	F2	F3	F4					
SA–316	(P–5)	F1	F2	F3	F4					
SA–298	(P–5)				F4					
SA–298	(P–6)	Stainless Chrome Electrodes			F4					
SA–298	(P–7)				F4					
SA–298	(P–8)	Stainless Chrome-Nickel Electrodes			F5					
SA–251	(P–1)						F6	F6	F6	F6
SA–251	(P–3)						F6	F6	F6	F6
SA–371	(P–5)					F7				
SA–371	(P–6)					F7				
SA–371	(P–7)					F7				
SA–371	(P–8)					F7				
SA–371	(P–8)					F7				

Note 1: The F-Number grouping of electrodes and welding rods in this table is based essentially on their usability characteristics which fundamentally determine the ability of a welder to make satisfactory welds with a given electrode. Qualification of a welder for any given F-Number, 1 through 4, provided the nominal alloy content does not exceed 6 per cent, automatically qualifies him for any lower F-Number, but not vice versa. F-Number 4 having a nominal alloy content over 6 per cent, and F-Numbers 5, 6, and 7, require independent qualification.

Note 2: E-45 series electrodes are not permitted for welding under these Standards.

Note 3: Covering and filler metal types not listed above may be used, but shall require separate qualification for performance.

6.15. Classification of Weld Metal Analyses for Procedure Qualification

Weld Metal Anal. No.	Equivalent P-Number Ref. for Plate or Pipe	Type of Weld Deposit	Cr (%)	Mo (%)	Ni (%)	Mn Max (%)	Si Max (%)
A-1	(P-1)	Mild steel	—	—	—	0.60	0.50
A-2	(P-3)	Carbon moly	0.50 Max	0.40–0.65	—	0.60	0.50
A-3	(P-4)	Chrome-moly (½–2% Cr)	0.50–2.00	0.40–0.65	—	1.00	1.00
A-3	(P-4)	Nickel-moly	—	0.30–1.00	1.50–3.75	1.00	1.00
A-4	(P-5)	Chrome-moly (2–10% Cr)	2.00–10.00	0.40–1.50	—	1.00	2.00
A-5	(P-6)	High-alloy martensitic	11.00–15.00	0.70 Max	—	2.00	1.00
A-6	(P-7)	High-alloy ferritic	11.00–30.00	1.00 Max	—	1.00	3.00
A-7	(P-8)	Chrome-nickel weld metals containing more than 1% ferrite	AISI Types 302–304–308–309–316–317–318 347–309 Mo–309 Cb				
A-8	(P-8)	Chrome-nickel weld metals which are fully austenitic	AISI Types 310–310 Cb–310 Mo–330				

Note 1: The carbon content of the above weld deposits shall not exceed 0.15 per cent except for Group A-8 where carbon contents up to 0.30 per cent may be used. Higher carbon weld metals shall require separate qualification.

Note 2: Qualification of a procedure with a weld metal type up to and including A-5 shall also qualify welding with any lower A-Number. Weld metals Numbers A-6, A-7, and A-8 require separate qualifications (see 2.9.1, V-2).

Note 3: Weld metals analyses which are not listed above but fall within the material analyses listed in the P-Number grouping shall fall within the A-Number corresponding to the P-Number listing in 6.15. Carbon restrictions (Note 1) shall apply (see 2.9.1, V-2).

Note 4: For submerged arc-welding, qualification with A-1 shall qualify the procedure for welding with A-2 analysis types and vice versa.

Note 5: Weld metals analyses not listed in the above table and of compositions other than those listed in P-1 to P-8 shall require separate qualification.

7. RECOMMENDED FORMS

7.1. Recommended Form For Manufacturer's Record of Qualification Test of Welding Procedures and Operators

Record of............................(Process or Operator)

Manufacturer.................... Address........................

Welding Operator: Name.......................................

Designating No................. Signature........................

Base Metal: Specification...........................

(ICC or ASME Desig.)

Tensile Strength ....

Thickness............inches Welding Position............

Welding Done in Accordance With Manufacturer's Welding Specification No....................

Dated...................

TYPE OF TEST	REMARKS
Tension (Transverse to Weld)	(tensile strength, psi)
	(location of initial failure)
Lap Joint Tension	
Tension (Longitudinal to Weld)	(elongation, per cent)
	(location of initial failure)
Flattening	
Standard Guided-bend Root-bend	
Standard Guided-bend Face-bend	
Alternate Guided-bend Root-bend	
Root-Break	

The undersigned manufacturer certifies that the statements made in this report are correct and that the test specimens were prepared, welded or brazed, and tested in accordance with "Standards for Welding and Brazing on Thin Walled Containers."

Date ....

Signed ....
(Manufacturer)

By ....

7.2. Recommended Form of Procedure Specification (Metal Arc Welding Process)

PROCEDURE SPECIFICATION FOR METAL ARC WELDING OF....................(state class of object to be welded).

SPECIFICATION NO............ DATE........................

PROCESS. The welding shall be done by the metal arc process.

BASE METAL. The base material shall conform to the specifications for............................(insert here references to standard ICC or other Code designations, or give the chemical analysis and physical properties).

FILLER METAL. The filler metal shall conform to Classification Number....................of the AWS-ASME Specification for....................(insert here the title of the desired specification).

POSITION. The welding shall be done in the........ ....(give the position or positions in which the welding will be done. See 6.1 or 6.2).

PREPARATION OF BASE MATERIAL. The edges of surfaces of the parts to be joined by welding shall be prepared by................................(state whether sheared, machined, ground, gas cut, etc.) as shown on the attached sketches and shall be cleaned of substantially all oil or grease and excessive amounts of scale or rust. (The sketches referred to should show the arrangement of parts to be welded with the spacing and details of the welding groove, if used. Such sketches should be comprehensive and cover the full range of material or base metal thicknesses to be welded.)

NATURE OF ELECTRIC CURRENT. The current used shall be........................(state whether direct or alternating, and if alternating give the frequency). The base material shall be on the........................... (state whether negative or positive) side of the line.

WELDING TECHNIQUE. The welding technique, electrode sizes, and mean voltages and current for each electrode shall be substantially as shown on the attached sketches.

APPEARANCE OF WELDING LAYERS. There shall be practically no undercutting on the side walls of the welding groove or the adjoining base material.

CLEANING. All cold slag or flux remaining on any bead of welding shall be removed before laying down the next successive bead.

DEFECTS. Any crack or major blow holes that appear on the surface of any bead of welding shall be removed by chipping, grinding, or gouging before depositing the next successive bead of welding.

PREHEATING. (This paragraph should describe any preheating that will be done.)

HEAT-TREATMENT. (This paragraph should describe any heat-treatment or stress-relieving that is given the welded parts before or after welding.)

....
(Name of manufacturer)

7.3. Recommended Form of Procedure Specification (Oxyacetylene Welding Process)

PROCEDURE SPECIFICATION FOR OXYACETYLENE WELDING OF............................(State class of object to be welded.)

SPECIFICATION NO.................DATE.........................

PROCESS. The welding shall be done by the oxy-acetylene process.

BASE METAL. The base material shall conform to the Specifications for............................(insert here references to standard ICC or other Code designations, or give the chemical analysis and physical properties).

FILLER METAL. The filler metal shall conform to Classification Number....................of the AWS-ASME Specification for...................(insert here the title of the desired specification).

POSITION. The welding shall be done in the ....(give the position or positions in which the welding will be done. See 6.1 or 6.2).

PREPARATION OF BASE MATERIAL. The edges or surfaces of the parts to be joined by welding shall be prepared by...................................(state whether sheared, machined, ground, gas cut, etc.) as shown on the attached sketches and shall be cleaned of substantially all oil or grease and excessive amounts of scale or rust. (The sketches referred to should show the arrangement of parts to be welded with the spacing and details of the welding groove, if used. Such sketches should be comprehensive and cover the full range of material or base metal thickness to be welded.)

SIZE OF WELDING TIP. The range in size of welding tips used shall be as shown on the attached sketch.

NATURE OF FLAME. The flame used for welding shall be............................(state whether a neutral flame or one with slight excess of acetylene is to be used).

METHOD OF WELDING. The method of welding used shall be that known as....................(describe whether "backhand" or "forehand").

SIZE OF WELDING ROD. The size of rod used for the various base material thicknesses shall be as shown on the attached sketch.

NUMBER OF LAYERS OF WELDING. The number of layers of welding used shall be as shown on the attached sketches.

CLEANING. All slag or flux remaining on any layer of welding shall be removed before laying down the next successive layer.

DEFECTS. Any crack or major blow holes that appear on the surface of any layer of welding shall be removed by chipping, grinding, or gouging before depositing the next successive bead of welding.

PREHEATING. (This paragraph should describe any preheating that will be done.)

HEAT-TREATMENT. (This paragraph should describe any heat-treatment or stress-relieving that is given the welded parts before or after welding.)

....

(Name of manufacturer)

7.4. Recommended Form of Procedure Specification (Submerged Arc-Welding Process)

PROCEDURE SPECIFICATION FOR SUBMERGED ARC-WELDING OF................................(State class of object to be welded.)

SPECIFICATION NO.............DATE.............................

PROCESS. The welding shall be done by the ....(machine or semi-automatic; state which) submerged arc process.

BASE METAL. The base material shall conform to the specifications for....................(insert here references to standard ICC or other Code designations, or give the chemical analysis and physical properties).

FILLER METAL. The filler metal shall conform to the following chemical requirements................... (insert here the chemical composition range or trade designation).

FLUX. The flux shall conform to the following chemical requirements............................(insert here the chemical composition range or trade designation).

POSITION. The welding shall be done in the ....(give the position or positions in which the welding will be done. See 6.1 or 6.2).

PREPARATION OF BASE MATERIAL. The edges or surfaces of the parts to be joined by welding shall be prepared by....................(state whether sheared, machined, ground, gas cut, etc.) as shown on the attached sketches and shall be cleaned of substantially all oil or grease and excessive amounts of scale or rust. (The sketches referred to should show the arrangement of parts to be welded with the spacing and details of the welding groove, if used. Such sketches should be comprehensive and cover the full range of material or base metal thickness to be welded.)

NATURE OF ELECTRIC CURRENT. The current used shall be....................(state whether direct or alternating and if alternating give the frequency). The base material shall be on the............................ (state whether negative or positive) side of the line.

JOINT WELDING PROCEDURE. The welding technique such as electrode sizes, travel speeds, and mean voltages and currents for each electrode shall be substantially as shown on the attached sketches.

APPEARANCE OF WELDING LAYERS. There shall be practically no undercutting on the side walls of the welding groove or the adjoining base material.

CLEANING. All cold slag or flux remaining on any bead of welding shall be removed before laying down the next successive bead.

DEFECTS. Any cracks or major blow holes that appear on the surface of any bead of welding shall be removed by chipping, grinding, or gouging before depositing the next successive bead of welding.

PREHEATING AND TEMPERATURE CONTROL. (This paragraph should describe any preheating and control of temperature during and after welding that will be done.)

HEAT-TREATMENT. (This paragraph should describe any heat-treatment or stress-relieving that is given the welded parts before or after welding.)

....

(Name of manufacturer)

7.5. Recommended Form of Procedure Specification (Inert Gas Metal Arc-Welding Process)

PROCEDURE SPECIFICATION FOR INERT GAS

METAL ARC-WELDING OF....................(state class of object to be welded).

SPECIFICATION NO....................DATE..................

PROCESS. The welding shall be done by the........ ....(machine or semi-automatic; consumable or non-consumable electrode; state which)...................inert gas metal arc process.

BASE METAL. The base material shall conform to the specifications for........................(insert here references to standard ICC or other Code designations, or give the chemical analysis and physical properties).

FILLER METAL. The filler metal shall conform to the following chemical requirements........................ (insert here the chemical composition range or trade designation).

SHIELDING GAS............................(state type).

POSITION. The welding shall be done in the ....(give the position or positions in which the welding will be done. See 6.1 or 6.2).

PREPARATION OF BASE MATERIAL. The edges or surfaces of the parts to be joined by welding shall be prepared by....................(state whether sheared, machined, ground, gas cut, etc.) as shown on the attached sketches and shall be cleaned of substantially all oil or grease and excessive amounts of scale or rust. (The sketches referred to should show the arrangement of parts to be welded with the spacing and details of the welding groove, if used. Such sketches should be comprehensive and cover the full range of material or base metal thickness to be welded.)

NATURE OF ELECTRIC CURRENT. The current used shall be............................(state whether direct or alternating and if alternating give the frequency). The base material shall be on the........... ....(state whether negative or positive) side of the line.

JOINT WELDING PROCEDURE. The welding technique such as electrode sizes, travel speeds, and mean voltages and currents for each electrode shall be substantially as shown on the attached sketches.

APPEARANCE OF WELDING LAYERS. There shall be practically no undercutting on the side walls of the welding groove or the adjoining base material.

CLEANING. All cold slag or flux remaining on any bead of welding shall be removed before laying down the next successive bead.

DEFECTS. Any cracks or major blow holes that appear on the surface of any bead of welding shall be removed by chipping, grinding, or gouging before depositing the next successive bead of welding.

PREHEATING AND TEMPERATURE CONTROL. (This paragraph should describe any preheating and control of temperature during and after welding that will be done.)

HEAT-TREATMENT. (This paragraph should describe any heat-treatment or stress-relieving that is given the welded parts before or after welding.)

....

(Name of manufacturer)

SECTION H

Disposition of Unserviceable Cylinders

1. INTRODUCTION

1.1. From time to time Compressed Gas Association, Inc., receives inquiries for recommendations for the disposal of compressed gas cylinders which, for one reason or another, are no longer considered to be serviceable. Some of these unserviceable cylinders failed to qualify for further use under the maintenance requirements of the regulations of the Interstate Commerce Commission or those of the Board of Transport Commissioners for Canada. In other cases cylinders are occasionally found that appear to have been out of service for a long time, are inadequately marked, and are considered to be unsafe for further use. In the latter case the cylinders may either be empty or charged with gas.

1.2. The proper safe disposition of unserviceable compressed gas cylinders is important, as a very substantial potential hazard may exist that must be recognized and evaluated by those who attempt to dispose of them. Where the content of the cylinder is unknown and there is no ready means for identifying its properties, the hazard is especially great. Compressed gas cylinders may have a very high energy content, for some cylinders are charged to high pressures. For example, cylinders used for the common atmospheric gases are usually charged to 2200 psig at 70 F. Also, cylinders may be charged with flammable gases which, when released to the atmosphere, may form explosive mixtures with air, or may form explosive mixtures within the container if the disposal procedure is improper. Some cylinders may contain toxic or poisonous substances or materials that are highly corrosive, oxidizing or reactive.

2. SCOPE

2.1. In preparing these recommendations an attempt was made to anticipate practical considerations that might arise. They do not cover all possible circumstances, nor do they contain all remedies. It is hoped, however, that they will serve a useful purpose and will be of aid to those engaged in the compressed gas industries as well as others who may have an interest, such as the fire services, persons engaged in the scrap metal industries, various units of the military and organizations or individuals that may at some time or other be faced with the problems of safe disposition of unserviceable cylinders.

2.2. The information contained in these recommendations was developed from sources believed to be reliable. The safety suggestions are based upon the experience of members of Compressed Gas Association, Inc., and others. Neither the Association nor its members make any guarantee of results and assume no liability either jointly or severally in connection with the information or the safety suggestions contained herein. It should not be assumed that every acceptable safety procedure is contained or that abnormal or unusual circumstances may not warrant or require further or additional procedures. These recommendations are not to be confused with state, municipal, or insurance requirements or with accepted national safety codes.

3. SUMMARY OF DISPOSAL PROCEDURE

This section briefly summarizes the disposal procedures covered in greater detail in the sections that follow.

3.1. The recommended procedure for disposing of unserviceable cylinders is to discharge the content and cut the cylinder into two or more pieces so that it cannot be used again for pressure service. In order to accomplish this safely, it is necessary to identify the nature of the hazard and take precautions to guard against possible explosion, fire or exposure to toxic gases.

3.2. The first step is to identify the content of the cylinder. If the labeling on the cylinder is not legible, clues to the content may be obtained from the markings on the cylinder pertaining to pressure rating, owner identification, valve outlet, and other factors. If the content of the cylinder cannot be positively identified, further procedure should be carried out under the assumption that the worst possible conditions may exist.

3.3. After identifying the content the cylinder may be discharged or otherwise emptied, using appropriate safety measures to guard against the known hazard. It is desirable to attach a suitable needle valve to the discharge connection of the cylinder valve with a tube leading to a safe area so that the discharge rate can be controlled within safe limits. If the cylinder valve proves to be inoperative, it will be necessary to take extraordinary measures to safely discharge the content.

3.4. After the cylinder has been discharged, it will be necessary to thoroughly purge the inside before cutting unless the content is known to be inert such as nitrogen or carbon dioxide.

3.5. The final step is to cut the cylinder into two or more pieces with a cutting torch or other suitable means. It is recommended that the ICC, CRC or BTC markings on the cylinder be destroyed with a torch in the process.

4. MARKINGS ON COMPRESSED GAS CYLINDERS

4.1. There are certain markings on compressed gas cylinders constructed in accord with specifications of the Interstate Commerce Commission or the Board of Transport Commissions for Canada, which are required to be permanently affixed to the container. In addition there are certain markings or labels that are customarily used by shippers of compressed gases although such markings or labeling may not be required by any regulation now in effect. Product labeling is required by some authorities who have jurisdiction over the handling and use of some of the products of the compressed gas industry, but these requirements are usually related to the end use of the commodity rather than any initial requirement for packaging for transportation. An explanation of these markings follows:

4.1.1. Marking required by the Interstate Commerce Commission or the Board of Transport Commissioners for Canada:

4.1.1.1. All cylinders constructed in accord with the specifications of the Interstate Commerce Commission (ICC), or of the Board of Transport Commissioners for Canada (BTC), bear a mark stamped into the shoulder of the cylinder or otherwise applied in an approved manner, to indicate compliance with the specific specification; for example:

ICC-3A480; indicates construction in accord with Specification ICC-3A for a service pressure of 480 psig;

ICC-3A2015; indicates construction in accord with Specification ICC-3A for a service pressure of 2015 psig;

ICC-4B240; indicates construction in accord with Specification ICC-4B for a service pressure of 240 psig;

*CRC-3A2015 or BTC-3A2015; indicates construction in accord with Specification 3A of the Board of Transport Commissioners for Canada for a service pressure of 2015 lb.

*The Board of Transport Commissioners for Canada (BTC) was formerly known as the Canadian Railroad Commission (CRC). Cylinders constructed in accord with BTC Specifications prior to June 1, 1953 are marked "CRC"; those manufactured after that date require "BTC" marking.

4.1.2. Content Identification Marking:

4.1.2.1. The compressed gas industry has consistently recommended that the marking of cylinders to indicate content should be by written legend, preferably by use of the chemical name or a commonly accepted name of the material contained. This legend should be applied to the cylinder and maintained in a clear and legible manner. The standard for such marking is published as American Standard Method of Marking Portable Compressed Gas Containers to Identify the Material Contained, Z48.1. (Refer to Part III, Chapter 3, Section A.)

4.1.2.2. The Interstate Commerce Commission and the Board of Transport Commissioners for Canada both require that cylinders containing compressed gases, when offered for transportation, must be marked with the proper shipping name as shown in the commodity list which is incorporated in their respective shipping regulations unless specifically exempted by their regulations.

4.1.3. Labeling Required by Interstate Commerce Commission and the Board of Transport Commissioners for Canada:

4.1.3.1. The Interstate Commerce Commission and the Board of Transport Commissioners for Canada require that each cylinder containing a compressed gas, when offered for transportation, must be conspicuously labeled with a label complying with the specifications outlined in their respective regulations. For compressed gases: a "RED" label is required on cylinders of flammable gases; a "GREEN" label is required on cylinders of nonflammable gases; a "POISON GAS" label is required on cylinders containing a Class A poison; and a "POISON" label on cylinders containing a Class B poison. The definitions of the commodity classes noted are specifically contained in the ICC and BTC regulations.

4.1.4. Product Labeling:

4.1.4.1. In some instances manufacturers of compressed gases mark their cylinders with product names. These names may either be proper chemical names or a commonly accepted name, such as a trade name. Where such product labeling is used it is presumed that it is not in conflict with the requirements of the ICC, the BTC, or other authority that may have jurisdiction. Product labeling usually gives adequate information concerning the name of the firm familiar with the charging of the cylinder with gas.

4.1.5. Special Marking Requirements:

4.1.5.1. In some of the compressed gas industries, such as the medical gases industry, additional labeling requirements are imposed by authorities other than those concerned with transportation. The Federal Food, Drug, and Cosmetic Act specifies labeling requirements for certain of the compressed gases which are used medicinally. The United States Department of Agriculture, Live Stock Branch, Production and Marketing Administration also has labeling requirements for cylinders containing some gases. A few state and local authorities have promulgated regulations which require labeling of cylinders in addition to the requirements of the transportation authorities. It is not feasible to list here all of these requirements.

5. IDENTIFICATION OF CYLINDER OWNER OR CYLINDER CHARGER

5.1. It is highly desirable to have the content of any unserviceable gas cylinder disposed of by or under the direction of the firm that charged it. Where this is not feasible it should be done by one engaged in producing or distributing the same product. Such people would usually have or could obtain the detailed information necessary to carry out these operations safely and expeditiously. For this reason every effort should be made to determine the name of the firm that last charged the cylinder.

5.2. In addition to the cylinder marking noted in 4.1.1, the ICC and the BTC require that cylinders made in compliance with their respective specifications, shall be marked at the time of manufacture with a symbol that will identify either the cylinder manufacturer, the initial owner, or both. This symbol is registered with the Bureau of Explosives, New York, N. Y., which assists the Interstate Commerce Commission and the Board of Transport Commissioners for Canada. This symbol will not necessarily indicate ownership of the cylinder at all times, for ownership may change and there is no provision in either ICC or BTC regulations to require that the registered symbol must be changed when necessary to reflect ownership of the cylinder. In order to determine the identity of the firm familiar with the charging of gas into any particular cylinder it is probably more reliable to refer to the product labeling or similar markings on the cylinder.

6. IDENTIFICATION OF CYLINDER CONTENT

6.1. To identify the gas content of cylinders first refer to the marking or labeling described in 4.1.2; 4.1.3; 4.1.4; and 4.1.5. The cylinder should be marked with the chemical name or the commonly accepted name of the commodity contained. If content markings are not available, do not place reliance upon the color of the cylinder, or other color coding to determine the cylinder content. There is no universal color code for cylinder content identification for very good reasons which are given in some detail in a bulletin available from Compressed Gas Association, Inc. Where cylinders have adequate product labeling or labels such as those described in 4.1.5, it is reasonable to rely upon labels for the identity of the cylinder's content. In many instances the cylinders will be marked either by stenciling, stamping, or by the use of appropriate decalcomania containing the name of the gas.

6.2. Cylinders bearing the marks ICC-8, ICC-8AL, CRC-8, CRC-8AL, BTC-8, or BTC-8AL, are restricted to acetylene service. Also, cylinders marked ICC-3D, CRC-3D, or BTC-3D are restricted to poison gas service. Such cylinders may be readily identified by these markings.

6.3. In order to determine whether there

is gas in the cylinder under pressure, the valve should be opened slightly and immediately closed to determine by sound or soap suds whether or not gas is being released. This should only be done in a safe area and with the cylinder properly supported. This test should be made with caution, and only a minimum amount of gas should be permitted to escape. When opening the valve the outlet should be pointed away from anyone in the vicinity. Do not inhale the released gas. Failure of any gas to escape is no guarantee that the cylinder is empty as it is possible that the valve may be inoperative.

6.4. It may be necessary to determine whether a cylinder contains gas in the liquid phase. It is not always possible to make such a determination by sound. However, with some thin-walled cylinders it is sometimes possible to hear the liquid in the cylinder when it is shaken. A more exact determination can be made by laying the cylinder on its side (*Caution*—in a well ventilated area or with the workmen wearing gas masks with independent air supply) and opening the valve slightly and immediately closing it, noting the escape of gas in the liquid phase. Care should be taken to be sure that the valve outlet is pointed down and away from the operator when the valve is opened. Cylinders for some liquefied gases are often marked with a tare weight. Weighing of such cylinders may therefore establish the presence of gas in the cylinder.

6.5. Valve outlet thread dimensions have been standardized by the compressed gas industry. These dimensions are published by Compressed Gas Association, Inc., as Pamphlet V-1 and by American Standards Association as "American Standard Compressed Gas Cylinder Valve Outlet and Inlet Connections, B57.1." The Canadian Standards Association has also adopted these standards. A comparison of the valve outlet thread on a cylinder valve with the recommended standard as contained in CGA V-1, ASA-B57.1 or CSA-B96, may provide a good check in the determination of cylinder content. This method, however, is not conclusive as the same dimensional valve outlet standard is sometimes utilized for a number of gases. It is also possible that a cylinder may not be equipped with a valve using the recommended standard.

6.6. While it is recommended that one should not rely upon safety device design as a means for checking cylinder content, it may sometimes be used as an aid in classifying the probable content of a cylinder. The Safety Relief Device Standards of Compressed Gas Association, Inc., contain a table which indicates the types of devices recommended for use on ICC and BTC specification containers. (Refer to Part III, Chapter 2, Section A.). Unless very familiar with the use of safety devices on compressed gas cylinders one should not rely upon the type of safety device as an indication of the specific commodity contained in the cylinder.

7. DISPOSITION OF CYLINDER CONTENT

7.1. The content of an unserviceable cylinder should be discharged safely by burning flammable gases or by discharging noxious gases through vent tubing to a safe area and the cylinder purged of any flammable gas or gas mixture before any attempt is made to destroy the cylinder. It is desirable to have the content of cylinders removed only by qualified persons or under their supervision.

7.2. If the content of a cylinder must be removed by some method other than through a properly operating valve, care must be used to release the content slowly so that the released energy does not cause the cylinder to move out of control. Under such circumstances, it is recommended that the cylinder be firmly held in a vise or by other suitable means, so that any danger of the cylinder being tossed around is minimized. In addition, the discharge connection of the cylinder valve should be connected through tubing and a needle valve to prevent excessive discharge rates.

7.3. In the event that a cylinder valve is damaged, preventing the discharge of the commodity in a normal manner, it may be possible to release the pressure in the cylinder through the safety device. This procedure is not generally recommended. It should not be attempted where the gas content may be noxious or where the ejection of the safety relief device under pressure may be hazardous, without proper protection to personnel and discharging into a safe location. The use of this method of releasing cylinder content should be restricted to qualified persons familiar with gas cylinders and their safety devices.

7.4. Many of the compressed gases may be safely vented to the atmosphere. Obviously, the inert gases can be so vented without creating any undue hazard. If flammable gases or gases which may present a health hazard are vented to the atmosphere, it should be done very carefully and at an isolated location in such a manner and at such a rate as is necessary to assure safety from fire or contamination of the atmosphere.

7.5. As an extreme means of discharging the gas contained in an unserviceable cylinder where the valve is damaged to such an extent that it cannot be used for the release of gas, it is possible to accomplish release of pressure by perforating the cylinder with a high powered rifle bullet. This method should be used only as a last resort. It should only be attempted by qualified individuals at a suitable isolated place, with the cylinder located where possible in a depression below surrounding grade level. The rifle should be fired from a distance not less than 50 yd from the cylinder through a protection shield that is adequate to protect personnel from any flying fragments. Wind direction and intensity must be considered.

7.6. After releasing the pressure and discharging the content from an unserviceable cylinder, the cylinder should be purged, if it previously contained a flammable or toxic material, before any attempt is made to destroy it with a cutting torch. Purging can be accomplished with the use of inert gases, steam, or by filling the cylinder with water. Detailed instructions for purging containers that have held combustible materials are contained in a pamphlet published by the American Welding Society, 345 E. 47th St., New York, N. Y. entitled "Safe Practices For Welding And Cutting Containers That Have Held Combustibles." This pamphlet is available at a nominal charge.

8. DISPOSITION OF UNSERVICEABLE EMPTY CYLINDERS (OTHER THAN ACETYLENE)

8.1. Having removed the gas content of a cylinder and after purging it if it previously contained a flammable or toxic gas as mentioned in 7.6, the empty cylinder may be destroyed with a cutting torch, or by other appropriate means that will make it unusable as a pressure vessel. It is recommended that the ICC, CRC or BTC markings on the cylinder be destroyed with a torch and that the remainder of the cylinder be cut into two or more pieces, or that it be destroyed by other suitable means.

9. DISPOSITION OF UNSERVICEABLE ACETYLENE CYLINDERS

9.1. As noted in 6.2, cylinders marked ICC-8, ICC-8AL, CRC-8, CRC-8AL, BTC-8, or BTC-8AL are authorized for acetylene only. These cylinders are filled with a porous mass which serves as an absorbent for a solvent that is utilized to retain acetylene in solution. An unserviceable acetylene cylinder, therefore, may retain varying quantities of solvent and gas. Before attempting to destroy one of these cylinders it is important that every precaution be taken to de-energize the cylinder. It is always best to have this work done by personnel completely familiar with these cylinders. The following procedures should be observed:

9.1.1. The cylinder should be removed to an isolated location where escaping gas will present no hazard to personnel or property.

9.1.2. The cylinder valve should be opened and should remain open for at least 24 hours to permit the discharge of residual acetylene.

9.1.3. The cylinder valve should be removed

very carefully, making sure that all gas pressure has been released before completely unscrewing the valve. This operation should be performed with caution to prevent accident in the event of release of gas which may be retained in the cylinder if the valve happened to be clogged.

9.1.4. All safety devices should be removed from the cylinder. It should be noted that these devices may be in both ends of the cylinder.

9.1.5. The cylinder should be placed in a horizontal position. If there is more than one cylinder they should be stacked, one cylinder wide and not more than 5 cylinders high. Wood or other fuel should be placed around the cylinders and ignited. The cylinders should be kept in an intense fire for at least 6 hours. *Warning* is repeated that this operation should be performed in a location remote from any building or from any place where people may work or assemble.

9.1.6. After 6 hours in the fire, apply a flame to the valve spud so as to ignite any acetone vapors which may issue from the opening. This can be readily done by using a lighted cutting torch.

9.1.7. Using a cutting torch, cut out the valve spud and all markings on the head of the cylinder, including the registered symbol, serial number, and other identification marks.

9.1.8. Using a cutting torch, make a circumferential cut midway in the cylinder so the filler can be cracked and the cylinder broken in half.

9.1.9. Allow the cylinder halves to lie as long as acetone vapors continue to burn.

9.1.10. Never store the scrapped cylinder in a confined place.

10. DISPOSITION OF UNSERVICEABLE CHARGED CYLINDERS WHEN CONTENT CANNOT BE DISCHARGED

10.1. It may sometimes be necessary to dispose of a cylinder which cannot be de-energized safely by releasing the pressure and removing the content. Under such conditions disposition must be made in such a manner that there is no hazard to life or property at the time of disposition or later. The alternatives for disposing of such a cylinder are: (1) by dumping it into a large body of deep water when the cylinder and content are sufficiently heavy to assure the sinking of the cylinder; or (2) burying it in a place and at a sufficient depth to assure that it can do no harm to persons or property.

11. SUMMARY

11.1. Unserviceable compressed gas cylinders, which have been de-energized by release of all pressure and by removing the commodity, and provided that they have been purged of all flammable or noxious gases, should be scrapped by rendering the container unsuitable for further use as a pressure vessel. Permanent markings and identification symbols should be destroyed with a torch, and the cylinder cut into at least 2 pieces.

11.2. Unserviceable cylinders which contain gases should preferably be de-energized and prepared for destruction by those most familiar with these cylinders and the commodities contained in them. These individuals are usually the manufacturers or distributors of compressed gases.

11.3. Individuals or organizations not engaged in the compressed gas industry, who may have reason to dispose of an unserviceable compressed gas cylinder, should acquaint themselves with the names and addresses of the nearest manufacturers or distributors of the type of compressed gas for which the cylinder was intended and seek advice before attempting to destroy the cylinder. This applies to such organizations as scrap dealers, the fire services, military organizations, and others.

11.4. Compressed Gas Association, Inc., New York, N. Y., will assist inquirers through counsel and advice to the best of its ability to do so. Copies of ICC regulations and BTC regulations may be obtained for a nominal fee from the Bureau of Explosives, 63 Vesey St., New York, N. Y. and The Printer to the Queen's Most Excellent Majesty, Ottawa, Canada, respectively.

CHAPTER 4

Compressed Gas Cylinder Valve Connection Systems

Accidentally connecting a compressed gas cylinder valve outlet with equipment not designed for the gas contained in the cylinder may result in serious hazards. Because of this, standard valve outlet connections have been established for valves used with cylinders containing the different gases. These standard connections are made so that the valve connection for one gas will not fit the connections prescribed for other incompatible gases. Standard valve outlet connections thus help to protect personnel and equipment.

Standards for gas cylinder valve connection systems developed by Compressed Gas Association, Inc., are used throughout the United States and Canada. The major CGA connection standards are summarized in this chapter as follows:

Section A. American-Canadian Standard Cylinder Valve Outlet and Inlet Connections.

Section B. The Pin-Index Safety System for Medical Gas Flush-Type Connections.

Section C. The Diameter-Index Safety System for Low-Pressure Medical Gas Connections.

SECTION A

American-Canadian Standard Cylinder Valve Outlet and Inlet Connections

This section summarizes the uniform American-Canadian Standards for the outlet and inlet connections of compressed gas cylinder valves. Detailed drawings which give the full specifications for each standard connection and component in the system are published in Pamphlet V-1 of Compressed Gas Association, Inc., entitled "Compressed Gas Cylinder Valve Outlet and Inlet Connections," 1965 edition. The pamphlet presents American Standard B57.1-1965 and Canadian Standard B96-1965.

INTRODUCTION*

History

The first effective efforts to develop standards for compressed gas cylinder valve connections followed immediately after World War I, and were inspired by the difficulties encountered both by industry and the military

* This Introduction is for general information and is not a part of American and Canadian Standard Compressed Gas Cylinder Valve Outlet and Inlet Connections.

services because of the multiplicity of connections that were then in use.

Through the activity of the Gas Cylinder Valve Thread Committee of Compressed Gas Manufacturers' Association, Inc., material progress was made through the years that followed with the result that, when our country became involved in World War II, the gas industries themselves had materially improved this situation. Several of the compressed gas industries had achieved virtual standardization at tremendous cost for replacement of valve equipment. Their standards, however, were not completely formalized nor fully coordinated with other related standards. Much of the progress between World War I and World War II was the result of interest in this problem by the Federal Specifications Board.

The circumstances surrounding industrial and military users of compressed gases during World War II brought into clear focus the need for acceleration of the standardization project for cylinder valve connections. They created not only the necessity but also a splendid opportunity for the compressed gas industry, the military services, and other federal agencies to study cooperatively the standardizing problems of valve outlet threads. These studies resulted in closer definition and appreciation of each valve outlet and in a more balanced relationship between the many types and sizes.

When the standards associations representing Great Britain, Canada and the United States met in Ottawa in October, 1945, to consider unification of screw threads, a fairly well developed plan for standardization of compressed gas cylinder valve connections was presented to the Conference by the Valve Thread Standardization Committee of Compressed Gas Manufacturers' Association, Inc. These proposed standards represented the experience and knowledge of compressed gas manufacturers, valve manufacturers, and the needs and requirements of varied users of gas cylinder valves, including the military services and other federal agencies. Approval of these standards to the extent to which they were then developed was given by the U. S. Department of Commerce, the U. S. Army, and the U. S. Navy through the Interdepartmental Screw Thread Committee following a joint meeting with the representatives of CGMA in August, 1945. Much progress was made later in this year at the Canadian Section Meeting of CGMA tending to unify United States and Canadian practices. During January, 1946, through conference between representatives of CGMA Valve Thread Standardization Committee and the Interdepartmental Screw Thread Committee in Washington agreements were reached that resulted in final approval of considerable additional gas cylinder valve thread data for inclusion in the National Bureau of Standards' Handbook H-28.

The Compressed Gas Manufacturers' Association, Inc. changed its name in January, 1949, and its Valve Thread Standardization Committee became the Valve Standards Committee of Compressed Gas Association, Inc. During the interval between January, 1946 and February, 1949, this committee developed its standards sufficiently to present them to the American Standards Association and the Canadian Standards Association. They were accepted as standards in 1949, accomplishing an objective that was established some 30 years before. Since that date additional connections have been developed and have been included in subsequent editions of the standard. Similarly, alternate connections have been removed as they became obsolete.

Medical Gas Connections

As early as the spring of 1940, it was evident to various medical societies, as well as to the manufacturers of medical gases that a system should be devised to prevent the interchangeability of medical gas cylinders equipped with flush-type valves when used with medical gas administering apparatus. Various means for accomplishing this were studied. The most difficult obstacle to be overcome was that of devising a system that would permit the adjustment of existing apparatus without interfering with its use and without requiring that it be returned to the manufacturer for conversion. The system (see Section B of this chapter) contained in these standards, and known as "The Pin-Index Safety System for Flush-Type Cylinder Valves," is the result of the concerted efforts of the companies and organizations concerned. This standard was submitted to the International Organization for Standardization and has been adopted as an international standard.

Suggestions for Changes

Suggestions, based upon experience gained in its use, will be welcomed. Such suggestions should be sent to the American Standards Association, Incorporated, New York, N. Y., the Canadian Standards Association, Ottawa, Canada, or to Compressed Gas Association, Inc., New York, N. Y.

1. GENERAL

1.1. The Valve Standards Committee of Compressed Gas Association, Inc., knitting together the experience and knowledge of gas producers, valve manufacturers, military services, and other federal agencies with the uses and requirements of varied consumers, established detailed dimensions for all elements of valve outlet and inlet connections.

1.2. These standards represent the best existing American and Canadian practice and provide a coordinated plan for the future. Standard outlet connections for the respective gases are fully defined and complete in themselves. The relation of one outlet to another is fixed so as to minimize undesirable connections.

2. OUTLET CONNECTIONS

2.1. Basic Thread Divisions

2.1.1. The threaded outlets are separated into four basic divisions—internal and external, as well as right-hand and left-hand. Within each of the four divisions, further separation is made by varying the pitch and diameter of the threads. The diameters within each division are so spaced that adjoining sizes either will not enter or will not engage.

2.1.2. There is a separate division of flush outlets for medical gases which uses the Pin-Index Safety System (see Section B of this chapter).

2.1.3. As far as practicable, the design of connections and assignment of the connections to gases has been made so as to prevent the interchange of connections which may result in a hazard. With the exception of outlets having taper pipe threads which seal at the threads, each outlet provides for screw threads which do not seal but merely hold the nipple against its seat. These screw threads have the national form, but are not in the regular series.

2.1.4. Past practice has firmly established many outlet connections for specific gases or groups of gases and in many cases these connections were retained. Small differences in the threads and other elements of the same connection were reconciled into one form and size, properly recorded and defined. By adhering to existing outlets where practicable, it was possible to put the new standard system into effect without the inconvenience and expense of a cumbersome and costly change-over. Alternate and costandards have been established for some gases.

2.1.5. Keeping the established practice in mind when classifying and assigning the gases to their outlets, an effort was made to follow a plan whereby right-hand threads would be used for nonfuel gases and left-hand threads would be used for fuel gases. Left-hand threads are identified by a groove on the hexagon nut. An external thread is used on the valve in most cases, but some important groups of gases have an internal thread on the valve outlet.

2.2. Numbering System

2.2.1. The numbering system shown in Table 1 provides for definite symbols for the complete outlet connection, for the valve outlet only, for the mating assembly, as well as for the nipple, nut and washers. The last digit of the designating numbers for the outlet connec-

TABLE 1. NUMBERING SYSTEM FOR VALVE OUTLET CONNECTIONS

Conn. No.	001 Valve Outlet		002 Mating Assembly		003 Nipple	004 Nut or Con-verter	005 Washer
	No.	Thread	No.	Thread			
120	121	.373″—24NGO-RH-EXT	122	.375″—24NGO-RH-INT	123	124	—
160	161	⅛″—27NGT-RH-INT	162	⅛″—27NGT-RH-EXT	163	—	—
200	201	.625″—20NGO-RH-EXT	202	.628″—20NGO-RH-INT	203	204	—
240	241	⅜″—18NGT-RH-INT	242	⅜″—18NPT-RH-EXT	243	244	245
280	281	.745″—14NGO-RH-EXT	282	.750″—14NGO-RH-INT	283	284	—
290	291	.745″—14NGO-LH-EXT	292	.750″—14NGO-LH-INT	293	294	—
300	301	.825″—14NGO-RH-EXT (for conical nipple)	302	.830″—14NGO-RH-INT (conical nipple)	303	304	—
320	321	.825″—14NGO-RH-EXT (for flat nipple)	322	.830″—14NGO-RH-INT (flat nipple)	323	324	325
330	331	.825″—14NGO-LH-EXT (for flat nipple)	332	.830″—14NGO-LH-INT (flat nipple)	333	334	335
350	351	.825″—14NGO-LH-EXT (for round nipple)	352	.830″—14NGO-LH-INT (round nipple)	353	354	—
360	361	½″—14NGT-RH-EXT	362	½″—14NGT-RH-INT	—	364	365
400	401	.850″—14NGO-RH-EXT (for round nipple)	402	.855″—14NGO-RH-INT (round nipple)	403	404	—
410	411	.850″—14NGO-LH-EXT	412	.855″—14NGO-LH-INT	413	414	—
420	421	.850″—14NGO-RH-EXT (for flat nipple)	422	.855″—14NGO-RH-INT (flat nipple)	423	424	425
510	511	.885″—14NGO-LH-INT	512	.880″—14NGO-LH-EXT	513	514	—
520	521	.895″—18NGO-RH-EXT	522	.899″—18NGO-RH-INT	523	524	—
540	541	.903″—14NGO-RH-EXT	542	.908″—14NGO-RH-INT	543	544	—
580	581	.965″—14NGO-RH-INT	582	.960″—14NGO-RH-EXT	583	584	—
590	591	.965″—14NGO-LH-INT	592	.960″—14NGO-LH-EXT	593	594	—
620	621	1.030″—14NGO-RH-EXT (with groove)	622	1.035″—14NGO-RH-INT (flare converter)	—	624	625
640	641	1.030″—14NGO-RH-EXT (with ⅛″—27NGT-RH-INT)	642	1.035″—14NGO-RH-INT (NGT converter)	—	644	645
660	661	1.030″—14NGO-RH-EXT (without groove)	662	1.035″—14NGO-RH-INT (nut)	663	664	665
670	671	1.030″—14NGO-LH-EXT	672	1.035″—14NGO-LH-INT	673	674	675
800	801	⅜″—18NGT-RH-INT	802	Yoke connection	803	—	805
820	821	1.030″—14NGO-RH-EXT (without groove)	822	Yoke connection washer on outer face	823	—	825
840	841	1.030″—14NGO-RH-EXT (without groove)	842	Yoke connection washer inside of recess	843	—	845
860	861	Flush outlet for medical gases	862	Yoke connection	863	—	865
1310	1311	Yoke outlet for air	1312	Yoke connection	1313	—	1315
1320	1321	.825″—14NGO-RH-EXT (for small round nipple)	1322	.830″—14NGO-RH-INT (small round nipple)	1323	1324	—
1340	1341	.825″—14NGO-RH-EXT (for large round nipple)	1342	.830″—14NGO-RH-INT (large round nipple)	1343	1344	—

Each complete outlet connection is different. However, one or more of its components may be the same as those used on other connections, as follows:

VALVE OUTLETS 241 and 801 are identical.
661, 821 and 841 are identical.
621, 641 and 661 are similar.

NIPPLES 323, 333 and 423 are identical.
353, 403 and 413 are identical.
513, 583 and 593 are identical.
663 and 673 are identical.

NUTS 304, 324, 1324 and 1344 are identical.
334 and 354 are identical.
404 and 424 are identical.

WASHERS 325, 335 and 425 are identical.
625 and 645 are identical.
665 and 675 are identical.

tions and components has the following significance:

0—complete outlet connection,
1—valve outlet,
2—mating assembly (see Table 1),
3—nipple,
4—nut or converter,
5—washer,
9—plug.

2.3. NGO Thread Details

2.3.1. Most of the connecting threads are of the National Gas Outlet (NGO) type. The designation NGO is a separate and distinct symbol for valve outlet threads. This symbol was suggested and designated by the Interdepartmental Screw Thread Committee, representing the Federal services to provide for the peculiar needs of the industry. Detailed dimensions of NGO threads are given on the respective pages showing the complete outlet connection and in Table 3 of CGA Pamphlet V-1.

2.3.2. An allowance (minimum clearance) of from 0.0020 to 0.0050 in. between the mating parts is established for NGO threads to provide the desired looseness of fit at the threads, and to assure interchangeability between products of different manufacturers who lacked a common standard in the past. The tolerances are in the direction of greater looseness and are determined on the basis of NS-3 data, except for the major diameter of the external threads for which the tolerance is limited to 0.0050 in.

2.4. Other Thread Details and Gaging

2.4.1. In addition to the NGO threads, other types of threads are used on the outlet connections. The types of threads specified in the various outlet connections are listed in the following table along with references to information on limits of thread size and gaging.

Type of Thread	Limits of Size of Thread	Gages and Gaging
NGO	Table 3, p. 11	H28(I), §VI
NGT	Table 4, p. 53	Pages 53 and 54
NPSL	H28(II), Table VIII.6, p. 9	H28(II), §VII, ¶9, p. 11. ASA-B2.1
NPT	H28(II), Tables VII.1 & VII.2, pp. 4-5	H28(II), §VII, ¶8, p. 11. ASA-B2.1

Notes: H28 refers to National Bureau of Standards Handbook H28 (1957) "Screw Thread Standards for Federal Services." (I) and (II) indicate Parts I and II of H28.

ASA-B2.1 refers to American Standard Pipe Threads (except Dryseal), B2.1-1960.

Page numbers refer to CGA Pamphlet V-1 unless otherwise indicated.

2.5. Valve Clearance

2.5.1. The maximum radius of any part of the valve from its centerline has been specified to insure clearance for the smallest ($3\frac{1}{8}$ in.) standard cylinder valve protecting cap.

2.6. Small Valve Series

2.6.1. Some gases, besides being available in commercial cylinders of the more conventional size, are likewise supplied in small cylinders which incorporate a valve different from the one on the larger cylinders. Thus, acetylene connection No. 200 with a $\frac{3}{8}$ in. NGT inlet thread is used on motorcycle type cylinders holding 10 cu ft or less of gas; acetylene connection No. 520 with a $\frac{3}{8}$ in. NGT inlet thread is used on lighting type cylinders holding 40 cu ft or less of gas; phosgene connection No. 160 with a special $2\frac{3}{8}$ in. diameter inlet thread incorporates a valve with a gas-tight protection cap to close off any entrapped gas, and connection No. 120 is designed for very small cylinders in which rare atmospheric gases are handled.

3. INLET THREADS

3.1. Inlet threads on the valve and in the cylinder neck have also been standardized and are included in CGA Pamphlet V-1.

4. ADAPTERS

4.1. In the standardization of compressed gas valve outlet connections, more than one outlet is provided for some gases. To provide interchangeability of equipment for the same gas, adapters may be required. See Appendix of CGA Pamphlet V-1 for details.

TABLE 2. GASES AND CONNECTION SYMBOLS

Gas	Standard Connection USA and Canada	Alternate Standard Connection[1]	
		USA and Canada	Canada Only
Acetylene	510	300	410
Acetylene, small valve series	200, 520	—	—
Air, for human respiration	*950, 1310, 1340	—	400[3]
Air, industrial	590	—	400[3]
Ammonia, anhydrous	240, 800	—	—
Argon	580	—	—
Boron trifluoride	330	—	—
Bromochloromethane	620	—	—
Butadiene	510	—	—
Butane	510	300	—
Carbon dioxide	320, 940	—	420[4]
Carbon dioxide-ethylene oxide mixture	350	—	—
Carbon dioxide-oxygen mixture (CO_2 over 7%)	320, 940	—	—
Carbon monoxide	350	—	—
Chlorine	820	660, 840	—
Chlorine trifluoride (corrosive liquid)	670	—	—
Cyclopropane	510, 920	—	—
Dichlorodifluoromethane	620	—	—
Dichlorodifluoromethane-difluoroethane mixture	620	—	—
Difluorodibromoethane	620	—	—
Difluoroethane	660	—	—
Difluoromonochloroethane	660	—	—
Dimethylamine, anhydrous	240	—	—
Dimethyl ether	510	—	—
Ethane	350	—	—
Ethyl chloride (flammable liquid)	300	—	—
Ethylene	350, 900	—	—
Ethylene oxide (flammable liquid)	510	—	—
Fluorine	670	—	—
Helium	580, 930	350[2]	—
Helium-oxygen mixture (He over 80%)	580, 930	—	—
Hydrogen	350	—	—
Hydrogen chloride, anhydrous	330	—	—
Hydrogen cyanide, anhydrous (Class A poison)	160	—	—
Hydrogen fluoride, anhydrous (corrosive liquid)	660	—	—
Hydrogen sulphide	330	—	—
Isobutane	510	—	—
Krypton	580	—	—
Krypton, small valve series	120	—	—
Methane	350	—	—
Methyl bromide	620	—	—
Methyl chloride	620	360	—
Methyl mercaptan	330	—	—
Monochlorodifluoromethane	620	—	—

TABLE 2. GASES AND CONNECTION SYMBOLS—(*cont'd*)

Gas	Standard Connection USA and Canada	Alternate Standard Connection[1]	
		USA and Canada	Canada Only
Monochlorotetrafluoroethane	620	—	—
Monochlorotrifluoromethane	620	—	—
Monomethylamine, anhydrous	240	—	—
Neon	580	—	—
Neon, small valve series	120	—	—
Nitrogen	580	590	400[5]
Nitrous Oxide	320[6], 910, 1320	—	—
Oxygen	540, 870	—	—
Oxygen-carbon dioxide mixture (CO_2 not over 7%)	280, 880	—	—
Oxygen-helium mixture (He not over 80%)	280, 890	—	—
Phosgene (Class A poison)	640	—	—
Phosgene, small valve series	160	—	—
Propane	510	350[2], 300	—
Propylene	510	—	—
Sulfur dioxide	620	360	—
Sulfur hexafluoride	590	—	—
Tetrafluoroethylene	620	—	—
Trifluorobromomethane	620	—	—
Trifluorochloroethylene	620	—	—
Trimethylamine, anhydrous	240	—	—
Vinyl Chloride	290	—	—
Vinyl Methyl Ether	290	—	—
Xenon	580	—	—
Xenon, small valve series	120	—	—

*CGA standard only, not an American or Canadian Standard.
[1]The alternate connections are still used in certain segments of industry but are not generally acceptable to the U.S.A. Federal Services.
[2]Obsolete effective January 1, 1972.
[3]Obsolete effective January 1, 1967.
[4]Obsolete effective September 1, 1966.
[5]Obsolete effective January 1, 1966.
[6]Obsolete effective November 1, 1966.

SECTION B

The Pin-Index Safety System for Medical Gas Flush-type Connections

The possibility of error in attaching the flush-type valves with which gas cylinders and other sources of gas supply are equipped, to gas apparatus* with yoke connections, has long been recognized as a potential hazard. Color, labeling and other safeguards, while helpful, have not provided adequately against the human elements of carelessness, mental lapse, preoccupation, and the like.

To eliminate the hazard of accidental substitution of the wrong gas on apparatus utilizing yoke type connections, Compressed Gas Association, Inc., in cooperation with the American Society of Anesthesiologists, the American Hospital Association, and others, undertook the problem of developing a mechanical system to prevent interchangeability.

*The term "apparatus" as used herein refers to all machines, and equipment for the utilization of medical gases, having yoke connections.

A great number of suggestions were studied, many of them excellent. Much time was spent in the careful evaluation of the many factors involved. Among these were: the intricacies of adapting existing apparatus and valves to avoid obsolescence; the creation of a simple yet effective solution which would present a minimum of expense and confusion to owners of apparatus and valves alike; and, the desirability of devising a system that would offer reasonable possibility of acceptance as an international standard.

The Pin-index safety system, was accepted by all concerned as providing a practical, simple, and effective method. Successful field trials were conducted in a number of prominent hospitals. This system has been adopted as an American and Canadian standard, and included in "American Standard-Canadian Standard Compressed Gas Cylinder Valve Inlet and Outlet Connections." It has been adopted by Compressed Gas Association, Inc., the American Hospital Association, and the American Society of Anesthesiologists. It has also been accepted as an international standard.

THE PIN-INDEX SAFETY SYSTEM

The system consists of a combination of two pins projecting from the yoke assembly of the apparatus and so positioned as to fit into matching holes in the cylinder valve. Conversion of existing equipment was accomplished in the field by affixing to the yoke an adapter device bearing the pins. The Compressed Gas Association, Inc., recommends that all new gas apparatus should incorporate the Pin-index safety system.

The pin-index safety system provides for 10 combinations, each using 2 position holes on the cylinder valve and 2 corresponding pins on the yoke. Connections for eight medical gases or gas mixtures are in use at present leaving 2 position combinations available for future assignment. The recommendations of the medical profession were followed in selecting the various pin positions for the various gases, particularly in the case of the gas mixtures. With this 2-pin system it is impossible for a gas cylinder of one gas to be unintentionally attached to a yoke pin-indexed for any other use. For example, a carbon dioxide yoke has pins so spaced as to receive only a carbon dioxide cylinder valve which has been drilled with holes matched to the spacing of the carbon dioxide yoke pins.

Note: The pin-index safety system does not replace any of the means of identification now in use with medical gases such as labels, and markings. It provides an additional and positive safeguard.

FIG. 1. Typical adapter with pins, to be installed on the gas machine yoke. On the left is shown a cylinder valve correspondingly drilled.

FIG. 2. Yoke with adapter installed. These examples use nitrous oxide pin positioning. Observe identifying gas symbol on adapter.

PROGRAM FOR THE PIN-INDEX SAFETY SYSTEM

It is not hard to imagine the difficulties which would have been encountered if this program had not been carefully timed for the various steps required to convert existing apparatus and to incorporate the system in new equipment. Some manufacturers, for instance, were not able to convert as readily as others. This presented an embarassing situation in the case of a medical gas supplier who might not have been able to supply, soon enough, cylinders with drilled valves for use on converted apparatus. Furthermore, time had to be allowed for the manufacture of the necessary jigs and fixtures required to insure universal conformity with the pin-index safety system standards.

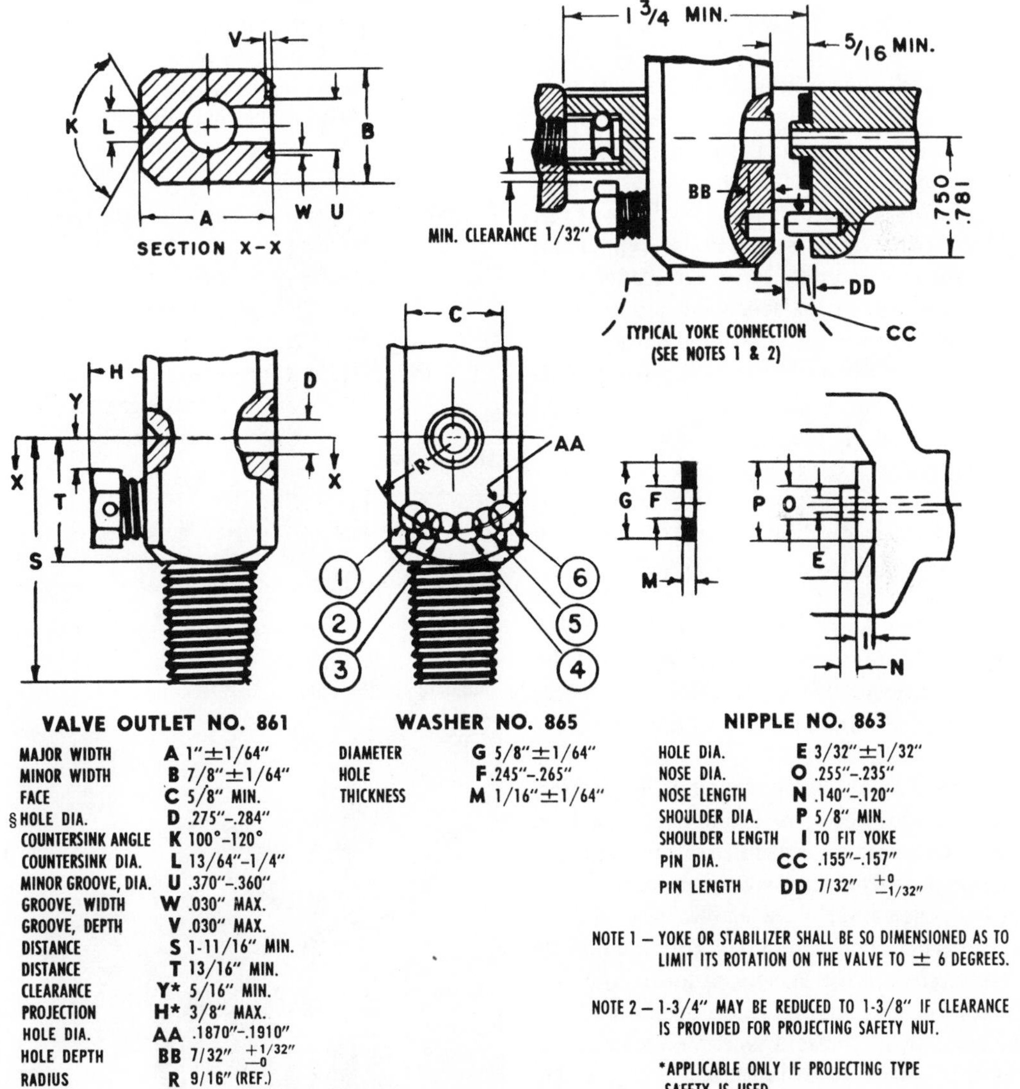

VALVE OUTLET NO. 861

MAJOR WIDTH	A	1"±1/64"
MINOR WIDTH	B	7/8"±1/64"
FACE	C	5/8" MIN.
§HOLE DIA.	D	.275"-.284"
COUNTERSINK ANGLE	K	100°-120°
COUNTERSINK DIA.	L	13/64"-1/4"
MINOR GROOVE, DIA.	U	.370"-.360"
GROOVE, WIDTH	W	.030" MAX.
GROOVE, DEPTH	V	.030" MAX.
DISTANCE	S	1-11/16" MIN.
DISTANCE	T	13/16" MIN.
CLEARANCE	Y*	5/16" MIN.
PROJECTION	H*	3/8" MAX.
HOLE DIA.	AA	.1870"-.1910"
HOLE DEPTH	BB	7/32" +1/32" −0
RADIUS	R	9/16" (REF.)

WASHER NO. 865

DIAMETER	G	5/8"±1/64"
HOLE	F	.245"-.265"
THICKNESS	M	1/16"±1/64"

NIPPLE NO. 863

HOLE DIA.	E	3/32"±1/32"
NOSE DIA.	O	.255"-.235"
NOSE LENGTH	N	.140"-.120"
SHOULDER DIA.	P	5/8" MIN.
SHOULDER LENGTH	I	TO FIT YOKE
PIN DIA.	CC	.155"-.157"
PIN LENGTH	DD	7/32" +0 −1/32"

NOTE 1 – YOKE OR STABILIZER SHALL BE SO DIMENSIONED AS TO LIMIT ITS ROTATION ON THE VALVE TO ± 6 DEGREES.

NOTE 2 – 1-3/4" MAY BE REDUCED TO 1-3/8" IF CLEARANCE IS PROVIDED FOR PROJECTING SAFETY NUT.

*APPLICABLE ONLY IF PROJECTING TYPE SAFETY IS USED

§Must be central within .010." Break sharp edge on outlet hole and gasket groove.

FIG. 3. Yoke outlet for medical gases. Standard flush outlet cylinder valve yoke connection. Basic dimensions for connection numbers 870 thru 940, inclusive.

The pin-index safety system is also utilized as an added safety factor in the charging of cylinders by the gas suppliers. This use required additional equipment and some modifications in present charging practices. All in all, simultaneous introduction of the new system on an industry-wide basis was an involved project and required the careful cooperation of everyone concerned.

This significant advance towards safer use of medical gases is a tribute to the cooperation of all, both users and manufacturers alike. Incidentally, previous standardization of basic dimensions for flush-type valves, achieved by Compressed Gas Association, Inc., has helped make possible the development of the pin-index safety system.

The ability of gas suppliers and apparatus manufacturers to adhere to a recommended conversion timetable depended largely upon the rate at which users returned empty cylinders with undrilled valves to the gas suppliers after drilled valves became available. A good housecleaning of "loafing" cylinders in out of the way places released a great many cylinders and provided a safeguard against costly unusable cylinder inventory when the conversion of gas apparatus began. It was essential that no apparatus be converted before a given date to avoid obsoleting full cylinders with undrilled valves. All users were requested to contact the manufacturer of their apparatus in sufficient time to ensure a conversion of equipment as soon as possible.

Figure 3 shows detail drawings illustrating the hole positions in flush-type valves in accordance with the pin-index safety system as standardized by Compressed Gas Association, Inc. For each gas, 2 of the 6 holes indicated are specified in the full standards. These standards were offered to the American and Canadian Standards Associations for adoption and for inclusion in the present American and Canadian Standards for Compressed Gas Cylinder Valve Inlet and Outlet Connections. Adoption by the American Standards Association and the Canadian Standards Association was concluded in 1953 (see A.S.A.-B57.1 and C.S.A.-B96). The standards have also been adopted by the International Organization for Standardization as an International Standard.

PIN-INDEX STANDARD CONNECTIONS SPECIFIED FOR MEDICAL GASES*

Connection Number	Specified for
870	Oxygen, medical
880	Oxygen-carbon dioxide mixture (CO_2 not over 7 per cent)
890	Oxygen-helium mixtures (He not over 80 per cent)
900	Ethylene
910	Nitrous oxide
920	Cyclopropane
930	Helium
	Helium-oxygen mixture (O_2 less than 20 per cent)
940	Carbon dioxide
	Carbon dioxide-oxygen mixture (CO_2 over 7 per cent)
950*	Air

*CGA standard only; when Connections 860-940 were approved as an International Standard (ISO), it was agreed that no nation would adopt either of the two remaining pin-index connections as a National Standard without referring the matter to the ISO. Adoption of Connection 950 as an American and Canadian standard and as an international standard is therefore awaiting action.

SECTION C

The Diameter-Index Safety System for Low-pressure Medical Gas Connections

In any field of human endeavor there are hazards which may be created by preoccupation, mental lapse, carelessness, and the like. The medical gas industry, organized within Compressed Gas Association, Inc., has long recognized its responsibility to the medical profession and to the general public by working, to the extent of mechanical practicability, toward eliminating the hazards of accidental substitution of the wrong medical gases by users of anesthetic, resuscitation and therapeutic administering equipment.

The diameter-index safety system was developed by the association to meet the need for a standard to provide noninterchangeable connections where removable exposed threaded connections are employed in conjunction with individual gas lines of medical gas administering equipment, at pressures of 200 psig or less, such as outlets from medical gas regulators and connectors for anesthesia, resuscitation and therapy apparatus. Removable threaded connections are those which are commonly and readily engaged or disengaged in routine use and service.

The diameter-index safety system supplements but does not replace:

(1) any of the means for medical gas identification now in use,

(2) the pin-index safety system,

(3) the existing threaded outlet standards for cylinder valves, or

(4) automatic quick coupler valves which also provide noninterchangeable connections for medical gas, air and suction equipment.

Compressed Gas Association, Inc., does not presume to designate specifically where the diameter-index safety system should find application on medical gas, air and suction equipment.

THE DIAMETER-INDEX SAFETY SYSTEM

This system presents, except for oxygen, a new concept in design for low pressure medical gas connections. The long established 9/16 in. 18 thread connection (see Connection No. 1240) has been retained for oxygen. For all other medical gas, air and suction equipment, noninterchangeable indexing is achieved by a series of increasing and decreasing diameters in the component parts of the connections. Before this standard was adopted, several proposals were evaluated. The prime objective was to develop a simple, effective system which would involve minimum expense and confusion to the equipment owners and manufacturers. After months of development and testing the diameter-index safety system, herein described, was approved as a standard by Compressed Gas Association, Inc.

The system recognizes the established use of suction and air in the immediate areas where connections to that equipment could accidentally be interchanged with medical gas equipment. Therefore, air and suction have been assigned indexed positions in this system. Provision has been made for the larger bore required for suction service (see Connection No. 1220).

The direction of flow through connections of the system is optional.

DESCRIPTION OF THE SYSTEM

Each connection of the diameter-index safety system consists of a body, nipple and nut. Except for oxygen, as stated above, this system is based on having two concentric and specific bores in the body and two concentric and specific shoulders on the nipple. To achieve noninterchangeability between different connections, the 2 diameters on each part vary in opposite directions so that as one diameter increases, the other decreases. In this way only the properly mated and intended parts fit with each other. Attempts to bring together unintended parts result in interference either at the large diameter or at the small diameter preventing thread engagement.

The critical diameters from one size to the next vary in basic steps of .012 in. Manufacturing tolerances on mating connections provide a clearance from .003 to .009 in. at each of the diameters. The interference of nonmating connections is a minimum of .003 in.

As shown in Fig. 1, the small bore in the body mates with the small diameter of the nipple, and the large bore in the body mates with the large diameter of the nipple. The four diameters marked (1) are for the set of conditions making a specific connection (Standard Connection No. 1020) thereby allowing the nut to engage the threads on the body.

For another set of conditions marked (2) (Standard Connection No. 1040) the small diameters are enlarged and the large diameters are reduced so that only these and no others will go together. It is impossible to assemble the body with any unintended nipple because of interference either at the large diameter or at the small diameter thereby preventing thread engagement.

The principle of increasing the small diameters and decreasing the large diameters continues until condition 11 (Connection No. 1220) is reached. Here the critical diameters become equal, forming single mating diameters for the bore and for the nipple. The nipple of this connection has the largest nose diameter, and is therefore designed with an extra large bore and assigned to suction. For mechanical reasons it was necessary to design the suction nut $\frac{1}{8}$ in. longer than the nuts used with other connections. This extra length made it possible to have thread engagement between nonmating parts if the suction nut was accidentally used in place of the shorter nut. To avoid this the hole in the suction nut was enlarged so that all other nipples, except the suction nipple, would pass through it.

Standard Connection No. 1240 retains the long established standard oxygen connection (9/16 in.-18 thread) and is quite different from the others which are based on diameter indexing.

Figure 2 shows a set of drawings showing the basic dimensions or minimum bore diameters of the various connections comprising the diameter-index safety system.

REFERENCES

"Regulator Connection Standards," Compressed Gas Association, Inc.

"Specification for Rubber Welding Hose," Compressed Gas Association, Inc.

"Standard Hose Connection Specifications," Compressed Gas Association, Inc.

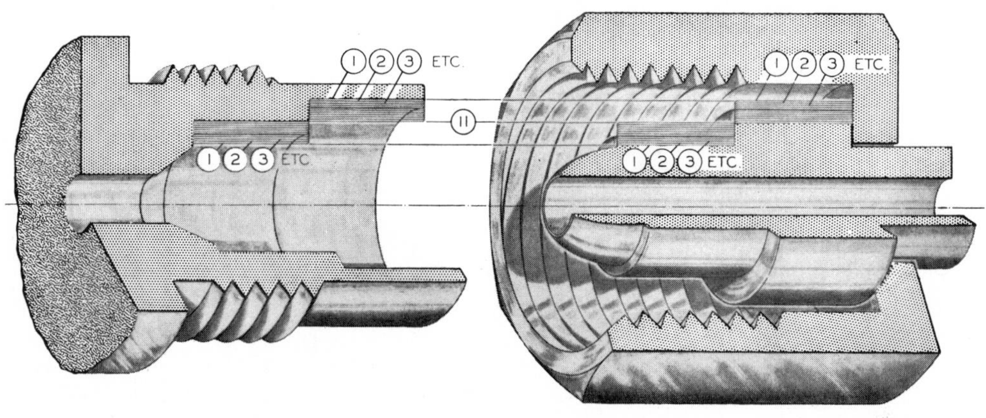

FIG. 1.

DIAMETER INDEX SAFETY SYSTEM

Standard Low-Pressure Connections for Medical Gases, Air and Suction.

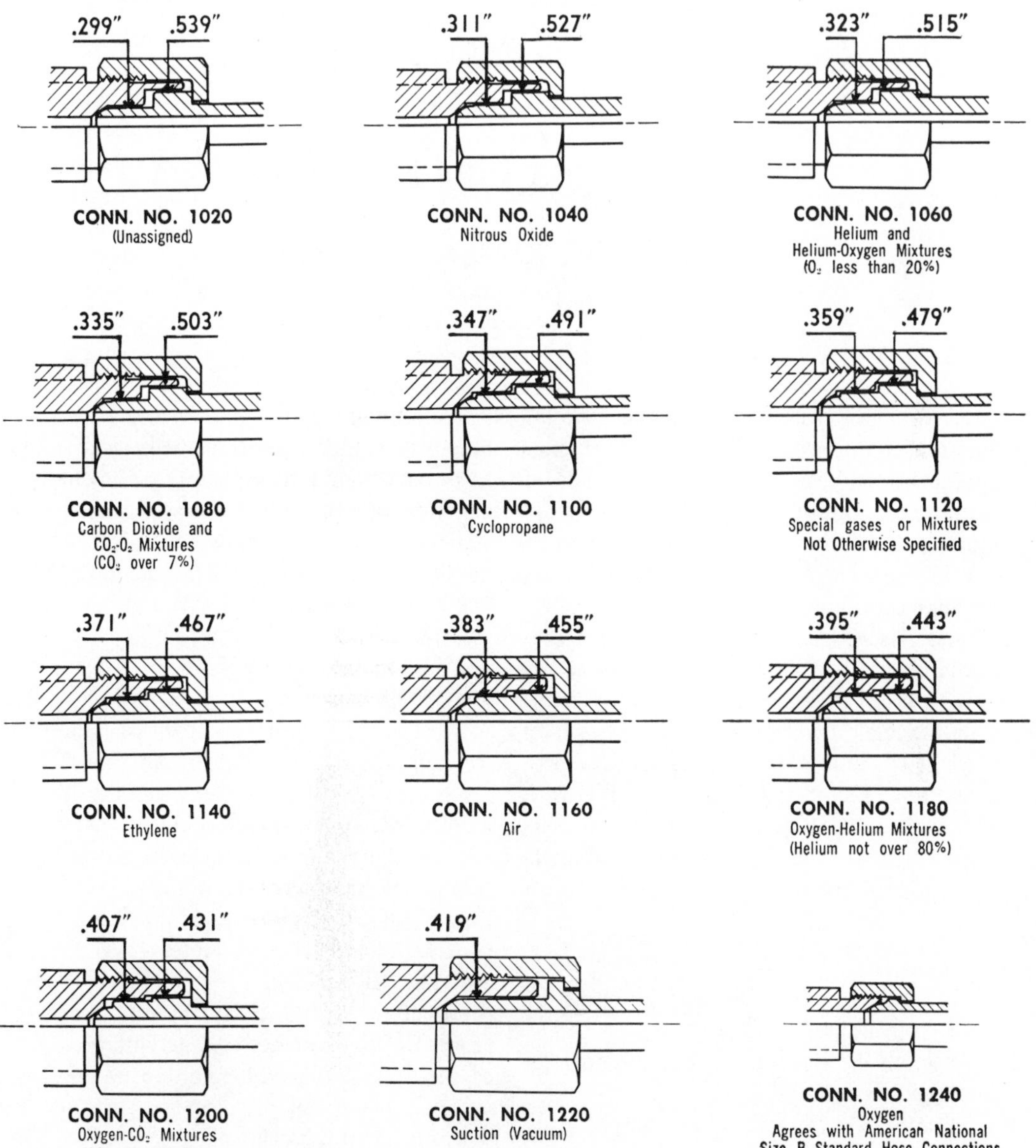

FIG. 2. Diameter-index safety system. Standard low-pressure connections for medical gases, air and suction.

Note: Dimensions shown are minimum bore diameters and not to be used for manufacture. Complete details and dimensions are contained in CGA Pamphlet V-5, "Diameter-Index Safety System," available from the Compressed Gas Association.

CHAPTER 5

Compressed Gas Shipping and Storage Containers

Intended primarily for persons unfamiliar with the compressed gas field, this chapter describes and illustrates the major kinds of containers used today for the compressed and liquefied gases (including those liquefied at very low or cryogenic temperatures). It does not represent standards published by Compressed Gas Association, Inc., as do the preceding chapters of Part III. The CGA, however, through its technical committees, continually reviews and recommends changes in existing ICC and BTC specifications, or proposes new specifications to keep pace with new developments in the industry. The chapter treats the various kinds of containers in sections as follows:

Section A. Cylinders and Small Containers
Section B. Containers for Shipping by Rail in Bulk
Section C. Containers for Shipping by Highway in Bulk
Section D. Containers for Shipping by Water in Bulk
Section E. Cryogenic Containers
Section F. Storage Containers

Safety relief devices for these different types of containers are not dealt with in this chapter because they are discussed in detail in the preceding Chapter 2. Similarly, readers should see the preceding Chapter 1 for information on the safe handling of cylinders and other containers; Chapter 3, for the labeling, requalifying, repair and disposition of cylinders; Chapter 4, for cylinder valve connection standards; and the following Chapter 6, for methods of unloading bulk shipments of liquefied compressed gas.

ICC, BTC AND ASME CODES FOR COMPRESSED GAS CONTAINERS

Most of the shipping and storage containers for compressed gas made in North America are built to comply with one or both of two major codes of detailed specifications. These are the identical codes adopted by the U. S. Interstate Commerce Commission and the Board of Transport Commissioners for Canada identified in Chapter 2 of Part I, and the "Boiler and Pressure Vessel Code" (particularly Section VIII, on unfired pressure vessels) of the American Society of Mechanical Engineers. (The 1965 edition of the code is available from the ASME at 345 E. 47th St., New York, N. Y., 10017.) The ICC and BTC codes are obligatory for containers shipped in interstate or interprovincial commerce, and apply mainly to shipping containers; the ASME code applies mainly to stationary storage vessels (though truck cargo tanks and portable tanks must conform to the ASME code under ICC and BTC specifications).

Inspectors examine compressed gas containers built to the respective codes as they are made, and approve them for container markings (usually stamped) which indicate that the marked container fulfills the code specifications identified in the marks. Inspectors authorized by the Bureau of Explosives of the Association of American Railroads are required for high-pressure cylinders. Inspectors are required to file a report on each individual container that is made under the specifications of these codes before the container goes into service, and each container made is identified and thus registered by serial number and manufacturer's symbol. Compliance with one or more of these codes is often stipulated in state and insurance company requirements as well as in federal regulations.

SECTION A

Cylinders and Small Containers

Cylinders for compressed gases are generally defined in the ICC and BTC specifications as containers having a maximum water capacity of 1000 lb (about 120 gal) or less, although the relatively new 3AX and 3AAX specifications permit the use of larger cylinders, a popular size having a water capacity of approximately 5000 lb. They are made in a wide variety of heights and proportions, and range in capacities from their authorized maximums down to a cubic foot or less. Cylinders broad and squat in proportions are generally

In left foreground, cryogenic tank car of the AAR-204W type for transporting nonflammable gases in the form of cryogenic fluids; the car carries the gases at pressures below 25 psig, and is not subject to ICC regulations. Cryogenic storage tanks of large and small capacity appear at the rear, and a truck cargo tank shows partly at the right, in this view of the shipping area of an air separation plant producing atmospheric gases.

made for low-pressure service, while cylinders tall and thin in proportions are intended primarily for high-pressure service. Cylinders are most often made with flat bottoms, or are fitted with foot-rings so that they may be stood on end and securely fastened in an upright position while their contents are being withdrawn. Some cylinders, though, have hemispherically rounded sealed ends and are designed for withdrawal of contents while securely fastened in a horizontal or slanted position. One end of the cylinder (or both ends, in some special cases) is tapered in to a neck which is tapped with screw threads for attachment of the cylinder valve.

Compressed gas cylinders of a great many of the different sizes and styles in which they are made are shown here, ranging from some of the smallest at bottom to some of the largest at top. The narrow, thin, seamless cylinders are generally those built for high-pressure service; the broader, thick welded cylinders are in general ones made for service at moderate or low pressures. Valve protection caps are in place on some of the cylinders shown, while others appear without valves screwed into the threaded openings for them. Some compressed gas cylinders are made with concave or dished heads, instead of the convex heads used for all the cylinders illustrated here.

Cylinders are the type of compressed gas containers most widely authorized for different means of shipment. They are the only container type accepted for air shipment in the case of most gases.

CYLINDER MANUFACTURE

Cylinders are made from seamless tubing, brazed or welded tubing, billets in the billet process, or flat sheets drawn to cylindrical shapes in large punch-press dies. Sealed ends are made either by spinning in a lathe under a flame at red or white heat, by forging, or by die-drawing. The sealed ends of some cylinders are closed by spinning or forging. In some instances the sealed end of such cylinders is drilled and then "plugged" with an additional metal piece.

Provision for Flexible Use of ICC and BTC Specification Cylinders

Under the ICC and BTC regulations, a cylinder meeting any one specification may be authorized for shipping a number of different gases, or any one gas may be authorized for shipment in a number of different specification cylinders. This differs from the usual European practice, in which a cylinder meeting a single set of cylinder specifications is authorized for shipping only a single designated gas. The one exception in North American practice concerns acetylene, which is authorized for shipment as a "dissolved" gas only in acetylene cylinders—those meeting specifications ICC-8 or ICC-8AL, BTC-8 or BTC-8AL. In turn, 8 or 8AL cylinders must not be charged with any gas but acetylene.

Required Cylinder Markings

Cylinders built to ICC or BTC specifications must be marked permanently with symbols like the following stamped into the shoulder (the part sloping up to the neck, or else into the top head or the neck itself):

ICC-3A 2015
462
XY
CGA

These marks give first the ICC or BTC specification to which the cylinder has been made (in the case above, ICC-3A specification). Immediately following is the service pressure for which the cylinder was built (2015 psig, in this case). Next appears the serial number of the individual cylinder ("462," above), and the official mark of the inspector ("XY"). The last mark is that of the cylinder owner ("CGA"). Such marks may instead be stamped in one line, as follows:

ICC-3A2015—462—XY—CGA

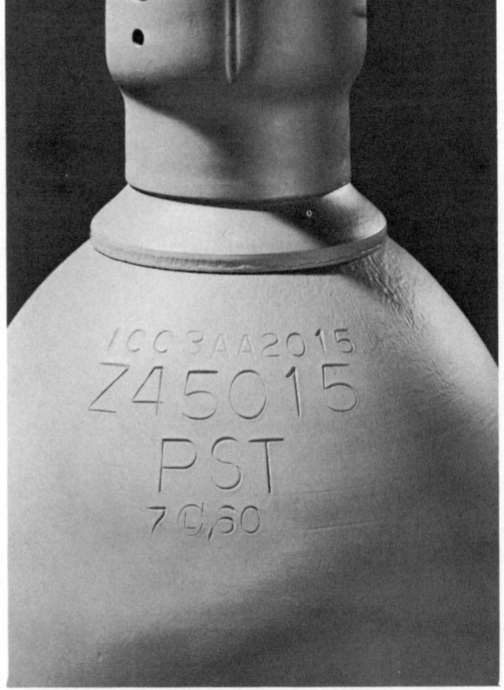

Permanent markings stamped into the shoulder or head of a new cylinder in accordance with ICC requirements, as shown by the cylinder above, include: the ICC specification to which the cylinder was made ("ICC-3AA"), followed by the service pressure in psig ("2015"); on the next line, the serial number of the individual cylinder, assigned by the manufacturer ("Z45015"); next, the identifying mark of the cylinder's manufacturer ("PST"); and last, the month and date of the qualification test required before the cylinder can go into service ("7, 60" for July 1960, with the "7" and "60" separated in this case by the optional identifying mark of the inspector who made the first test). Months and years of subsequent hydrostatic tests required periodically for requalification follow the date of the initial test, being stamped below or beside the first date on the shoulder.

Still other markings are required. Cylinders are tested as part of the final inspection in being manufactured, and the month and year of the initial qualifying test are required as stamped markings (appearing as in 5-66 for May, 1966); the date stamp is placed so that the dates of subsequent retestings for required requalification can be added later on the shoulder, top head or neck. Moreover, the ICC and BTC codes require that the word "SPUN" or "PLUG" be stamped near the specification mark when an end closure has been made by one or the other of these means. The manufacturer's name or symbol must also be part of the permanent markings. As a result, the full stamping to appear on a new cylinder might read as follows:

ICC-3A2015	SPUN
462	(M)
XY	5-66
CGA	

If a plus sign is given immediately after the test date marking of a specification 3A or 3AA cylinder, it means that the cylinder is authorized for charging up to 10 per cent in excess of the marked service pressure (up to 2200 psig at 70 F in this case) if the cylinder valve is equipped with an ICC-approved safety relief device and if used for certain gases as provided in 13.302 of ICC or BTC regulations.

For cylinders stamped "ICC-3" or "ICC-3E" without any marked service pressure following the specification number, the service pressure is 1800 psig.

Commonly Made Cylinder Types

Among types of cylinders manufactured in the largest quantities today are the following:

ICC-3A. These are seamless steel cylinders, and are produced usually for high-pressure service. Service pressures for which they are built, though, range from 150 to 15,000 psig.

ICC-3AA. These are also seamless, high-pressure cylinders, but they must be made from designated alloy steels heat-treated as specified so that the metal withstands higher working stresses than that of 3A cylinders. As a result, a 3AA cylinder is lighter than a 3A cylinder having the same service pressure.

ICC-3B. Also seamless cylinders like the 3A but built for lower service pressures (150 to 500 maximum psig).

ICC-3E. A small cylinder limited in size to 2-in. maximum diameter and 24-in. maximum length and built to the one required service pressure of 1800 psig; also a seamless cylinder.

ICC-8 or ICC-8AL. Seamless low-pressure cylinders for acetylene service only built to the single authorized service pressure of 250 psig; heads attached by welding or by brazing by the dipping process; ICC-8 specification cylinders may have a longitudinal seam if it is forge lap welded; welded circumferential body seam authorized if body has no longitudinal seam. These cylinders also contain a permanent porous inner filling which is saturated with the acetylene solvent, usually acetone.

Small Containers Exempt from Cylinder Requirements

Under the ICC or BTC code, many compressed gases are authorized for rail and highway shipment in small containers that are exempt from the specification, packaging, marking and labeling requirements set in the code for authorized cylinders. In water carrier shipment, though, the contents of these small containers must be marked on the outside of boxes in which they are packed. The only gases which may not be shipped in small containers exempt from cylinder requirements are those designated "poison gas" or "no exception" in the ICC or BTC regulations. Gases for which exemptions are allowed may be shipped under any one of the following conditions in small containers as indicated:

Any exempt gas—in containers of not more than 4 fluid oz water capacity (7.22 cu in.).

In refillable containers, for any nonflammable, nonliquefied exempt gas—in metal containers not over one quart in capacity if charged to not over 170 psig at 70 F, or not over 30 gal in capacity if charged to not over 75 psig at 70 F (and also only if tested to three times the filled pressure at each refilling).

In nonrefillable containers, for any exempt gas—in inside (boxed) metal containers not over 32 cu in. in capacity and if charged to not over 55 psig at 70 F (and meeting other requirements).

In nonrefillable containers, for any exempt gas that is nonflammable and nonliquefied—in inside metal containers not over 31.83 cu in. (17.6 fluid oz) in capacity and if charged to not over 140 psia at 130 F (and meeting other requirements).

Other specific exemptions are also provided in the regulations for compressed gases packaged in foodstuffs, soap, cosmetics, beverages, biologicals, electronic tubes, audible fire alarm systems, and fire extinguishers.

SECTION B

Containers for Shipping by Rail in Bulk

Bulk rail shipment of compressed gases is authorized under the ICC and BTC codes in containers of three major kinds: single-unit tank cars; multi-unit (or "ton multi-unit") tank cars; and specification 107A tank cars that consists of clustered and fixed cylindrical tubes that extend the length of the car and are manifolded in a common header. Adaptations of existing tank car specifications, and proposed specifications for new types of cars, must be approved by the Committee on Tank Cars of the Association of American Railroads.

Single-unit Tank Cars

Single-unit tank cars authorized for compressed gas shipment by the ICC or the BTC carry the gas in a single large pressure tank which is permanently mounted to the car frame. These cars resemble, generally, the familiar railroad tank cars used for oil and other nonpressurized liquids. Single-unit pressure tank cars for use with some of the cooler liquefied gases (such as sulfur dioxide or carbon dioxide) are often made with insulation (lagging) between inner and outer metal shells. Tank car types frequently required for compressed gases are those complying with ICC or BTC specifications 105A, 112A, and 114A. (The number following a designation like 105A, as in "105A500-W," is the marked test pressure of the tank—500 psig, in this example. The highest test pressure authorized for a single-unit tank car in current regulations is 600 psig.)

Single-unit tank cars that had been built under ICC or BTC specifications until a few years ago had been made with capacities ranging up to about 85,000 lb of water or some 10,000 gal.

Insulated single-unit tank cars for shipping compressed gas by rail stretch far into the distance in this train of 54 cars of conventional size (some 10,000 gal water capacity per car); a train this long is needed to carry each day's full production of the large-scale anhydrous ammonia plant appearing in the background.

Oversize Single-unit Tank Cars

A number of oversize single-unit tank cars, popularly called "jumbo" or "pregnant whale" cars, have been built and put into compressed gas service under ICC specifications or special permits in the last few years. These cars have been introduced primarily for liquefied gases that are shipped in very large quantities, like propane and ammonia, and have capacities in exceptional cases as large as 50,000 or even 60,000 gal per car. However, a fairly common capacity for jumbo cars today is 30,000 or 32,000 gal.

Extremely large tank car capacities have been obtained within the maximum height and width allowances for railroad cars by making the biggest oversize cars far longer than usual and swelling out the lower tank contours in long "belly" bulges. One design enlarges capacity through a so-called "figure-eight" cross section that consists of two intersecting circles, one above the other.

The largest oversize tank cars are usually restricted to main lines in order to avoid possible weight and other problems they might encounter in traversing spur lines not built for fast, heavy-duty service.

Multi-unit or "TMU" Tank Cars

Multi-unit tank cars consist of a kind of flatbed railroad car that carries 15 large cylindrical pressure tanks crosswise on the car. The filled tanks are lifted onto or off the car by crane or hoist. Specifications 106A and 110A in the ICC and BTC codes are the principal ones for multi-unit tanks, and the codes limit their water capacity to a minimum of 1500 lb and a maximum of 2600 lb (about 120 to 215 gal). 106A tanks have forged-welded heads formed convex to pressure. Present authorized test pressures are 500 and 800 psig for both types plus 1000 psig for the 110A type.

Their popular name, "ton containers" or "ton multi-unit tanks", (which is used throughout the "Handbook") stems from the

An oversize single-unit tank car that is uninsulated; its capacity is some 33,000 gal.

One of the largest oversize single-unit tank cars made, this uninsulated giant has a capacity of some 50,000 gal.

fact that these tanks were first introduced to transport a ton of liquefied chlorine apiece.

TMU tanks are uninsulated.

107A Tank Cars made up of Clustered Tubular Tanks

Tank car specification 107A in the ICC and BTC codes provides for cars carrying clustered sets of long tubular tanks advantageous for transporting bulk quantities of nonliquefied gases (like helium, argon, or air) at high pressure. The multiple tubular tanks are either hollow forged, drawn or seamless tubing. They are permanently mounted on the car, and are connected or manifolded to a common header at one end of the car. A common diameter for the tubes is 30 in., and as many as 30 or more tubes may be installed on a single car frame. Test pressures for the 107A tank cars range up to 3500 psig, and capacities up to a quarter-million standard cubic feet of gas (in the case of helium).

"Piggyback" Rail Transport

Cargo tanks on truck semi-trailers and tube trailers are also often transported "piggyback" on railroad flatcars in bulk shipment of compressed gases by rail.

A multi-unit tank car for rail transport of compressed gas, as above, carries 15 tanks of at least 1,000 lb water capacity each; the tanks are popularly called "ton multi-unit" or "TMU" tanks because they were first introduced to carry one ton of liquefied chlorine per tank. TMU tanks are usually loaded and unloaded with a hoist, and the cars are almost always shipped with the full 15 tanks in place because the rail shipping charge per car is the same regardless of the number of tanks carried. TMU tanks are also authorized for the highway shipment of some compressed gases.

The trailer-mounted truck cargo tank at left and the tube trailer at right aboard a railroad flatcar illustrate how railroad "piggy back" shipment can be used for combined rail and highway transport of compressed gases.

SECTION C

Containers for Shipping by Highway in Bulk

Three major types of containers are authorized for the bulk shipment of compressed gases by highway under the ICC code: truck or truck semi-trailer cargo tanks; portable tanks; and tube trailers.

Cargo Tanks

Cargo tanks are large-capacity tanks permanently mounted on truck bodies or semi-trailer bodies. The specifications most often authorized for compressed gas cargo tanks in the ICC code are MC-330 and MC-331. Tanks of each type must comply with the ASME pressure vessel code, and must have a design pressure of not less than 100 psig nor more than 500 psig. Either type may be insulated for use with such liquefied gases as carbon dioxide or nitrous oxide, and either may be fitted with refrigerating and heating coils. Cargo tank capacities range up to as large as 10,000 gal or more per tank; no minimum capacity is given for them in the specifications. In some instances, truck tractors draw double tank trailers of compressed gas.

A relatively small compressed gas cargo tank mounted directly on a truck body (instead of on a semi-trailer, the common mounting for larger tanks) for local deliveries of LP-gas. The tank is uninsulated.

This moderately large, insulated cargo tank illustrates semi-trailer mounting widely used for highway shipment of compressed gas; the tank shown is typical of those built primarily for liquefied carbon dioxide.

Portable Tanks

Portable tanks complying with specification ICC-51 are also authorized for shipping many compressed gases. The ICC-51 specification provides for steel tanks of at least 1000 lb water capacity (about 120 gal) with service pressures of not less than 100 psig nor more than 500 psig. The cylindrical ICC-51 tanks are often made with flat skid mountings attached, and are commonly called "skid tanks." Portable tanks are shipped primarily by truck, but they also are used to some extent in rail shipment. U. S. water shipment regulations limit the maximum gross weight of full portable tanks to 20,000 lb.

TMU Tanks on Trucks

For some gases, ICC regulations authorize shipment of TMU or multi-unit tanks by motor vehicle as well as by rail. TMU tanks shipped on trucks must be securely chocked or clamped while in transit, and adequate facilities must be present for handling tanks where transfer in transit is necessary.

Tube Trailers

High-pressure nonliquefied gases like oxygen and nitrogen at atmospheric temperatures are often shipped by highway in tube trailers. These are truck semi-trailers on which a number of very long gas cylinders have been mounted and manifolded in a common header. Tube trailer service pressures are as high as 2000 psig or more. The tubular cylinders of the trailers are often made according to 3A or 3AA cylinder specifications, or to cylinder specification 3AX or 3AAX (3AX and 3AAX specifications provide for containers some 22 in. in diameter instead of the customary 9⅝ in. diameter of 3A and 3AA cylinders; also, 3AX and 3AAX cylinders have a minimum size of 1000 lb water capacity under the ICC code). Tube trailers have been built to carry as much as 45,000 standard cu ft of oxygen or 128,000 standard cu ft of helium.

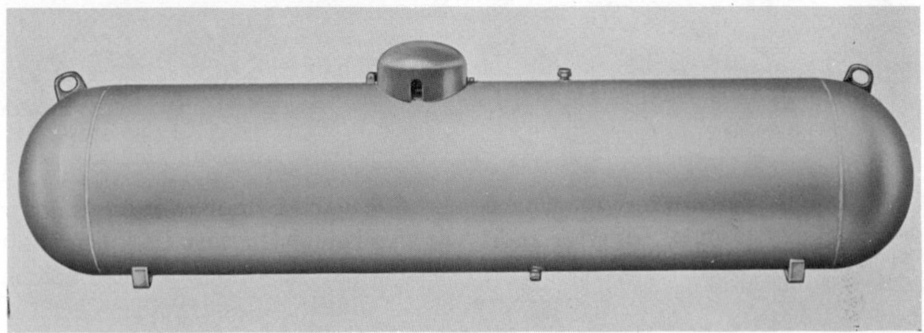

Portable tanks used for highway and rail shipment of compressed gas are made in many different sizes above the specified minimum of 1,000 lb water capacity (about 120 gal) and in a wide variety of shapes, as this unusually long portable tank suggests. Their outfitting with flat skid mountings, as illustrated above, has led to the common practice of calling them also, "skid tanks."

SECTION D

Containers for Shipping by Water in Bulk

Practically all types of containers authorized for shipping compressed gases on land are also authorized under some conditions for water shipment; for example, even single-unit tank cars are approved for cargo vessels or railroad car ferry vessels, and tank trailers are approved for cargo vessels and trailerships, in the cases of some gases.

Tankships and Tank Barges

Some tankships and many tank barges are built with fixed pressure tanks primarily for

Fixed pressure tanks built into tank barges like the one above are used for inland waterway shipment of increasing numbers of gases. Among gases shipped in bulk by barge are anhydrous ammonia, inhibited butadiene, chlorine, anhydrous dimethylamine, liquefied hydrogen, LP-gas methyl chloride, and vinyl chloride.

Tankship with fixed pressure tanks was built to carry 7.5 million gallons (14,000 long tons) of liquefied natural gas ("LNG") at −259 F in trans-Atlantic service; the vessel is also equipped for ocean shipment of ammonia.

bulk water transport of compressed and liquefied gases. Regulations of the U. S. Coast Guard and of the Canadian Department of Transport (see Chapter 2, Part I) set forth detailed requirements for fixed tanks or barges used for shipping anhydrous ammonia and chlorine. Special permission of these regulatory agencies nas authorized tank vessels for other compressed gases; the first of any importance was a ship equipped to transport LP-gases in intercoastal services; and more recently ocean-going tankers that carry liquefied methane in insulated and refrigerated tanks from North America to Europe. Among other gases that are shipped in substantial quantities by tank barge are inhibited butadiene, anhydrous dimethylamine, liquefied hydrogen, methyl chloride and vinyl chloride.

SECTION E

Cryogenic Containers

A wide variety of containers has been developed in recent years for shipping gases liquefied at very low or cryogenic temperatures—ones ranging from about −250 F down to the neighborhood of absolute zero, or −459.69 F. The containers usually have high-efficiency insulation, and some of them dissipate heat absorbed in the contained cryogenic fluid by venting small amounts of vapor.

Containers for cryogenic fluids include specification 4L cylinders of the ICC and BTC codes; small non-regulated shipping containers of various kinds; the ICC and BTC specification 113A tank car, a single unit car designed primarily for liquefied hydrogen; a non-regulated tank car complying with Association of American Railroads specification AAR-204W, intended mainly for liquefied oxygen, nitrogen, or argon; various kinds of non-regulated cargo tank trucks and tank semi-trailers; and the liquefied methane tankers and liquefied hydrogen tank barges mentioned in the preceding section, which are authorized by special permit. (Non-regulated

The row of individual small-volume containers for gases liquefied at cryogenic temperatures is made up of insulated spherical flasks with bright-metal outside casings and, fourth from left in the row, a cylinder of the ICC-4L type. Larger cryogenic containers include the insulated, dolly-mounted portable tank at right and larger truck cargo tanks in background.

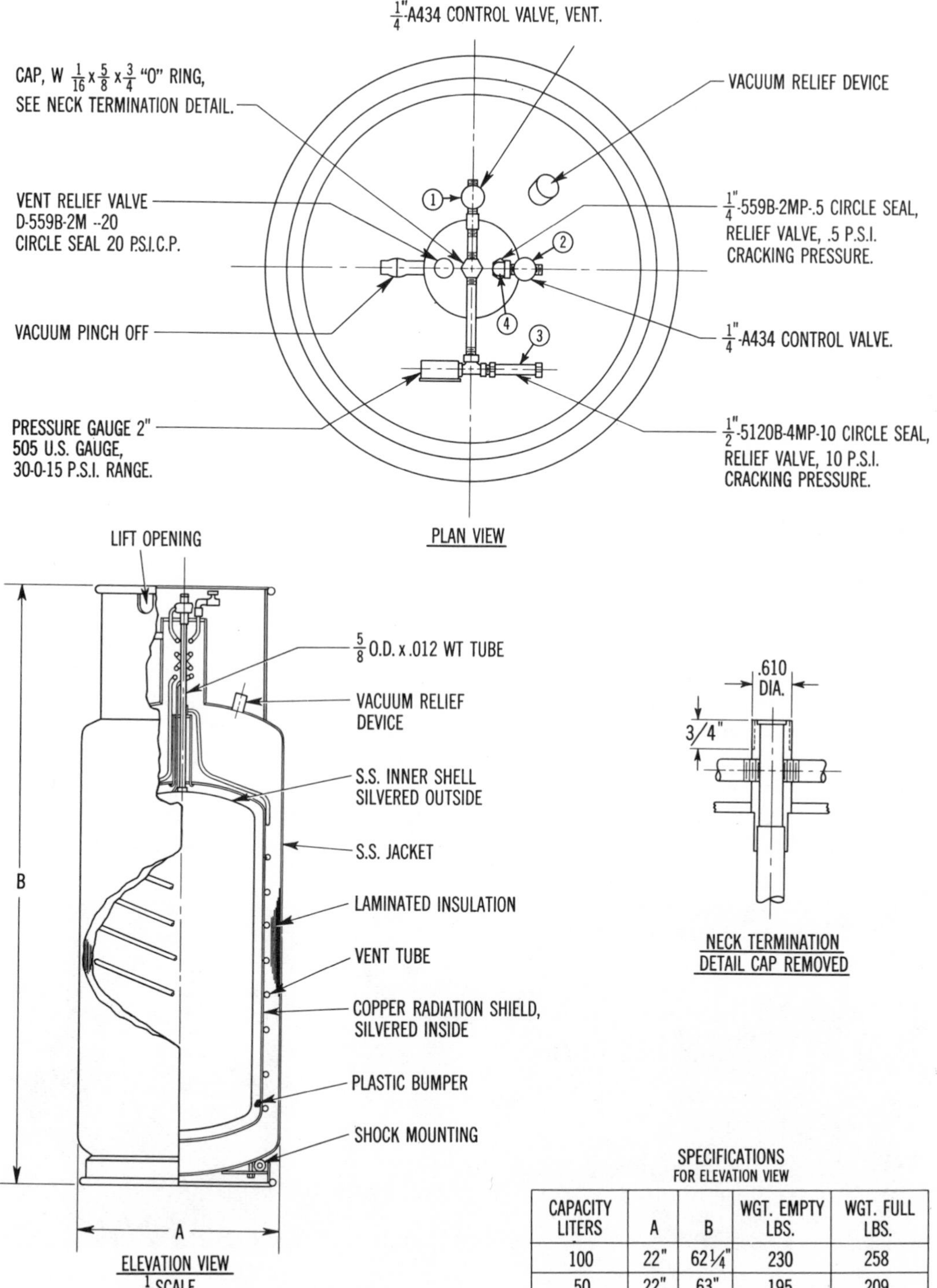

SPECIFICATIONS
FOR ELEVATION VIEW

CAPACITY LITERS	A	B	WGT. EMPTY LBS.	WGT. FULL LBS.
100	22"	62¼"	230	258
50	22"	63"	195	209

Cross-section view of a cryogenic cylinder of the ICC-4L type designed for use with liquid helium at −452 F the lowest boiling point of any known substance. The cylinder is insulated both by a vacuum and a laminated, high-efficiency insulating material.

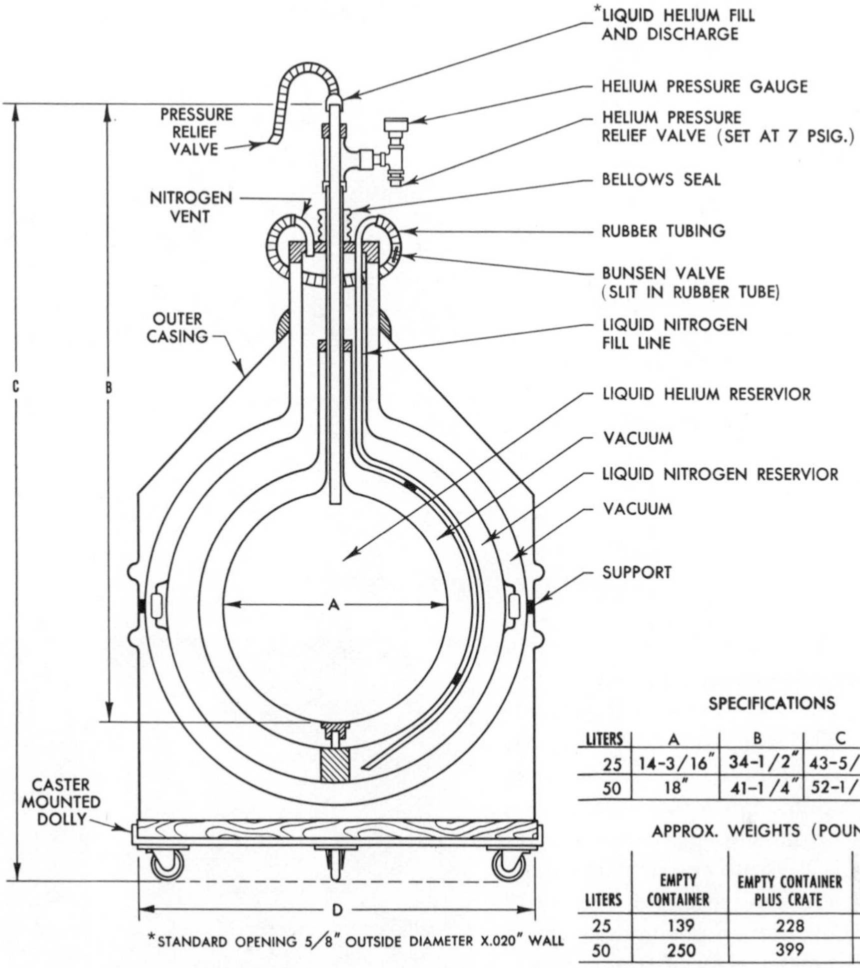

*LIQUID HELIUM FILL AND DISCHARGE
HELIUM PRESSURE GAUGE
HELIUM PRESSURE RELIEF VALVE (SET AT 7 PSIG.)
BELLOWS SEAL
RUBBER TUBING
BUNSEN VALVE (SLIT IN RUBBER TUBE)
LIQUID NITROGEN FILL LINE
LIQUID HELIUM RESERVIOR
VACUUM
LIQUID NITROGEN RESERVIOR
VACUUM
SUPPORT
PRESSURE RELIEF VALVE
NITROGEN VENT
OUTER CASING
C
B
A
CASTER MOUNTED DOLLY
D
*STANDARD OPENING 5/8" OUTSIDE DIAMETER X.020" WALL

SPECIFICATIONS

LITERS	A	B	C	D
25	14-3/16"	34-1/2"	43-5/8"	20-1/4"
50	18"	41-1/4"	52-1/8"	25-1/8"

APPROX. WEIGHTS (POUNDS)

LITERS	EMPTY CONTAINER	EMPTY CONTAINER PLUS CRATE	EMPTY CONTAINER, CRATE AND LQN_2 & LQHe
25	139	228	286
50	250	399	518

Cross-section view of a spherical, small-volume flask for liquid helium which is insulated by liquid nitrogen at −320 F and by double vacuum layers.

LIQUEFIED GASES
LIQUEFIED HYDROGEN
HYDX 704

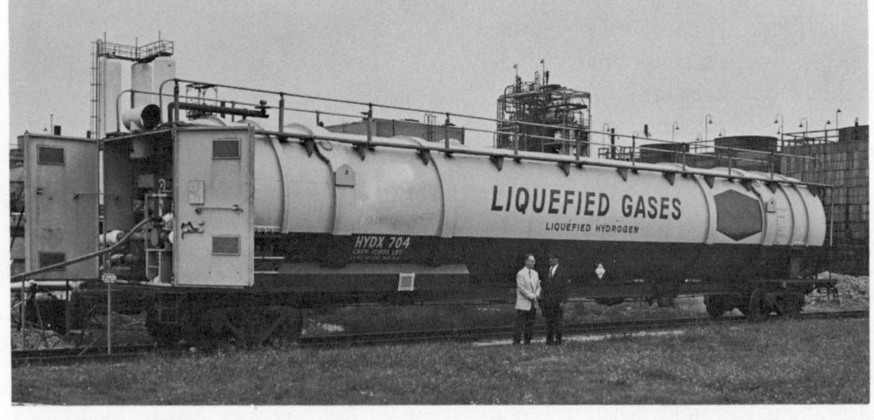

Cryogenic tank car of the ICC-113A type for transporting liquefied hydrogen at −423 F.

in the sense used here means not subject to ICC or BTC regulations, because the containers indicated as non-regulated are ones for nonflammable and very cold liquefied gases carried at pressures below 25 psig. A number of the regulated and non-regulated shipping containers are described in more detail in the Part II sections on gases which are often shipped as cryogenic fluids, such as helium, hydrogen, oxygen, nitrogen and argon).

Among proposed specifications for additional cryogenic containers that are being developed by Compressed Gas Association, Inc., as this volume goes to press are MC-341, for truck cargo tanks; and 51L, for portable tanks.

Heavily insulated cryogenic cargo tank for highway shipment of liquefied helium.

SECTION F

Storage Containers

Compressed and liquefied gases are often stored by users of smaller quantities in the shipping containers—in banks or storerooms of cylinders, in TMU tanks and portable tanks, in high-pressure tube trailers or 107A tank cars.

Stationary storage tanks into which gases are transferred from shipping containers are most often steel pressure vessels conforming to the ASME code. Storage tanks for relatively low-pressure liquefied or nonliquefied gases are made in a great variety of sizes, designs, and materials.

Gases formerly shipped only in nonliquefied, high-pressure form are today increasingly shipped and stored as compact cryogenic fluids. Cryogenic storage systems often include vaporizing units for converting the very cold fluid to gas at atmospheric temperatures and pressures.

Among the largest and most striking types of storage tanks now used for liquefied gases are huge underground caverns or refrigerated, lined and capped pits dug into the ground. These are developed most often for such fuel gases as propane-butane. One such pit storage facility being developed in New Jersey for liquefied natural gas will hold, when full, one billion standard cubic feet of gas.

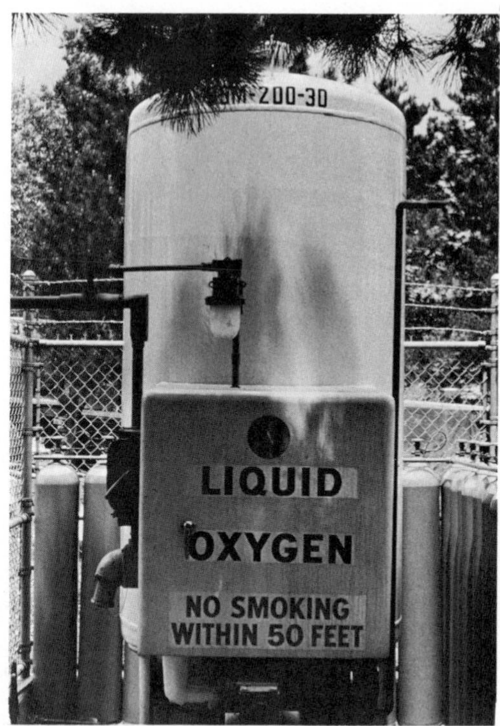

Compact liquid oxygen storage system typical of those at hospitals is located outdoors and has standby supply of higher-pressure gaseous oxygen in manifolded cylinders.

Aerial view of tanks for typical storage installation for propane or ammonia in open farm country; siding and unloading gear for receiving large shipments by rail appear at left bottom, while truck-loading pumps to provide for local deliveries show at right center.

Low, earth-covered dome caps 160,000-barrel frozen pit for storing LP-gas liquefied at moderately low temperature and pressure. The large storage pit, for which the surrounding earth serves as insulation, is located in the Western United States. A similar storage pit for liquefied natural gas in the Northeast has a capacity of one billion cubic feet of gas at standard conditions.

CHAPTER 6

How to Receive and Unload Liquefied Compressed Gases*

F. R. FETHERSTON

Liquefied compressed gases commonly shipped to users in bulk by rail or over the highway in cargo tank trucks are: anhydrous ammonia, liquefied petroleum gas (LP-gas), chlorine, methyl chloride, sulfur dioxide and, broadly, liquefied fluorinated hydrocarbons.

In liquid form, carbon dioxide and nitrous oxide are liquefied compressed gases, and these are so shipped. However, due to special considerations, these two gases are not covered in this explanation.

All subject gases are shipped in single-unit tank cars. There are several ways to unload them. Some are also shipped in bulk by rail in multi-unit (TMU) tank cars. In such shipments a number of large containers, usually 15, are transported on a car tank underframe for removal therefrom and subsequent movement to point of use. Whichever gas these containers hold, the method of removing them from the car underframe is identical.

The method of withdrawing their content at point of use is the same except with sulfur dioxide and liquefied fluorinated hydrocarbons, which may require heat to effect transfer.

Never use direct heat in any form to unload containers holding the other compressed gases. Where you need heat to transfer the contents of sulfur dioxide or liquified fluorinated hydrocarbon containers, use it only as prescribed by the manufacturers.

"Cargo tank truck" is an industry term used to designate a truck on which a tank is mounted to transport liquids or gases. "Cargo tank truck," "tank truck," and "transport tank truck," as used in this explanation, are synonymous. In Interstate Commerce Commission (ICC) regulations, all are defined as motor vehicles on which cargo tanks are mounted as a permanent part of the vehicle itself. The capacity of the tanks range up to approximately 10,800 gal.

Safe unloading of single-unit tank cars, safe removal of containers from TMU tank cars, unloading of these containers at point of use, and safe unloading of cargo tank trucks is a matter of applying known safety procedures.

Liquefied compressed gases can be safely handled and unloaded only:

(1) when their physical and chemical properties are understood;

(2) when regulations and standards governing their handling are complied with fully;

(3) when you realize that accidents are the results of human failure.

*Reprinted with permission from *Chemical Engineering*, Nov. 2, 1959. Copyright 1959 by McGraw-Hill Publishing Co., Inc., 330 W. 42nd St., New York, N. Y. 10036.

Accidents have happened during unloading operations. Results, in some cases, have been property damage, personal injuries and loss of life.

Compressed Gas Association, Inc., which is and has been active for almost 50 years in developing procedures for safe and efficient transportation, storage and handling of all compressed gases, as well as the related standards, adamantly takes the position that there can be no leeway, short cuts, laxity, nor carelessness on the part of handling personnel during any stage of the unloading operations.

Safe procedures are spelled out precisely in regulations, standards, the literature of technical associations, and in the instructions of the compressed gas manufacturers. These must be followed to the letter.

EQUIPMENT AND TECHNIQUES

Start with careful selection of personnel charged with the responsibility for unloading. Never send a boy to do a man's job. Choose reliable, intelligent men with a high sense of responsibility preferably from the chemical operator level and arrange to train them thoroughly.

Training must be continuous. Failure to keep personnel well trained is a source of danger. Include a medical officer, safety supervisor and fire protection engineer (if they are available) in training class.

The compressed gas manufacturer is the key to a good training program. He will supply technical data describing the physical and chemical properties of the compressed gas in question, knowledge concerning its characteristics, information about protective equipment, measures to take to prevent an accident before it happens and the corrective and first-aid steps in case of an accident.

Make it a *must* for personnel to study these data until they are completely familiar.

Some manufacturers of certain compressed gases furnish qualified technical representatives to train personnel in every step of the unloading procedure. Take advantage of this service where it is available.

Bear in mind that employees leave a company, or they may be promoted or transferred to other departments or locales, and so are relieved of their unloading duties. In such circumstances, be sure the replacement is thoroughly trained in unloading procedures before he is allowed to participate in the operation. Many unloading accidents have been traced to failure to observe this obvious precaution.

Responsibility for Unloading Operations

With single-unit tank cars and TMU cars, full responsibility for unloading them safely rests with you, the consignee. In unloading cargo tank trucks, where equipment for making the transfer to the consignee's storage facilities is a part of the vehicle itself (photograph on p. 369) responsibility for safe unloading rests with the carrier or the transfer owner.

In cases where equipment for transferring a compressed gas from a cargo tank truck is provided by the consignee, the consignee then has a vested interest in the transfer operations and to this extent shares responsibility for safety. There are exceptions to this: some users own their cargo tank trucks and these users are responsible for unloading operations.

Tank Cars are Similar

Except for capacities and specifications governing fabrication of their tanks, single-unit tank cars used in liquefied compressed gas service are similar.

Single-unit tank cars for transporting anhydrous ammonia and LP-gas are insulated or uninsulated while those used to transport the other liquefied compressed gases are insulated to protect against atmospheric temperature differentials. The cars are never filled liquid full, but have a vapor space, termed outage, the volume of which is calculated for each specific gas.

Arrangement of their valves and fittings is identical (Fig. 1). These are located on a manhole cover plate on top of the car, not in a dome as such, but in a protective housing. The

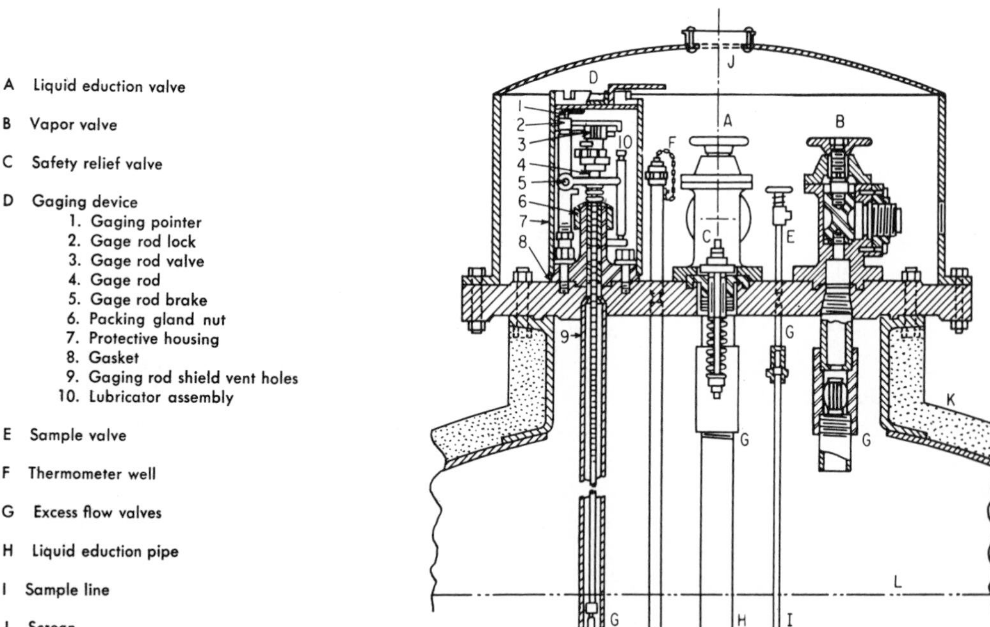

FIG 1. Tankcars for liquefied compressed gases have identical valve arrangement.

manhole cover is never removed for unloading. Unloading is only through the manually operated valves.

Usually there are 4 such valves. Two are in line with the track (towards the ends of the car) and are connected to liquid eduction pipes that extend to the bottom of the car. These are the liquid valves.

Two valves (in some cars only 1) are mounted towards the sides of the car, and these terminate in the car's vapor space. These are the vapor valves.

In cases where cars have only 1 vapor valve, a liquid level gage may be installed in the space otherwise occupied by the second vapor valve. Such cars usually have a thermometer well.

Excess flow check valves are installed on the liquid eduction lines of some single-unit tank cars. These operate when the liquid valves are opened too quickly; the unloading rate is too high; or a break or rupture occurs in the liquid unloading lines. When the difficulty has been corrected, reopening the check valves is accompanied by an audible sound.

Single-unit tank cars in liquefied compressed gas service are protected from excessive internal pressure by a spring-loaded safety relief valve located in the center of the manhole cover plate. Valve setting depends upon the specification tank car used.

At high temperatures, such as might be encountered in a fire, the safety relief valve prevents buildup of dangerous internal pressures. Unloading procedures must be such that the pressure developed does not cause the safety relief valve to blow.

Removing TMU Containers

Liquefied compressed gases transported in multi-unit (TMU) tank cars are: anhydrous ammonia, chlorine, methyl chloride, sulfur dioxide and liquefied fluorinated hydro-carbons.

Although LP-gas generally is not shipped by TMU cars, ICC regulations provide for their use.

In TMU shipments, these gases are contained in large cylinders commonly known as "ton containers." This names comes from their capacity in terms of chlorine. They are shipped on a car underframe; shipments usually consisting of 15 containers. Fewer than 15 units may be shipped as a TMU car, but this arrangement is impractical since freight charge is figured for the full 15 containers at prevailing carload rates.

Before removing containers from the car frame, it is necessary to observe the same ICC regulations concerning personnel, setting brakes, blocking wheels of the car and posting caution placards as in unloading single-unit cars.

Full containers are removed from the underframe by a lifting hook in combination with a hoist on a trolley (see accompanying photo-

TMU containers being removed with hook, hoist, and trolley.

graph) or a jib boom. Never use a lifting magnet, nor a rope or chain sling, to unload or otherwise handle these containers.

Containers must be protected from shock which might damage valves, fuse plugs or the container itself. Keep valve protecting hoods in place at all times when containers are being moved.

When containers are trucked from the rail siding to the storage area, they must be placed on saddles on the truck. Preferably, clamp them down to prevent shifting and rolling.

Valves and Safety Relief Devices

Each container has 2 valves, each equipped with a $\frac{1}{2}$-in. eduction pipe. When the container is positioned horizontally with a slight downward pitch and one valve is directly above the other, the top eduction pipe ends in the vapor phase and the eduction pipe of the bottom valve in the liquid phase.

Either liquid or vapor can be withdrawn by connecting to the appropriate valve.

These two valves are protected by hoods. Keep hoods in place at all times except when the container is connected for withdrawing its content.

Each container is protected by fusible plugs in each head. The fusible metal is designed to soften or melt at 157 to 165 F. Do not tamper with the fusible plugs under any circumstances. It is recommended that containers never be allowed to reach a temperature exceeding 125 F.

ICC regulations provide that TMU containers may be transported under certain conditions on trucks or semitrailers. They must be chocked or clamped on the truck to prevent shifting. You must have adequate facilities available when transfer in transit is necessary.

Primary Precautions

Comply with all rules set forth in Section 74.561 through Section 74.563 of ICC regulations.

Once caution signs are posted, do not allow the tank cars so protected to be coupled or moved.

Do not permit other cars to be placed on the same track except after notifying the person who placed the signs, and provided there is a standard derail properly set and locked in the derailing position between the car being unloaded and cars to be set on the same track.

Never permit a compressed gas to be loaded into a container that holds or has held another compressed gas.

Unloading valves must not be opened until caution signs are posted, nor the signs removed, even temporarily, until valves are closed and lines disconnected.

When tank cars are placed on a dead-end siding, protect them on the switch end by a derail; if on open-end siding, at both ends.

Follow faithfully all procedures recommended by your compressed gas manufacturer.

In the case of flammable gases, establish a firm "No Smoking" rule for all unloading personnel. Make it mandatory that the rule is observed during the entire unloading operation.

Have your personnel check the area for all open flames and have flames extinguished before starting operations. Instruct your personnel to prevent all persons in the vicinity from smoking, carrying any flame or lighted cigar, pipe or cigarette.

Make sure the car is level so it can be gaged correctly and unloaded completely. The compressed gas manufacturer will supply data for sampling procedures and for gaging tank cars. The American Petroleum Institute publishes a tentative code for gaging tank cars containing compressed gases derived from petroleum.[2]

Check These Regulations Before Starting Any Unloading Operation

In the U. S. the regulations of the Interstate Commerce Commission (ICC) govern unloading of compressed gas shipped in interstate commerce. In Canada, the regulations of the Board of Transport Commissioners for Canada (BTC) govern the operation, and these parallel the regulations of the ICC for rail transportation.

In addition, there may be local, state or provincial regulations. Consult your compressed gas manufacturer as to the existence of the latter.

Interstate Commerce Commission regulations are published in Agent T. C. George's Tariff No. 15, reissues thereof, and supplements thereto.[1] Obtain a copy. Make sure that excerpts from these regulations, as they pertain to compressed gas, are placed in the hands of unloading personnel for study.

If you are building a new plant, Section 74.560 of the ICC regulations tells the circumstances under which you must own (or lease) your rail siding. This section also states the exceptions and circumstances under which compressed gas tank cars—particularly TMU cars for shipping anhydrous ammonia or chlorine, and single-unit tank cars for shipping anhydrous ammonia, liquefied hydrocarbons or liquefied petroleum gas—may be unloaded from carrier tracks.

Tank Cars.

Section 74.561 gives the regulations governing unloading of all compressed gases, including liquefied compressed gases, from tank cars. Refer to it for full requirements. Significantly, this section emphasizes the importance of personnel by placing this subject first on the list.

Following items are excerpts from this section:

"Unloading operations should be performed only by reliable persons properly instructed and made responsible for careful compliance with this part.

"Brakes must be set and wheels blocked on all cars being unloaded.

"Caution signs must be so placed on the track or car as to give necessary warning to persons approaching car from open end or ends of siding and must be left up until after car is unloaded and disconnected from discharge connection. Signs must be of metal or other suitable material, at least 12 by 15 in. in size and bear the words 'STOP—Tank Car Connected,' or 'STOP—Men at Work,' the word 'STOP' being in letters at least 4 in. high and the other words in letters at least 2 in. high. The letters must be white on a blue background . . .

". . . care must be taken to avoid spilling any of the contents over tank or car . . .

"Unloading connections must be securely attached to unloading pipes . . . before discharge valves are opened . . .

"Tank cars must not be allowed to stand with unloading connections attached after unloading is completed, and throughout the entire period of unloading, or while car is connected to unloading device, the car must be attended by unloader."

Cargo Tank Trucks.

Where responsibility for unloading cargo tank trucks rests with the common carrier or transport owner, his operator will be trained in all phases of the operation. He will be thoroughly familiar with the regulations, both federal and local.

However, since the operation is performed within the battery limits of your (the consignee) plant, you have an interest.

As a matter of information, general regulations governing cargo tank truck unloading are stated here. Check these regulations and make sure you follow them.

Subpart B of Part 77 of ICC regulations applies to unloading compressed gas cargo tank trucks. In part they provide that:

There shall be no smoking on or about the vehicle during unloading. Extreme care shall be taken to keep fire away and to prevent persons in the vicinity from smoking, lighting matches, carrying any flame or lighted cigar, pipe or cigarette.

The handbrake of the transport must be securely set and other reasonable precautions taken to prevent any motion of the vehicle.

No tools which are likely to damage the tank shall be used.

The transport must be attended at all times. Under ICC regulations the delivery hose, when attached to the vehicle, is declared a part thereof.

Unless the engine of the transport truck is

to be used for operation of a transfer pump of the vehicle, the engine must be stopped during unloading; and, in any event, care should be exercised to prevent the ignition of vapors. Unless the delivery hose is equipped with a shutoff valve at its discharge end, the engine of the motor vehicle shall be stopped at the finish of the unloading operation while the discharge connection is disconnected.

No tank motor vehicle shall be moved, coupled or uncoupled when loading connections are attached to the vehicle, nor shall any semitrailer or trailer be left without the power unit unless chocked or equivalent means are provided to prevent motion.

General Precautions

Phone or wire the compressed gas manufacturer immediately for assistance and instructions if:

(1) A tank car is received in bad order. Meanwhile, do not attempt to unload it.

(2) In event of failure of fittings or a leak that cannot be readily repaired by simple adjustment or tightening of the fitting. Meanwhile, evacuate the area around the car and permit only properly instructed and protected personnel to enter the area.

(3) If the tank car cannot be unloaded after following all instructions.

(4) In event of an accident of any kind.

In addition to notifying the compressed gas manufacturer concerning the above, also notify the nearest ICC inspector. He is usually very helpful in correcting these problems. You can get his name, address and phone number by writing to: Interstate Commerce Commission, Washington, D. C.

It is illegal to ship a defective or leaking tank car. If a car is in bad order or if there is an accident, telephone the car owner. He can lend some valuable assistance.

Insist that personnel observe the following:

(1) Never introduce a compressed gas into a tank car or container of any kind that holds or has held another compressed gas or chemical.

(2) Never introduce water or other material into a tank car or any other compressed gas container.

(3) Do not remove tank car valves under any conditions.

(4) Never hammer tank car fittings in an attempt to open excess flow check valves, nor for any other reason.

(5) Never enter empty tank cars for any purpose whatsoever.

(6) Do not attempt to correct leaks at unions or other fittings in a line with a wrench while the line is under pressure.

(7) Never fill the storage tank liquid full; the consequences are very serious. Observe readings on gaging devices and fill only to allowable maximum filling level.

(8) Where compressor, pump or piping, handling a heavier-than-air compressed gas, is located in a trench or pit, caution the operator to check these low spots for vapors since they can occupy an enclosure or trench by displacing the air.

(9) When all unloading connections seem to be tight, pressurize the system and check for any leaks.

(10) Never break a hose coupling under pressure.

Because of their low boiling points, the subject gases must be liquefied under pressure. On release at normal temperatures and atmospheric pressure the liquids vaporize rapidly.

In contact with skin surfaces this evaporation of liquids causes severe burns (frostbite) and when large quantities of the liquid are involved, deep and severe freezing of the area in contact with the liquid results.

For this reason, it is important for unloading personnel to wear protective gloves to prevent coming into contact with liquid when breaking hose connections or when venting the transfer hose. Large-lensed spectacles or goggles should be worn to prevent liquid from contacting the eyes and causing possible injury due to freezing of the moisture in the eyes.

As concerns gas masks, protective clothing, fire extinguishers and related safety equipment, make certain these are of approved types, always in perfect condition and stored in areas that are accessible in case of a leak. See to it that personnel are thoroughly instructed in their use and keep personnel alert and continuously trained by occasional "dry runs."

Tank Car Transfer Lines

Single-unit tank cars in liquefied compressed gas service are always unloaded from the top.

Transfer lines and their flexible hoses, or unloading coils, must be of materials suitable for transferring the particular gas being handled as recommended by the compressed gas manufacturer. When not in use, plug the hoses or coils to prevent accumulation of moisture and dirt.

Different methods of unloading require special valves and equipment. In all cases, the liquid unloading line is equipped with a pressure gage and shutoff valve.

With the exception of chlorine, liquid and vapor lines are vented to atmosphere through small valves. While bleeder valves are installed in chlorine transfer lines, chlorine is not vented to the atmosphere but to a lime or caustic disposal system.

In all cases, never break a hose coupling or connection while the transfer line is under pressure.

Mechanical Unloading Methods

Liquefied compressed gases are transferred from single unit tank cars to storage (in case of chlorine, direct to process) by pressure differential. Before unloading, check the pressure in the tank car and in the storage tank to determine which is higher. If pressure in the tank car is the higher of the two, then mechanical means will not be required to transfer the liquid. However, as pressure between the car and storage equalizes, mechanical methods are required to restore it.

Methods and gases to which they must be applied are:

By compressor—anhydrous ammonia, LP-gas, methyl chloride, sulfur dioxide and liquefied fluorinated hydrocarbons.

By vaporizer—anhydrous ammonia. (This method is not commonly used.)

By direct-acting liquid pump—LP-gas.

By air or gas repressuring—LP-gas.

By air-padding—chlorine and sulfur dioxide.

All equipment used must be designed for the particular gas being transferred.

Weighing.

This is often the most desirable method for determining when a car is empty of liquid. Some transfer systems include a sight glass in the liquid discharge line. When vapor is observed in this glass, it indicates the car is emptied of liquid. Where cars are equipped with a sample valve, you can check indication given by the sight glass by opening the sample valve slowly.

Unloading With Compressor.

The suction side of the compressor is connected to the vapor line of the storage tank: the discharge side is connected to one or both vapor valves on the tank car through suitable transfer hoses.

The compressor withdraws and compresses vapor from the storage tank and transfers it to the tank car's vapor space. In this way, the desired pressure differential is created which transfers liquid to storage from the tank car through a hose connected to the liquid line on the tank car and to a vapor line from the storage tank. A typical compressor unloading setup is shown in Fig. 2.

To withdraw vapors from the tank car after it is empty of liquid, the car's liquid valve is closed, and the compressor's reversing valve turned. The compressor then withdraws most of the vapor in the tank car and returns it to storage. If the compressor does not have a reversing valve, its suction and discharge connections are reversed to accomplish this result.

Allow some vapor to remain in the car to prevent a vacuum from developing that may damage the car.

When using a compressor, take precautions to prevent lubricating oil from contaminating the tank car or storage tank. Do this by installing a suitable oil trap or mist extractor on both

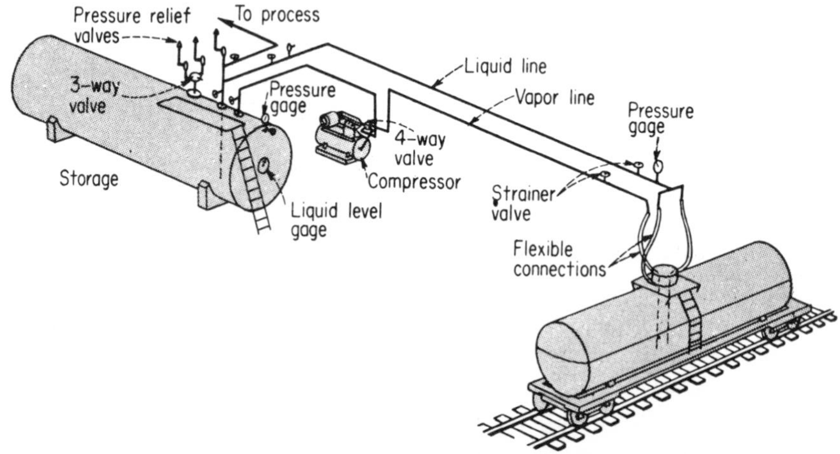

FIG. 2. Using a compressor to unload a single-unit tankcar.

the suction and discharge sides of the compressor. Drain the trap or mist extractor before each unloading.

Unloading With Vaporizer.

Liquid from storage is charged into a vaporizer, sometimes termed an evaporator, and its vapor section is connected to one or both vapor valves on the tank car. When heat is turned into the vaporizer, the vapor exerts pressure on top of the liquid in the tank car causing it to flow to storage.

By this method, all of the liquid can be transferred to storage but the vapor remains in the car. A steam vaporizer method of tank car unloading is illustrated in Fig. 3.

Unloading With Pump.

When a direct-acting liquid pump is used for transfer, an equalizing hose assembly is connected to the storage tank vapor line and to the tank car vapor valves. This equalizing hose is used if storage tank pressure is greater than tank car pressure and to save wear and tear on the pump and motor.

If the tank car pressure is higher than storage tank pressure, allow the pump's bypass to remain open as long as the rate of liquid flow from the tank car to storage is satisfactory.

Unloading by Repressure.

An air dryer must be used to unload by this method.

Connect the repressuring hose assembly to the tank car vapor valve and to the compressed air or gas supply line. This line must have a check valve and an atmospheric vent.

There should be a pressure gage on the repressurizing line near the tank car vapor valve to permit readings on the tank car pressure. If the difference in pressure is less than 10 psig, no restriction of flow is necessary. If a long unloading line is involved, this differential may be increased slightly.

While air (or any fluid) is passing through a pipe at high velocity, the pressure gage mounted directly in the pipeline reads much lower than the pressure existing at the source of the pressure. Where the air or gas supply is at a much higher pressure than that desired in the tank car, there have been cases of relief valves on the tank car releasing while the pressure gage in the air or gas line indicated a safe pressure.

By either placing the pressure gage directly in the line to the tank car vapor valve or stopping the flow of air or gas through the pipeline momentarily, you can get an accurate reading and overcome false readings.

Maintain from 5 to 10 psig more pressure on the tank car than on the storage tank.

Open the repressuring line valve just enough to maintain the desired differential unless a differential controller is used. Avoid building up too great a differential or else the

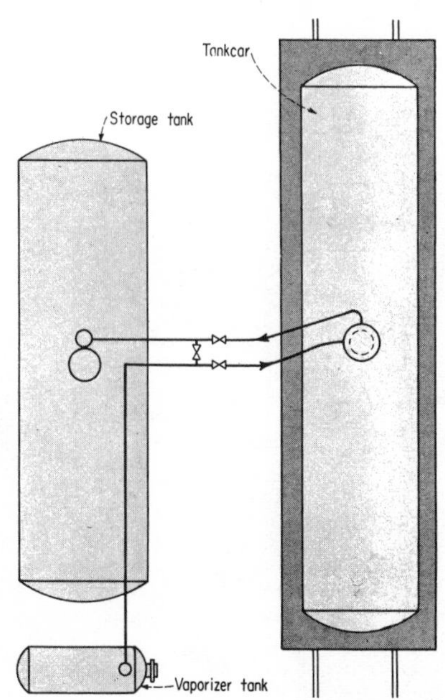

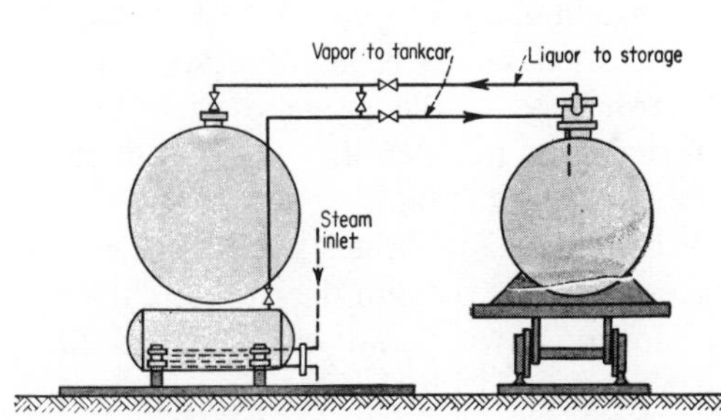

FIG. 3. Unloading a single-unit tankcar by steam vaporizer.

tank car excess flow valves will close.

As in the compressor method, a flow of gas instead of liquid through the sight glass, if one is installed in the system, indicates the car is empty. When the tank car is empty, first shut off the air or gas repressuring supply, and then proceed in the same way as when unloading with a compressor.

Do not use untreated artificial gas for repressure or the shipment or the tank car may be contaminated. If properly cleaned, coal gas, blue water gas, coke oven gas, producer gas or natural gas, may be used. Also, butane-air or propane-air mixtures may be used to unload by gas repressure.

Repressuring gas or air should be as dry as possible to eliminate the addition of water to the product as well as to prevent freezing troubles during cold weather. Unloading by air or gas is impractical for propane or high propane content mixtures because of the pressure involved.

Unloading With Air-Padding.

When air-padding is used, it is imperative that clean, oil free, cooled, dry compressed air be introduced into the tank car's vapor space through its vapor valve to transfer the car's liquid to storage (or as in the case of chlorine, direct to process).

The setup consists of an air compressor and its tank equipped with a pressure regulator and an air dryer with appropriate valves and gages, as shown in Fig. 4. A dependable check valve must be incorporated in the air-padding line as also shown. Never use a plant air system for air-padding since the possibility exists that vapors can be drawn into the plant air system.

Preparing for Unloading

When the tank car is protected as required by ICC regulations, break the seal on the protective cover housing the unloading valves and related equipment, and check (by the means prescribed later for the separate gases) the various valves for leaks which may have occured in transit.

If any working parts of the tank car are in bad repair, notify the supplier at once. Do not tamper with the car in an effort to correct the trouble. Instead, phone or wire the compressed gas manufacturer, state the trouble, and follow his instructions.

Make the car ready for unloading by carefully and cautiously removing the plugs from

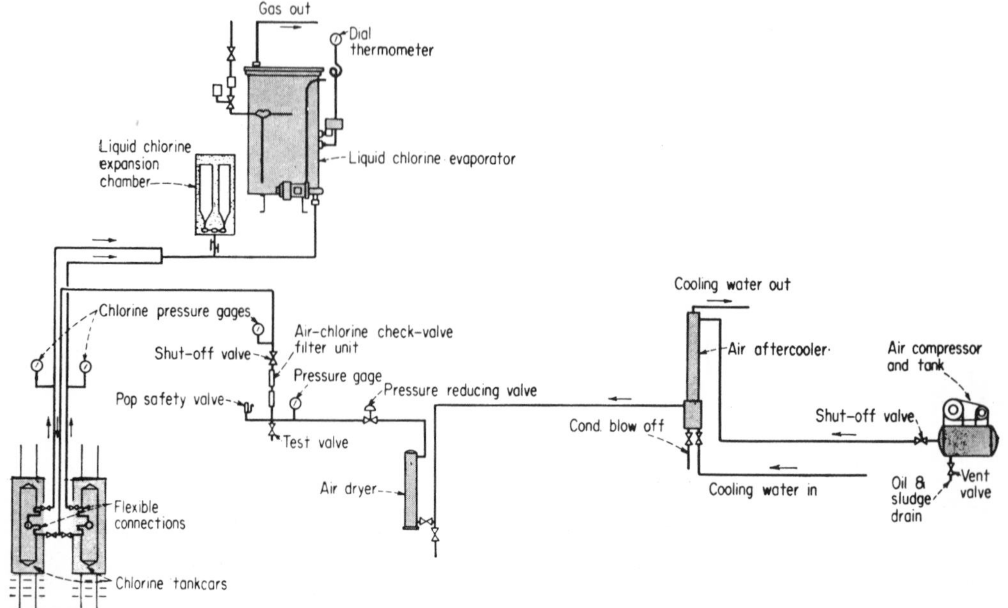

FIG. 4. Unloading two single-unit tankcars by air padding.

the car's liquid and vapor valves. The liquid transfer assembly is connected to the tank car's liquid valve, or valves. The vapor line transfer assembly is connected to the tank car's vapor valve or valves.

If the pressure in the tank car is higher than that in the storage tank, start the unloading operation by opening the tank car's liquid valves (only one may be used, if desired) and then open all other valves on the liquid line toward the storage tank.

When pressure between tank car and storage tank nears equalization, restore pressure in the tank car by the method appropriate for the liquefied compressed gas being transferred. This proceeds by working from the source of the pressure to the tank car.

If a compressor, vaporizer or a pump is used, open the vapor valve on the storage tank as well as all valves on the intake and discharge sides of the vapor transfer equipment.

Place the equipment in operation and open the vapor valve, or valves, on the tank car.

If air or gas repressuring or air padding is used, place the equipment in operation, and open all valves leading to the tank car with the tank car's vapor valve or valves opened last. As in the case of liquid transfer, open these valves slowly.

The sight glass, when one is installed in the liquid discharge line, is observed by the operator, as are the pressure gages in the transfer lines, to determine when all liquid is removed from the tank car.

When this is established, prepare the car for disconnecting (except in the case of chlorine) by closing down liquid transfer lines first. Then, starting with the storage tank's inlet valve, close all valves up to and including the tank car's liquid valve.

If you use a compressor on the vapor side and want to recover the vapors in the tank car, shut down the compressor and turn its plugcock, or reverse the vapor transfer lines on the compressor. Then, place the compressor in operation again to withdraw vapors from the tank car to storage.

In shutting down the vapor transfer lines, close the tank car's vapor valve first, and then successively close the other valves by working from the tank car to the source of pressure. After bleeding off all transfer lines to atmospheric pressure through the small bleed and purge valves, located in the liquid and vapor lines, disconnect the flexible hoses or coils from the tank car either by breaking the flanged connection or by opening up a pipe union.

Plug the unloading lines, as well as all tank car valve openings and place the tank car fittings in proper position.

Lower the tank car's protective housing carefully and secure it in place with an iron pin. When signs protecting the car have been removed and all other regulations required by the ICC observed, inform the railroad agent that the car is empty and furnish him with return billing instructions.

Precautions When Unloading Tank Cars.

(1) Never introduce a compressed gas into a tank car or container of any kind that holds or has held another compressed gas or chemical.

(2) Never introduce water of another material into a tank car or any other compressed gas container.

(3) Do not remove tank car valves—under any conditions.

(4) Never hammer tank car fittings in an attempt to open excess flow check valves, nor for any other reason.

(5) Never enter empty tank cars for any purpose whatsoever.

(6) Do not attempt to correct leaks at unions, or other fittings in a line, with a wrench while the line is under pressure.

(7) Never fill your storage tank full; the consequences are very serious. Observe readings on gaging devices and fill only to allowable maximum level.

(8) Where compressor, pump or piping, handling a heavier-than-air compressed gas, is located in a trench or pit, caution the operator to check these low spots for vapors since they can occupy an enclosure or trench by displacing the air.

(9) When all unloading connections seem to be tight, pressurize the system and check for any leaks.

(10) Never break a hose coupling under pressure.

Handling TMU Containers

With the exception of LP-gas, as previously noted, all subject gases are shipped in TMU containers. These containers may be removed from the car underframe when the car is spotted on carrier tracks provided the ICC specification regarding the presence of handling equipment is complied with. The cylinders are then transported by truck, as previously described, to the user's plant where similar handling equipment must be available. Cautions concerning the handling of these cylinders as earlier described must be strictly observed.

Ton containers should be stored on the user's premises in a cool, dry place and protected against heat sources. You can make a convenient storage rack by supporting the containers at each end on a railroad rail or an I-beam. When containers are not being used, keep the valve protective hoods in place at all times.

Store full and empty containers in different places to avoid confusion in handling. It is good practice to tag empties. Do not store containers near elevators or gangways or in locations where heavy objects may fall and strike them.

Never store containers near combustible or flammable materials. When containers hold flammable gases, keep them away from all ignition sources. Keep the storage room well ventilated and so arranged that any container can be removed with a minimum of handling of other containers. When practical, the storage room should be fireproof. Avoid storage in subsurface locations.

Make sure containers stored outdoors are free of debris, tall grass and away from public access. They should be kept clean and inspected regularly.

Never store containers holding a flammable gas or a gas that affects the respiratory system. Never place such containers where fumes can enter a ventilating system, or where wind can carry fumes to populated areas.

Use containers in the order received. Observe precautions stated under "Valves and Safety Devices" as well as the general and specific rules for unloading tank cars.

Except when recommended by the compressed gas manufacturer, TMU containers should not be manifolded to withdraw contents from two or more containers simultaneously.

To discharge the content of TMU container, place it so that it is in a nearly horizontal position with a slight downward pitch (about 1 in. over-all) toward the valves. When the container is placed so that one valve is directly above the other, vapor can be withdrawn from the upper valve or liquid from the lower.

Assembly for withdrawing a container's content usually consists of a transfer line equipped with a pressure reducing valve, rotameter, control valve and diffuser. In chlorine service the system may include only a pressure and control valve.

For all subject gages, test TMU container connections and transfer piping for leaks as later described under handling of each specific gas.

Sometimes, it is necessary to use heat to help the flow of either gas or liquid from sulfur dioxide or liquefied fluorinated hydrocarbon ton containers. When heat is used, the method approved by the compressed gas manufacturer must be followed. Exercise great care, as fusible plugs in ton containers melt at 157 to 165 F. Never allow containers to reach a temperature above 125 F. Never apply blow torches or steam hoses nor use an open flame from any source or heat them.

Unloading Cargo Tank Trucks

All subject liquefied compressed gases are, or may be, shipped in cargo tank trucks conforming to ICC Specification MC-330.* While it is permissible to ship chlorine in cargo tank trucks of this specification in the U. S. and Canada, no shipments of this kind have been made to date. Similarly, methyl chloride is being transported in MC-330 transports, but only in a few isolated cases and usually in a cargo tank truck owned by the user of the product.

Anhydrous ammonia is unloaded from cargo tank trucks by a compressor or liquid pump; LP-gas by a compressor or liquid pump; and methyl chloride and sulfur dioxide by a compressor. Liquefied fluorinated hydrocarbons are usually unloaded from cargo tank trucks by a turbine pump although you can use a compressor. ICC regulations for unloading cargo tank trucks and the precautions stated herein are to be observed.

Procedures are generally those described in unloading single-unit tank cars. Even where responsibility for unloading cargo tank trucks rests with the carrier or transport owner, the consignee should be as familiar with the entire unloading procedure as is the transport's operator in charge of the transfer.

*Subsequent to the publication of this article, ICC specification MC-330 was superseded by a new specification MC-331.

PROBLEMS WITH SPECIFIC GASES

Anhydrous Ammonia

Anhydrous ammonia is classified by the ICC as a nonflammable gas.

In transferring ammonia from containers, including single-unit tank cars, TMU containers and cargo tank trucks, never use compressed air as it will contaminate ammonia.

The continuous presence of the sharp, irritating odor of ammonia is evidence of a leak. Leaks can be located in ammonia by allowing fumes from an open bottle or hydrochloric acid (from a squeeze bottle of sulfuric acid, or from a sulfur dioxide aerosol container) to come in contact with ammonia vapor. This produces a dense fog. Leaks may also be detected with moist phenolphthalein or litmus paper. Sulfur tapers for detecting ammonia leaks are not recommended.

When there is a leak around an ammonia

container valve stem, it usually can be corrected by tightening the packing gland nut which has a left-hand thread.

When a leak occurs in a congested area where atmospheric dissipation is not feasible, absorb the ammonia in water. Its high solubility in water maybe utilized to control escape of ammonia vapor. Applying a large volume of water from a fog or spray nozzle lessens vaporization, as the vapor pressure of ammonia in water is much less than that of liquid ammonia.

Do not neutralize liquid ammonia with acid —heat generated by the reaction may increase the fumes.

Only an authorized person should attempt to stop a leak, and if there is any question as to the seriousness of the leak, a gas mask of the type approved by the U. S. Bureau of Mines for use with ammonia must be worn. Have all persons not equipped with such masks leave the affected area until the leak is stopped.

Also, provide personnel subject to exposure to ammonia with a hat, gloves, suit and boots, all garments of rubber. Garments worn beneath rubber outer clothing should be of cotton. Some protection to the skin may be obtained by applying protecting oils before exposure to ammonia. Supply approved eye goggles if the eyes are not protected by a full face mask.

Although ammonia is flammable only within the narrow limits of 16 to 25 per cent by volume, the mixture of oil with ammonia broadens this range. Therefore, take every precaution to keep sources of flame or sparks from areas that have ammonia storage or use.

In the event a fire does break out in an area containing ammonia, make every effort to remove portable containers from the premises. If they cannot be removed, inform the firemen of their location.

For data concerning the physiological effects of ammonia, protective equipment and first aid measures, obtain a copy of "Pamphlet G-2, Anhydrous Ammonia" published by Compressed Gas Association, Inc.[3]

LP-Gases

All LP-gases are classified by the ICC as flammable gases and must be treated as such. Obtain a copy of the National Fire Protection Association "Pamphlet 58, Standard for the Storage and Handling of Liquefied Petroleum Gases."[4]

Liquid LP-gas leaks in transfer piping are indicated by frost at the point of leakage due to the low boiling point of the material.

Extremely small leaks and leaks in vapor transfer piping can be detected by applying soap suds or a similar material to the suspected area.

Under no circumstances should a flame be used to detect a leak.

The most important safety considerations in unloading LP-gas are: avoid unnecessary release of the product; keep any open flames and other sources of ignition away from the unloading area; and make sure suitable first-aid type fire extinguishers are available (dry chemical or carbon dioxide types are suitable for LP-gas fires).

It is important for personnel to be familiar with the characteristics of LP-gas. You must realize the importance of not extinguishing a fire unless by doing so you stop the source of the leakage (as by closing a valve). This is so because if you extinguish the fire and leakage is allowed to continue, unburned vapor could accumulate and possibly result in a more serious hazard than if you allowed the escaping gas to burn.

If a fire is in progress and an LP-gas storage tank is exposed, it is of prime importance to keep the container cool by applying hose streams of water until the fire is properly extinguished. By the same token, if fire threatens a single-unit tank car, a TMU car or a cargo transport truck, immediately remove these from the area. If this isn't possible, apply hose streams of water, as in protecting storage tanks.

Unloading From Tank Cars.

If unloading LP-gas from single-unit tank cars with a compressor and vapors are being recovered from the tank car's tank, stop the compressor and close all vapor lines when the tank car vapor pressure is reduced to about 15 psig. Bleed the vapor lines and replace all plugs in tank car valve opening and transfer hoses.

To unload by gas or air repressure, after the plugs are removed from the tank car liquid education valves:

(1) Connect the unloading hose and storage assembly line which may be a riser, or part of the storage tank manhole assembly.

(2) Connect the repressuring hose assembly.

(3) Connect the repressuring hose assembly to the tank car vapor valve and to the compressed air (or gas) supply line. This line *must* have a check valve and an atmospheric vent.

(4) Open the tank car liquid eduction valves slowly but completely.

(5) If an unloading riser is used, then slowly open the liquid line valve.

Chlorine

In both its liquid and gaseous form, chlorine is neither flammable nor explosive. It is classified as a nonflammable compressed gas by the ICC. Its principal hazard arises from inhalation from leaks. For data describing its physiological effects, handling, employee training and protection, and its chemical characteristics and physical properties, obtain a copy of The Chlorine Institute's manual[5].

Also, consult with the chlorine manufacturer concerning these matters, as well as the equipment required for handling chlorine, specifications and maintenance of the equipment and the special safety precautions and safety equipment required.

All chlorine containers, including single-unit tank cars, are intended for use by you (the customer) to deliver chlorine direct to process. Thus, differing from the other liquefied compressed gases, there is no need for intermediate storage tanks between the shipping container and the process. Moreover, the use of chlorine storage tanks between the container and process is frequently hazardous.

Usually, chlorine tank cars are filled at low temperature and pressure. Normally, the inherent pressure of the vapor in the tank car is sufficient to accomplish withdrawal of liquid chlorine to process, but sometimes, especially

Tank truck for LP-gas service, above, contains its own unloading equipment.

during the winter, the car is air-padded to accomplish liquid withdrawal.

In Fig. 4 is shown a typical arrangement for unloading liquid chlorine for tank cars. Liquid flow is through the tank car's liquid eduction valve to a liquid evaporator within battery limits, then direct to process. Whenever liquid chlorine can be trapped between two valves, the line must be protected by a heated expansion chamber.

When air-padding is required, introduce only clean, oil-free, cooled, dry compressed air into the tank car through its vapor valve.

Absorbing Chlorine in a Liquid.

If chlorine is to be absorbed in a liquid, there is a tendency for the liquid to suck back into the container when the container becomes empty due to the creation of a partial vacuum. Avoid this as it has resulted in numerous accidents.

As soon as the container is empty and its pressure has dropped to zero, shut the container valve. Then, vent air into the line leading from the container, after the valve has been shut off to prevent liquid from "sucking back" into the line. For this purpose, install a "vacuum break" valve or loop on the chlorinator well line.

Rules for Unloading.

In general, rules for unloading all chlorine containers apply to the tank car operation. These include:

Safety devices must never be tampered with.

Open container valves slowly.

Make sure that threaded connections are the same as those on the container valve outlets. Never force connections that don't fit. The outlet threads on valves of TMU containers are not tapered pipe threads.

Containers or valves should never be altered or repaired by unauthorized personnel.

Gas leaks around the valve stem may usually be checked by tightening the packing nut.

Test for chlorine leaks by attaching a cloth to one end of a stick, soak the cloth with strong ammonia-water, and apply to the suspected area. A white cloud of ammonium chloride results if there is any chlorine gas leakage. Do not use the usual household ammonia—it is not strong enough.

Use only reducing valves and gages designed for chlorine. Consult the chlorine producer for details.

In addition to complying with ICC regulations governing unloading tank cars, the following special rules are recommended where chlorine is being handled in tank car quantities:

Switch-tracks on which chlorine tank cars are placed for unloading should be devoted solely to this purpose.

When chlorine tank cars are located on a dead-end siding, protect the cars on the switch end by a locked derail; if on an open-end siding, at both ends. Keys should be in charge of a designated responsible person.

Unloading single-unit chlorine tank cars should be done through flexible metal connections that compensate for the vertical rise of the tank car as its springs decompress during unloading. For all details relating to the unloading line, consult the chlorine producer. However, do not depend upon flexible connections as a safety factor in case of a bump by switching operations during unloading.

Place derails as stated previously and use a blue lantern, suitably placed, if unloading at night.

Perform unloading, connecting and disconnecting operations only in well-lighted places by reliable persons who are properly instructed, responsible for operations and equipped with chlorine gas masks.

In shutting off the flow of liquid chlorine, be careful not to leave the line full of liquid chlorine with valves closed at both ends. Otherwise, the liquid remaining in the line may warm up and the resulting expansion of the liquid burst the pipeline by the hydrostatic pressure thus developed.

Close the tank car valve first and then allow the discharge line to empty by continuing to use the chlorine remaining in the line. After this, close both ends of the line.

Never apply heat directly to tank cars. It may, however, be desirable in some locations subject to low winter temperatures, to unload the tank cars in a shed maintained at 70 to 75 F.

In certain cases, particularly in winter weather with its low outside temperature, the pressure in single-unit cars may not be adequate and it is then necessary to increase it by adding dried compressed air to secure the desired rate of discharge.

Use only thoroughly dried clean air for this purpose. Have clean air supplied by a separate air compressor used only for this service. Never connect an existing compressed air system to the chlorine tank car.

The air dryer must be of suitable capacity to supply the full requirements. It must be maintained in proper operating condition. Consult the chlorine producer for all details relating to the application of dry air to single-unit tank cars. A specification for adding dry air to single-unit tank cars can also be obtained from The Chlorine Institute, Inc.

If chlorine is to be used at an elevation substantially above that of the tank car, high pressure is required in the car to force the liquid to that higher elevation. It is much safer to locate a chlorine evaporator at the elevation of the tank car and pipe the gaseous chlorine to the higher elevation.

With single-unit tank cars used for all other liquefied compressed gases, the content is usually withdrawn to storage. It is desirable to unload them as quickly as possible and at a rate consistent with good practice by connecting both liquid eduction valves to the transfer lines. However, the rate of withdrawal of liquid chlorine from tank cars is governed by the requirements of the process, and withdrawal may not be continuous. Accordingly, it may be desirable to unload through only one liquid valve. Open this valve, as is the case when two are used, slowly but completely.

When air-padding is necessary, connection of the compressed air line is made to one or both vapor valves. When a tank car has been "padded" with air, the pressure may become excessively high if the car is allowed to warm up when no chlorine is being withdrawn. Consequently, check the tank car pressure periodically during prolonged shutdowns so that excessive pressure may be vented carefully into caustic soda solution or into milk of lime in a large tank before the safety valve releases.

Chlorine can be withdrawn in gaseous form by unloading through the vapor valve or valves. However, this procedure is quite unusual.

The pipe connected to a tank car discharge line should be no longer than 18 in. and should be screwed in with a pipe wrench not larger than an 18-in. size. Use of longer connecting lines and larger wrenches may result in tilting the tank car valve on its gasket, causing a chlorine leak.

Extra-heavy black iron or steel pipe is recommended for chlorine service. All threads should be clean and sharp, preferably cut with new dies.

If night operations are necessary, adequately light the area where the tank car is hooked up.

Gas Masks.

Canister gas masks of a type approved by the U. S. Bureau of Mines for chlorine service should always be readily available when chlorine is unloaded, stored, transported or used. Locate gas masks outside the probable area of contamination so that it is possible to reach them in case of emergency.

Canister-type gas masks do not supply oxygen; they absorb the chlorine content present in the air leaving clean air to breathe. Where the chlorine content is greater than 1 per cent, use a self-contained oxygen breathing apparatus or fresh air hose mask. Each chlorine consumer should have at least one such device available in his plant. All personnel who may be required to use gas masks should be properly instructed in their application and use. Obtain a copy of National Safety Council Pamphlet No. 64, "Respiratory Protective Equipment."[6]

A poor gas mask is worse than no gas mask at all. Since the active materials in a canister become inactive when exposed to chlorine or air, keep canisters sealed. Renew them after each use.

Emergency Measures.

In an emergency, telephone the chlorine manufacturer.

In case of fire, remove chlorine containers from the fire zone immediately.

As soon as there is any indication of the presence of chlorine in the air, take steps to correct the condition. Chlorine leaks never get better; they always get worse if not corrected promptly.

Keep on the windward side of the leak and higher than the leak. Since gaseous chlorine is approximately $2\frac{1}{2}$ times as heavy as air, it tends to lie close to the ground.

When a chlorine leak occurs, authorized, trained personnel equipped with gas masks should investigate. All other persons should be kept away from the affected area until cause of the leak is discovered and corrected. If the leak is extensive, warn all persons in the path of the fumes.

Do not spray water on a chlorine leak. To do so makes the leak worse because of the corrosive action of wet chlorine. Heat supplied by even cold water causes liquid chlorine to change to gas at a faster rate.

When a leak occurs in equipment in which chlorine is being used, immediately close the chlorine container valve.

If a chlorine container is leaking in such a position that chlorine is escaping as a liquid, turn the container so that chlorine gas escapes. The quantity of chlorine escaping from a gas leak is about 1/15 the amount that escapes from a liquid leak through the same size hole.

If a chlorine leak occurs in transit in a congested area, it is recommended that the vehicle keep moving, if possible, until it reaches an open area where the escaping gas will be less hazardous. If a chlorine leak occurs in transit and the conveying vehicle is wrecked, the container or containers should be shifted so that gaseous chlorine, rather than liquid, is escaping. If possible, transfer the containers to a suitable vehicle and take it to open country.

Leaks at valve stems are often stopped by tightening the valve packing nuts or closing the valve.

The severity of a chlorine leak can be lessened by reducing the pressure on the leaking container. This may be done by absorbing chlorine gas, from the container, in caustic soda solution. Evaporation of some of the liquid chlorine cools the remaining liquid, reducing its pressure.

At regular points of storage and use, make emergency preparations for disposing of chlorine from leaking cylinders or ton containers. Chlorine may be absorbed in caustic soda, soda ash or hydrated lime solution. Caustic soda solution is preferred as it absorbs chlorine most readily.

A suitable container to hold the solution should be provided in a convenient location. Chlorine may be passed into the solution through an iron pipe or rubber hose properly weighted to hold it under the surface. Do not immerse the container in the solution.

Methyl Chloride

Methyl chloride is classified as a flammable gas. It burns feebly but forms explosive mixtures with air. The end product of high-temperature decomposition may be toxic. For data concerning the chemical and physical properties of methyl chloride, its physiological effects, protective equipment, etc., obtain a copy of the "Methyl Chloride Data Sheet"[7] published by the Manufacturing Chemists' Association, Inc.

Throughout the entire single-unit tank car unloading operation, particularly while connecting and disconnecting, take great caution to make sure the working area is free of heated surfaces, flames, static electricity, railroad locomotives, gasoline tractors and all other sources of ignition. Tank cars and TMU containers should be grounded electrically.

Under no circumstances should water or other materials be introduced into tank cars which contain, or have contained methyl chloride.

In testing for leaks, use soapy water; in freezing weather or around very cold pipes or equipment, use glycerine. Never test for leaks with an open flame. This is prohibited, and it applies not only to tank cars but also to all containers holding methyl chloride.

When unloading single-unit tank cars after all of the liquid has been transferred, the greater part (but not all) of the methyl chloride vapors may be recovered by creating a slight pressure differential. Observe caution in this operation, as a slight residual pressure must always remain on the tank car so that no air is drawn into it to form explosive mixtures when the pipes are disconnected.

When unloading tank cars, if a leak occurs which cannot be readily repaired by simple adjustment or tightening of the fittings, telephone the methyl chloride manufacturer at once for instructions, and evacuate the area around the car immediately. Permit only properly protected and instructed personnel to enter the contaminated area.

In case of TMU container leaks, all sources of ignition must be removed from the area at once, and if the leak cannot be stopped, transfer the methyl chloride to another container. As to the procedure, obtain and follow the detailed instructions of the methyl chloride manufacturer.

Fire and Explosion Hazards.

Avoid all sources of ignition, of whatever nature, when unloading single-unit tank cars and TMU containers holding methyl chloride.

Methyl chloride fires are gas fires. Most effective method of extinguishing them is to shut off the flow of vapor by closing the valves.

Carbon dioxide or dry chemical may be used to extinguish the flame to permit access to shutoff valves. If the valve is in the area of the fire, attack it if possible, close it, and then attack and extinguish the secondary fire which consists of other burning material ignited by the gas fire.

Circumstances may make it impossible to attack the valve, and in such a case, the flame may be allowed to continue burning while the surrounding area and objects are cooled with water spray.

Provide employees engaged in extinguishing fires with gas masks to protect them from methyl chloride vapors and the toxic combustion products formed.

Sulfur Dioxide

Sulfur dioxide is classified by the ICC as a nonflammable gas. In both its gaseous and liquid form it is neither flammable nor explosive. It is a respiratory and a skin and eye irritant. Obtain a copy of the "Sulfur Dioxide" pamphlet published by Compressed Gas Association, Inc.,[8] for data concerning its chemical and physical properties, physiological effects, protective equipment, etc.

The pressure differential necessary to unload sulfur dioxide from a single-unit tank car into a storage tank may be controlled in several ways. In some instances it is only necessary to release the pressure on the storage tank by actually using sulfur dioxide from the storage tank in the regular way while unloading the tank car.

Where you have 2 or more storage tanks, the car may be unloaded by building the pressure up in one of the storage tanks to 25 or 30 psig higher than the pressure in the tank to be filled. This higher pressure is then applied to the tank car to force the liquid into the storage tank.

Probably the best method of unloading a single-unit tank car is by installing a compressor with the suction side connected to the top of the storage tank and the discharge side connected to one of the gas valves of the tank car.

The least desirable method of unloading is to apply air pressure to the tank car through the vapor valve. If this method is used, exercise great care to insure that the supply of air is clean and dry and at a sufficiently high pressure to prevent back flow of sulfur dioxide into the air line.

Feeding SO_2 into a Solution.

Where it is desirable to feed sulfur dioxide

gas into a solution, the pipe leading from the valve on the container will frost on the outside and may even cause the solution to freeze at the point where the sulfur dioxide is being absorbed. If this happens you may correct the trouble by steam jacketing a small section of the feed line.

When gas is being fed to a solution, the evaporation of the liquid in the container refrigerates the entire content. As a result, the pressure might be reduced to a point where there is little or no flow of gas. In this case it is necessary to apply heat to the container or to connect a number of containers in parallel. This retards the rate of evaporation in each one.

Because of the corrosive nature of sulfurous acid, exercise great care to prevent the solution from drawing back into the upper valve chambers when the feed valve is shut off at the container or storage tank. To prevent this from happening, it is imperative that the feed line is vented or that a stainless steel ball check or check valve is installed in the line.

Leaks.

The occurrence of sulfur dioxide leaks are indicated by the pungent odor of the gas. Locate by using a wad of cloth soaked in ammonium hydroxide. This produces white fumes near the point of the leak.

Leaks which might develop are ordinarily not serious and can be readily controlled. Where leaks do occur, shut off the supply of sulfur dioxide by closing the appropriate valve. Leaks at unions or other fittings are often eliminated by tightening the connection.

Do not work on a line while it is under pressure. If corrosion is indicated, take care to empty the lines before working on them. A broken fitting might lead to a serious loss of sulfur dioxide before the supply valves can be shut off.

Although serious leaks rarely occur, careless handling sometimes results in this condition. Tank car and TMU container valves are made of brass and if struck by a heavy object, may be broken.

Leaks may also occur from carelessness in heating TMU containers, causing the fusible plug to melt and discharge the contents. Care in handling eliminates those dangers. Occasionally, leaks may develop in the valve packing, but these can be checked by tightening the packing nut.

In the event of a leaking container, wherever possible, move the container to an open area where the hazard due to escaping sulfur dioxide is minimized.

If a container is discharging too freely to permit movement, it should be arranged, if possible, in such a position that the leak is at the top, thus discharging gaseous sulfur dioxide and not liquid.

If large quantities of gas can be withdrawn rapidly into equipment, or satisfactorily vented, the evaporation often lowers the pressure in the container to such a point that you can move it to the open without difficulty; or, possibly, the leak can be repaired by persons provided with suitable masks.

If the leak still prevents removing the container to an open area, gaseous or liquid sulfur dioxide can be vented into a solution of lime, caustic soda or other alkaline material. One lb of sulfur dioxide is equivalent to about 2 lb of lime or $1\frac{1}{2}$ lb of caustic soda.

When a leak does occur, only an authorized employee should attempt to stop it, and if there is any question as to the seriousness of a leak, a suitable gas mask should be worn. In general, where leaks are serious, the employee even when equipped with a suitable gas mask, should remain in the contaminated area only long enough to make emergency adjustments.

Provisions for an Emergency.

All employees handling sulfur dioxide should be impressed with the potential danger it represents and should be trained in its safe handling. In addition, provide them with personal protective equipment for use in an emergency, and drill them until they are familiar with its use.

This protective equipment should include a gas mask of a type approved by the Bureau of

Mines for sulfur dioxide service. Take care to assure that masks are kept in proper working order and that they are stored so as to be readily available in case of need.

Canister-type masks are unsafe for high concentrations of sulfur dioxide. Warn employees of possible failure of this type of mask in the event of a really serious leak. Self-contained oxygen breathing apparatus or a mask with a long air hose and outside source of air may be required under extreme conditions.

Other protective equipment provided should include goggles or large-lensed spectacles to eliminate the possibility of liquid sulfur dioxide coming in contact with the eyes and causing possible injury.

If sulfur dioxide should be released, the irritating effect of the gas will force personnel to leave the area before they have long been exposed to dangerous concentrations. To facilitate their rapid evacuation, provide sufficient, well-marked, easily accessible exits.

Since sulfur dioxide neither burns nor supports combustion, there is no danger of fire or explosion due to igniting gas or liquid.

If a fire breaks out due to some other cause in an area containing sulfur dioxide, make every effort to remove the containers from the area to prevent overheat which would lead to melting of the fuse plugs. If they cannot be removed, inform the firemen of their location.

Fluorocarbons

Liquefied fluorinated hydrocarbons are classified as nonflammable gases by the ICC. Those presently shipped in single-unit tank cars, TMU containers and cargo tank trucks are: dichlorodifluoromethane, trichloromonofluoromethane, monochlorodifluoromethane, dichlorotetrafluoromethane and trichlorotrifluoromethane.

These gases are odorless, and leaks cannot be detected by sense of smell. Frosting is evidence of a large leak while smaller leaks may be located by means of a halide torch.

Avoid contact with the liquid and excessive inhalation of vapor. In case of a severe leak, persons entering area of dense concentration of vapor should wear an air gas mask of the type approved by the U. S. Bureau of Mines for liquefied fluorinated hydrocarbon service.

Check Regulations and Industry Standards

ICC regulations, and such local regulations as may exist in your area as well as industry standards are the "Bible" for handling and unloading compressed gases. Statements made herein are intended only to complement these regulations and standards and not to supplant them.

Obviously, within the limits of this explanation, it is not possible to set forth all the precautions to be taken when handling and unloading a particular gas. It is possible only to mention some of the more important. In all cases, the compressed gas manufacturer must be consulted concerning specific handling and transfer problems, and he will advise in full concerning them.

Acknowledgments. A special committee to prepare this statement was appointed by the Executive Board of Compressed Gas Association, Inc. The committee consisted of D. J. Barday, "Freon" Products Division, E. I. du Pont de Nemours & Co., Inc.; F. J. Heller, Phillips Petroleum Co.; E. C. Perrine, Nitrogen Division, Allied Chemical Corp.; James S. Walker, Hooker Chemical Corp.; George T. Wrenn, Jr., Virginia Chemicals Inc., and F. R. Fetherston, Compressed Gas Association, Inc., who acted as chairman and whose name appears in the byline. The chairman acknowledges the contributions and efforts of each committee member in the preparation of this statement.

REFERENCES

1. "Agent T. C. George's Tariff No. 15 Publishing Interstate Commerce Commission Regulations for Transportation of Explosives and Other Dangerous Articles by Land and Water in Rail Freight Service and by Motor Vehicle (Highway) and Water Including Specifications for Shipping Containers," Bureau of Explosives, 63 Vesey St., New York, N. Y., 10007. Nominal charge.

2. "Standard 1202 (Feb. 1960), Measuring, Sampling and Calculating Tank Car Quantities and Calibrating Tank Car Tanks (Pressure Type Tank Cars)," American Petroleum Institute, 1271 Avenue of the Americas, New York, N. Y., 10020. Nominal charge.
3. "Anhydrous Ammonia (Pamphlet G-2)," Compressed Gas Association, Inc., 500 Fifth Avenue, New York, N. Y. 10036. Nominal charge.
4. "Standard for the Storage and Handling ot Liquefied Petroleum Gases, NFPA Pamphlef No. 58," National Fire Protection Association International, 60 Batterymarch St., Boston 10, Mass. 02110. Nominal charge.
5. "Chlorine Manual," The Chlorine Institute, Inc., 342 Madison Ave., New York, N. Y., 10017. Nominal charge.
6. "Respiratory Protective Equipment," National Safety Council, 425 N. Michigan Ave., Chicago, Ill. Nominal charge.
7. "Safety Data Sheet SD-40, Methyl Chloride, Manufacturing Chemists' Association, 1825 Connecticut Ave., N. W., Washington, D. C. 2009. Nominal charge.
8. "Sulfur Dioxide" (Pamphlet G-3), Compressed Gas Association, Inc. Nominal charge.

Appendices

The following appendices to the *Handbook* present supplementary information important to users, distributors, producers, and others concerned with the compressed and liquefied gases.

Appendix A summarizes, in three tables, certain significant features of compressed gas regulation (by states) of the United States. Appendix B consists of a complete list of the present service publications of Compressed Gas Association, Inc. The CGA regularly adds to and revises these publications, and current information about them may be obtained on request to the Association.*

The summary information given in Appendix A on regulations of the states is necessarily general and subject to change subsequent to the time of its collection. As a result, in its application to any specific situation, it should be taken as indicative, rather than conclusive. The CGA secured the data from sources believed to be fully authoritative, but the Association and the publisher can assume no responsibility in connection with use of the data. However, the summary outlined in the tables represents the most comprehensive collection of information on state regulation that has yet been made available to the public. The major purposes that the three tables of the appendix are designed to serve are as follows:

Table 1, on the use by states of compressed gas standards developed by the National Fire Protection Association, is intended to aid persons responsible for the location, design and installation of systems, including piping, containers, instruments, and equipment to handle compressed gases;

Table 2, on state laws pertaining to the Boiler and Pressure Vessel Code, Section VIII, of the American Society of Mechanical Engineers (the Unfired Pressure Vessel Code), should prove helpful primarily to persons responsible for compressed gas storage installations (because compressed gas storage tanks are commonly built to the requirements of this ASME Code);

Table 3, on state adoptions of the ICC regulations for shipping "explosives and other dangerous articles," and of the ICC "motor carrier safety regulations," is conceived as a general guide primarily for persons responsible for the transportation of compressed gases within state borders.

* 500 Fifth Avenue, New York, New York 10036

APPENDIX A

Summary of Selected State Regulations and Codes Concerning Compressed Gases

TABLE 1. Use of Gas Standards of the National Fire Protection Association by Fire Marshalls and Other Officials of States (and Cities)*

Note: The titles of the N.F.P.A. Standards identified by number in the table are:

51—Standard for the Installation and Operation of Oxygen-Fuel Gas Systems for Welding and Cutting.
54—Standard for the Installation of Gas Appliances and Gas Piping.
56—Code for the Use of Flammable Anesthetics.
565—Standard for Nonflammable Medical Gas Systems.
566—Standard for the Installation of Bulk Oxygen Systems at Consumer Sites.
567—Standard for Gaseous Hydrogen Systems at Consumer Sites.
58—Standard for the Storage and Handling of Liquefied Petroleum Gases.
59—Standard for the Storage and Handling of Liquefied Petroleum Gases at Utility Gas Plants.

State	No. 51	No. 54	No. 56	No. 565	No. 566	No. 567	No. 58	No. 59
Ala.	Yes	Yes	Yes	Yes	Yes	Yes	Yes	Yes
Alaska	Yes	Yes	Yes	Yes	Yes	Yes	Yes	Yes
Ariz. (Phoenix only)	Yes	Yes	Yes	Yes	Yes	Yes	Yes	Yes
Ark.	—	—	—	—	—	—	—	—
Calif.	Yes	Yes	Yes	Yes	Yes	No	Yes	No
Colo.	—	—	—	—	—	—	—	—
Conn.	Informal	Yes	Informal	Informal	Informal	Informal	Yes	Yes
Del.	No	Yes	Informal	Informal	Informal	No	Yes	Yes
D.C.	—	—	—	—	—	—	—	—

* *Explanatory notes to Table 1:*

1. "Yes" entered in the table means that the standard is either adopted by reference or reproduced in a law, regulation or similar vehicle. In a number of instances, standards are included by virtue of adoption of the National Fire Codes of the N.F.P.A., of which they are part.
2. "Informal" in the table means that the standard is used informally under the broad jurisdictional powers of the regulating official or agency concerned.
3. "No" in the table means that no use of the standard was reported when the information given in the table was requested. A "No" response may reflect a misunderstanding of the question, or reservation on one or another specific point in a standard rather than on the entire standard.
4. "Only" after the name of a city in a state listed in the table's first column means that the use of N.F.P.A. standards indicated is the use by the regulating officials of that city rather than the officials of that state.
5. Information in the table does not indicate which edition of the standard is used by the respective states.
6. Not included in the table but of general interest are those states which have adopted the Fire Prevention Code of the National Board of Fire Underwriters (now named the American Insurance Association), in full or in part, namely: Del., Fla., Hawaii, Ky., and Md.

TABLE 1—*continued*

State	No. 51	No. 54	No. 56	No. 565	No. 566	No. 567	No. 58	No. 59
Fla.	Informal	Yes	Informal	Informal	Informal	Informal	Yes	Informal
Ga.	Yes	Yes	No	No	Informal	No	Yes	Yes
Hawaii	Yes	Informal	Yes	Informal	Informal	Informal	Yes	Yes
Idaho	—	—	—	—	—	—	—	—
Ill.	Yes	Yes	Informal	Informal	Informal	Informal	Yes	Yes
Ind.	—	—	—	—	—	—	—	—
Iowa	Informal	Yes	Informal	Informal	Informal	Informal	Yes	Yes
Kans. (Kans. City only)	Yes	Yes	Yes	Yes	Yes	Yes	Yes	Yes
Ky.	Yes	Yes	Yes	No	Yes	Yes	Yes	Yes
La. (Shreveport only)	Yes	Yes	Yes	Yes	Yes	Yes	Yes	Yes
Me.	—	—	—	—	—	—	—	—
Md.	Informal	Yes	Informal	Informal	Informal	Informal	Yes	Yes
Mass. (Boston only)	Informal	Yes	Informal	Informal	Informal	Informal	Yes	Informal
Mich.	Yes	Yes	Yes	Informal	Yes	Informal	Yes	Yes
Minn.	Yes	Yes	Informal	Informal	Informal	Informal	Yes	Yes
Miss.	—	—	—	—	—	—	—	—
Mo. (St. Louis only)	Yes	Yes	Yes	Yes	Yes	Yes	Yes	Yes
Mont.	Informal	Informal	Yes	Yes	Informal	Informal	Informal	Informal
Neb. (Omaha only)	Yes	Yes	Yes	Yes	Yes	Yes	Yes	Yes
Nev.	—	—	—	—	—	—	—	—
N. H.	Informal	Informal	Informal	Informal	Informal	Informal	Yes	Informal
N. J. (Jersey City only)	Yes	No	Yes	Yes	Yes	Yes	Yes	Yes
N. Mex.	No	No	No	No	No	No	Yes	No
N. Y. (N.Y.C. only)	Informal	No	Informal	Informal	Informal	Informal	Informal	No
(Syracuse only)	Yes	Yes	Yes	Yes	Yes	Yes	Yes	Yes
(Buffalo only)	Yes	Yes	Yes	Yes	Yes	Yes	Yes	Yes
N. Car.	Informal	Yes	Yes	Yes	Yes	Informal	Yes	Informal
N. Dak.	Informal	Informal	Informal	Informal	No	No	Yes	Yes
Ohio	Informal	Informal	Informal	Informal	Informal	Informal	Yes	Informal
(Toledo only)	Yes	Yes	Informal	Yes	Informal	Informal	Yes	Yes
Okla. (Tulsa only)	No	Yes	Yes	No	Informal	Informal	Yes	No
Oregon	Informal	Yes	Informal	Informal	Informal	Informal	Yes	Yes
Pa. (Philadelphia only)	Yes	Informal	Yes	No	No	Informal	Yes	Yes
R. I.	Informal	Informal	Informal	No	Informal	No	Yes	Informal
S. Car. (Greenville only)	Yes	Yes	Informal	Informal	Informal	Informal	Yes	Informal
S. Dak.	Informal	Yes	Informal	Informal	Informal	Informal	Yes	Yes
Tenn. (Memphis only)	Informal	No	Yes	Yes	Informal	No	No	Yes

TABLE 1—*continued*

State	No. 51	No. 54	No. 56	No. 565	No. 566	No. 567	No. 58	No. 59
Texas (Dallas only)	Yes	Yes	Yes	Yes	Yes	Yes	Yes	Yes
Utah	Informal	Yes	Informal	Informal	No	No	Yes	No
Vt.	—	—	—	—	—	—	—	—
Va.	No	Yes	No	No	No	No	Yes	No
Wash. (Seattle only)	Yes	Yes	Yes	Yes	Yes	Yes	Yes	Yes
W. Va.	Yes	Yes	Yes	Yes	Yes	Yes	Yes	Yes
Wisc. (Milwaukee only)	Informal	Yes	Informal	Informal	Informal	Informal	Yes	Yes
Wyo. (Cheyenne only)	Informal	Informal	Informal	Informal	Informal	Informal	Informal	Informal

TABLE 2. States (and Cities and Counties within States) and Provinces that have Adopted Section VIII of the ASME Boiler and Pressure Vessel Code, Or Other Codes for Unfired Pressure Vessels, with Name and Address of Enforcement Officials.

Key to symbols in second column: A—Law requires ASME construction
O—Have own construction code
(L)—Operator's license required

Explanatory Note:

Section VIII of the Boiler and Pressure Vessel Code of the ASME (American Society of Mechanical Engineers) concerns unfired pressure vessels, and the standards of Compressed Gas Association, Inc., recommend its use for compressed gas storage containers. The information below is reprinted with permission from the "1966 Data Sheet" of the Uniform Boiler and Pressure Vessel Laws Society (57 Pratt St., Hartford, Conn. 06103), a nonprofit, nonpartisan technical body supported by voluntary contributions.

State	Adoption of Code	Enforcement Official
Alaska	A	R. D. Molt, Dept. of Labor, Box 2141, Juneau, Alaska.
Ariz.	—	—
Phoenix	A	S. A. Antignano, 301 Municipal Bldg., Phoenix, Ariz.
Tucson	A	R. C. Higginbotham, Inspec. Div., City Hall, Tucson, Ariz.
Ark.	A	J. T. Crosby, Blr. Inspec. Div., Box 1797, Little Rock, Ark.
Calif.	A	A. I. Snyder, Dept. Indus. Rel., Box 603, San Fransisco, Calif.
Los Angeles	A	James Cameron, Rm. 200, City Hall, Los Angeles, Calif.
San Fran.	A	John T. Edson, Rm. 103, 450 McAllister St., San Fransisco, Calif.
Canal Zone	A	M. B. Nickel, Indus. Div., Panama Canal Co., Cristobal, C.Z.
Colo.	—	—
Denver	A	H. O. McIntosh, Bldg. Dept., City & County Bldg., Denver, Colo.
D. C.	A(L)	R. T. Clark, Smoke & Blr. Sec., 14th & E Sts., N.W., Washington, D.C.
Fla.	—	—
Miami	A	H. O'Banion, P.O. Box 708, Coconut Grove Station, Miami, Fla.
Miami Beach	A	Div. of Blr. Inspec., 1351 NW 12th St., Miami, Fla.
No. Miami Beach	A(L)	Dept. of Blr. Inspec., 17011 NE 19th Ave., No. Miami Beach, Fla.
Tampa	A(L)	W. A. Cooper, Blr. Bur., 301 N. Florida Ave., Tampa, Fla.
Dade County	A	W. R. Angleton, Justice Bldg., 1351 NW 12th St., Miami, Fla.
Guam	A	—

TABLE 2—*continued*

State	Adoption of Code	Enforcement Official
Idaho	A	W. L. Robison, Dept. of Labor, 317 Main St., Boise, Idaho.
Ill.	—	—
Chicago	A	John F. Hanley, Rm. 300, 320 N. Clark St., Chicago, Ill.
East St. Louis	A(L)	Joseph Iwasyszyn, City Hall, East St. Louis, Ill.
Ind.	A	Charles Harris, Jr., 512 State Office Bldg., Indianapolis, Ind.
Iowa	A	Arthur Parkhurst, Bur. of Labor, Blr. Div., Des Moines, Ia.
Kans.	A	C. A. Shriver, Dept. of Labour, 401 Topeka Ave., Topeka, Kans.
La.	—	—
New Orleans	A	Leo J. Vivien, Rm. 7E04, City Hall, New Orleans, La.
Jefferson Parish	A	H. Schouest, Jr., Dept. of Inspec., 3301 Metairie Rd., Metairie, La.
Mass.	A	Thomas Dickson, Div. of Inspec., 1010 Commonwealth Ave., Boston, Mass.
Mich.	A	Royal Beckwith, Dept. of Labor, Lansing, Mich.
Dearborn	A	T. J. O'Neill, Bldg. & Safety, City Hall, Dearborn, Mich.
Detroit	A	F. J. Drogosch, 408 City-County Bldg., Detroit, Mich.
Minn.	A	Indus. Commission, St. Paul, Minn.
Mo.	—	—
Kansas City	A(L)	E. S. Bybee, Div. of Bldg. & Inspec., Kansas City, Mo.
St. Louis	A	E. J. Robbins, Pub. Safety Dept., City Hall, St. Louis, Mo.
St. Louis County	A	R. A. Jauer, Chf. Mech. Inspector, County Court House, Clayton, Mo.
Neb.	A	Emery R. Shaw, Dept. of Labor, Lincoln, Neb.
Omaha	A	A. L. Rono, 104 City Hall, Omaha, Neb.
Nev.	A	A. C. Lamb, Dept. of Indus. Safety, 347 S. Wells Ave., Reno, Nev.
N. J.	A	J. L. Sullivan, Labor & Indus. Bldg., P.O. Box V, Trenton, N. J.
N. Y.	—	—
Buffalo	A	Thos. A. Hearn, 2501 City Hall, Buffalo, N. Y.
N. Car.	A	E. L. Clodfelter, Dept. of Labor, Raleigh, N. Car.
Greensboro	A(L)	Lloyd H. Doolittle, Inspec. Dept., Drawer W-2, Greensboro, N. Car.
Ohio	O(L)	Philip Meseroll, Dept. of Indus. Rel., 220 S. Parsons Ave., Columbus, Ohio.
Okla.	—	—
Oklahoma City	A	E. T. Garlock, 107–8 City Hall, Oklahoma City, Okla.
Tulsa	A	Robert B. McGill, City Hall, 4th & Cinn., Tulsa, Okla.
Ore.	A	D. R. Bartosch, 115 Labor & Industries Bldg., Salem, Ore.
Pa.	A	John Riddiough, Labor & Indus. Bldg., Blr. Div., Harrisburg, Pa.
Puerto Rico	A	Jose A. Rosado Ortiz, Dept. of Labor, San Juan, Puerto Rico.
Tenn.	A	C. W. Allison, Dept. of Labor, C-1, 120 Cordell Hull Bldg., Nashville, Tenn.
Memphis	A	Otis R. Kyle, Rm. 406, City Hall, 125 N. Main St., Memphis, Tenn.
Utah	A	Fred S. Thomas, Indus. Commission, Salt Lake City, Utah.
Vt.	A	Albert A. Fraser, Dept. of Ind. Rel., Montpelier, Vt.
Va.	A	C. S. Mullen, Jr., P.O. Box 1157, Richmond, Va.
Richmond	A	J. G. Williams, Jr., 501 N. 9th St., Rm. 130, Richmond, Va.
Arlington County	A	Elmer Shrout, Chf. Mech. Inspector, Court House, Arlington, Va.
Wash.	A	Stuart T. Viggers, 1601 2nd Ave., Seattle, Wash.
Seattle	A	S. B. Voris, Dept. of Bldgs., Rm. 503, 600 4th Ave., Seattle, Wash.
Spokane	A	R. R. Reese, Supt. of Bldg., 451 City Hall, Spokane, Wash.
Tacoma	A	LeRoy Galbraith, 432 County-City Bldg., Tacoma, Wash.
Wisc.	A	R. Ostrem, Indus. Commission, 4802 Sheboygan Ave., Madison, Wisc.
Milwaukee	A	G. M. Kuetemeyer, Municipal Bldg., Milwaukee, Wisc.

TABLE 2—*continued*

Canadian Province	Adoption of Code	Enforcement Official
Alberta	A	A. J. Rees, Dept. of Labour, Terrace Bldg., Edmonton, Alberta.
Brit. Col.	A	S. Smith, 501 W. 12th Ave., Vancouver, Brit. Col.
Manitoba	A(L)	W. L. Garvin, 611 Norquay Bldg., Winnipeg, Manitoba.
New Brunswick	A	J. L. Sisk, P.O. Box 580, Fredericton, New Brunswick.
Newfoundland and Labrador	A	T. A. Stewart, Dept. of Labour, St. Johns, Newfoundland.
Northwest Territory	A(L)	S. M. Hodgson, Dep. Com. of NW Territory, Ottawa, Ontario.
Nova Scotia	A	W. F. Urquhart, Johnston Bldg., Dept. of Labour, Halifax, Nova Scotia.
Ontario	A(L)	L. J. Hutchinson, 74 Victoria St., Toronto, Ontario.
Quebec	A	W. A. Berriman, Rm. 410, 355 McGill St., Montreal, Quebec.
Saskatchewan	A	Joseph Taylor, New Telephone Bldg., Regina, Saskatchewan.
Yukon Terr.	A	J. E. Halsall, Box 2373, Vancouver, Brit. Col.

TABLE 3. State Adoptions of ICC Regulations on Transportation of Explosives and Other Dangerous Articles ("EODA"; with State Agencies Administering), and of ICC Motor Carrier Safety Regulations ("MCSR")

State	EODA Adopted	State Agency Administering	MCSR Adopted
Ala.	Partial	Pub. Ser. Commission	No
Alaska	No	—	Partial
Ariz.	Yes	Ariz. Corp. Commission	Yes
Ark.	Yes	Ark. Commerce Commission	Yes
Calif.	Partial	Calif. Highway Patrol	Partial
Colo.	Partial	Pub. Utilities Commission	Partial
Conn.	No	—	No
Del.	No	—	Partial
D. C.	No	—	Partial
Fla.	Yes	Pub. Utilities Commission	Partial
Ga.	No	—	Partial
Hawaii	—	—	—
Idaho	Yes	Pub. Utilities Commission	Yes
Ill.	No	—	Partial
Ind.	No	—	Partial
Iowa	No	—	Partial
Kans.	Partial	State Fire Marshal, State Board of Agric., Dept. of Revenue	Partial
Ky.	Partial	Dept. of Pub. Safety, Div. of Fire Prevention	Partial
La.	No	—	No
Me.	Partial	Pub. Utilities Commission	Partial
Md.	Partial	Pub. Ser. Commission, Dept. of Motor Vehicles, Baltimore Harbor Tunnel, Toll Facilities Div., State Roads Div.	Partial

TABLE 3—*continued*

State	EODA Adopted	State Agency Administering	MCSR Adopted
Mass.	No	—	No
Mich.	No	—	Partial
Minn.	Partial	Railroad & Warehouse Commission, State Fire Marshal	No
Miss.	No	—	Partial
Mo.	Yes	Pub. Ser. Commission, Bus & Truck Dept.	Partial
Mont.	Partial	Board of Railroad Commissioners	Yes
Neb.	No	—	Partial
Nev.	Yes	Pub. Ser. Commission	Partial
N. H.	No	—	No
N. J.	No	—	Partial
N. Mex.	Partial	State Corp. Commission	Partial
N. Y.	Yes	Pub. Ser. Commission, Thruway Authority	Partial
N. Car.	Yes	Utilities Commission	Partial
N. Dak.	Partial	Fire Marshal	Partial
Ohio	No	—	Partial
Okla.	Partial	Okla. Highway Patrol, LPG Administration	Partial
Ore.	Yes	Pub. Utilities Commission	Partial
Pa.	Yes	—	Partial
R. I.	No	—	Partial
S. Car.	Partial	Pub. Ser. Commission, Traffic Bureau	Partial
S. Dak.	Partial	State Fire Marshal	Partial
Tenn.	Yes	Pub. Ser. Commission	No
Tex.	No	—	Partial
Utah	Partial	Pub. Ser. Commission	Yes
Vt.	No	—	No
Va.	Partial	Corp. Comm., State Police	Partial
Wash.	Yes	State Police	Partial
W. Va.	No	—	Partial
Wisc.	Partial	Pub. Ser. Commission	Partial
Wyo.	Yes	Pub. Ser. Commission	Yes

APPENDIX B

Service Publications of Compressed Gas Association

PAMPHLET SERIES

Cylinders

C–1. Methods for Hydrostatic Testing of Compressed Gas Cylinders.

C–2. Disposition of Unserviceable Compressed Gas Cylinders.

C–3. Standards for Welding and Brazing on Thin-Walled Containers.

C–4. American Standard Method of Marking Portable Compressed Gas Containers to Identify the Material Contained (American Standard Z48.1–1954).

C–5. Cylinder Service Life—Seamless High-Pressure Cylinder Specifications ICC–3, ICC–3A, ICC–3AA.

C–6. Standards for Visual Inspection of Compressed Gas Cylinders.

C–7. A Guide for the Preparation of Labels for Compressed Gas Containers.

C–8. Standard for Requalification of ICC–3HT Cylinders.

Gases

G–1. Acetylene (see also "Oxy-Acetylene," below)

- *G–1.2.* Recommendations for Chemical Acetylene Metering.
- *G–1.3.* Acetylene Transmission for Chemical Synthesis.
- *G–1.4.* Standard for Acetylene Cylinder Charging Plants.

Oxy-Acetylene

- *OA*–1. Oxy-Acetylene Cutting.
- *OA*–2. Oxy-Acetylene Welding and Its Applications.
- *OA–4.* Braze Welding of Iron and Steel by the Oxy-Acetylene Process.
- *OA–5.* Safe Practices for Installation and Operation of Oxy-Acetylene Welding and Cutting.
- *OA–6.* Flame-Hardening by the Oxy-Acetylene Process.
- *OA–7.* Hard-Facing by the Oxy-Acetylene Process.
- *OA–8.* Carbide Lime—Its Value and Its Uses.

G–2. Anhydrous Ammonia

- *G–2.1.* American Standard Safety Requirements for the Storage and Handling of Anhydrous Ammonia (American Standard K–61.1–1966).

G–3. Sulfur Dioxide.

G–4. Oxygen (see also "Oxy-Acetylene," above).

- *G–4.1.* Equipment Cleaned for Oxygen Service.
- *G–4.2.* Standard for Bulk Oxygen Systems at Consumer Sites (also National Fire Protection Association No. 566).

G–5. Hydrogen

- *G–5.1.* Standard for Gaseous Hydrogen Systems at Consumer Sites.
- *G–5.2.* Standard for Liquefied Hydrogen Systems at Consumer Sites.

G–6. Carbon Dioxide.

- *G–6.1T.* Tentative Standard for Low Pressure Carbon Dioxide Systems at Consumer Sites.

G–7. Tentative Standard for Compressed Air for Human Respiration (revision in preparation).

G–8.1. Standard for Nitrous Oxide Systems at Consumer Sites.

Protection and Safe Handling

P–1. Safe Handling of Compressed Gases.

P–2. Characteristics and Safe Handling of Medical Gases.

P–2.1. Standard for Medical Vacuum Systems in Hospitals (revision in preparation).

P–2.2T. Tentative Standard for the Use of Inhalant and Resuscitative Equipment and Agents.

P–3. Standards for Solid Ammonium Nitrate (Nitrous Oxide Grade).

P–4. Safe Handling of Cylinders by Emergency Rescue Squads.

P–5. Care of High Pressure Air Cylinders for Underwater Breathing.

P–6. Standard Density Data, Atmospheric Gases and Hydrogen.

Safety Relief Devices

S–1.1 (*S–1, Part 1*). Safety Relief Device Standards for Compressed Gas Cylinders.

S–1.2 (*S–1, Part 2*). Safety Relief Device Standards for Cargo and Portable Tanks.

S–1.3 (*S–1, Part 3*). Safety Relief Device Standards for Compressed Gas Storage Containers (with design pressures exceeding 15 psig).

S–3. Frangible Disc Safety Device Assembly.

S–4. Recommended Practice for the Manufacture of Fusible Plugs.

Valves

V–1. American-Canadian Standard Compressed Gas Cylinder Valve Connections (American Standard B57.1–1965; Canadian Standard B96–1965).

V–2. Pin-Index Safety System for Flush-Type Cylinder Valves.

V–5. Diameter-Index Safety System for Non-Interchangeable Low Pressure Connections for Medical Gases, Air and Suction.

SAFETY BULLETIN SERIES

(single-sheet suitable for posting)

SB–1. Hazards of Refilling Compressed Refrigerant (Halogenated Hydrocarbons) Gas Cylinders.

SB–2. Oxygen Deficient Atmospheres.

GAS SPECIFICATION SERIES

Air
Helium
Oxygen

IN PREPARATION

Gas Specification Series

Argon
Halogenated Hydrocarbon Refrigerants
Hydrogen
Nitrogen

Other Series

S–1.4 (*S–1, Part 4*). Safety Relief Device Standards for Compressed Gas Storage Containers (with design pressures of 15 psig or less).

Standard for hose and regulator connections for welding and cutting equipment.

Standard for visual inspection of cargo tank trucks.

Research report on low-pressure storage of anhydrous ammonia.

Index